Studies in Big Data

Volume 160

The series "Studies in Big Data" (SBD) publishes new developments and advances in the various areas of Big Data- quickly and with a high quality. The intent is to cover the theory, research, development, and applications of Big Data, as embedded in the fields of engineering, computer science, physics, economics, and life sciences. The books of the series refer to the analysis and understanding of large, complex, and/or distributed data sets generated from recent digital sources coming from sensors or other physical instruments as well as simulations, crowd sourcing, social networks or other internet transactions, such as emails or video click streams and other. The series contains monographs, lecture notes and edited volumes in Big Data spanning the areas of computational intelligence including neural networks, evolutionary computation, soft computing, fuzzy systems, as well as artificial intelligence, data mining, modern statistics and Operations research, as well as self-organizing systems. Of particular value to both the contributors and the readership are the short publication timeframe and the world-wide distribution, which enable both wide and rapid dissemination of research output.

The books of this series are reviewed in a single blind peer review process.

Indexed by SCOPUS, EI Compendex, SCIMAGO and zbMATH.

All books published in the series are submitted for consideration in Web of Science.

Kerven Durdymyradov · Mikhail Moshkov ·
Azimkhon Ostonov

Decision Trees Versus Systems of Decision Rules

A Rough Set Approach

Springer

Kerven Durdymyradov
Computer, Electrical and Mathematical Sciences and Engineering Division
King Abdullah University of Science and Technology
Thuwal, Saudi Arabia

Mikhail Moshkov
Computer, Electrical and Mathematical Sciences and Engineering Division
King Abdullah University of Science and Technology
Thuwal, Saudi Arabia

Azimkhon Ostonov
Computer, Electrical and Mathematical Sciences and Engineering Division
King Abdullah University of Science and Technology
Thuwal, Saudi Arabia

ISSN 2197-6503 ISSN 2197-6511 (electronic)
Studies in Big Data
ISBN 978-3-031-71585-3 ISBN 978-3-031-71586-0 (eBook)
https://doi.org/10.1007/978-3-031-71586-0

This Springer imprint is published by the registered company Springer Nature Switzerland AG
The registered company address is: Gewerbestrasse 11, 6330 Cham, Switzerland

To our families

Preface

Decision trees and systems of decision rules are widely used as means of representing knowledge, as classifiers that predict decisions for new objects, as well as algorithms for solving various problems of fault diagnosis, combinatorial optimization, etc. Decision trees and systems of decision rules are among the most interpretable models of knowledge representation and classification. Investigating the relationships between these two models is an important task in computer science.

This book explores the complexity of decision trees and decision rule systems, and the relationships between them, within the framework of rough set theory, for problems over information systems, for decision tables from closed classes, and for problems involving formal languages. The possibilities of transforming decision rule systems into decision trees are being studied in detail.

Instead of decision rule systems, this book often uses nondeterministic decision trees, which can be thought of as representations of decision rule systems, sometimes in shortened form. It is possible to say that the book is devoted to studying the relationships between deterministic and nondeterministic decision trees.

The results obtained in the book may be useful for researchers using decision trees and decision rule systems in data analysis, especially in rough set theory, logical analysis of data, and test theory. This book can also be used to create courses for graduate students.

Thuwal, Saudi Arabia
May 2024

Kerven Durdymyradov
Mikhail Moshkov
Azimkhon Ostonov

Acknowledgements We are greatly indebted to King Abdullah University of Science and Technology for the immense support.

We are thankful to Professor Andrzej Skowron for stimulating discussions.

We extend an expression of gratitude to Prof. Janusz Kacprzyk, to Dr. Thomas Ditzinger, and to the Series Studies in Big Data staff at Springer for their support in making this book possible.

Competing Interests This study was supported by King Abdullah University of Science and Technology.

Contents

Chapter 1
Introduction

Decision trees [3, 4, 13, 52, 53, 66, 69] and decision rule systems [10, 11, 17, 23, 56, 57, 61, 62] are widely used as a means for knowledge representation, as classifiers that predict decisions for new objects, and as algorithms for solving various problems of fault diagnosis, combinatorial optimization, etc. Decision trees and rules are among the most interpretable models for classifying and representing knowledge [36]. The study of relationships between these two models is an important task of computer science.

This book is devoted to the investigation, within the framework of rough set theory [60–62, 71, 72], of the complexity of decision trees and decision rule systems and relationships between them including problem of transformation of a decision rule system into a decision tree.

We begin this chapter with a preliminary discussion of the book, then describe four directions of research, consider some conclusions, and finally discuss the use of the book.

1.1 Preliminary Discussion

In the present book, we study decision problems: for a given input, we should return a decision corresponding to this input. We consider only problems each of which can be solved if we know values of attributes from a finite set of attributes. To solve these problems, we use algorithms in which each elementary operation (query) consists in computation of the value of an attribute for the given input. We consider two types of such algorithms: deterministic decision trees and decision rule systems. We assume that each rule from the system is true for the considered problem and rules cover all inputs. For example, in Fig. 1.1, a deterministic decision tree and a decision rule

K. Durdymyradov et al., *Decision Trees Versus Systems of Decision Rules*,
Studies in Big Data 160, https://doi.org/10.1007/978-3-031-71586-0_1

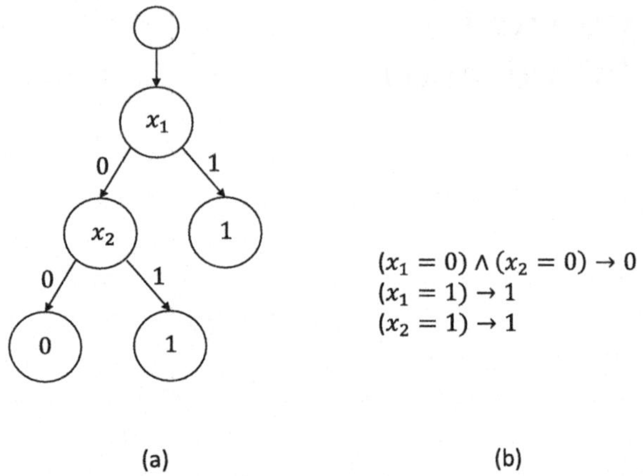

Fig. 1.1 Deterministic decision tree and decision rule system for disjunction $x_1 \vee x_2$

system for the disjunction $x_1 \vee x_2$ are depicted. As attributes, we use variables x_1 and x_2.

In this book, instead of decision rule systems, we often consider nondeterministic decision trees represented them. For example, in Fig. 1.2, two nondeterministic decision trees are depicted, which represent the decision rule system shown in Fig. 1.1b. The tree depicted in Fig. 1.2 (a) represents this rule system as it is. The tree depicted in Fig. 1.2b "compresses" the considered decision rule system.

For problems with only two types of decisions 0 and 1, we can also study strongly nondeterministic decision trees. These trees can be considered as representations of systems of true rules that cover all inputs with the decision 1. For example, in

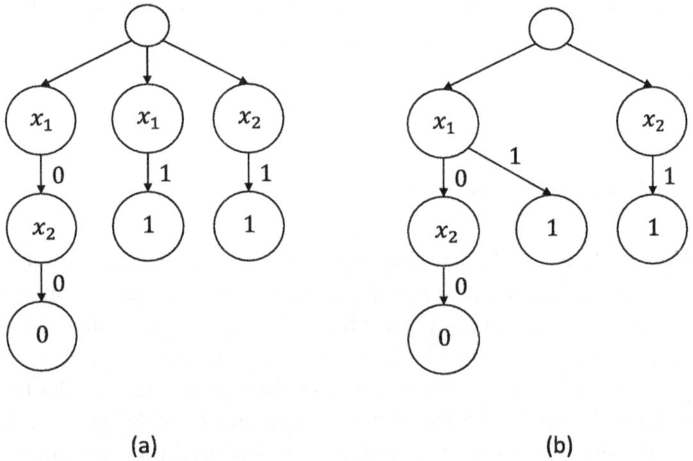

Fig. 1.2 Nondeterministic decision trees for disjunction $x_1 \vee x_2$

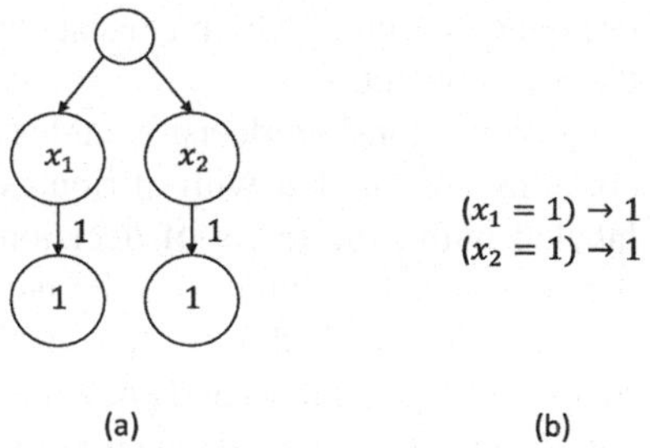

Fig. 1.3 Strongly nondeterministic decision tree for disjunction $x_1 \vee x_2$ and decision rule system represented by this tree

Fig. 1.3a, a strongly nondeterministic decision tree for the disjunction $x_1 \vee x_2$ is depicted. This tree represents the decision rule system shown in Fig. 1.3b.

We study time complexity of decision trees, in particular, the depth of decision trees, which is equal to the maximum number of nodes labeled with attributes in a path from the root to a terminal node of the tree. For example, the depth of decision trees depicted in Figs. 1.1a and 1.2 is 2 and the depth of the decision tree depicted in Fig. 1.3a is 1. Besides depth, we study general complexity measures and two types of these measures: limited and bounded complexity measures.

Let F be the set of attributes under consideration and F^* be the set of all finite words over the alphabet F including the empty word λ. A (general) complexity measure is any mapping $\psi : F^* \to \omega$, where ω is the set of nonnegative integers. Let Γ be a decision tree and ξ be a path in Γ from the root to a terminal node. We correspond to it a word $\varphi(\xi) \in F^*$ in which letters are attributes attached to nodes of ξ in the order from the root to the terminal node. Then the complexity of Γ is equal to $\max(\psi(\varphi(\xi)))$, where the maximum is taken among all paths ξ from the root to a terminal node of Γ.

A complexity measure ψ is called limited if, for any $\alpha_1, \alpha_2, \alpha_3 \in F^*$, $\psi(\alpha_1\alpha_2) \le \psi(\alpha_1) + \psi(\alpha_2)$, $\psi(\alpha_1\alpha_2\alpha_3) \ge \psi(\alpha_1\alpha_3)$, and $\psi(\alpha_1) \ge |\alpha_1|$, where $|\alpha_1|$ is the length of α_1. A complexity measure ψ is called bounded if, for any $\alpha_1, \alpha_2 \in F^*$, $\psi(\alpha_1) = 0$ if and only if $\alpha_1 = \lambda$, $\psi(\alpha_1) = \psi(\alpha_1')$ for any word α_1' obtained from α_1 by permutation of letters, $\psi(\alpha_1) \le \psi(\alpha_1\alpha_2)$, $\psi(\alpha_1\alpha_2) \le \psi(\alpha_1) + \psi(\alpha_2)$, and $\psi(\alpha_1) \ge |\alpha_1|$. It is easy to see that the depth is a bounded complexity measure and each bounded complexity measure is a limited complexity measure.

We investigate different models of decision problems. We begin with problems over information systems [59]. An information system is a pair consisting of a set of objects called the universe and a set of functions from the universe to a finite set. These functions are called attributes. A problem with many-valued decisions over this information system is defined by a finite number of attributes that divide the universe into finite number of domains in which the considered attributes have fixed values and a mapping, which corresponds to each domain a finite set of decisions. For a given object from the universe, we should find a decision from the set attached to the domain to which this object belongs. We consider also conventional problems for which a decision is attached to each domain. In this case, for a given object, we should return the decision attached to the domain to which this object belongs. We use queries each of which consists in the computation of the value of an attribute for the given object.

Next we consider decision tables. A decision table is a rectangular table in which columns are labeled with attributes, rows are pairwise different and each row is labeled with a finite set of decisions. Rows are interpreted as tuples of values of the considered attributes. For a given row, we should return a decision from the set of decisions attached to it. Besides the considered tables called decision tables with many-valued decisions and known also as multi-label decision tables [12, 76, 80], we study conventional decision tables in which rows are labeled with arbitrary decisions and decision tables with 0-1-decisions in which rows are labeled with decisions 0 or 1. In the last two cases, for a given row, we should recognize the decision attached to it. We use queries each of which consists in the recognition of the value of an attribute in the given row. In this book, we study decision tables from classes of tables closed under the operations of changing of decisions and removal of attributes (columns), and sometimes some other operations.

The last type of decision problems is related to the formal languages over finite alphabets. We fix length n of words from the language and consider two problems. The problem of recognition: for a given word of the length n from the language, we should recognize it. The problem of membership: for a given word of the length n over the considered alphabet, we should recognize if this word belongs to the language. To solve these problems we can use the following queries. For $i \in \{1, \dots, n\}$, we can ask what is the ith letter in the given word.

For the considered decision problems, we study relationships among three parameters: the minimum complexity of a deterministic decision tree, the minimum complexity of a nondeterministic decision tree, and the complexity of problem description. In particular, for the depth, the complexity of problem description is the number of attributes in it. For decision tables with 0-1-decisions, instead of nondeterministic decision trees, we consider strongly nondeterministic decision trees. For problems over information systems, we also study time-space trade-off for deterministic and nondeterministic decision trees using as time complexity the depth and as space complexity the number of nodes.

It is easy to transform a deterministic decision tree into a decision rule system. The inverse transformation is a more difficult task. In this book, we study unimprovable upper and lower bounds on the minimum depth of decision trees derived from decision rule systems depending on the various parameters of these systems. We also investigate the complexity of constructing decision trees and acyclic decision graphs representing decision trees from decision rule systems. Finally we study opportunities not to build the entire decision tree, but to describe the computation path in this tree for the given input.

1.2 Directions of Research

There are four main directions of the research corresponding to the four parts of the book.

1.2.1 *Problems over Information Systems*

The first part of the book is devoted to the study of problems over information systems. Most results on deterministic decision trees and decision rules is obtained for finite information systems represented often by decision tables [13, 17, 23, 61, 66, 69]. The results related to infinite information systems were achieved initially in the study of deterministic trees over linear and algebraic information systems. These results include lower [8, 21, 22, 24, 37, 73, 78, 79] and upper [20, 33, 38] bounds on the minimum depth of deterministic decision trees. To the best of our knowledge, the problems of complexity of decision trees over arbitrary infinite information systems were not considered prior to [40, 41] for deterministic and prior to [43, 44] for nondeterministic decision trees.

We developed two approaches to the study of deterministic and nondeterministic decision trees over arbitrary information systems: local, when the decision trees solving a problem can use only attributes from the problem description, and global, when the decision trees solving a problem can use arbitrary attributes from the considered information system [4, 40, 44, 50–53, 57]. In this part, we study decision trees within both frameworks and consider two topics of the research.

The first topic is related to the comparative analysis of the three parameters of problems with many-valued decisions over arbitrary information system: the complexity of a problem description, the minimum complexity of a deterministic decision tree solving this problem, and the minimum complexity of a nondeterministic decision tree solving the problem. The complexity of a problem description is equal to the complexity of a deterministic decision tree solving this problem, which sequentially computes values of all attributes from its description. We consider both general complexity measures and limited complexity measures including depth and weighted depth of decision trees.

We fix an information system and a complexity measure and, for each ordered pair of parameters, we consider two functions that give us upper and lower bounds for the first parameter depending on upper and lower bounds, respectively, for the second parameter. The resulting 18 functions well describe the relationships between the three parameters under consideration, but it is most likely impossible to list all such tuples of 18 functions. Instead of these functions, we study their types from the set $\{\alpha, \beta, \gamma, \delta, \epsilon\}$. Let f be a partial function from $\omega = \{0, 1, 2, \ldots\}$ to ω. If the function f has an infinite domain and it is bounded from above, then its type is equal to α. If the function f has an infinite domain, is not bounded from above, and the inequality $f(n) \geq n$ holds only for a finite set of numbers $n \in \omega$, then its type is equal to β, etc. Thus, we obtain 18-tuples of types of functions.

In the case of the global approach, we listed all possible seven 18-tuples of types of functions for general complexity measures and six 18-tuples of types of functions for limited complexity measures. In the case of the local approach, we listed all possible seven 18-tuples of types of functions for general complexity measures and five 18-tuples of types of functions for limited complexity measures. The obtained results allow us to understand not only types of relationships for each ordered pair of parameters, but also joint behavior of types of relationships for all ordered pairs of the considered three parameters.

Note that the book [53] includes similar studies but with different types of functions. Moreover, for functions characterizing the lower and upper bounds of the parameters, different sets of types of functions are used.

The second topic is related to the study of relationships between time and space complexity for deterministic and nondeterministic decision trees solving conventional problems over infinite binary information systems. In these systems, attributes have values from the set $\{0, 1\}$. As time and space complexity, we use the depth and the number of nodes in decision trees.

For the global approach, in the worst case, with the growth of the number of attributes in the problem description, (i) the minimum depth of deterministic decision trees grows either almost as logarithm or linearly, (ii) the minimum depth of nondeterministic decision trees either is bounded from above by a constant or grows linearly, (iii) the minimum number of nodes in deterministic decision trees has either polynomial or exponential growth, and (iv) the minimum number of nodes in nondeterministic decision trees has either polynomial or exponential growth. Thus, each of the considered four parameters has two types of growth. Based on these results, we divide the set of all infinite binary information systems into five nonempty complexity classes in which each parameter has fixed type of growth and tuples of types for different classes are different.

For these complexity classes, we study issues related to time-space trade-off for deterministic decision trees. To this end, we investigate pairs of functions (φ, ψ) such that, for any problem over the considered information system, there exists a deterministic decision tree over this system, which solves the problem and for which the depth is at most $\varphi(n)$ and the number of nodes is at most $\psi(n)$, where n is the number of attributes in the problem description. We also study similar pairs of functions (φ, ψ) for nondeterministic decision trees.

We are interested in minimizing the functions φ and ψ. The main questions we study are: is it possible to simultaneously minimize the functions under consideration, and if not, how close can one get to the pair of functions $(\varphi^{\min}, \psi^{\min})$, obtained as a result of independent minimization of the functions φ and ψ.

For the global approach, for four out of the five complexity classes and all information systems from these classes, we can minimize the functions φ and ψ for deterministic decision trees simultaneously. One complexity class contains both information systems for which we can minimize the functions φ and ψ for deterministic decision trees simultaneously and information systems for which it is impossible.

However, for each such information system, we found a pair of functions (φ, ψ) for deterministic decision trees, which is close to the pair $(\varphi^{\min}, \psi^{\min})$.

For two out of the five complexity classes and all information systems from these classes, we can minimize the functions φ and ψ for nondeterministic decision trees simultaneously. For the rest three classes, all information systems from these classes do not allow simultaneous minimization of functions φ and ψ for nondeterministic decision trees. For some information systems, we found pairs of functions (φ, ψ) for nondeterministic decision trees, which are close to the pairs $(\varphi^{\min}, \psi^{\min})$. For the rest of information systems, there are no pairs of functions (φ, ψ) for nondeterministic decision trees that are close to the pairs $(\varphi^{\min}, \psi^{\min})$.

For the local approach, we are doing similar research. The main differences related to the complexity classes are as follows: (i) the minimum depth of deterministic decision trees grows either as a logarithm or linearly and (ii) we divide the set of all infinite binary information systems into three nonempty complexity classes in which each of the considered four parameters has fixed type of growth and tuples of types for different classes are different.

For the local approach, for each complexity class and all information systems from this class, we can minimize the functions φ and ψ for deterministic decision trees simultaneously. The same situation is with two complexity classes and nondeterministic decision trees. For each information system from the third complexity class, we cannot minimize the functions φ and ψ for nondeterministic decision trees simultaneously. However, for each such information system, we found a pair of functions (φ, ψ) for nondeterministic decision trees, which is close to the pair $(\varphi^{\min}, \psi^{\min})$.

Both for the global and the local approaches, the conducted research allows us to understand joint behavior of types of growth of the considered four parameters and to obtain nontrivial results related to time-space trade-off for decision trees.

This part consists of Chaps. 2–5: two chapters on each topic (one related to the global approach, the other to the local one).

1.2.2 Decision Tables from Closed Classes

The second part of the book examines decision tables, which are a well-known way of presenting the information needed to make decisions. These tables are used, in particular, in data analysis, including classification problems, in modeling and studying combinatorial optimization problems, fault diagnosis, computational geometry, etc. [10, 17, 23, 28, 52, 57, 61, 63, 69]. Note that finite information systems with a selected decision attribute, data sets with a selected class attribute, and partially defined Boolean functions studied in various fields of data analysis as representations of decision problems can naturally be interpreted as decision tables.

In this part, we study decision tables from classes of tables closed relative to the operations of changing of decisions and removal of attributes (columns), and sometimes some other operations. The most natural examples of such classes are closed classes of decision tables generated by information systems, where each problem over the information system is transformed into a decision table. Note that the family of all closed classes is essentially wider than the family of closed classes generated by information systems. In particular, the union of two closed classes generated by two information systems is a closed class. However, generally, there is no an information system that generates this class.

The main aims of this part are to generalize results obtained earlier for problems over information systems in [51] to the case of decision tables from closed classes, obtain for closed classes of decision tables results similar to some statements proved in [53] for information systems, and obtain completely new results related to the closed classes.

Various classes of objects that are closed under different operations are intensively studied. Among them, in particular, are classes of Boolean functions closed under the operation of superposition [64] and minor-closed classes of graphs [68]. Decision tables represent an interesting mathematical object deserving mathematical research, in particular, the study of closed classes of decision tables.

This part continues the study of closed classes of decision tables that started by work [39] and frozen for various reasons for many years. In [39], we studied the dependence of the minimum depth of deterministic decision trees and the depth of deterministic decision trees constructed by a greedy algorithm on the number of attributes (columns) for conventional decision tables from classes closed under operations of removal of columns and changing of decisions. In the present part, we consider four topics of research.

The first topic is related to the investigation of classes of decision tables with many-valued decisions closed under operations of removal of columns, changing of decisions, permutation of columns, and duplication of columns. We consider both general complexity measures and limited complexity measures including depth and weighted depth of decision trees. We study relationships among the three parameters of these tables: the complexity of the decision table, the minimum complexity of a deterministic decision tree, and the minimum complexity of a nondeterministic decision tree. The complexity of a decision table is equal to the complexity of a deterministic decision tree for this table, which sequentially computes values of all attributes attached to its columns.

We fix a closed class of decision tables and a complexity measure (we call this pair a class-measure-pair or cm-pair in short) and, for each ordered pair of parameters, we consider two functions that give us upper and lower bounds for the first parameter depending on upper and lower bounds, respectively, for the second parameter. Instead of the resulting 18 functions, we study their types from the set $\{\alpha, \beta, \gamma, \delta, \varepsilon\}$ and obtain 18-tuples of types of functions.

We listed all possible seven 18-tuples of types of functions for general complexity measures and five 18-tuples of types of functions for limited complexity measures. The obtained results allow us to understand not only types of relationships for each ordered pair of parameters, but also joint behavior of types of relationships for all ordered pairs of the considered three parameters. These results generalize the results obtained in the first part of the book in Chap. 3. We also define the notion of a union of two cm-pairs and study the 18-tuple of types of functions for the resulting cm-pair depending on 18-tuples of types of functions for the initial cm-pairs.

The second topic is related to the investigation of classes of decision tables with many-valued decisions closed relative to removal of attributes (columns) and changing of decisions assigned to rows. We consider so called bounded complexity measures including depth and weighted depth of decision trees. For tables from an arbitrary closed class, we study three functions that characterize (i) the growth in the worst case of the minimum complexity of deterministic decision trees with the growth of the complexity of the set of attributes attached to columns, (ii) the growth in the worst case of the minimum complexity of nondeterministic decision trees with the growth of the complexity of the set of attributes attached to columns, and (iii) the growth in the worst case of the minimum complexity of deterministic decision trees with the growth of the minimum complexity of nondeterministic decision trees. Note that for bounded complexity measures, the complexity of the set of attributes attached to columns of a decision table is equal to the complexity of this table.

We proved that the first function is either bounded from above by a constant, or grows as a logarithm, or grows almost linearly (as a function depending on n, it is bounded from above by n and is equal to n for infinitely many n). The second function is either bounded from above by a constant or grows almost linearly. We indicated a criterion for the third function to be defined everywhere and described a wide range of behavior of this function, when it is defined everywhere. This function (as a function depending on n) is either bounded from above by a constant or is greater than or equal to n for infinitely many n. In particular, for any nondecreasing function φ such that $\varphi(n) \geq n$ and $\varphi(0) = 0$, the third function can grow between $\varphi(n)$ and $\varphi(n) + n$. We also indicated the conditions under which the third function is bounded from above by a polynomial.

The third topic is similar to the second, but instead of closed classes of decision tables with many-valued decisions, we consider classes of conventional decision tables closed relative to removal of attributes (columns) and changing of decisions assigned to rows. In this topic, we study three functions similar to functions studied in the second topic. The first two functions have the same types of behavior as in the second topic. For the third function, we have only a rough classification of its behavior, when it is everywhere defined: this function (as a function depending on n) is either bounded from above by a constant or is greater than or equal to n for infinitely many n. We also consider two more functions that characterize the growth in the worst case of the complexity of deterministic and nondeterministic decision trees constructed by some algorithms with the growth of the complexity of the set of attributes attached to columns. For these functions, we study their types of behavior.

The fourth topic is related to the investigation of classes of decision tables with 0-1-decisions closed relative to removal of attributes (columns) and changing of decisions assigned to rows. We consider bounded complexity measures including depth and weighted depth of decision trees. For tables from an arbitrary closed class, we study four functions that characterize (i) the growth in the worst case of the minimum complexity of deterministic decision trees with the growth of the complexity of the set of attributes attached to columns, (ii) the growth in the worst case of the minimum complexity of deterministic decision trees with the growth of the minimum complexity of tests (a test is a set of attributes whose values are sufficient to determine a decision), (iii) the growth in the worst case of the minimum complexity of strongly nondeterministic decision trees with the growth of the complexity of the set of attributes attached to columns, and (iv) the growth in the worst case of the minimum complexity of deterministic decision trees with the growth of the minimum complexity of strongly nondeterministic decision trees.

We proved that the first and the second functions are either bounded from above by a constant, or grow as a logarithm, or grow almost linearly. The third function is either bounded from above by a constant or grows almost linearly. We found a criterion for the fourth function to be defined everywhere and described a wide range of behavior of this function, when it is everywhere defined. This function (as a function depending on n) is either bounded from above by a constant or is greater than or equal to n for infinitely many n. In particular, the fourth function can grow as an arbitrary nondecreasing function φ such that $\varphi(n) \geq n$ and $\varphi(0) = 0$. We also indicated conditions for this function to be bounded from above by a polynomial.

Some results and proofs obtained in this part for decision tables with many-valued decisions, for conventional decision tables, and for decision tables with 0-1-decisions are similar. However, we decided to present these results with full proofs. The reason is that generally decision tables of different kinds can differ significantly. As an example, compare so-called complete decision tables with many-valued decisions and complete conventional decision tables. A decision table with n columns filled with zeros and ones is called complete if it has 2^n pairwise different rows. One can show that, for a conventional complete decision table, the minimum depth of a deterministic decision tree does not exceed the square of the minimum depth of a nondeterministic decision tree. This is a simple generalization of the results obtained for Boolean functions in [9, 26, 75] (see also [15]). In Chap. 7, we construct an infinite sequence of complete decision tables with many-valued decisions for which the minimum depth of deterministic decision trees does not bounded from above by a constant but the minimum depth of nondeterministic decision trees is at most 3.

This part consists of Chaps. 6–9, corresponding to the four topics discussed.

1.2.3 Recognition and Membership Problems for Formal Languages

In the third part, we study formal languages over a finite alphabet Σ. Let L be a language over the alphabet Σ and $L(n)$ be the set of words from this language of the length n. For this set, we investigate the minimum depth of deterministic and nondeterministic decision trees solving the recognition and the membership problems. In the case of recognition problem, for a given word from $L(n)$, we should recognize it using queries (attributes) each of which, for some $i \in \{1, \ldots, n\}$, returns the ith letter of the word. In the case of membership problem, for a given word over the alphabet Σ of the length n, we should recognize if it belongs to $L(n)$ using the same queries.

A well-known approach to evaluate complexity of an infinite language L over a finite alphabet Σ is to study its so-called combinatorial complexity (known also as counting function) $f_L(n)$ that is the number of words of the length n in L [19, 70]. For many years, one of the authors has been introducing new parameters of languages into scientific use: the minimum depth of deterministic and nondeterministic decision trees for the recognition and membership problems related to the language [45, 47, 52, 54, 55]. This way is more complicated, but can give more detailed classification of languages. It is easy to describe two languages (see introduction to Chap. 11) such that for both languages, the counting function grows linearly. For the first language, the minimum depth of deterministic decision trees solving the problem of recognition grows as a logarithm, but for the second language, the minimum depth of deterministic decision trees solving the problem of recognition grows linearly. The present part continues this line of research. In this part, we consider two topics.

The first topic is related to the study of arbitrary binary languages (languages over the alphabet $E = \{0, 1\}$) that are subword-closed: if a word $w_1 u_1 w_2 \cdots w_m u_m w_{m+1}$ belongs to a language, then the word $u_1 \cdots u_m$ belongs to this language. Subword-closed languages have attracted the attention of researchers in the field of formal languages for many years [5, 14, 25, 27, 58].

We proved that, for an arbitrary binary subword-closed language, with the growth of n, the minimum depth of deterministic decision trees solving the problem of recognition is either bounded from above by a constant, or grows as a logarithm, or linearly. For other types of trees and problems (nondeterministic decision trees solving the problem of recognition, and deterministic and nondeterministic decision trees solving the membership problem), with the growth of n, the minimum depth of decision trees is either bounded from above by a constant or grows linearly. We also studied joint behavior of the minimum depths of the considered four types of decision trees and described five complexity classes of binary subword-closed languages.

The second topic is related to the study of arbitrary regular factorial languages over a finite alphabet Σ, which are well known in the field of formal languages [6, 7, 18]. A factorial language satisfies the following condition: if a word $w_1 u w_2$ belongs to the language, then the word u also belongs to it. For a given problem (problem of recognition or membership problem) and type of decision trees solving the problem (deterministic or nondeterministic), instead of the minimum depth $h(n)$ of a decision tree of the considered type solving the problem for $L(n)$, we study the smoothed minimum depth $H(n) = \max\{h(m) : m \leq n\}$.

For an arbitrary regular factorial language, with the growth of n, the smoothed minimum depth of deterministic decision trees solving the problem of recognition is either bounded from above by a constant, or grows as a logarithm, or linearly. These results follow immediately from more general, obtained in [47] for arbitrary regular languages.

For other cases (nondeterministic decision trees solving the problem of recognition, and deterministic and nondeterministic decision trees solving the membership problem), with the growth of n, the smoothed minimum depth of decision trees is either bounded from above by a constant, or grows linearly. In the conference paper [45], a classification of arbitrary regular languages depending on the smoothed minimum depth of nondeterministic decision trees solving the problem of recognition was announced without proofs. In the present part, we consider simpler classification for regular factorial languages with full proofs.

As corollaries of the obtained results, we studied joint behavior of the smoothed minimum depths of decision trees for the considered four cases and described five complexity classes of regular factorial languages. We also investigated the class of regular factorial languages over the alphabet $E = \{0, 1\}$ each of which is given by one forbidden word.

It is clear that each subword-closed language is a factorial language. Moreover, each subword-closed language over a finite alphabet is a regular language [25]. One can show that the language $L(00)$ over the alphabet E given by one forbidden word 00 is a regular factorial language, which is not subword-closed. Therefore the class of subword-closed languages over the alphabet E is a proper subclass of the class of regular factorial languages over the alphabet E.

The main difference between the first and the second topics of research is that, in the first topic, we do not assume that the subword-closed languages are given by deterministic finite automata. Instead of this, we describe simple criteria (based on the presence in the language of words of special types) for the behavior of the minimum depths of deterministic and nondeterministic decision trees solving the problem of recognition. Differently formulated criteria for the behavior of the minimum depth of decision trees solving the recognition problem require very different proofs. One more difference is that in the first topic we directly consider the minimum depth of decision trees, not the smoothed minimum depth.

This part consists of Chaps. 10 and 11 corresponding to the two considered topics. Note that in these chapters, instead of talking about deterministic or nondeterministic decision trees solving a problem, we talk about decision trees solving a problem deterministically or nondeterministically.

1.2.4 Transforming Decision Rule Systems Into Deterministic Decision Trees

Regular methods for extracting decision rules from deterministic decision trees have been known for a long time [65–67]: first, each path in the decision tree from the root to a terminal node is assigned a decision rule, and then the resulting rules are simplified. In the fourth part, we study the more complex problem of transforming a decision rule system into a deterministic decision tree. There are at least three directions of research related to this problem:

- In a number of papers [1, 2, 29–32, 34, 35, 74], it was proposed to build decision trees or their generalizations, known as decision structures, in two stages: first, build decision rules based on the input data, and then build decision trees or decision structures based on the constructed rules. There are various methodological considerations and experimental results that justify this approach.

- The relationship between the depth of deterministic decision trees and nondeterministic decision trees (representing systems of decision rules) has been studied for both decision tables and problems over finite and infinite information systems [44, 48, 50, 51, 53]. The most famous results in this direction are related to decision trees for computing Boolean functions [9, 26, 42, 75].
- One of the authors of this book begun in [46, 49] the development of the so-called syntactic approach to the study of the considered problem, which assumes that we do not know input data but only have a system of decision rules that must be transformed into decision tree. This part continues work in this direction. The results obtained in [46, 49] are described in this part.

Let there be a system S of decision rules of the form

$$(a_{i_1} = \delta_1) \wedge \cdots \wedge (a_{i_m} = \delta_m) \rightarrow \sigma \ ,$$

where $a_{i_1}, \ldots, a_{i_m}$ are attributes, $\delta_1, \ldots, \delta_m$ are values of these attributes, and σ is a decision. We study three algorithmic problems associated with this system:

- For a given input (a tuple of values of all attributes included in S), it is necessary to find at least one rule that is realizable for this input (having a true left-hand side) or show that there are no such rules.
- For a given input, it is necessary to find all the rules that are realizable for this input, or show that there are no such rules.
- For a given input, it is necessary to find all the right-hand sides of rules that are realizable for this input (we should return realizable rules with these right-hand sides), or show that there are no such rules.

For each problem, we consider two its variants. The first assumes that in the input each attribute can have only those values that occur for this attribute in the system S. In the second case, we assume that in the input any attribute can have any value.

Our goal is to minimize the number of queries for attribute values in the given input. For this purpose, deterministic decision trees are studied as algorithms for solving the described six problems. In this part, we consider two topics of research.

The first topic is related to the study of unimprovable upper and lower bounds on the minimum depth of decision trees depending on three parameters of the decision rule system S: the total number $n(S)$ of different attributes in the rules belonging to the system S, the maximum length $d(S)$ of a decision rule, and the maximum number $k(S)$ of attribute values.

To illustrate the results obtained, consider the problem $AR(S)$ for a system of decision rules S: for a given input containing only values of attributes that occur in the system S, it is necessary to find all the rules from S that are realizable for this input.

For this problem, we study lower $h_{AR}(n, d, k)$ and upper $H_{AR}(n, d, k)$ unimprovable bounds on the minimum depth of decision trees solving the problem $AR(S)$ for systems of decision rules S such that $n(S) = n$, $d(S) = d$, and $k(S) = k$. Let $n, d, k \in \omega \setminus \{0\}$ and $d \leq n$. We prove that $H_{AR}(n, d, k) = n$, and $h_{AR}(n, d, k) = n$ if $k = 1$ or $d = 1$ and $\max\left\{d, \frac{n(k-1)}{k^d}\right\} \leq h_{AR}(n, d, k) \leq d + \frac{n}{k^{d-1}}$ if $k \geq 2$ and $d \geq 2$.

We show that not only for the problem $AR(S)$, but for each of the considered problems, there are systems of decision rules for which the minimum depth of the decision trees that solve the problem is much less than the total number of attributes in the rule system. For such systems of decision rules, it is advisable to use decision trees.

To illustrate the process of transformation of decision rule systems into decision trees, we generalize well known inequality $h(f) \leq C(f)^2$ [9, 26, 75], where $h(f)$ is the minimum depth of a deterministic decision tree computing Boolean function f and $C(f)$ is the certificate complexity of f (see details in [15]), to the case of functions of k-valued logic. Note that the certificate complexity $C(f)$ of the Boolean function f is equal to the minimum depth of a nondeterministic decision tree computing this function.

The second topic is related to the problems of constructing deterministic decision trees and decision graphs representing these trees, and to the opportunities not to build the entire deterministic decision tree, but to describe the computation path in this tree for a given input.

To illustrate the obtained results, we consider the problem $AR(S)$. First, we study possibilities to derive decision trees from decision rule systems using polynomial time algorithms. We denote by $P_{DT}(AR)$ the set of pairs $(k, d) \in (\omega \setminus \{0\})^2$ satisfying the following condition: there exists a polynomial time algorithm that, for an arbitrary decision rule system S with $k(S) = k$ and $d(S) = d$, constructs a decision tree solving the problem $AR(S)$. We prove that $P_{DT}(AR) = \{(1, d) : d \in \omega \setminus \{0\}\}$.

Besides decision trees, we study acyclic decision graphs that are defined similarly to the decision trees, but instead of finite directed trees with roots, finite directed graphs with roots that do not have directed cycles are considered. We denote by $P_{DG}(AR)$ the set of pairs $(k, d) \in (\omega \setminus \{0\})^2$ satisfying the following condition: there exists a polynomial time algorithm that, for an arbitrary decision rule system S with $k(S) = k$ and $d(S) = d$, constructs an acyclic decision graph solving the problem $AR(S)$. We prove that $P_{DG}(AR) = \{(1, d) : d \in \omega \setminus \{0\}\}$, i.e., for the problem $AR(S)$ (unlike some other problems), the use of acyclic decision graphs cannot improve the situation with the existence of polynomial time algorithms.

These difficulties can be circumvented by considering acyclic decision graphs with writing, which, in addition to nodes labeled with attributes and one terminal node, have writing nodes labeled each with a decision rule. We begin the work with the empty set of rules and each time, when we come to a writing node that is labeled with a decision rule, we add this rule to the set under construction. When we reach the terminal node, the set of rules formed by this moment is the result of the work of the considered acyclic decision graph with writing.

We denote by $P_{DGW}(AR)$ the set of pairs $(k, d) \in (\omega \setminus \{0\})^2$ satisfying the following condition: there exists a polynomial time algorithm that, for an arbitrary decision rule system S with $k(S) = k$ and $d(S) = d$, constructs an acyclic decision graph with writing solving the problem $AR(S)$. We prove that $P_{DGW}(AR) = (\omega \setminus \{0\})^2$.

It would be possible to move on to the study of acyclic decision graphs and acyclic decision graphs with writing. However, when constructing them, one should simultaneously try to make the depth as small as possible without allowing an excessive increase in the number of nodes. The possibilities of such bi-criteria optimization are the subject of a special study in the future.

In the second topic, we consider another approach: instead of constructing the entire decision tree, we restrict ourselves to the consideration of polynomial time algorithms that, for a given tuple of attribute values, describe the work of the decision tree on this tuple. To this end, we study bounds on the depth of decision trees and algorithms for the description of the decision tree work based on node covers for the hypergraph corresponding to the considered decision rule system (nodes of this hypergraph are attributes from the rule system and edges correspond to rules).

We mention here only one result for the problem $AR(S)$. We developed a polynomial time algorithm, which models the operation of a decision tree that solves the problem $AR(S)$ for a given tuple of attribute values. The depth of this tree does not exceed the cube of the minimum depth of a decision tree solving the problem $AR(S)$. For the remaining problems considered, we are conducting similar studies.

We also described dynamic programming algorithms for the computation of the minimum depth of decision trees.

This part consists of Chaps. 12 and 13 corresponding to the two topics discussed.

1.3 Some Conclusions

Several topics run through most of the book. Such topics include the study of the behavior of functions that characterize the complexity of decision trees, separately for each function. In the previous sections of the introduction we paid enough attention to this. In the present section, we look at another topic: how these functions behave together. The conclusions obtained are important for the purposes of this book.

In the first three parts of the book, we study mainly three parameters of decision problems represented either as problems over information systems, or as decision tables from closed classes, or as problems related to formal languages: the complexity of problem description, the minimum complexity of a deterministic decision tree solving the problem, and the minimum complexity of a nondeterministic decision tree solving the problem. The complexity of problem description is equal to the complexity of the deterministic decision tree that solves the problem by sequential computation of values of all attributes from the problem description. In the case of the depth, this parameter is equal to the number of attributes in the problem description.

A distinctive feature of this book is the study of the joint behavior of functions characterizing relationships between pairs of these parameters. In this section, we consider the joint behavior of the following three most studied functions: the function $\mathscr{D}$, which characterizes the growth in the worst case of the minimum complexity of deterministic decision trees with the growth of the complexity of problem description, the function $\mathscr{N}$, which characterizes the growth in the worst case of the minimum complexity of nondeterministic decision trees with the growth of the complexity of the problem description, and the function $\mathscr{H}$, which characterizes the growth in the worst case of the minimum complexity of deterministic decision trees with the growth of the minimum complexity of nondeterministic decision trees (the generalized notation $\mathscr{D}$, $\mathscr{N}$, and $\mathscr{H}$ are used only in this section). First, we consider the joint behavior of the functions $\mathscr{D}$ and $\mathscr{N}$.

In Chaps. 2, 3, and 6, we looked at types of functions instead of functions. In particular, for a function $f \in \{\mathscr{D}, \mathscr{N}\}$,

- The type α means that the function f has infinite domain and is bounded from above by a constant.
- The type β means that the function f has infinite domain, is not bounded from above by a constant, the inequality $f(n) \leq n$ is true for all n, and the equality $f(n) = n$ holds only for a finite number of n.
- The type γ means that the function f has infinite domain, the inequality $f(n) \leq n$ is true for all n, and the equality $f(n) = n$ holds for infinite number of n.

For general complexity measures, for problems over information systems studied in the frameworks of global and local approaches (see Chaps. 2 and 3) and for decision tables from closed classes (see Chap. 6), all possible pairs of types of the functions $\mathscr{D}$ and $\mathscr{N}$ are listed in Table A depicted in Fig. 1.4.

Let us consider now limited complexity measures. For problems over information systems studied in the framework of the global approach (see Chap. 2), all possible pairs of types of the functions $\mathscr{D}$ and $\mathscr{N}$ are listed in Table A depicted in Fig. 1.4. For problems over information systems studied in the framework of the local approach (see Chap. 3) and for decision tables from closed classes (see Chap. 6), all possible pairs of types of the functions $\mathscr{D}$ and $\mathscr{N}$ are listed in Table B depicted in Fig. 1.4. The same results hold for the depth.

In Chaps. 4 and 5, we study the depth of decision trees for problems over infinite binary information systems in the frameworks of global and local approaches, respectively. In the case of the global approach, all possible pairs of types of the functions $\mathscr{D}$ and $\mathscr{N}$ coincide with the pairs listed in Table A depicted in Fig. 1.4 with the exception of the first pair $\alpha\alpha$. The type β means here that, for the considered function $f \in \{\mathscr{D}, \mathscr{N}\}$, $f(n) = \Omega(\log(n))$ and $f(n) = O(\log(n)^{1+\varepsilon})$, where ε is an arbitrary positive real number, and the type γ means that $f(n) = n$ for any n. In

$A =$

$\mathscr{D}$	$\mathscr{N}$
α	α
β	α
γ	α
γ	γ

$B =$

$\mathscr{D}$	$\mathscr{N}$
α	α
β	α
γ	γ

Fig. 1.4 Types of joint behavior of functions $\mathscr{D}$ and $\mathscr{N}$ for problems over information systems and decision tables from closed classes

the case of the local approach, all possible pairs of types of the functions $\mathscr{D}$ and $\mathscr{N}$ coincide with the pairs listed in Table B depicted in Fig. 1.4 with the exception of the first pair $\alpha\alpha$. The type β means here that $f(n) = \Theta(\log n)$ and the type γ means that $f(n) = n$ for any n. Both for the global and the local approaches, the pair $\alpha\alpha$ is possible if we add to the consideration the finite binary information systems.

In Chaps. 7 and 8, we study closed classes of decision tables and bounded complexity measures. We show that all possible pairs of types of the functions $\mathscr{D}$ and $\mathscr{N}$ are listed in Table B depicted in Fig. 1.4. The type β means here that, for the considered function $f \in \{\mathscr{D}, \mathscr{N}\}$, $f(n) = \Theta(\log n)$. For the depth, the type γ means that $f(n) = n$ for any n.

In Chap. 10, for binary subword-closed languages, we investigate the depth of decision trees solving the recognition and the membership problems. The possible joint behavior of the functions $\mathscr{D}$ and $\mathscr{N}$ for the recognition problem is described in Table A depicted in Fig. 1.5. For the membership problem, the possible joint behavior of the functions $\mathscr{D}$ and $\mathscr{N}$ is described in Table B depicted in Fig. 1.5.

As we see, for the membership problem, it is impossible that the function $\mathscr{D}$ is not bounded from above by a constant, but the function $\mathscr{N}$ is bounded from above by a constant. It's easy to explain. The decision table corresponding to the membership problem represents a Boolean function. For each Boolean function, the minimum depth of a deterministic decision tree does not exceed the square of the minimum depth of a nondeterministic decision tree [9, 26, 75]. Therefore, if the function $\mathscr{N}$ is bounded from above by a constant, then the function $\mathscr{D}$ is bounded from above by a constant too.

Now let us briefly characterize the joint behavior of all three functions $\mathscr{D}$, $\mathscr{N}$, and $\mathscr{H}$ for problems over information systems and for decision tables from closed classes. If both functions $\mathscr{D}$ and $\mathscr{N}$ are bounded from above by constants (have types $\alpha\alpha$), then the function $\mathscr{H}$ is also bounded from above by a constant. If the function $\mathscr{N}$ is bounded from above by a constant but the function $\mathscr{D}$ is not bounded from above by a constant ($\mathscr{D}$ and $\mathscr{N}$ have types $\beta\alpha$ or $\gamma\alpha$), then the function $\mathscr{H}$ has a finite domain. If both functions $\mathscr{D}$ and $\mathscr{N}$ are not bounded from above by constants (have types $\gamma\gamma$), then the function $\mathscr{H}$ has either finite or infinite domain. In the latter case, the inequality $\mathscr{H}(n) \geq n$ holds for infinitely many n.

If the function $\mathscr{H}$ has an infinite domain, then its behavior can be very diverse. In particular, for any nondecreasing function $\varphi : \omega \to \omega$ such that $\varphi(n) \geq n$ and $\varphi(0) = 0$, the function $\mathscr{H}(n)$ can grow between $\varphi(n)$ and $\varphi(n) + n$. This result is obtained in Chap. 7, where we study closed classes of decision tables with many-valued decisions and bounded complexity measures. In this chapter, we also found the

$A =$

$\mathscr{D}$	$\mathscr{N}$
$O(1)$	$O(1)$
$\Theta(\log n)$	$O(1)$
$\Theta(n)$	$\Theta(n)$

$B =$

$\mathscr{D}$	$\mathscr{N}$
$O(1)$	$O(1)$
$\Theta(n)$	$\Theta(n)$

Fig. 1.5 Joint behavior of functions $\mathscr{D}$ and $\mathscr{N}$ for problems related to formal languages

condition for the function $\mathscr{H}(n)$ to have an infinite domain and indicated conditions for this function to be bounded from above by a polynomial on n.

1.4 Use of Book

This book covers a variety of material with the general goal of understanding the relationship between deterministic decision trees and systems of decision rules, often represented by nondeterministic decision trees. To help readers cope with this diversity, we have included a fairly detailed introduction and made the chapters largely independent, sometimes at the cost of repeating definitions and statements.

The results obtained in the book may be useful to researchers working with decision trees and decision rule systems in data analysis, especially in rough set theory [61, 62], logical analysis of data [10, 11], and test theory [16, 77]. The book can also be used to create courses for graduate students.

References

1. Abdelhalim, A., Traoré, I., Nakkabi, Y.: Creating decision trees from rules using RBDT-1. Comput. Intell. **32**(2), 216–239 (2016)
2. Abdelhalim, A., Traoré, I., Sayed, B.: RBDT-1: A new rule-based decision tree generation technique. In: G. Governatori, J. Hall, A. Paschke (eds.) Rule Interchange and Applications, International Symposium, RuleML 2009, Las Vegas, Nevada, USA, November 5-7, 2009. Proceedings, Lecture Notes in Computer Science, vol. 5858, pp. 108–121. Springer (2009)
3. AbouEisha, H., Amin, T., Chikalov, I., Hussain, S., Moshkov, M.: Extensions of Dynamic Programming for Combinatorial Optimization and Data Mining. Intelligent Systems Reference Library, vol. 146. Springer (2019)
4. Alsolami, F., Azad, M., Chikalov, I., Moshkov, M.: Decision and Inhibitory Trees and Rules for Decision Tables with Many-valued Decisions. Intelligent Systems Reference Library, vol. 156. Springer (2020)
5. Atminas, A., Lozin, V.V.: Deciding atomicity of subword-closed languages. In: Diekert, V., Volkov, M.V. (eds.) Developments in Language Theory - 26th International Conference, DLT 2022, Tampa, FL, USA, May 9-13, 2022, Proceedings, Lecture Notes in Computer Science, vol. 13257, pp. 69–77. Springer (2022)
6. Atminas, A., Lozin, V.V., Moshkov, M.: WQO is decidable for factorial languages. Inf. Comput. **256**, 321–333 (2017)
7. Avgustinovich, S.V., Frid, A.E.: Canonical decomposition of a regular factorial language. In: Grigoriev, D., Harrison, J., Hirsch, E.A. (eds.) Computer Science - Theory and Applications, First International Symposium on Computer Science in Russia, CSR 2006, St. Petersburg, Russia, June 8-12, 2006, Proceedings, Lecture Notes in Computer Science, vol. 3967, pp. 18–22. Springer (2006)
8. Ben-Or, M.: Lower bounds for algebraic computation trees (preliminary report). In: 15th Annual ACM Symposium on Theory of Computing, STOC 1983, pp. 80–86 (1983)
9. Blum, M., Impagliazzo, R.: Generic oracles and oracle classes (extended abstract). In: 28th Annual Symposium on Foundations of Computer Science, Los Angeles, California, USA, 27-29 October 1987, pp. 118–126. IEEE Computer Society (1987)

10. Boros, E., Hammer, P.L., Ibaraki, T., Kogan, A.: Logical analysis of numerical data. Math. Program. **79**, 163–190 (1997)
11. Boros, E., Hammer, P.L., Ibaraki, T., Kogan, A., Mayoraz, E., Muchnik, I.B.: An implementation of logical analysis of data. IEEE Trans. Knowl. Data Eng. **12**(2), 292–306 (2000)
12. Boutell, M.R., Luo, J., Shen, X., Brown, C.M.: Learning multi-label scene classification. Pattern Recognit. **37**(9), 1757–1771 (2004)
13. Breiman, L., Friedman, J.H., Olshen, R.A., Stone, C.J.: Classification and Regression Trees. Wadsworth and Brooks (1984)
14. Brzozowski, J.A., Jirásková, G., Zou, C.: Quotient complexity of closed languages. Theory Comput. Syst. **54**(2), 277–292 (2014)
15. Buhrman, H., de Wolf, R.: Complexity measures and decision tree complexity: a survey. Theor. Comput. Sci. **288**(1), 21–43 (2002)
16. Chegis, I.A., Yablonskii, S.V.: Logical methods of electric circuit control. Trudy MIAN SSSR **51**, 270–360 (1958). (in Russian)
17. Chikalov, I., Lozin, V.V., Lozina, I., Moshkov, M., Nguyen, H.S., Skowron, A., Zielosko, B.: Three Approaches to Data Analysis - Test Theory, Rough Sets and Logical Analysis of Data. Intelligent Systems Reference Library, vol. 41. Springer (2013)
18. Crochemore, M., Mignosi, F., Restivo, A.: Automata and forbidden words. Inf. Process. Lett. **67**(3), 111–117 (1998)
19. D'Alessandro, F., Intrigila, B., Varricchio, S.: On the structure of the counting function of sparse context-free languages. Theor. Comput. Sci. **356**(1–2), 104–117 (2006)
20. Dobkin, D.P., Lipton, R.J.: Multidimensional searching problems. SIAM J. Comput. **5**(2), 181–186 (1976)
21. Dobkin, D.P., Lipton, R.J.: A lower bound of the $(1/2)n^2$ on linear search programs for the knapsack problem. J. Comput. Syst. Sci. **16**(3), 413–417 (1978)
22. Dobkin, D.P., Lipton, R.J.: On the complexity of computations under varying sets of primitives. J. Comput. Syst. Sci. **18**(1), 86–91 (1979)
23. Fürnkranz, J., Gamberger, D., Lavrac, N.: Foundations of Rule Learning. Springer, Cognitive Technologies (2012)
24. Grigoriev, D., Karpinski, M., Vorobjov, N.: Improved lower bound on testing membership to a polyhedron by algebraic decision trees. In: 36th Annual Symposium on Foundations of Computer Science, FOCS 1995, pp. 258–265 (1995)
25. Haines, L.H.: On free monoids partially ordered by embedding. J. Comb. Theory **6**, 94–98 (1969)
26. Hartmanis, J., Hemachandra, L.A.: One-way functions, robustness, and the non-isomorphism of NP-complete sets. In: Proceedings of the Second Annual Conference on Structure in Complexity Theory, Cornell University, Ithaca, New York, USA, June 16-19, 1987. IEEE Computer Society (1987)
27. Hospodár, M.: Power, positive closure, and quotients on convex languages. Theor. Comput. Sci. **870**, 53–74 (2021)
28. Humby, E.: Programs from Decision Tables, Computer Monographs, vol. 19. Macdonald, London and American Elsevier, New York (1973)
29. Imam, I.F., Michalski, R.S.: Learning decision trees from decision rules: A method and initial results from a comparative study. J. Intell. Inf. Syst. **2**(3), 279–304 (1993)
30. Imam, I.F., Michalski, R.S.: Should decision trees be learned from examples or from decision rules? In: Komorowski, H.J., Ras, Z.W. (eds.) Methodologies for Intelligent Systems, 7th International Symposium, ISMIS '93, Trondheim, Norway, June 15-18, 1993, Proceedings, Lecture Notes in Computer Science, vol. 689, pp. 395–404. Springer (1993)
31. Imam, I.F., Michalski, R.S.: Learning for decision making: the FRD approach and a comparative study. In: Ras, Z.W., Michalewicz, M. (eds.) Foundations of Intelligent Systems, 9th International Symposium, ISMIS '96, Zakopane, Poland, June 9-13, 1996, Proceedings, Lecture Notes in Computer Science, vol. 1079, pp. 428–437. Springer (1996)

32. Kaufman, K.A., Michalski, R.S., Pietrzykowski, J., Wojtusiak, J.: An integrated multi-task inductive database VINLEN: initial implementation and early results. In: Dzeroski, S., Struyf, J. (eds.) Knowledge Discovery in Inductive Databases, 5th International Workshop, KDID 2006, Berlin, Germany, September 18, 2006, Revised Selected and Invited Papers, Lecture Notes in Computer Science, vol. 4747, pp. 116–133. Springer (2006)
33. Meyer auf der Heide, F.: A polynomial linear search algorithm for the n-dimensional knapsack problem. In: 15th Annual ACM Symposium on Theory of Computing, STOC 1983, pp. 70–79 (1983)
34. Michalski, R.S., Imam, I.F.: Learning problem-oriented decision structures from decision rule: The AQDT-2 system. In: Ras, Z.W., Zemankova, M. (eds.) Methodologies for Intelligent Systems, 8th International Symposium, ISMIS '94, Charlotte, North Carolina, USA, October 16-19, 1994, Proceedings, Lecture Notes in Computer Science, vol. 869, pp. 416–426. Springer (1994)
35. Michalski, R.S., Imam, I.F.: On learning decision structures. Fundam. Inform. **31**(1), 49–64 (1997)
36. Molnar, C.: Interpretable Machine Learning. A Guide for Making Black Box Models Explainable, 2 edn. (2022). https://christophm.github.io/interpretable-ml-book/
37. Morávek, J.: A localization problem in geometry and complexity of discrete programming. Kybernetika **8**(6), 498–516 (1972)
38. Moshkov, M.: On conditional tests. Sov. Phys. Dokl. **27**, 528–530 (1982)
39. Moshkov, M.: On depth of conditional tests for tables from closed classes. In: Markov, A.A. (ed.) Combinatorial-Algebraic and Probabilistic Methods of Discrete Analysis, pp. 78–86. Gorky University Press, Gorky (1989). (in Russian)
40. Moshkov, M.: Decision Trees. Theory and Applications. Nizhny Novgorod University Publishers, Nizhny Novgorod (1994). (in Russian)
41. Moshkov, M.: Optimization problems for decision trees. Fundam. Inform. **21**(4), 391–401 (1994)
42. Moshkov, M.: About the depth of decision trees computing Boolean functions. Fundam. Inform. **22**(3), 203–215 (1995)
43. Moshkov, M.: Two approaches to investigation of deterministic and nondeterministic decision trees complexity. In: 2nd World Conference on the Fundamentals of Artificial Intelligence, WOCFAI 1995, pp. 275–280 (1995)
44. Moshkov, M.: Comparative analysis of deterministic and nondeterministic decision tree complexity. Global approach. Fundam. Inform. **25**(2), 201–214 (1996)
45. Moshkov, M.: Complexity of deterministic and nondeterministic decision trees for regular language word recognition. In: Bozapalidis, S. (ed.) Proceedings of the 3rd International Conference Developments in Language Theory, DLT 1997, Thessaloniki, Greece, July 20–23, 1997, pp. 343–349. Aristotle University of Thessaloniki (1997)
46. Moshkov, M.: Some relationships between decision trees and decision rule systems. In: Polkowski, L., Skowron, A. (eds.) Rough Sets and Current Trends in Computing, First International Conference, RSCTC'98, Warsaw, Poland, June 22-26, 1998, Proceedings, Lecture Notes in Computer Science, vol. 1424, pp. 499–505. Springer (1998)
47. Moshkov, M.: Decision trees for regular language word recognition. Fundam. Inform. **41**(4), 449–461 (2000)
48. Moshkov, M.: Deterministic and nondeterministic decision trees for rough computing. Fundam. Inform. **41**(3), 301–311 (2000)
49. Moshkov, M.: On transformation of decision rule systems into decision trees. In: Proceedings of the Seventh International Workshop Discrete Mathematics and its Applications, Moscow, Russia, January 29 – February 2, 2001, Part 1, pp. 21–26. Center for Applied Investigations of Faculty of Mathematics and Mechanics, Moscow State University (2001). (in Russian)
50. Moshkov, M.: Classification of infinite information systems depending on complexity of decision trees and decision rule systems. Fundam. Inform. **54**(4), 345–368 (2003)
51. Moshkov, M.: Comparative analysis of deterministic and nondeterministic decision tree complexity. Local approach. In: Peters, J.F., Skowron, A. (eds.) Transactions on Rough Sets IV, Lecture Notes in Computer Science, vol. 3700, pp. 125–143. Springer (2005)

52. Moshkov, M.: Time complexity of decision trees. In: Peters, J.F., Skowron, A. (eds.) Transactions on Rough Sets III, Lecture Notes in Computer Science, vol. 3400, pp. 244–459. Springer (2005)
53. Moshkov, M.: Comparative Analysis of Deterministic and Nondeterministic Decision Trees, Intelligent Systems Reference Library, vol. 179. Springer (2020)
54. Moshkov, M.: Decision trees for regular factorial languages. Array **15**, 100,203 (2022)
55. Moshkov, M.: Decision trees for binary subword-closed languages. Entropy **25**(2), 349 (2023)
56. Moshkov, M., Piliszczuk, M., Zielosko, B.: Partial Covers, Reducts and Decision Rules in Rough Sets - Theory and Applications, Studies in Computational Intelligence, vol. 145. Springer (2008)
57. Moshkov, M., Zielosko, B.: Combinatorial Machine Learning - A Rough Set Approach, Studies in Computational Intelligence, vol. 360. Springer (2011)
58. Okhotin, A.: On the state complexity of scattered substrings and superstrings. Fundam. Inform. **99**(3), 325–338 (2010)
59. Pawlak, Z.: Information systems theoretical foundations. Inf. Syst. **6**(3), 205–218 (1981)
60. Pawlak, Z.: Rough sets. Int. J. Parallel Prog. **11**(5), 341–356 (1982)
61. Pawlak, Z.: Rough Sets - Theoretical Aspects of Reasoning about Data, Theory and Decision Library: Series D, vol. 9. Kluwer (1991)
62. Pawlak, Z., Skowron, A.: Rudiments of rough sets. Inf. Sci. **177**(1), 3–27 (2007)
63. Pollack, S.L., Hicks, H.T., Harrison, W.J.: Decision Tables: Theory and Practice. Wiley (1971)
64. Post, E.: Two-valued Iterative Systems of Mathematical Logic, Annals of Math. Studies, vol. 5. Princeton University Press, Princeton-London (1941)
65. Quinlan, J.R.: Generating production rules from decision trees. In: McDermott, J.P. (ed.) Proceedings of the 10th International Joint Conference on Artificial Intelligence. Milan, Italy, August 23-28, 1987, pp. 304–307. Morgan Kaufmann (1987)
66. Quinlan, J.R.: C4.5: Programs for Machine Learning. Morgan Kaufmann (1993)
67. Quinlan, J.R.: Simplifying decision trees. Int. J. Hum Comput Stud. **51**(2), 497–510 (1999)
68. Robertson, N., Seymour, P.D.: Graph minors. XX. Wagner's conjecture. J. Comb. Theory, Ser. B **92**(2), 325–357 (2004)
69. Rokach, L., Maimon, O.: Data Mining with Decision Trees - Theory and Applications, Series in Machine Perception and Artificial Intelligence, vol. 69. World Scientific (2007)
70. Shur, A.M.: Combinatorial complexity of regular languages. In: Hirsch, E.A., Razborov, A.A., Semenov, A.L., Slissenko, A. (eds.) Computer Science - Theory and Applications, Third International Computer Science Symposium in Russia, CSR 2008, Moscow, Russia, June 7-12, 2008, Proceedings, Lecture Notes in Computer Science, vol. 5010, pp. 289–301. Springer (2008)
71. Skowron, A., Jankowski, A., Swiniarski, R.W.: Foundations of rough sets. In: Kacprzyk, J., Pedrycz, W. (eds.) Springer Handbook of Computational Intelligence, Springer Handbooks, pp. 331–348. Springer (2015)
72. Skowron, A., Rauszer, C.: The discernibility matrices and functions in information systems. In: Slowinski, R. (ed.) Intelligent Decision Support - Handbook of Applications and Advances of the Rough Sets Theory, Theory and Decision Library, vol. 11, pp. 331–362. Springer (1992)
73. Steele, J.M., Yao, A.C.: Lower bounds for algebraic decision trees. J. Algorithms **3**(1), 1–8 (1982)
74. Szydlo, T., Sniezynski, B., Michalski, R.S.: A rules-to-trees conversion in the inductive database system VINLEN. In: Klopotek, M.A., Wierzchon, S.T., Trojanowski, K. (eds.) Intelligent Information Processing and Web Mining, Proceedings of the International IIS: IIPWM'05 Conference held in Gdansk, Poland, June 13-16, 2005, Advances in Soft Computing, vol. 31, pp. 496–500. Springer (2005)
75. Tardos, G.: Query complexity, or why is it difficult to separate $NP^A \cap coNP^A$ from P^A by random oracles A? Comb. **9**(4), 385–392 (1989)
76. Vens, C., Struyf, J., Schietgat, L., Dzeroski, S., Blockeel, H.: Decision trees for hierarchical multi-label classification. Mach. Learn. **73**(2), 185–214 (2008)

77. Yablonskii, S.V., Chegis, I.A.: On tests for electric circuits. Uspekhi Matematicheskikh Nauk **10**(4), 182–184 (1955). (in Russian)
78. Yao, A.C.: Algebraic decision trees and Euler characteristics. In: 33rd Annual Symposium on Foundations of Computer Science, FOCS 1992, pp. 268–277 (1992)
79. Yao, A.C.: Decision tree complexity and Betti numbers. In: 26th Annual ACM Symposium on Theory of Computing, STOC 1994, pp. 615–624 (1994)
80. Zhou, Z., Zhang, M., Huang, S., Li, Y.: Multi-instance multi-label learning. Artif. Intell. **176**(1), 2291–2320 (2012)

Part I
Problems Over Information Systems

The first part of the book is devoted to the investigation of problems over information systems. In this part, we study decision trees for these problems within both global and local frameworks and consider two topics of the research. The first topic is related to the comparative analysis of the three parameters of problems with many-valued decisions over arbitrary information systems: the complexity of a problem description, the minimum complexity of a deterministic decision tree solving this problem, and the minimum complexity of nondeterministic decision tree solving the problem. The second topic is related to the study of relationships between time and space complexity for deterministic and nondeterministic decision trees solving conventional problems over infinite binary information systems. As time and space complexity, we consider the depth and the number of nodes in decision trees. This part consists of the four chapters.

Chapters 2 and 3 are devoted to the comparative analysis of the three parameters of problems with many-valued decisions over arbitrary information systems in the frameworks of the global and local approaches, respectively.

Chapters 4 and 5 are devoted to the study of relationships between time and space complexity for deterministic and nondeterministic decision trees solving conventional problems over infinite binary information systems in the frameworks of the global and local approaches, respectively.

Chapter 2
Comparative Analysis of Deterministic and Nondeterministic Decision Tree Complexity. Global Approach

2.1 Introduction

In this chapter, we consider arbitrary (finite and infinite) information systems and study problems with many-valued decisions over these information systems. An information system [5] consists of a set of objects called universe and a set of attributes that are functions from the universe to a finite set. Any problem is specified by a number of attributes that divide the universe into domains in which these attributes have fixed values. A nonempty finite set of decisions is attached to each domain. For a given object from the universe, it is required to find a decision from the set attached to the domain containing this object.

Problems with many-valued decisions arise naturally in such areas as combinatorial optimization, fault diagnosis, computational geometry. For example, let $g_1, \ldots, g_m$ be functions from $\mathbb{R}^n$ to $\mathbb{R}$, where $\mathbb{R}$ is the set of real numbers. For a given $\bar{a} \in \mathbb{R}^n$, it is required to find a number from the set

$$D(\bar{a}) = \{i : i \in \{1, \ldots, m\}, g_i(\bar{a}) = \min\{g_1(\bar{a}), \ldots, g_m(\bar{a})\}\} .$$

If, for some $\bar{a}$, the set $D(\bar{a})$ contains at least two numbers, then we deal with a problem with many-valued decisions, which can be formulated as a problem over an appropriate information system. It can be, for example, an information system with the universe equal to $\mathbb{R}^n$ and the set of attributes containing function $\mathrm{sign}(g_i - g_j)$ for any i, j such that $1 \leq i < j \leq m$.

As algorithms for problem solving, we consider deterministic and nondeterministic decision trees over the information system. One can interpret nondeterministic decision trees solving a problem as a way for representation of arbitrary complete (applicable to any object) systems of true decision rules for the problem.

For a given information system, we consider both general complexity measures and limited complexity measures including depth and weighted depth of decision trees. The depth of a decision tree is the maximum number of nodes labeled with attributes in a path from the root to a terminal node of the tree.

K. Durdymyradov et al., *Decision Trees Versus Systems of Decision Rules*,
Studies in Big Data 160, https://doi.org/10.1007/978-3-031-71586-0_2

There are two approaches to decision tree investigation: the local approach, where for a problem, the decision trees are considered using only attributes from the problem description, and the global one, where for problem solving, all attributes from the considered information system can be used in decision trees. In this chapter, decision trees are investigated within the frameworks of the global approach.

In the present chapter, we study so-called sm-pairs (U, ψ), where U is an information system and ψ is a complexity measure over U. The sm-pair is called limited if ψ is a limited complexity measure.

We have the three parameters $\psi^i_{Ug}(z)$, $\psi^d_{Ug}(z)$ and $\psi^a_{Ug}(z)$ (index g refers to the global approach) for any problem z from the set $\mathrm{Probl}(U)$ of all problems over U:

- $\psi^i_{Ug}(z)$—the complexity of the problem z description. This parameter is equal to the complexity of a deterministic decision tree solving the problem z, which sequentially computes values of all attributes from the description of z.
- $\psi^d_{Ug}(z)$—the minimum complexity of a deterministic decision tree solving the problem z.
- $\psi^a_{Ug}(z)$—the minimum complexity of a nondeterministic decision tree solving the problem z.

We will investigate the relationships between any two such parameters for problems from $\mathrm{Probl}(U)$. Let us consider, for example, the parameters $\psi^i_{Ug}(z)$ and $\psi^d_{Ug}(z)$. Let $n \in \omega = \{0, 1, 2, \ldots\}$. We will study relations of the kind $\psi^i_{Ug}(z) \le n \Rightarrow \psi^d_{Ug}(z) \le u$, which are true for any problem $z \in \mathrm{Probl}(U)$. The minimum value of u is most interesting for us. This value (if exists) is equal to

$$\mathscr{U}^{di}_{U\psi g}(n) = \max\{\psi^d_{Ug}(z) : z \in \mathrm{Probl}(U), \psi^i_{Ug}(z) \le n\} .$$

We will also study relations of the kind $\psi^i_{Ug}(z) \ge n \Rightarrow \psi^d_{Ug}(z) \ge l$. In this case, the maximum value of l is most interesting for us. This value (if exists) is equal to

$$\mathscr{L}^{di}_{U\psi g}(n) = \min\{\psi^d_{Ug}(z) : z \in \mathrm{Probl}(U), \psi^i_{Ug}(z) \ge n\} .$$

The two functions $\mathscr{U}^{di}_{U\psi g}$ and $\mathscr{L}^{di}_{U\psi g}$ describe how the behavior of the parameter $\psi^d_{Ug}(z)$ depends on the behavior of the parameter $\psi^i_{Ug}(z)$.

There are 18 similar functions for all ordered pairs of parameters $\psi^i_{Ug}(z)$, $\psi^d_{Ug}(z)$ and $\psi^a_{Ug}(z)$. The resulting 18 functions well describe the relationships between the three parameters under consideration, but it is most likely impossible to list all such tuples of 18 functions. In this chapter, instead of functions we will study types of functions. With any function, we will associate its type from the set $\{\alpha, \beta, \gamma, \delta, \epsilon\}$. For example, if a function f has an infinite domain and it is bounded from above, then its type is equal to α. If the function f has an infinite domain, is not bounded from above, and the inequality $f(n) \ge n$ holds for a finite number of n, then its type is equal to β, etc. Thus, we will list 18-tuples of types of functions. These tuples will be represented in tables called the global types of sm-pairs. We will prove that there are only seven realizable global types of sm-pairs and only six realizable global types of limited sm-pairs.

First, we will study 9-tuples of types of functions $\mathcal{U}^{bc}_{U\psi g}$, $b, c \in \{i, d, a\}$. These tuples will be represented in tables called global upper types of sm-pairs. We will list all realizable global upper types of sm-pairs and limited sm-pairs. After that, we will extend the results obtained for global upper types of sm-pairs to the case of global types of sm-pairs.

The obtained results allow us to understand not only types of relationships for each ordered pair of parameters, but also joint behavior of types of relationships for all ordered pairs of the considered three parameters.

Note that in this chapter, instead of deterministic or nondeterministic decision trees solving a problem, it will be convenient for us to talk about decision trees solving a problem deterministically or nondeterministically.

This chapter is based on the papers [2–4]. It consists of six sections. In Sect. 2.2, basic definitions and results are discussed. Sections 2.3–2.6 contain proofs of both auxiliary and main results.

2.2 Basic Definitions and Results

2.2.1 Decision Trees

Let $\omega = \{0, 1, 2, \ldots\}$, $E_k = \{0, 1, \ldots, k-1\}$, $k \geq 2$, A be a nonempty set and F be a nonempty set of functions from A to E_k. Functions from F will be called *attributes*, and the pair $U = (A, F)$ will be called an *information system*. Denote by F^* the set of all finite words over the alphabet F including the empty word λ.

A node in a finite directed tree is called the *root* if it is the only node, which has no entering edges. A tree, which has such a node, will be called a *finite directed tree with root*. Nodes without leaving edges will be called *terminal* nodes. Nodes, which are neither the root nor terminal, will be called *working* nodes. A *complete path* in a finite directed tree with root is any sequence $\xi = v_0, d_0, \ldots, v_m, d_m, v_{m+1}$ of nodes and edges of the tree such that v_0 is the root, v_{m+1} is a terminal node and, for $i = 0, \ldots, m$, the edge d_i leaves the node v_i and enters the node v_{i+1}.

Definition 2.1 *A decision tree over the information system U* is a marked finite directed tree with root, which has at least two nodes and possesses the following properties:

- The root and the edges leaving the root are not labeled.
- Each working node is labeled with an attribute from the set F.
- Each edge leaving a working node is labeled with a number from E_k.
- Each terminal node is labeled with a number from ω.

Definition 2.2 A decision tree is called *deterministic* if it satisfies the following conditions:

- Exactly one edge leaving the root.
- Edges leaving a working node are labeled with pairwise different numbers.

The set of decision trees over U will be denoted by $\mathrm{Tree}(U)$. Let $\Gamma \in \mathrm{Tree}(U)$. Denote by $\mathrm{At}(\Gamma)$ the set of attributes attached to working nodes of Γ. By $\mathrm{Path}(\Gamma)$, we denote the set of complete paths in Γ. Let $\xi = v_0, d_0, \dots, v_m, d_m, v_{m+1}$ be a complete path in Γ. We denote by $\tau(\xi)$ the number attached to the node v_{m+1}. Define a word $\varphi(\xi)$ from F^* and a subset $\mathscr{A}(\xi)$ of the set A associated with ξ. If $m = 0$, then $\varphi(\xi) = \lambda$ and $\mathscr{A}(\xi) = A$. Let $m > 0$ and, for $j = 1, \dots, m$, the node v_j be labeled with the attribute f_j, and the edge d_j be labeled with the number δ_j. Then $\varphi(\xi) = f_1 \cdots f_m$ and $\mathscr{A}(\xi) = \{a : a \in A, f_1(a) = \delta_1, \dots, f_m(a) = \delta_m\}$. Denote $\mathscr{S}(\xi) = \{f_1(x) = \delta_1, \dots, f_m(x) = \delta_m\}$.

2.2.2 Problems

The set of nonempty finite subsets of the set ω will be denoted by $\mathscr{P}(\omega)$.

Definition 2.3 A *problem over* U is any $(n+1)$-tuple $z = (\nu, f_1, \dots, f_n)$, where $n \in \omega \setminus \{0\}$, $\nu : E_k^n \to \mathscr{P}(\omega)$ and $f_1, \dots, f_n \in F$. The problem z may be interpreted as a problem of searching for at least one number from the set $z(a) = \nu(f_1(a), \dots, f_n(a))$ for a given $a \in A$.

We denote by $\mathrm{Probl}(U)$ the set of problems over the information system U. Let $z = (\nu, f_1, \dots, f_n) \in \mathrm{Probl}(U)$. Denote $\mathrm{At}(z) = \{f_1, \dots, f_n\}$. Let $\Gamma \in \mathrm{Tree}(U)$.

Definition 2.4 We will say that the tree Γ solves the problem z *nondeterministically* if the following conditions hold:

- $\bigcup_{\xi \in \mathrm{Path}(\Gamma)} \mathscr{A}(\xi) = A$.
- For any $a \in A$ and $\xi \in \mathrm{Path}(\Gamma)$, if $a \in \mathscr{A}(\xi)$, then $\tau(\xi) \in z(a)$.

We can also say that Γ is a nondeterministic decision tree solving the problem z.

Definition 2.5 We will say that the tree Γ solves the problem z *deterministically* if Γ is a deterministic decision tree, which solves z nondeterministically. We can also say that Γ is a deterministic decision tree solving the problem z.

2.2.3 Complexity Measures

Definition 2.6 A *complexity measure over* U is any mapping $\psi : F^* \to \omega$.

Definition 2.7 The complexity measure ψ will be called *limited* if it possesses the following properties:

(a) $\psi(\alpha_1\alpha_2) \le \psi(\alpha_1) + \psi(\alpha_2)$ for any $\alpha_1, \alpha_2 \in F^*$.
(b) $\psi(\alpha_1\alpha_2\alpha_3) \ge \psi(\alpha_1\alpha_3)$ for any $\alpha_1, \alpha_2, \alpha_3 \in F^*$.
(c) For any $\alpha \in F^*$, the inequality $\psi(\alpha) \ge |\alpha|$ holds, where $|\alpha|$ is the length of α.

We now consider some examples of complexity measures. Let $w : F \to \omega \setminus \{0\}$. We define the function $\psi^w : F^* \to \omega$ in the following way: $\psi^w(\alpha) = 0$ if $\alpha = \lambda$ and $\psi^w(\alpha) = \sum_{i=1}^{m} w(f_i)$ if $\alpha = f_1 \cdots f_m$. The function ψ^w is a limited complexity measure over U and it is called a *weighted depth*. If $w \equiv 1$, then the function ψ^w is called the *depth* and is denoted by h.

We extend an arbitrary complexity measure ψ onto the set Tree (U) in the following way: $\psi(\Gamma) = \max\{\psi(\varphi(\xi)) : \xi \in \mathrm{Path}(\Gamma)\}$ for any $\Gamma \in \mathrm{Tree}(U)$. The value $\psi(\Gamma)$ will be called the *complexity of the decision tree* Γ.

Let ψ be a complexity measure over U and $z = (\nu, f_1, \ldots, f_n) \in \mathrm{Probl}(U)$. The value $\psi^i_{Ug}(z) = \psi(f_1 \cdots f_n)$ will be called the *complexity of the problem z description*. We denote by $\psi^d_{Ug}(z)$ the minimum complexity of a decision tree $\Gamma \in \mathrm{Tree}(U)$, which solves the problem z deterministically. We denote by $\psi^a_{Ug}(z)$ the minimum complexity of a decision tree $\Gamma \in \mathrm{Tree}(U)$, which solves the problem z nondeterministically.

2.2.4 Global Types of SM-Pairs

Definition 2.8 A pair (U, ψ), where U is an information system and ψ is a complexity measure over U, will be called a *system-measure-pair* (or, *sm-pair*, in short). If ψ is a limited complexity measure, then sm-pair (U, ψ) will be called a *limited sm-pair*.

Let (U, ψ) be a sm-pair. We have the three parameters $\psi^i_{Ug}(z)$, $\psi^d_{Ug}(z)$ and $\psi^a_{Ug}(z)$ for any problem $z \in \mathrm{Probl}(U)$. We now define functions that describe relationships among these parameters.

Definition 2.9 Let $b, c \in \{i, d, a\}$. We define partial functions $\mathscr{U}^{bc}_{U\psi g} : \omega \to \omega$ and $\mathscr{L}^{bc}_{U\psi g} : \omega \to \omega$ in the following way:

$$\mathscr{U}^{bc}_{U\psi g}(n) = \max\{\psi^b_{Ug}(z) : z \in \mathrm{Probl}(U),\ \psi^c_{Ug}(z) \le n\}\ ,$$
$$\mathscr{L}^{bc}_{U\psi g}(n) = \min\{\psi^b_{Ug}(z) : z \in \mathrm{Probl}(U),\ \psi^c_{Ug}(z) \ge n\}\ .$$

If the value $\mathscr{U}^{bc}_{U\psi g}(n)$ is definite, then it is the unimprovable upper bound on the values $\psi^b_{Ug}(z)$ for problems $z \in \mathrm{Probl}(U)$ satisfying $\psi^c_{Ug}(z) \le n$. If the value $\mathscr{L}^{bc}_{U\psi g}(n)$ is definite, then it is the unimprovable lower bound on the values $\psi^b_{Ug}(z)$ for problems $z \in \mathrm{Probl}(U)$ satisfying $\psi^c_{Ug}(z) \ge n$.

Let g be a partial function from ω to ω. We denote by $\mathrm{Dom}(g)$ the domain of g. Denote $\mathrm{Dom}^+(g) = \{n : n \in \mathrm{Dom}(g), g(n) \ge n\}$ and $\mathrm{Dom}^-(g) = \{n : n \in \mathrm{Dom}(g), g(n) \le n\}$.

Definition 2.10 We now define the value $\text{typ}(g) \in \{\alpha, \beta, \gamma, \delta, \epsilon\}$ called the *type* of g.

- If $\text{Dom}(g)$ is an infinite set and g is a bounded from above function, then $\text{typ}(g) = \alpha$.
- If $\text{Dom}(g)$ is an infinite set, $\text{Dom}^+(g)$ is a finite set, and g is an unbounded from above function, then $\text{typ}(g) = \beta$.
- If both sets $\text{Dom}^+(g)$ and $\text{Dom}^-(g)$ are infinite, then $\text{typ}(g) = \gamma$.
- If $\text{Dom}(g)$ is an infinite set and $\text{Dom}^-(g)$ is a finite set, then $\text{typ}(g) = \delta$.
- If $\text{Dom}(g)$ is a finite set, then $\text{typ}(g) = \epsilon$.

Definition 2.11 We denote by $\text{typ}_g(U, \psi)$ a table with three rows and three columns in which rows from top to bottom and columns from the left to the right are labeled with indices i, d, a. The pair $\text{typ}(\mathcal{L}^{bc}_{U\psi g})\ \text{typ}(\mathcal{U}^{bc}_{U\psi g})$ is in the intersection of the row with index $b \in \{i, d, a\}$ and the column with index $c \in \{i, d, a\}$. The table $\text{typ}_g(U, \psi)$ will be called the *global type of sm-pair* (U, ψ).

2.2.5 Basic Results

The main problem investigated in this chapter is to find all global types of sm-pairs as well as of limited sm-pairs. The solution of this problem describes all possible (in terms of functions $\mathcal{U}^{bc}_{U\psi g}$, $\mathcal{L}^{bc}_{U\psi g}$ types, $b, c \in \{i, d, a\}$) relationships among the complexity of problem description, the complexity of problem solving by deterministic decision trees, and the complexity of problem solving by nondeterministic decision trees.

We now define seven tables:

$T_1 =$

	i	d	a
i	$\epsilon\alpha$	$\epsilon\alpha$	$\epsilon\alpha$
d	$\epsilon\alpha$	$\epsilon\alpha$	$\epsilon\alpha$
a	$\epsilon\alpha$	$\epsilon\alpha$	$\epsilon\alpha$

$T_2 =$

	i	d	a
i	$\gamma\gamma$	$\epsilon\epsilon$	$\epsilon\epsilon$
d	$\alpha\alpha$	$\epsilon\alpha$	$\epsilon\alpha$
a	$\alpha\alpha$	$\epsilon\alpha$	$\epsilon\alpha$

$T_3 =$

	i	d	a
i	$\gamma\gamma$	$\delta\epsilon$	$\epsilon\epsilon$
d	$\alpha\beta$	$\gamma\gamma$	$\epsilon\epsilon$
a	$\alpha\alpha$	$\alpha\alpha$	$\epsilon\alpha$

$T_4 =$

	i	d	a
i	$\gamma\gamma$	$\gamma\epsilon$	$\epsilon\epsilon$
d	$\alpha\gamma$	$\gamma\gamma$	$\epsilon\epsilon$
a	$\alpha\alpha$	$\alpha\alpha$	$\epsilon\alpha$

$T_5 =$

	i	d	a
i	$\gamma\gamma$	$\gamma\epsilon$	$\gamma\epsilon$
d	$\alpha\gamma$	$\gamma\gamma$	$\gamma\gamma$
a	$\alpha\gamma$	$\gamma\gamma$	$\gamma\gamma$

$T_6 =$

	i	d	a
i	$\gamma\gamma$	$\gamma\epsilon$	$\gamma\epsilon$
d	$\alpha\gamma$	$\gamma\gamma$	$\gamma\delta$
a	$\alpha\gamma$	$\beta\gamma$	$\gamma\gamma$

$T_7 =$

	i	d	a
i	$\gamma\gamma$	$\gamma\epsilon$	$\gamma\epsilon$
d	$\alpha\gamma$	$\gamma\gamma$	$\gamma\epsilon$
a	$\alpha\gamma$	$\alpha\gamma$	$\gamma\gamma$

Theorem 2.1 *The relation* $\text{typ}_g(U, \psi) \in \{T_1, T_2, T_3, T_4, T_5, T_6, T_7\}$ *holds for any sm-pair* (U, ψ). *For any* $i \in \{1, 2, 3, 4, 5, 6, 7\}$, *there exists a sm-pair* (U, ψ) *such that* $\text{typ}_g(U, \psi) = T_i$.

Theorem 2.2 *The relation* $\text{typ}_g(U, \psi) \in \{T_2, T_3, T_4, T_5, T_6, T_7\}$ *holds for any limited sm-pair* (U, ψ). *For any* $i \in \{2, 3, 4, 5, 6, 7\}$, *there exists a limited sm-pair* (U, h) *such that* $\text{typ}_g(U, h) = T_i$.

Example 2.1 Let $n \in \omega \setminus \{0\}$ and $U(n) = (\mathbb{R}^n, L_n)$, where $\mathbb{R}$ is the set of real numbers, $L_n = \{s(\sum_{i=1}^{n} a_i x_i + a_{n+1}) : a_i \in \mathbb{R}, 1 \leq i \leq n+1\}$, $s : \mathbb{R} \to \{0, 1\}$ and $s(a) = 0$ if and only if $a < 0$ for any $a \in \mathbb{R}$. Using results from [1], one can prove that $\text{typ}_g(U(n), h) = T_3$.

Seven similar examples with full proofs can be found in Sect. 2.4 (see Lemmas 2.7–2.13).

2.3 Possible Global Upper Types of SM-Pairs

Definition 2.12 Let (U, ψ) be a sm-pair. We denote by $\text{typ}_{gu}(U, \psi)$ a table with three rows and three columns in which rows from top to bottom and columns from the left to the right are labeled with indices i, d, a and the value $\text{typ}(\mathcal{U}_{U\psi g}^{bc})$ is in the intersection of the row with index $b \in \{i, d, a\}$ and the column with index $c \in \{i, d, a\}$. The table $\text{typ}_{gu}(U, \psi)$ will be called the *global upper type of sm-pair* (U, ψ).

In this section, all possible global upper types of sm-pairs are listed. We now define seven tables:

$t_1 =$

	i	d	a
i	α	α	α
d	α	α	α
a	α	α	α

$t_2 =$

	i	d	a
i	γ	ϵ	ϵ
d	α	α	α
a	α	α	α

$t_3 =$

	i	d	a
i	γ	ϵ	ϵ
d	β	γ	ϵ
a	α	α	α

$t_4 =$

	i	d	a
i	γ	ϵ	ϵ
d	γ	γ	ϵ
a	α	α	α

$t_5 =$

	i	d	a
i	γ	ϵ	ϵ
d	γ	γ	γ
a	γ	γ	γ

$t_6 =$

	i	d	a
i	γ	ϵ	ϵ
d	γ	γ	δ
a	γ	γ	γ

$t_7 =$

	i	d	a
i	γ	ϵ	ϵ
d	γ	γ	ϵ
a	γ	γ	γ

Proposition 2.1 *The relation* $\text{typ}_{gu}(U, \psi) \in \{t_1, t_2, t_3, t_4, t_5, t_6, t_7\}$ *holds for any sm-pair* (U, ψ).

Proposition 2.2 *The relation* $\text{typ}_{gu}(U, \psi) \in \{t_2, t_3, t_4, t_5, t_6, t_7\}$ *holds for any limited sm-pair* (U, ψ).

First, we prove some auxiliary statements.

Lemma 2.1 *Let* (U, ψ) *be a sm-pair and* $z \in \text{Probl}(U)$. *Then the inequalities* $\psi_{Ug}^{a}(z) \leq \psi_{Ug}^{d}(z) \leq \psi_{Ug}^{i}(z)$ *hold.*

Proof Let $z = (\nu, f_1, \ldots, f_n)$. It is not difficult to construct a decision tree $\Gamma_0 \in \mathrm{Tree}(U)$ solving the problem z deterministically by sequential finding values of the attributes $f_1, \ldots, f_n$. Evidently, $\psi(\Gamma_0) = \psi^i_{Ug}(z)$. Therefore $\psi^d_{Ug}(z) \le \psi^i_{Ug}(z)$. If a decision tree $\Gamma \in \mathrm{Tree}(U)$ solves the problem z deterministically, then the decision tree Γ solves the problem z nondeterministically. Therefore $\psi^a_{Ug}(z) \le \psi^d_{Ug}(z)$. □

Let (U, ψ) be a sm-pair, $n \in \omega$ and $b, c \in \{i, d, a\}$. A notation $\mathscr{U}^{bc}_{U\psi g}(n) = \infty$ means that the set $\{\psi^b_{Ug}(z) : z \in \mathrm{Probl}(U), \psi^c_{Ug}(z) \le n\}$ is infinite. Evidently, if $\mathscr{U}^{bc}_{U\psi g}(n) = \infty$, then $\mathscr{U}^{bc}_{U\psi g}(n+1) = \infty$. It is not difficult to prove the next statement.

Lemma 2.2 *Let (U, ψ) be a sm-pair and $b, c \in \{i, d, a\}$. Then*
(a) *If there exists $n \in \omega$ such that $\mathscr{U}^{bc}_{U\psi g}(n) = \infty$, then* $\mathrm{typ}(\mathscr{U}^{bc}_{U\psi g}) = \epsilon$.
(b) *If there is no $n \in \omega$ such that $\mathscr{U}^{bc}_{U\psi g}(n) = \infty$, then* $\mathrm{Dom}(\mathscr{U}^{bc}_{U\psi g}) = \{n : n \in \omega, n \ge n_0\}$, *where* $n_0 = \min\{\psi^c_{Ug}(z) : z \in \mathrm{Probl}(U)\}$.

Let (U, ψ) be a sm-pair and $b, c, e, f \in \{i, d, a\}$. A notation $\mathscr{U}^{bc}_{U\psi g} \lhd \mathscr{U}^{ef}_{U\psi g}$ means that, for any $n \in \omega$, the following statements hold:
(a) If the value $\mathscr{U}^{bc}_{U\psi g}(n)$ is defined, then either $\mathscr{U}^{ef}_{U\psi g}(n) = \infty$ or the value $\mathscr{U}^{ef}_{U\psi g}(n)$ is defined and the inequality $\mathscr{U}^{bc}_{U\psi g}(n) \le \mathscr{U}^{ef}_{U\psi g}(n)$ holds.
(b) If $\mathscr{U}^{bc}_{U\psi g}(n) = \infty$, then $\mathscr{U}^{ef}_{U\psi g}(n) = \infty$.
Let $\preceq$ be a linear order on the set $\{\alpha, \beta, \gamma, \delta, \epsilon\}$ and $\alpha \preceq \beta \preceq \gamma \preceq \delta \preceq \epsilon$.

Lemma 2.3 *Let (U, ψ) be a sm-pair. Then* $\mathrm{typ}(\mathscr{U}^{bi}_{U\psi g}) \preceq \mathrm{typ}(\mathscr{U}^{bd}_{U\psi g}) \preceq \mathrm{typ}(\mathscr{U}^{ba}_{U\psi g})$ *and* $\mathrm{typ}(\mathscr{U}^{ab}_{U\psi g}) \preceq \mathrm{typ}(\mathscr{U}^{db}_{U\psi g}) \preceq \mathrm{typ}(\mathscr{U}^{ib}_{U\psi g})$ *for any $b \in \{i, d, a\}$.*

Proof From the definition of the functions $\mathscr{U}^{bc}_{U\psi g}$, $b, c \in \{i, d, a\}$, and from Lemma 2.1 it follows that $\mathscr{U}^{bi}_{U\psi g} \lhd \mathscr{U}^{bd}_{U\psi g} \lhd \mathscr{U}^{ba}_{U\psi g}$ and $\mathscr{U}^{ab}_{U\psi g} \lhd \mathscr{U}^{db}_{U\psi g} \lhd \mathscr{U}^{ib}_{U\psi g}$ for any $b \in \{i, d, a\}$. Using these relations and Lemma 2.2, we obtain the statement of the lemma. □

Lemma 2.4 *Let (U, ψ) be a sm-pair and $b, c \in \{i, d, a\}$. Then*
(a) $\mathrm{typ}(\mathscr{U}^{bc}_{U\psi g}) = \alpha$ *if and only if the function ψ^b_{Ug} is bounded from above on the set* $\mathrm{Probl}(U)$.
(b) *If the function ψ^b_{Ug} is unbounded from above on* $\mathrm{Probl}(U)$, *then* $\mathrm{typ}(\mathscr{U}^{bb}_{U\psi g}) = \gamma$.

Proof The statement (a) of lemma is obvious. (b) Let the function ψ^b_{Ug} be unbounded from above on $\mathrm{Probl}(U)$. One can show that in this case the equality $\mathscr{U}^{bb}_{U\psi g}(n) = n$ holds for infinitely many $n \in \omega$. Therefore $\mathrm{typ}(\mathscr{U}^{bb}_{U\psi g}) = \gamma$. □

Corollary 2.1 *Let (U, ψ) be a sm-pair and $b \in \{i, d, a\}$. Then* $\mathrm{typ}(\mathscr{U}^{bb}_{U\psi g}) \in \{\alpha, \gamma\}$.

Lemma 2.5 *Let (U, ψ) be a sm-pair and* $\mathrm{typ}(\mathscr{U}^{ii}_{U\psi g}) \ne \alpha$. *Then*

$$\mathrm{typ}(\mathscr{U}^{id}_{U\psi g}) = \mathrm{typ}(\mathscr{U}^{ia}_{U\psi g}) = \epsilon\ .$$

Proof Using Lemma 2.4, we conclude that the function ψ^i_{Ug} is unbounded from above on $\mathrm{Probl}(U)$. Let $m \in \omega$. Then there exists a problem $z = (\nu, f_1, \ldots, f_n) \in \mathrm{Probl}(U)$ such that $\psi^i_{Ug}(z) \geq m$. Let us consider a problem $z' = (\nu', f_1, \ldots, f_n)$, where $\nu' \equiv \{0\}$. It is clear that $\psi^i_{Ug}(z') \geq m$. Let Γ be a decision tree, which consists of the root, the terminal node labeled with 0 and the edge connecting these two nodes. One can show that the tree Γ solves the problem z' deterministically. Therefore $\psi^a_{Ug}(z') \leq \psi^d_{Ug}(z') \leq \psi(\Gamma) = \psi(\lambda)$. Taking into account that m is an arbitrary number from ω, we conclude that $\mathscr{U}^{id}_{U\psi g}(\psi(\lambda)) = \infty$ and $\mathscr{U}^{ia}_{U\psi g}(\psi(\lambda)) = \infty$. Using Lemma 2.2, we obtain $\mathrm{typ}(\mathscr{U}^{id}_{U\psi g}) = \mathrm{typ}(\mathscr{U}^{ia}_{U\psi g}) = \epsilon$. □

Lemma 2.6 *Let* (U, ψ) *be a sm-pair. Then* $\mathrm{typ}(\mathscr{U}^{ai}_{U\psi g}) \in \{\alpha, \gamma\}$.

Proof Let $U = (A, F)$ and $f : A \to E_k$ for any $f \in F$. Using Lemma 2.3 and Corollary 2.1, we obtain $\mathrm{typ}(\mathscr{U}^{ai}_{U\psi g}) \in \{\alpha, \beta, \gamma\}$. Assume that $\mathrm{typ}(\mathscr{U}^{ai}_{U\psi g}) = \beta$. Then there exists $m \in \omega \setminus \{0\}$ such that $\mathscr{U}^{ai}_{U\psi g}(n) < n$ for any $n \in \omega$, $n > m$. We prove by induction on n that, for any problem $z \in \mathrm{Probl}(U)$, if $\psi^i_{Ug}(z) \leq n$, then $\psi^a_{Ug}(z) \leq m_0$, where $m_0 = \max\{m, \psi(\lambda)\}$. From Lemma 2.1 it follows that under the condition $n \leq m$ the considered statement holds. Let it hold for some n, $n \geq m$. We show that this statement holds for $n + 1$ too. Let $z \in \mathrm{Probl}(U)$ and $\psi^i_{Ug}(z) \leq n + 1$. Since $n + 1 > m$, we obtain $\psi^a_{Ug}(z) \leq n$. Let $\Gamma \in \mathrm{Tree}(U)$, $\psi(\Gamma) = \psi^a_{Ug}(z)$ and Γ solves the problem z nondeterministically. Assume that in Γ there exists a complete path ξ in which there are no working nodes. In this case, a decision tree, which consists of the root, the terminal node labeled with $\tau(\xi)$ and the edge connecting these two nodes, solves the problem z nondeterministically. Therefore $\psi^a_{Ug}(z) \leq \psi(\lambda) \leq m_0$. Assume now that each complete path in the decision tree Γ contains a working node. Let $\xi \in \mathrm{Path}(\Gamma)$, $\xi = v_0, d_0, \ldots, v_p, d_p, v_{p+1}$, an attribute f_i be assigned to the node v_i and δ_i be a number assigned to the edge d_i, $i = 1, \ldots, p$. Let us consider a problem $z_\xi = (\nu_\xi, f_1, \ldots, f_p)$, where $\nu_\xi(\delta_1, \ldots, \delta_p) = \{\tau(\xi)\}$ and $\nu_\xi(\bar{\sigma}) = \{\tau(\xi) + 1\}$ for any p-tuple $\bar{\sigma} \in E^p_k$ such that $\bar{\sigma} \neq (\delta_1, \ldots, \delta_p)$. It is clear that $\psi^i_{Ug}(z_\xi) \leq n$. Using the inductive hypothesis, we obtain that there exists a decision tree $\Gamma_\xi \in \mathrm{Tree}(U)$, which has the following properties: Γ_ξ solves the problem z_ξ nondeterministically and $\psi(\Gamma_\xi) \leq m_0$. Let $\mathscr{A}(\xi) \neq \emptyset$. We denote by $\tilde{\Gamma}_\xi$ a tree obtained from Γ_ξ by the removal all nodes and edges that satisfy the following condition: there is no a complete path ξ' in Γ_ξ, which contains this node or edge and for which $\tau(\xi') = \tau(\xi)$. Let $\{\xi : \xi \in \mathrm{Path}(\Gamma), \mathscr{A}(\xi) \neq \emptyset\} = \{\xi_1, \ldots, \xi_r\}$. Let us identify the roots of the trees $\tilde{\Gamma}_{\xi_1}, \ldots, \tilde{\Gamma}_{\xi_r}$. We denote by G the obtained tree. It is not difficult to show that $G \in \mathrm{Tree}(U)$, $\psi(G) \leq m_0$ and the decision tree G solves the problem z nondeterministically. Thus, the considered statement holds. Using Lemma 2.4, we have $\mathrm{typ}(\mathscr{U}^{ai}_{U\psi g}) = \alpha$. The obtained contradiction shows that $\mathrm{typ}(\mathscr{U}^{ai}_{U\psi g}) \in \{\alpha, \gamma\}$. □

Proof of Proposition 2.1 Let (U, ψ) be a sm-pair. Using Corollary 2.1, we conclude that $\mathrm{typ}(\mathscr{U}^{ii}_{U\psi g}) \in \{\alpha, \gamma\}$. Using Corollary 2.1 and Lemma 2.3, we obtain $\mathrm{typ}(\mathscr{U}^{di}_{U\psi g}) \in \{\alpha, \beta, \gamma\}$. From Lemma 2.6 it follows that $\mathrm{typ}(\mathscr{U}^{ai}_{U\psi g}) \in \{\alpha, \gamma\}$.

(a) Let $\mathrm{typ}(\mathscr{U}^{ii}_{U\psi g}) = \alpha$. Using Lemmas 2.3 and 2.4, we obtain $\mathrm{typ}_{gu}(U, \psi) = t_1$.

(b) Let $\mathrm{typ}(\mathcal{U}^{ii}_{U\psi g}) = \gamma$ and $\mathrm{typ}(\mathcal{U}^{di}_{U\psi g}) = \alpha$. Using Lemmas 2.3, 2.4 and 2.5, we obtain $\mathrm{typ}_{gu}(U, \psi) = t_2$.

(c) Let $\mathrm{typ}(\mathcal{U}^{ii}_{U\psi g}) = \gamma$ and $\mathrm{typ}(\mathcal{U}^{di}_{U\psi g}) = \beta$. Using Lemma 2.5, we conclude that $\mathrm{typ}(\mathcal{U}^{id}_{U\psi g}) = \mathrm{typ}(\mathcal{U}^{ia}_{U\psi g}) = \epsilon$. Hence from Lemmas 2.3 and 2.6 it follows that $\mathrm{typ}(\mathcal{U}^{ai}_{U\psi g}) = \alpha$. Using this equality and Lemma 2.4, we obtain that $\mathrm{typ}(\mathcal{U}^{ad}_{U\psi g}) = \mathrm{typ}(\mathcal{U}^{aa}_{U\psi g}) = \alpha$. Using the equality $\mathrm{typ}(\mathcal{U}^{di}_{U\psi g}) = \beta$, Lemma 2.3 and Corollary 2.1, we have $\mathrm{typ}(\mathcal{U}^{dd}_{U\psi g}) = \gamma$. From the equalities $\mathrm{typ}(\mathcal{U}^{dd}_{U\psi g}) = \gamma$, $\mathrm{typ}(\mathcal{U}^{aa}_{U\psi g}) = \alpha$ and from Lemmas 2.2 and 2.4 it follows that $\mathrm{typ}(\mathcal{U}^{da}_{U\psi g}) = \epsilon$. Thus, $\mathrm{typ}_{gu}(U, \psi) = t_3$.

(d) Let $\mathrm{typ}(\mathcal{U}^{ii}_{U\psi g}) = \mathrm{typ}(\mathcal{U}^{di}_{U\psi g}) = \gamma$ and $\mathrm{typ}(\mathcal{U}^{ai}_{U\psi g}) = \alpha$. Using Lemma 2.5, we obtain $\mathrm{typ}(\mathcal{U}^{id}_{U\psi g}) = \mathrm{typ}(\mathcal{U}^{ia}_{U\psi g}) = \epsilon$. From Lemma 2.4 it follows that $\mathrm{typ}(\mathcal{U}^{ad}_{U\psi g}) = \mathrm{typ}(\mathcal{U}^{aa}_{U\psi g}) = \alpha$. Using Lemma 2.3 and Corollary 2.1, we obtain $\mathrm{typ}(\mathcal{U}^{dd}_{U\psi g}) = \gamma$. Hence taking also into account the equality $\mathrm{typ}(\mathcal{U}^{aa}_{U\psi g}) = \alpha$, and Lemmas 2.2 and 2.4, we obtain that $\mathrm{typ}(\mathcal{U}^{da}_{U\psi g}) = \epsilon$. Thus, $\mathrm{typ}_{gu}(U, \psi) = t_4$.

(e) Let $\mathrm{typ}(\mathcal{U}^{ii}_{U\psi g}) = \mathrm{typ}(\mathcal{U}^{di}_{U\psi g}) = \mathrm{typ}(\mathcal{U}^{ai}_{U\psi g}) = \gamma$. Using Lemma 2.5, we conclude that $\mathrm{typ}(\mathcal{U}^{id}_{U\psi g}) = \mathrm{typ}(\mathcal{U}^{ia}_{U\psi g}) = \epsilon$. From Lemma 2.3 and Corollary 2.1 it follows that $\mathrm{typ}(\mathcal{U}^{dd}_{U\psi g}) = \mathrm{typ}(\mathcal{U}^{ad}_{U\psi g}) = \mathrm{typ}(\mathcal{U}^{aa}_{U\psi g}) = \gamma$. Using Lemma 2.3, we have $\mathrm{typ}(\mathcal{U}^{da}_{U\psi g}) \in \{\gamma, \delta, \epsilon\}$. Therefore $\mathrm{typ}_{gu}(U, \psi) \in \{t_5, t_6, t_7\}$.

Proof of Proposition 2.2 Let (U, ψ) be a limited sm-pair. Taking into account that the complexity measure ψ has the property (c) and using Lemma 2.4, we obtain that $\mathrm{typ}(\mathcal{U}^{ii}_{U\psi g}) \neq \alpha$. Therefore $\mathrm{typ}_{gu}(U, \psi) \neq t_1$. From this relation and Proposition 2.1 it follows that the statement of Proposition 2.2 holds.

2.4 Realizable Global Upper Types of SM-Pairs

In this section all realizable global upper types of sm-pairs are listed.

Proposition 2.3 *For any* $i \in \{1, 2, 3, 4, 5, 6, 7\}$, *there exists a sm-pair* (U, ψ) *such that*

$$\mathrm{typ}_{gu}(U, \psi) = t_i \ .$$

Proposition 2.4 *For any* $i \in \{2, 3, 4, 5, 6, 7\}$, *there exists a limited sm-pair* (U, h) *such that*

$$\mathrm{typ}_{gu}(U, h) = t_i \ .$$

We now prove some auxiliary statements.

Let us define a sm-pair (U_1, π) as follows: $U_1 = (\omega, F_1)$, where $F_1 = \{f\}$, $f \equiv 0$, and $\pi \equiv 0$.

Lemma 2.7 $\mathrm{typ}_{gu}(U_1, \pi) = t_1$.

Proof Using Lemma 2.4, we obtain $\text{typ}(\mathscr{U}^{ii}_{U_1\pi g}) = \alpha$. From this equality and from Proposition 2.1 it follows that $\text{typ}_{gu}(U_1, \pi) = t_1$. □

Let us define a sm-pair (U_2, h) as follows: $U_2 = (\omega, F_2)$, where $F_2 = F_1$.

Lemma 2.8 $\text{typ}_{gu}(U_2, h) = t_2$.

Proof It is not difficult to show that $h^d_{U_2 g}(z) = 0$ for any problem $z \in \text{Probl}(U_2)$. It is clear that the function $h^i_{U_2 g}$ is unbounded from above on $\text{Probl}(U_2)$. Using Lemma 2.4, we conclude that $\text{typ}(\mathscr{U}^{ii}_{U_2 hg}) = \gamma$ and $\text{typ}(\mathscr{U}^{dd}_{U_2 hg}) = \alpha$. From these equalities and from Proposition 2.1 it follows that $\text{typ}_{gu}(U_2, h) = t_2$. □

Let us define a sm-pair (U_3, h) as follows: $U_3 = (\omega, F_3)$, where $F_3 = \{l_i : i \in \omega \setminus \{0\}\}$ and, for any $i \in \omega \setminus \{0\}$, $j \in \omega$, if $j \le i$, then $l_i(j) = 0$, and if $j > i$, then $l_i(j) = 1$.

Lemma 2.9 $\text{typ}_{gu}(U_3, h) = t_3$.

Proof Using Corollary 2.1 and Lemma 2.3, we obtain $\text{typ}(\mathscr{U}^{di}_{U_3 hg}) \in \{\alpha, \beta, \gamma\}$. We now prove that $\text{typ}(\mathscr{U}^{di}_{U_3 hg}) \ne \gamma$. Let $n \in \omega \setminus \{0\}$, $z = (\nu, l_{i_1}, \dots, l_{i_m}) \in \text{Probl}(U_3)$ and $m \le n$. Using "dichotomous" approach to the construction of decision trees, it is not difficult to prove that $h^d_{U_3 g}(z) \le 1 + \log_2 n$. Taking into account that z is an arbitrary problem over U_3 such that $h^i_{U_3 g}(z) \le n$, we conclude that $\mathscr{U}^{di}_{U_3 hg}(n) \le 1 + \log_2 n$. Therefore $\text{typ}(\mathscr{U}^{di}_{U_3 hg}) \ne \gamma$.

We now prove that $\text{typ}(\mathscr{U}^{di}_{U_3 hg}) = \beta$. Assume the contrary: $\text{typ}(\mathscr{U}^{di}_{U_3 hg}) = \alpha$. Using Lemma 2.4, we obtain that there exists a number $p \in \omega$ such that $h^d_{U_3 g}(z) \le p$ for any problem $z \in \text{Probl}(U_3)$. Therefore there exists a number $r \in \omega \setminus \{0\}$, which satisfies the following condition: for any problem $z \in \text{Probl}(U_3)$, there exists a decision tree $\Gamma \in \text{Tree}(U_3)$ that solves the problem z deterministically and has at most r terminal nodes. Let us consider a problem $z' = (\nu, l_1, \dots, l_r)$ from $\text{Probl}(U_3)$, where $\nu : E_2^r \to \mathscr{P}(\omega)$ and $\nu(\bar{\delta}_1) \cap \nu(\bar{\delta}_2) = \emptyset$ for any $\bar{\delta}_1, \bar{\delta}_2 \in E_2^r$ such that $\bar{\delta}_1 \ne \bar{\delta}_2$. Let Γ be an arbitrary decision tree over U_3, which solves the problem z' deterministically. It is not difficult to show that the tree Γ must have at least $r + 1$ terminal nodes. We obtained a contradiction. Therefore $\text{typ}(\mathscr{U}^{di}_{U_3 hg}) = \beta$. From this equality and from Proposition 2.1 it follows that $\text{typ}_{gu}(U_3, h) = t_3$. □

Let us define a sm-pair (U_4, h) as follows: $U_4 = (\omega, F_4)$, where F_4 is the set of all possible functions $f : \omega \to E_2$.

Lemma 2.10 $\text{typ}_{gu}(U_4, h) = t_4$.

Proof We now prove that $\text{typ}(\mathscr{U}^{ai}_{U_4 hg}) = \alpha$. Let $z \in \text{Probl}(U_4)$, Γ be a decision tree over U_4, which solves the problem z deterministically, and $\xi \in \text{Path}(\Gamma)$. Let us denote by f_ξ the attribute from F_4, which has the following property: $f_\xi(a) = 1$ if and only if $a \in \mathscr{A}(\xi)$ for any $a \in \omega$. We denote by Γ_ξ a decision tree over U_4, which consists of the root v_0, the working node v_1 labeled with f_ξ, the terminal node v_2 labeled with

$\tau(\xi)$, the edge d_0 connecting v_0 and v_1, and the edge d_1 labeled with 1 and connecting v_1 and v_2. Let us identify the roots of the trees Γ_ξ, $\xi \in \text{Path}(\Gamma)$. We denote by G the obtained tree. It is not difficult to show that $G \in \text{Tree}(U_4)$, $h(G) = 1$ and the decision tree G solves the problem z nondeterministically. Thus, $h^a_{U_4 g}(z) \le 1$ for any problem $z \in \text{Probl}(U_4)$. From here and from Lemma 2.4 it follows that $\text{typ}(\mathcal{U}^{ai}_{U_4 hg}) = \alpha$.

Let us prove that $\text{typ}(\mathcal{U}^{di}_{U_4 hg}) = \gamma$. Using Lemmas 2.2 and 2.3 and Corollary 2.1, we obtain $\text{Dom}(\mathcal{U}^{di}_{U_4 hg}) = \omega \setminus \{0\}$. Let $n \in \omega$ and $n \ge 1$. It is not difficult to show that there exist $f_1, \dots, f_n \in F_4$, which have the following property: for any $\delta_1, \dots, \delta_n \in E_2$ the system of equations

$$\{f_1(x) = \delta_1, \dots, f_n(x) = \delta_n\}$$

is compatible on ω. Let us consider a problem $z = (\nu, f_1, \dots, f_n)$, where $\nu : E_2^n \to \mathcal{P}(\omega)$ and $\nu(\bar{\delta}_1) \cap \nu(\bar{\delta}_2) = \emptyset$ for any $\bar{\delta}_1, \bar{\delta}_2 \in E_2^n$ such that $\bar{\delta}_1 \ne \bar{\delta}_2$. It is clear that $h^i_{U_4 g}(z) = n$. Let Γ be a decision tree over U_4, which solves the problem z deterministically and for which $h(\Gamma) = h^d_{U_4 g}(z)$. It is clear that the decision tree Γ must have at least 2^n terminal nodes. Therefore $h(\Gamma) \ge n$ and $h^d_{U_4 g}(z) \ge n$. Thus $\mathcal{U}^{di}_{U_4 hg}(n) \ge n$. Using Lemma 2.1, we conclude that $\mathcal{U}^{di}_{U_4 hg}(n) = n$. Taking into account that n is an arbitrary number from $\omega \setminus \{0\}$, we obtain that $\text{typ}(\mathcal{U}^{di}_{U_4 hg}) = \gamma$. From this equality, equality $\text{typ}(\mathcal{U}^{ai}_{U_4 hg}) = \alpha$ and Proposition 2.1 it follows that $\text{typ}_{gu}(U_4, h) = t_4$. □

Let us define a sm-pair (U_5, h) as follows: $U_5 = (\omega, F_5)$, where $F_5 = \{f_i : i \in \omega \setminus \{0\}\}$ and, for any $i \in \omega \setminus \{0\}$, $j \in \omega$, if $i = j$, then $f_i(j) = 1$, and if $i \ne j$, then $f_i(j) = 0$.

Lemma 2.11 $\text{typ}_{gu}(U_5, h) = t_5$.

Proof Let $z \in \text{Probl}(U_5)$. We now show that $h^d_{U_5 g}(z) = h^a_{U_5 g}(z)$. Let Γ be a decision tree over U_5, which solves the problem z nondeterministically and for which $h(\Gamma) = h^a_{U_5 g}(z)$. Let ξ be a complete path in the tree Γ such that $0 \in \mathcal{A}(\xi)$. If in the complete path ξ there are no working nodes, then, as it is not difficult to show, $h^d_{U_5 g}(z) = h^a_{U_5 g}(z) = 0$. Let we have $m > 0$ working nodes in the path ξ and $\mathcal{S}(\xi) = \{f_{i_1}(x) = \delta_1, \dots, f_{i_m}(x) = \delta_m\}$. Since $0 \in \mathcal{A}(\xi)$, $\delta_1 = \dots = \delta_m = 0$. It is clear that, for any $p \in \mathcal{A}(\xi)$, the relation $\tau(\xi) \in z(p)$ holds.

Let us describe a decision tree Γ_1 over U_5, which solves the problem z deterministically and for which $h(\Gamma_1) \le m$. For an arbitrary $p \in \omega$, the decision tree Γ_1 finds the values $f_{i_1}(p), \dots, f_{i_m}(p)$. If $f_{i_1}(p) = \dots = f_{i_m}(p) = 0$, then $p \in \mathcal{A}(\xi)$ and, hence, the problem z is solved since we know that $\tau(\xi) \in z(p)$. Let, for some $j \in \{1, \dots, m\}$, the equality $f_{i_j}(p) = 1$ holds. Then $p = i_j$ and, hence, the problem z is solved too since we know all the set $z(p)$.

It is clear that $h(\Gamma) \ge m$. Therefore $h(\Gamma_1) \le h^a_{U_5 g}(z)$ and $h^d_{U_5 g}(z) \le h^a_{U_5 g}(z)$. By Lemma 2.1, $h^d_{U_5 g}(z) \ge h^a_{U_5 g}(z)$ and, hence, $h^d_{U_5 g}(z) = h^a_{U_5 g}(z)$. From this equality it follows that there is no $n \in \omega$ for which $\mathcal{U}^{da}_{U_5 hg}(n) = \infty$. Using Lemma 2.2, we conclude that $\mathrm{Dom}(\mathcal{U}^{da}_{U_5hg}) = \{n : n \in \omega, n \geq n_0\}$ for some $n_0 \in \omega$ and $\mathcal{U}^{da}_{U_5hg}(n) \leq n$ for any $n \in \mathrm{Dom}(\mathcal{U}^{da}_{U_5hg})$.

Let $n \in \mathrm{Dom}(\mathcal{U}^{da}_{U_5hg})$ and $n \geq 1$. We now show that $\mathcal{U}^{da}_{U_5hg}(n) = n$. Let $z = (\nu, f_1, \ldots, f_n)$ be a problem from $\mathrm{Probl}(U_5)$ for which $\nu(\bar{\delta}_1) \cap \nu(\bar{\delta}_2) = \emptyset$ for any $\bar{\delta}_1, \bar{\delta}_2 \in E_2^n$ such that $\bar{\delta}_1 \neq \bar{\delta}_2$. Let Γ be a decision tree over U_5, which solves the problem z nondeterministically and for which $h(\Gamma) = h^a_{U_5g}(z)$. Let ξ be a complete path in the tree Γ such that $0 \in \mathcal{A}(\xi)$. It is clear that in the complete path ξ there is at least one working node. Let $\mathcal{S}(\xi) = \{f_{i_1}(x) = \delta_1, \ldots, f_{i_m}(x) = \delta_m\}$. Since $0 \in \mathcal{A}(\xi), \delta_1 = \ldots = \delta_m = 0$. It is clear that $i \notin \mathcal{A}(\xi)$ for any $i \in \{1, \ldots, n\}$. Therefore, for any $i \in \{1, \ldots, n\}$, the equation $f_i(x) = 0$ is contained in the system $\mathcal{S}(\xi)$. Consequently, $h(\Gamma) \geq n$ and $h^a_{U_5g}(z) \geq n$. It is clear that $h^i_{U_5g}(z) = n$. Using Lemma 2.1, we obtain $h^a_{U_5g}(z) = n$ and $h^d_{U_5g}(z) = n$. Therefore $\mathcal{U}^{da}_{U_5hg}(n) \geq n$ and, as proved above, $\mathcal{U}^{da}_{U_5hg}(n) = n$. Taking into account that n is an arbitrary number from ω such that $n \geq \max(n_0, 1)$, we obtain $\mathrm{Dom}^-(\mathcal{U}^{da}_{U_5hg})$ and $\mathrm{Dom}^+(\mathcal{U}^{da}_{U_5hg})$ are infinite sets and $\mathrm{typ}(\mathcal{U}^{da}_{U_5hg}) = \gamma$. Using Proposition 2.1, we obtain $\mathrm{typ}_{gu}(U_5, h) = t_5$. □

Let us define a sm-pair (U_6, h) as follows: $U_6 = (\omega, F_6)$, where $F_6 = F_5 \cup G_6$, $G_6 = \{g_{2i+1} : i \in \omega\}$ and, for any $i \in \omega$, $j \in \omega$, if $j \in \{2i+1, 2i+2\}$, then $g_{2i+1}(j) = 1$, and if $j \notin \{2i+1, 2i+2\}$, then $g_{2i+1}(j) = 0$.

Lemma 2.12 $\mathrm{typ}_{gu}(U_6, h) = t_6$.

Proof Let $z \in \mathrm{Probl}(U_6)$. We now show that $h^d_{U_6g}(z) \leq h^a_{U_6g}(z) + 1$. Let Γ be a decision tree over U_6, which solves the problem z nondeterministically and for which $h(\Gamma) = h^a_{U_6g}(z)$. Let ξ be a complete path in the tree Γ such that $0 \in \mathcal{A}(\xi)$. If in the complete path ξ there are no working nodes, then, as it is not difficult to show, $h^d_{U_6g}(z) = h^a_{U_6g}(z) = 0$. Let there be $m > 0$ working nodes in the path ξ and $\mathcal{S}(\xi) = \{q_1(x) = \delta_1, \ldots, q_m(x) = \delta_m\}$, where $q_i \in F_6, 1 \leq i \leq m$. Since $0 \in \mathcal{A}(\xi)$, $\delta_1 = \ldots = \delta_m = 0$. It is clear that, for any $p \in \mathcal{A}(\xi)$, the inclusion $\tau(\xi) \in z(p)$ holds.

Let us describe a decision tree Γ_1 over U_6, which solves the problem z deterministically and for which $h(\Gamma_1) \leq m + 1$. For an arbitrary $p \in \omega$, the decision tree Γ_1 finds the values $q_1(p), \ldots, q_m(p)$. If $q_1(p) = \ldots = q_m(p) = 0$, then $p \in \mathcal{A}(\xi)$ and, hence, the problem z is solved since we know that $\tau(\xi) \in z(p)$. Let, for some $j \in \{1, \ldots, m\}$, the equality $q_j(p) = 1$ holds. If $q_j = f_i$ for some $i \in \omega \setminus \{0\}$, then $p = i$ and, hence, the problem z is solved too since we know all the set $z(p)$. Let $q_j = g_{2i+1}$ for some $i \in \omega$. Then we obtain that $p \in \{2i+1, 2i+2\}$ and we find the value $f_{2i+1}(p)$. If $f_{2i+1}(p) = 1$, then $p = 2i+1$, and if $f_{2i+1}(p) = 0$, then $p = 2i+2$. In these cases, the problem z is solved too.

It is clear that $h(\Gamma) \geq m$. Therefore $h(\Gamma_1) \leq h^a_{U_6g}(z) + 1$ and $h^d_{U_6g}(z) \leq h^a_{U_6g}(z) + 1$. From this inequality it follows that there is no $n \in \omega$ for which $\mathcal{U}^{da}_{U_6hg}(n) = \infty$. Using Lemma 2.2, we conclude that $\mathrm{Dom}(\mathcal{U}^{da}_{U_6hg}) = \{n : n \in \omega, n \geq n_0\}$ for some $n_0 \in \omega$.

Let $n \in \mathrm{Dom}(\mathcal{U}^{da}_{U_6hg})$ and $n \geq 1$. We now show that $\mathcal{U}^{da}_{U_6hg}(n) \geq n + 1$. Let $z = (\nu, g_1, f_1, f_2, \ldots, g_{2n-1}, f_{2n-1}, f_{2n})$ be a problem from $\mathrm{Probl}(U_6)$ such that

$\nu(\bar{\delta}_1) \cap \nu(\bar{\delta}_2) = \emptyset$ for any $\bar{\delta}_1, \bar{\delta}_2 \in E_2^{3n}, \bar{\delta}_1 \neq \bar{\delta}_2$. It is not difficult to show that $h^a_{U_6 g}(z) \leq n$. Let us prove that $h^d_{U_6 g}(z) \geq n+1$. Let Γ be a decision tree over U_6, which solves the problem z deterministically and for which $h(\Gamma) = h^d_{U_6 g}(z)$. Let ξ be a complete path in the tree Γ such that $0 \in \mathscr{A}(\xi)$. It is clear that in the complete path ξ there is at least one working node. Let $\mathscr{S}(\xi) = \{q_1(x) = \delta_1, \ldots, q_m(x) = \delta_m\}$, where $q_i \in F_6, 1 \leq i \leq m$. Since $0 \in \mathscr{A}(\xi)$, $\delta_1 = \ldots = \delta_m = 0$. It is clear that $i \notin \mathscr{A}(\xi)$ for any $i \in \{1, \ldots, 2n\}$. Therefore the equation $g_{2i-1}(x) = 0$ or both equations $f_{2i-1}(x) = 0$ and $f_{2i}(x) = 0$ must belong to the system $\mathscr{S}(\xi)$ for any $i \in \{1, \ldots, n\}$. If the number of working nodes in the path ξ is greater than n, then the considered statement holds. Let the number of working nodes in the path ξ be equal to n. Then $\mathscr{S}(\xi) = \{g_1(x) = 0, \ldots, g_{2n-1}(x) = 0\}$. Let v be the last working node in the path ξ and let the attribute g_{2i-1} be assigned to the node v. Let us consider the complete path ξ' in the tree Γ such that $2i - 1 \in \mathscr{A}(\xi')$. Since Γ is a deterministic decision tree, the path ξ' contains the node v. Assume that in the path ξ' there are exactly n working nodes. Then $\mathscr{S}(\xi') = \{g_{2 \cdot 1-1}(x) = 0, \ldots, g_{2(i-1)-1}(x) = 0, g_{2i-1}(x) = 1, g_{2(i+1)-1}(x) = 0, \ldots, g_{2n-1}(x) = 0\}$ and $\mathscr{A}(\xi') = \{2i-1, 2i\}$, but this is impossible since $z(2i-1) \cap z(2i) = \emptyset$. Consequently, in the path ξ' there are at least $n+1$ working nodes and $h(\Gamma) \geq n+1$. Therefore $h^d_{U_6 g}(z) \geq n+1$ and $\mathscr{U}^{da}_{U_6 hg}(n) \geq n+1$. Taking into account that n is an arbitrary number from ω such that $n \geq \max(n_0, 1)$, we obtain $\mathrm{Dom}^-(\mathscr{U}^{da}_{U_6 hg})$ is a finite set and $\mathrm{Dom}^+(\mathscr{U}^{da}_{U_6 hg})$ is an infinite set. Consequently, $\mathrm{typ}(\mathscr{U}^{da}_{U_6 hg}) = \delta$. Using Proposition 2.1, we obtain $\mathrm{typ}_{gu}(U_6, h) = t_6$. □

Let us define a sm-pair (U_7, h) as follows: $U_7 = (\mathbb{Z}, F_7)$, where $\mathbb{Z}$ is the set of integers, $F_7 = G_7 \cup L_7, G_7 = \{f_i : i \in \omega \setminus \{0\}\}$ and $L_7 = \{l_i : i \in \omega \setminus \{0\}\}$. For any $i \in \omega \setminus \{0\}$, $j \in \mathbb{Z}$, if $j = -i$, then $f_i(j) = 1$, and if $j \neq -i$, then $f_i(j) = 0$. For any $i \in \omega \setminus \{0\}$, $j \in \mathbb{Z}$, if $j \leq i$, then $l_i(j) = 0$, and if $j > i$, then $l_i(j) = 1$.

Lemma 2.13 $\mathrm{typ}_{gu}(U_7, h) = t_7$.

Proof We now prove that $\mathrm{typ}(\mathscr{U}^{da}_{U_7 hg}) = \epsilon$. Let $n \in \omega \setminus \{0\}$ and $z_n = (\nu_n, l_1, \ldots, l_n)$ be the problem from $\mathrm{Probl}(U_7)$, where $\nu : E_2^n \to \mathscr{P}(\omega)$ and $\nu(\bar{\delta}_1) \cap \nu(\bar{\delta}_2) = \emptyset$ for any $\bar{\delta}_1, \bar{\delta}_2 \in E_2^n$ such that $\bar{\delta}_1 \neq \bar{\delta}_2$. It is not difficult to show that, for any compatible on $\mathbb{Z}$ system of equations

$$\{l_1(x) = \delta_1, \ldots, l_n(x) = \delta_n\},$$

where $\delta_1, \ldots, \delta_n \in E_2$, there exists a subsystem, which has the same set of solutions and which contains at most two equations. Using this fact, it is not difficult to show that $h^a_{U_7 g}(z_n) \leq 2$. Let us prove that there is no $m \in \omega$ such that $h^d_{U_7 g}(z_n) \leq m$ for any $n \in \omega \setminus \{0\}$. Assume the contrary. Then there exists a number $t \in \omega \setminus \{0\}$ satisfying the following condition: for any $n \in \omega \setminus \{0\}$, there exists a decision tree $\Gamma \in \mathrm{Tree}(U_7)$, which solves the problem z_n deterministically and has at most t terminal nodes. Let us consider a problem z_t. Let Γ be an arbitrary decision tree over U_7, which solves the problem z_t deterministically. It is not difficult to show that the tree Γ must have at least $t + 1$ terminal nodes. We obtained a contradiction. Thus, $h^a_{U_7g}(z_n) \le 2$ for any $n \in \omega \setminus \{0\}$ and there is no $m \in \omega$ such that $h^d_{U_7g}(z_n) \le m$ for any $n \in \omega \setminus \{0\}$. Therefore $\mathcal{U}^{da}_{U_7hg}(2) = \infty$. Using Lemma 2.2, we obtain $\mathrm{typ}(\mathcal{U}^{da}_{U_7hg}) = \epsilon$.

We now prove that the function $h^a_{U_7g}$ is unbounded from above on the set $\mathrm{Probl}(U_7)$. Assume the contrary. Then there exists $m \in \omega$ such that $h^a_{U_7g}(z) \le m$ for any $z \in \mathrm{Probl}(U_7)$. Let $z = (\nu, f_1, \ldots, f_{m+1})$ be a problem from $\mathrm{Probl}(U_7)$ for which $\nu(\bar{\delta}_1) \cap \nu(\bar{\delta}_2) = \emptyset$ for any $\bar{\delta}_1, \bar{\delta}_2 \in E_2^{m+1}$ such that $\bar{\delta}_1 \ne \bar{\delta}_2$. Let Γ be a decision tree over U_7, which solves the problem z nondeterministically and for which $h(\Gamma) = h^a_{U_7g}(z)$. Let ξ be a complete path in the tree Γ such that $0 \in \mathcal{A}(\xi)$. It is clear that in the complete path ξ there is at least one working node. Let $\mathcal{S}(\xi) = \{q_1(x) = \delta_1, \ldots, q_t(x) = \delta_t\}$, where $q_1, \ldots, q_t \in F_7$. Since $0 \in \mathcal{A}(\xi)$, $\delta_1 = \ldots = \delta_t = 0$. It is clear that $i \notin \mathcal{A}(\xi)$ for any $i \in \{-1, \ldots, -m - 1\}$. Therefore, for any $i \in \{1, \ldots, m + 1\}$, the equation $f_i(x) = 0$ is contained in the system $\mathcal{S}(\xi)$. Consequently, $h(\Gamma) \ge m + 1$ and $h^a_{U_5g}(z) \ge m + 1$. We obtained a contradiction. Thus, the function $h^a_{U_7g}$ is unbounded from above on the set $\mathrm{Probl}(U_7)$. By Lemma 2.4, $\mathrm{typ}(\mathcal{U}^{aa}_{U_7hg}) = \gamma$. Using the equality $\mathrm{typ}(\mathcal{U}^{da}_{U_7h}g) = \epsilon$ and Proposition 2.1, we obtain that $\mathrm{typ}_{gu}(U_7, h) = t_7$. □

Proof of Proposition 2.3 The statement of the proposition follows from Lemmas 2.7–2.13.

Proof of Proposition 2.4 The statement of the proposition follows from Lemmas 2.8–2.13.

2.5 Auxiliary Statements

We now prove some auxiliary statements, which will help us to analyze the relationships between global upper types and global types of sm-pairs.

Let X be a nonempty set, $f : X \to \omega$ and $g : X \to \omega$. We define partial functions $\mathcal{U}^{fg} : \omega \to \omega$ and $\mathcal{L}^{gf} : \omega \to \omega$ as follows: if $n \in \omega$, then

$$\mathcal{U}^{fg}(n) = \max\{f(x) : x \in X, g(x) \le n\} ,$$
$$\mathcal{L}^{gf}(n) = \min\{g(x) : x \in X, f(x) \ge n\} .$$

The notation $\mathcal{U}^{fg}(n) = \infty$ means that $\{f(x) : x \in X, g(x) \le n\}$ is an infinite set. Evidently, if $\mathcal{U}^{fg}(n) = \infty$, then $\mathcal{U}^{fg}(n + 1) = \infty$.

It is not difficult to prove

Lemma 2.14 *The following statements hold:*

(a) *If there exists $n \in \omega$ such that $\mathcal{U}^{fg}(n) = \infty$, then* $\mathrm{typ}(\mathcal{U}^{fg}) = \epsilon$.

(b) *If there is no $n \in \omega$ such that $\mathcal{U}^{fg}(n) = \infty$, then* $\mathrm{Dom}(\mathcal{U}^{fg}) = \{n : n \in \omega, n \ge n_g\}$, *where* $n_g = \min\{g(x) : x \in X\}$.

(c) *If the function f is bounded from above on X, then* $\mathrm{typ}(\mathcal{L}^{gf}) = \epsilon$.

(d) *If the function f is unbounded from above on X, then* $\mathrm{Dom}(\mathcal{L}^{gf}) = \omega$.

Lemma 2.15 *The following statements hold:*
(a) $\text{typ}(\mathcal{U}^{fg}) = \alpha$ *if and only if* $\text{typ}(\mathcal{L}^{gf}) = \epsilon$.
(b) $\text{typ}(\mathcal{U}^{fg}) = \epsilon$ *if and only if* $\text{typ}(\mathcal{L}^{gf}) = \alpha$.

Proof (a) Let $\text{typ}(\mathcal{U}^{fg}) = \alpha$. Using Lemma 2.14, one can prove that the function f is bounded from above on X and $\text{typ}(\mathcal{L}^{gf}) = \epsilon$. Let $\text{typ}(\mathcal{L}^{gf}) = \epsilon$. From Lemma 2.14 it follows that the function f is bounded from above on X and $\text{typ}(\mathcal{U}^{fg}) = \alpha$.

(b) Let $\text{typ}(\mathcal{U}^{fg}) = \epsilon$. By Lemma 2.14, the function f is unbounded from above on X and there exists $m \in \omega$ such that $\mathcal{U}^{fg}(m) = \infty$. Therefore $\text{Dom}(\mathcal{L}^{gf}) = \omega$ and $\mathcal{L}^{gf}(n) \leq m$ for any $n \in \omega$. Consequently, $\text{typ}(\mathcal{L}^{gf}) = \alpha$. Let $\text{typ}(\mathcal{L}^{gf}) = \alpha$. Using Lemma 2.14, we have $\text{Dom}(\mathcal{L}^{gf}) = \omega$ and there exists $m \in \omega$ such that $\mathcal{L}^{gf}(n) \leq m$ for any $n \in \omega$. Therefore $\mathcal{U}^{fg}(m) = \infty$ and $\text{typ}(\mathcal{U}^{fg}) = \epsilon$. □

Lemma 2.16 *Let* $\text{typ}(\mathcal{U}^{fg}) \neq \alpha$ *and* $\text{typ}(\mathcal{U}^{fg}) \neq \epsilon$. *Then*
(a) $|\text{Dom}^-(\mathcal{U}^{fg})| < \infty$ *if and only if* $|\text{Dom}^+(\mathcal{L}^{gf})| < \infty$.
(b) $|\text{Dom}^+(\mathcal{U}^{fg})| < \infty$ *if and only if* $|\text{Dom}^-(\mathcal{L}^{gf})| < \infty$.

Proof Using Lemmas 2.14 and 2.15, we conclude that $\text{Dom}(\mathcal{U}^{fg}) = \{n : n \in \omega, n \geq n_g\}$ and $\text{Dom}(\mathcal{L}^{gf}) = \omega$, where $n_g = \min\{g(x) : x \in X\}$.

(a) Let $|\text{Dom}^-(\mathcal{U}^{fg})| < \infty$. Then there exists $m \in \omega, m \geq n_g$, such that $\mathcal{U}^{fg}(n) > n$ for any $n \in \omega, n \geq m$. Let $n \geq m$. Then there exists an element $x_0 \in X$ such that $g(x_0) \leq n$ and $f(x_0) \geq n + 1$. Therefore $\mathcal{L}^{gf}(n + 1) \leq n$. Consequently, $\mathcal{L}^{gf}(n) < n$ for any $n \in \omega, n \geq m + 1$, and $|\text{Dom}^+(\mathcal{L}^{gf})| < \infty$.

Let $|\text{Dom}^+(\mathcal{L}^{gf})| < \infty$. Then there exists $m \in \omega$ such that $\mathcal{L}^{gf}(n) < n$ for any $n \in \omega, n \geq m$. Let $n \in \omega$ and $n \geq m$. Then there exists an element $x_0 \in X$ such that $g(x_0) \leq n - 1$ and $f(x_0) \geq n$. Therefore $\mathcal{U}^{fg}(n - 1) \geq n$. Consequently, $\mathcal{U}^{fg}(n) > n$ for any $n \in \omega, n \geq m - 1$, and $|\text{Dom}^-(\mathcal{U}^{fg})| < \infty$.

(b) Let $|\text{Dom}^+(\mathcal{U}^{fg})| < \infty$. Then there exists $m \in \omega, m \geq n_g$, such that $\mathcal{U}^{fg}(n) < n$ for any $n \in \omega, n \geq m$. Let $n \geq m$. Then, for any element $x \in X$ such that $g(x) \leq n$, the inequality $f(x) < n$ holds. Therefore, for any element $x \in X$ such that $f(x) \geq n$, the inequality $g(x) > n$ holds. Consequently, $\mathcal{L}^{gf}(n) > n$ and $|\text{Dom}^-(\mathcal{L}^{gf})| < \infty$.

Let $|\text{Dom}^-(\mathcal{L}^{gf})| < \infty$. Then there exists $m \in \omega$ such that $\mathcal{L}^{gf}(n) > n$ for any $n \in \omega, n \geq m$. Let $n \in \omega$ and $n \geq \max(m, n_g)$. Then, for any element $x \in X$ such that $f(x) \geq n$, the inequality $g(x) > n$ holds. Therefore, for any element $x \in X$ such that $g(x) \leq n$, the inequality $f(x) < n$ holds. Consequently, $\mathcal{U}^{fg}(n) < n$ and $|\text{Dom}^+(\mathcal{U}^{fg})| < \infty$. □

Lemma 2.17 *The following statements hold:*
(a) $\text{typ}(\mathcal{U}^{fg}) = \beta$ *if and only if* $\text{typ}(\mathcal{L}^{gf}) = \delta$.
(b) $\text{typ}(\mathcal{U}^{fg}) = \gamma$ *if and only if* $\text{typ}(\mathcal{L}^{gf}) = \gamma$.
(c) $\text{typ}(\mathcal{U}^{fg}) = \delta$ *if and only if* $\text{typ}(\mathcal{L}^{gf}) = \beta$.

Proof (a) Let $\text{typ}(\mathcal{U}^{fg}) = \beta$. Using Lemma 2.15, we obtain $\text{Dom}(\mathcal{L}^{gf})$ is an infinite set. From Lemma 2.16 it follows that $|\text{Dom}^-(\mathcal{L}^{gf})| < \infty$. Thus, $\text{typ}(\mathcal{L}^{gf}) = \delta$. Let $\text{typ}(\mathcal{L}^{gf}) = \delta$. Using Lemma 2.15, we conclude that $\text{Dom}(\mathcal{U}^{fg})$ is an infinite set and the function $\mathcal{U}^{fg}$ is unbounded from above. From Lemma 2.16 it follows that $|\operatorname{Dom}^+(\mathcal{U}^{fg})| < \infty$. Therefore $\operatorname{typ}(\mathcal{U}^{fg}) = \beta$.

(b) Let $\mathcal{U}^{fg} = \gamma$. Using Lemma 2.16, we have $\operatorname{typ}(\mathcal{L}^{gf}) = \gamma$. Let $\operatorname{typ}(\mathcal{L}^{gf}) = \gamma$. By Lemma 2.15, $\operatorname{typ}(\mathcal{U}^{fg}) \neq \alpha$ and $\operatorname{typ}(\mathcal{U}^{fg}) \neq \epsilon$. From here and from Lemma 2.16 it follows that $\operatorname{typ}(\mathcal{U}^{fg}) = \gamma$.

(c) Using Lemma 2.15 and statements (a), (b) of Lemma 2.17, we obtain $\operatorname{typ}(\mathcal{U}^{fg}) = \delta$ if and only if $\operatorname{typ}(\mathcal{L}^{gf}) = \beta$. □

Let us define a function $\rho : \{\alpha, \beta, \gamma, \delta, \epsilon\} \to \{\alpha, \beta, \gamma, \delta, \epsilon\}$ as follows: $\rho(\alpha) = \epsilon, \rho(\beta) = \delta, \rho(\gamma) = \gamma, \rho(\delta) = \beta, \rho(\epsilon) = \alpha$.

Proposition 2.5 *Let X be a nonempty set, $f : X \to \omega$, $g : X \to \omega$, $\mathcal{U}^{fg}(n) = \max\{f(x) : x \in X, g(x) \leq n\}$ and $\mathcal{L}^{gf}(n) = \min\{g(x) : x \in X, f(x) \geq n\}$ for any $n \in \omega$. Then $\operatorname{typ}(\mathcal{L}^{gf}) = \rho(\operatorname{typ}(\mathcal{U}^{fg}))$.*

Proof The statement of the proposition follows from Lemmas 2.15 and 2.17. □

2.6 Proofs of Theorems 2.1 and 2.2

Using Proposition 2.5, we obtain the following statement.

Proposition 2.6 *Let (U, ψ) be a sm-pair and $b, c \in \{i, d, a\}$. Then*

$$\operatorname{typ}(\mathcal{L}^{cb}_{U\psi g}) = \rho(\operatorname{typ}(\mathcal{U}^{bc}_{U\psi g}))\ .$$

Proof of Theorem 2.1 The statement of the theorem follows from Propositions 2.1, 2.3 and 2.6.

Proof of Theorem 2.2 The statement of the theorem follows from Propositions 2.2, 2.4 and 2.6.

References

1. Moshkov, M.: Decision trees with quasilinear checks. Proc. Inst. Math. SB RAS **27**, 108–141 (1994). (in Russian)
2. Moshkov, M.: Two approaches to investigation of deterministic and nondeterministic decision trees complexity. In: 2nd World Conference on the Fundamentals of Artificial Intelligence, WOCFAI 1995, pp. 275–280 (1995)
3. Moshkov, M.: Comparative analysis of deterministic and nondeterministic decision tree complexity. Global approach. Fundam. Inform. **25**(2), 201–214 (1996)
4. Moshkov, M.: Local and global approaches to comparative analysis of complexity of deterministic and nondeterministic decision trees. In: Actual Problems of Modern Mathematics 2, pp. 110–118. NII MIOO NGU Publishers, Novosibirsk (1996). (in Russian)
5. Pawlak, Z.: Information systems theoretical foundations. Inf. Syst. **6**(3), 205–218 (1981)

Chapter 3
Comparative Analysis of Deterministic and Nondeterministic Decision Tree Complexity. Local Approach

3.1 Introduction

In this chapter, we consider arbitrary (finite and infinite) information systems and study problems with many-valued decisions over these information systems. An information system [5] consists of a set of objects called universe and a set of attributes that are functions from the universe to a finite set. Any problem is specified by a number of attributes that divide the universe into domains on which these attributes have fixed values. A nonempty finite set of decisions is attached to each domain. For a given object from the universe, it is required to find a decision from the set attached to the domain containing this object. Problems with many-valued decisions arise naturally in such areas as combinatorial optimization, fault diagnosis, computational geometry.

They arise also in problems related to data analysis. Let us consider a data set represented as an information system with finite universe and finite set of attributes. One of these attributes is distinguished as the decision one, all other are considered as the conditional attributes. The conditional attributes divide the universe into domains on which these attributes have fixed values. Our aim is to find the value of the decision attribute using only values of conditional attributes. Let us consider an arbitrary domain. It is clear that using only values of conditional attributes, we will obtain the same value of the decision attribute for all objects from the domain. If the decision attribute is constant on this domain, then we can find its exact value for all objects from the domain. Otherwise, we can find exact value only for some objects. If we want to minimize the number of incorrect answers, we must choose a value of the decision attribute, which is true for the maximum number of objects from the considered domain. It is possible that there exists more than one such value. In this case, we deal with a problem with many-valued decisions.

As algorithms for problem solving, we consider deterministic and nondeterministic decision trees over the information system. One can interpret nondeterministic decision trees solving a problem as a way for representation of arbitrary complete (applicable to any object) systems of true decision rules for the problem.

K. Durdymyradov et al., *Decision Trees Versus Systems of Decision Rules*, Studies in Big Data 160, https://doi.org/10.1007/978-3-031-71586-0_3

For a given information system, we consider both general complexity measures and limited complexity measures including depth and weighted depth of decision trees. The depth of a decision tree is the maximum number of nodes labeled with attributes in a path from the root to a terminal node of the tree.

There are two approaches to decision tree investigation: the local approach, where for a problem, the decision trees are considered using only attributes from the problem description, and the global one, where for problem solving, all attributes from the considered information system can be used in decision trees. In this chapter, decision trees are studied within the frameworks of the local approach.

In the present chapter, we study so-called sm-pairs (U, ψ), where U is an information system and ψ is a complexity measure over U. The sm-pair is called limited if ψ is a limited complexity measure.

We have the three parameters $\psi^i_{Ul}(z)$, $\psi^d_{Ul}(z)$ and $\psi^a_{Ul}(z)$ (index l refers to the local approach) for any problem z from the set $\mathrm{Probl}(U)$ of all problems over U:

- $\psi^i_{Ul}(z)$—the complexity of the problem z description. This parameter is equal to the complexity of a deterministic decision tree solving the problem z, which sequentially computes values of all attributes from the description of z.
- $\psi^d_{Ul}(z)$—the minimum complexity of a deterministic decision tree solving the problem z and using only attributes from this problem description.
- $\psi^a_{Ul}(z)$—the minimum complexity of a nondeterministic decision tree solving the problem z and using only attributes from this problem description.

We will investigate the relationships between any two such parameters for problems from $\mathrm{Probl}(U)$. Let us consider, for example, the parameters $\psi^i_{Ul}(z)$ and $\psi^d_{Ul}(z)$. Let $n \in \omega = \{0, 1, 2, \ldots\}$. We will study relations of the kind $\psi^i_{Ul}(z) \leq n \Rightarrow \psi^d_{Ul}(z) \leq m$, which are true for any $z \in \mathrm{Probl}(U)$. The minimum value of m is most interesting for us. This value (if exists) is equal to

$$\mathscr{U}^{di}_{U\psi l}(n) = \max\{\psi^d_{Ul}(z) : z \in \mathrm{Probl}(U), \psi^i_{Ul}(z) \leq n\} .$$

We will also study relations of the kind $\psi^i_{Ul}(z) \geq n \Rightarrow \psi^d_{Ul}(z) \geq m$. In this case, the maximum value of m is most interesting for us. This value (if exists) is equal to

$$\mathscr{L}^{di}_{U\psi l}(n) = \min\{\psi^d_{Ul}(z) : z \in \mathrm{Probl}(U), \psi^i_{Ul}(z) \geq n\} .$$

The two functions $\mathscr{U}^{di}_{U\psi l}$ and $\mathscr{L}^{di}_{U\psi l}$ describe how the behavior of the parameter $\psi^d_{Ul}(z)$ depends on the behavior of the parameter $\psi^i_{Ul}(z)$.

There are 18 similar functions for all ordered pairs of parameters $\psi^i_{Ul}(z)$, $\psi^d_{Ul}(z)$ and $\psi^a_{Ul}(z)$. The resulting 18 functions well describe the relationships between the three parameters under consideration, but it is most likely impossible to list all such tuples of 18 functions. In this chapter, instead of functions we will study types of functions. With any function, we will associate its type from the set $\{\alpha, \beta, \gamma, \delta, \epsilon\}$. For example, if a function f has an infinite domain and it is bounded from above, then its type is equal to α. If the function f has an infinite domain, is not bounded from above, and the inequality $f(n) \geq n$ holds for a finite number of n, then its type is equal to β, etc. Thus, we will list 18-tuples of types of functions. These tuples will be represented in tables called the local types of sm-pairs. We will prove that there are only seven realizable local types of sm-pairs and only five realizable local types of limited sm-pairs.

First, we will study 9-tuples of types of functions $\mathcal{U}^{bc}_{U\psi l}$, $b, c \in \{i, d, a\}$. These tuples will be represented in tables called local upper types of sm-pairs. We will list all realizable local upper types of sm-pairs and limited sm-pairs. After that, we will extend the results obtained for local upper types of sm-pairs to the case of local types of sm-pairs.

The obtained results allow us to understand not only types of relationships for each ordered pair of parameters, but also joint behavior of types of relationships for all ordered pairs of the considered three parameters.

Note that in this chapter, instead of deterministic or nondeterministic decision trees solving a problem, it will be convenient for us to talk about decision trees solving a problem deterministically or nondeterministically.

This chapter is based on the papers [1–4]. It consists of five sections. In Sect. 3.2, basic definitions and results are discussed. Sections 3.3–3.5 contains proofs of both auxiliary and main results.

3.2 Basic Definitions and Results

3.2.1 Decision Trees

Let $\omega = \{0, 1, 2, \ldots\}$, $E_k = \{0, 1, \ldots, k-1\}$, $k \geq 2$, A be a nonempty set and F be a nonempty set of functions from A to E_k. Functions from F will be called *attributes*, and the pair $U = (A, F)$ will be called an *information system*. Denote by F^* the set of all finite words over the alphabet F including the empty word λ.

A node in a finite directed tree is called the *root* if it is the only node, which has no entering edges. A tree, which has such a node will be called a *finite directed tree with root*. Nodes without leaving edges will be called *terminal* nodes. Nodes, which are neither the root nor terminal, will be called *working* nodes. A *complete path* in a finite directed tree with root is any sequence $\xi = v_0, d_0, \ldots, v_m, d_m, v_{m+1}$ of nodes and edges of the tree such that v_0 is the root, v_{m+1} is a terminal node and, for $i = 0, \ldots, m$, the edge d_i leaves the node v_i and enters the node v_{i+1}.

Definition 3.1 A *decision tree over information system U* is a marked finite directed tree with root, which has at least two nodes and possesses the following properties:

- The root and the edges leaving the root are not labeled.
- Each working node is labeled with an attribute from the set F.
- Each edge leaving a working node is labeled with a number from E_k.
- Each terminal node is labeled with a number from ω.

Definition 3.2 A decision tree is called *deterministic* if it satisfies the following conditions:

- Exactly one edge leaving the root.
- Edges leaving a working node are labeled with pairwise different numbers.

The set of decision trees over U will be denoted by Tree(U). Let $\Gamma \in \text{Tree}(U)$. Denote by At($\Gamma$) the set of attributes attached to working nodes of Γ. By Path(Γ), we denote the set of complete paths in Γ. Let $\xi = v_0, d_0, \ldots, v_m, d_m, v_{m+1}$ be a complete path in Γ. We denote by $\tau(\xi)$ the number attached to the node v_{m+1}. Define a word $\varphi(\xi)$ from F^* and a subset $\mathscr{A}(\xi)$ of the set A associated with ξ. If $m = 0$, then $\varphi(\xi) = \lambda$ and $\mathscr{A}(\xi) = A$. Let $m > 0$ and, for $j = 1, \ldots, m$, the node v_j be labeled with the attribute f_j, and the edge d_j be labeled with the number δ_j. Then $\varphi(\xi) = f_1 \cdots f_m$ and $\mathscr{A}(\xi) = \{a : a \in A, f_1(a) = \delta_1, \ldots, f_m(a) = \delta_m\}$. Denote $\mathscr{S}(\xi) = \{f_1(x) = \delta_1, \ldots, f_m(x) = \delta_m\}$.

3.2.2 Problems

The set of nonempty finite subsets of the set ω will be denoted by $\mathscr{P}(\omega)$.

Definition 3.3 A *problem over* U is any $(n + 1)$-tuple $z = (\nu, f_1, \ldots, f_n)$, where $n \in \omega \setminus \{0\}$, $\nu : E_k^n \to \mathscr{P}(\omega)$ and $f_1, \ldots, f_n \in F$. The problem z may be interpreted as a problem of searching for at least one number from the set $z(a) = \nu(f_1(a), \ldots, f_n(a))$ for a given $a \in A$.

We denote by Probl(U) the set of problems over the information system U. Let $z \in \text{Probl}(U)$. Denote $\text{At}(z) = \{f_1, \ldots, f_n\}$. Let $\Gamma \in \text{Tree}(U)$.

Definition 3.4 We will say that the tree Γ solves the problem z *nondeterministically* if the following conditions hold:

- $\bigcup_{\xi \in \text{Path}(\Gamma)} \mathscr{A}(\xi) = A$.
- For any $a \in A$ and $\xi \in \text{Path}(\Gamma)$, if $a \in \mathscr{A}(\xi)$, then $\tau(\xi) \in z(a)$.

We can also say that Γ is a nondeterministic decision tree solving the problem z.

Definition 3.5 We will say that the tree Γ solves the problem z *deterministically* if Γ is a deterministic decision tree, which solves z nondeterministically. We can also say that Γ is a deterministic decision tree solving the problem z.

3.2.3 Complexity Measures

Definition 3.6 A *complexity measure over* U is any mapping $\psi : F^* \to \omega$.

Definition 3.7 The complexity measure ψ will be called *limited* if it possesses the following properties:

(a) $\psi(\alpha_1\alpha_2) \leq \psi(\alpha_1) + \psi(\alpha_2)$ for any $\alpha_1, \alpha_2 \in F^*$.
(b) $\psi(\alpha_1\alpha_2\alpha_3) \geq \psi(\alpha_1\alpha_3)$ for any $\alpha_1, \alpha_2, \alpha_3 \in F^*$.
(c) For any $\alpha \in F^*$, the inequality $\psi(\alpha) \geq |\alpha|$ holds, where $|\alpha|$ is the length of α.

We now consider some examples of complexity measures. Let $w : F \to \omega \setminus \{0\}$. We define the function $\psi^w : F^* \to \omega$ in the following way: $\psi^w(\alpha) = 0$ if $\alpha = \lambda$ and $\psi^w(\alpha) = \sum_{i=1}^{m} w(f_i)$ if $\alpha = f_1 \cdots f_m$. The function ψ^w is a limited complexity measure over U and it is called a *weighted depth*. If $w \equiv 1$, then the function ψ^w is called the *depth* and is denoted by h.

We extend an arbitrary complexity measure ψ onto the set $\text{Tree}(U)$ in the following way: $\psi(\Gamma) = \max\{\psi(\varphi(\xi)) : \xi \in \text{Path}(\Gamma)\}$ for any $\Gamma \in \text{Tree}(U)$. The value $\psi(\Gamma)$ will be called the *complexity of the decision tree* Γ.

Let ψ be a complexity measure over U and $z = (\nu, f_1, \ldots, f_n) \in \text{Probl}(U)$. The value $\psi^i_{Ul}(z) = \psi(f_1 \cdots f_n)$ will be called the *complexity of the problem z description*. We denote by $\psi^d_{Ul}(z)$ the minimum complexity of a decision tree $\Gamma \in$ Tree (U), which solves the problem z deterministically and satisfies the condition $\text{At}(\Gamma) \subseteq \text{At}(z)$. We denote by $\psi^a_{Ul}(z)$ the minimum complexity of a decision tree $\Gamma \in \text{Tree}(U)$, which solves the problem z nondeterministically and satisfies the condition $\text{At}(\Gamma) \subseteq \text{At}(z)$.

3.2.4 Local Types of SM-Pairs

Definition 3.8 A pair (U, ψ), where U is an information system and ψ is a complexity measure over U, will be called a *system-measure-pair* (or, *sm-pair*, in short). If ψ is a limited complexity measure, then sm-pair (U, ψ) will be called a *limited sm-pair*.

Let (U, ψ) be a sm-pair. We have the three parameters $\psi^i_{Ul}(z)$, $\psi^d_{Ul}(z)$ and $\psi^a_{Ul}(z)$ for any problem $z \in \text{Probl}(U)$. We now define functions that describe relationships among these parameters.

Definition 3.9 Let $b, c \in \{i, d, a\}$. We define partial functions $\mathcal{U}^{bc}_{U\psi l} : \omega \to \omega$ and $\mathcal{L}^{bc}_{U\psi l} : \omega \to \omega$ in the following way:

$$\mathcal{U}^{bc}_{U\psi l}(n) = \max\{\psi^b_{Ul}(z) : z \in \text{Probl}(U), \psi^c_{Ul}(z) \leq n\} ,$$
$$\mathcal{L}^{bc}_{U\psi l}(n) = \min\{\psi^b_{Ul}(z) : z \in \text{Probl}(U), \psi^c_{Ul}(z) \geq n\} .$$

If the value $\mathcal{U}^{bc}_{U\psi l}(n)$ is definite, then it is the unimprovable upper bound on the values $\psi^{b}_{Ul}(z)$ for problems $z \in \text{Probl}(U)$ satisfying $\psi^{c}_{Ul}(z) \leq n$. If the value $\mathcal{L}^{bc}_{U\psi l}(n)$ is definite, then it is the unimprovable lower bound on the values $\psi^{b}_{Ul}(z)$ for problems $z \in \text{Probl}(U)$ satisfying $\psi^{c}_{Ul}(z) \geq n$.

Let g be a partial function from ω to ω. We denote by $\text{Dom}(g)$ the domain of g. Denote $\text{Dom}^{+}(g) = \{n : n \in \text{Dom}(g), g(n) \geq n\}$ and $\text{Dom}^{-}(g) = \{n : n \in \text{Dom}(g), g(n) \leq n\}$.

Definition 3.10 We now define the value $\text{typ}(g) \in \{\alpha, \beta, \gamma, \delta, \epsilon\}$ called the *type* of g.

- If $\text{Dom}(g)$ is an infinite set and g is a bounded from above function, then $\text{typ}(g) = \alpha$.
- If $\text{Dom}(g)$ is an infinite set, $\text{Dom}^{+}(g)$ is a finite set, and g is an unbounded from above function, then $\text{typ}(g) = \beta$.
- If both sets $\text{Dom}^{+}(g)$ and $\text{Dom}^{-}(g)$ are infinite, then $\text{typ}(g) = \gamma$.
- If $\text{Dom}(g)$ is an infinite set and $\text{Dom}^{-}(g)$ is a finite set, then $\text{typ}(g) = \delta$.
- If $\text{Dom}(g)$ is a finite set, then $\text{typ}(g) = \epsilon$.

Definition 3.11 We denote by $\text{typ}_l(U, \psi)$ a table with three rows and three columns in which rows from top to bottom and columns from the left to the right are labeled with indices i, d, a. The pair $\text{typ}(\mathcal{L}^{bc}_{U\psi l})\,\text{typ}(\mathcal{U}^{bc}_{U\psi l})$ is in the intersection of the row with index $b \in \{i, d, a\}$ and the column with index $c \in \{i, d, a\}$. The table $\text{typ}_l(U, \psi)$ will be called the *local type of sm-pair* (U, ψ).

3.2.5 *Basic Results*

The main problem investigated in this chapter is to find all local types of sm-pairs as well as of limited sm-pairs. The solution of this problem describes all possible (in terms of functions $\mathcal{U}^{bc}_{U\psi l}$, $\mathcal{L}^{bc}_{U\psi l}$ types, $b, c \in \{i, d, a\}$) relationships among the complexity of problem description, the complexity of problem solving by deterministic decision trees, and the complexity of problem solving by nondeterministic decision trees.

We now define seven tables:

$T_1 =$

	i	d	a
i	$\epsilon\alpha$	$\epsilon\alpha$	$\epsilon\alpha$
d	$\epsilon\alpha$	$\epsilon\alpha$	$\epsilon\alpha$
a	$\epsilon\alpha$	$\epsilon\alpha$	$\epsilon\alpha$

$T_2 =$

	i	d	a
i	$\gamma\gamma$	$\epsilon\epsilon$	$\epsilon\epsilon$
d	$\alpha\alpha$	$\epsilon\alpha$	$\epsilon\alpha$
a	$\alpha\alpha$	$\epsilon\alpha$	$\epsilon\alpha$

$T_3 =$

	i	d	a
i	$\gamma\gamma$	$\delta\epsilon$	$\epsilon\epsilon$
d	$\alpha\beta$	$\gamma\gamma$	$\epsilon\epsilon$
a	$\alpha\alpha$	$\alpha\alpha$	$\epsilon\alpha$

$T_4 =$

	i	d	a
i	$\gamma\gamma$	$\gamma\epsilon$	$\epsilon\epsilon$
d	$\alpha\gamma$	$\gamma\gamma$	$\epsilon\epsilon$
a	$\alpha\alpha$	$\alpha\alpha$	$\epsilon\alpha$

$T_5 =$

	i	d	a
i	$\gamma\gamma$	$\gamma\epsilon$	$\gamma\epsilon$
d	$\alpha\gamma$	$\gamma\gamma$	$\gamma\gamma$
a	$\alpha\gamma$	$\gamma\gamma$	$\gamma\gamma$

$T_6 =$

	i	d	a
i	$\gamma\gamma$	$\gamma\epsilon$	$\gamma\epsilon$
d	$\alpha\gamma$	$\gamma\gamma$	$\gamma\delta$
a	$\alpha\gamma$	$\beta\gamma$	$\gamma\gamma$

$T_7 =$

	i	d	a
i	$\gamma\gamma$	$\gamma\epsilon$	$\gamma\epsilon$
d	$\alpha\gamma$	$\gamma\gamma$	$\gamma\epsilon$
a	$\alpha\gamma$	$\alpha\gamma$	$\gamma\gamma$

Theorem 3.1 *For any sm-pair* (U, ψ), *the relation* $\text{typ}_l(U, \psi) \in \{T_1, T_2, T_3, T_4, T_5, T_6, T_7\}$ *holds. For any* $i \in \{1, 2, 3, 4, 5, 6, 7\}$, *there exists a sm-pair* (U, ψ) *such that* $\text{typ}_l(U, \psi) = T_i$.

Theorem 3.2 *For any limited sm-pair* (U, ψ), *the relation* $\text{typ}_l(U, \psi) \in \{T_2, T_3, T_5, T_6, T_7\}$ *holds. For any* $i \in \{2, 3, 5, 6, 7\}$, *there exists a limited sm-pair* (U, h) *such that* $\text{typ}_l(U, h) = T_i$.

Example 1 Let $n \in \omega \setminus \{0\}$, $U(n) = (\mathbb{R}^n, L_n)$, $L_n = \{s(\sum_{i=1}^{n} a_i x_i + a_{n+1}) : a_i \in \mathbb{R}, 1 \leq i \leq n+1\}$, $s : \mathbb{R} \to \{0, 1\}$ and $s(a) = 0$ if and only if $a < 0$. One can prove that $\text{typ}_l(U(1), h) = T_3$ and $\text{typ}_l(U(n), h) = T_7$ for any $n, n \geq 2$.

Seven similar examples with full proofs are included in Sect. 3.4 (see Lemmas 3.8–3.14).

3.3 Possible Local Upper Types of SM-Pairs

Definition 3.12 Let (U, ψ) be a sm-pair. We denote by $\text{typ}_{lu}(U, \psi)$ a table with three rows and three columns in which rows from top to bottom and columns from the left to the right are labeled with indices i, d, a. The value $\text{typ}(\mathscr{U}^{bc}_{U\psi l})$ is in the intersection of the row with index $b \in \{i, d, a\}$ and the column with index $c \in \{i, d, a\}$. The table $\text{typ}_{lu}(U, \psi)$ will be called the *local upper type of sm-pair* (U, ψ).

In this section, all possible local upper types of sm-pairs are listed. We now define seven tables:

$t_1 =$

	i	d	a
i	α	α	α
d	α	α	α
a	α	α	α

$t_2 =$

	i	d	a
i	γ	ϵ	ϵ
d	α	α	α
a	α	α	α

$t_3 =$

	i	d	a
i	γ	ϵ	ϵ
d	β	γ	ϵ
a	α	α	α

$t_4 =$

	i	d	a
i	γ	ϵ	ϵ
d	γ	γ	ϵ
a	α	α	α

$t_5 =$

	i	d	a
i	γ	ϵ	ϵ
d	γ	γ	γ
a	γ	γ	γ

$t_6 =$

	i	d	a
i	γ	ϵ	ϵ
d	γ	γ	δ
a	γ	γ	γ

$t_7 =$

	i	d	a
i	γ	ϵ	ϵ
d	γ	γ	ϵ
a	γ	γ	γ

Proposition 3.1 *For any sm-pair* (U, ψ), *the relation* $\text{typ}_{lu}(U, \psi) \in \{t_1, t_2, t_3, t_4, t_5, t_6, t_7\}$ *holds.*

Proposition 3.2 *For any limited sm-pair* (U, ψ), *the relation* $\text{typ}_{lu}(U, \psi) \in \{t_2, t_3, t_5, t_6, t_7\}$ *holds.*

We divide the proofs of the propositions into a sequence of lemmas.

Lemma 3.1 *Let* (U, ψ) *be a sm-pair and* $z \in \text{Probl}(U)$. *Then the inequalities* $\psi^a_{Ul}(z) \leq \psi^d_{Ul}(z) \leq \psi^i_{Ul}(z)$ *hold.*

Proof Let $z = (\nu, f_1, \ldots, f_n)$. It is not difficult to construct a decision tree $\Gamma_0 \in \mathrm{Tree}(U)$, which solves the problem z deterministically by sequential computation values of the attributes $f_1, \ldots, f_n$. Evidently, $\psi(\Gamma_0) = \psi^i_{Ul}(z)$ and $\mathrm{At}(\Gamma_0) \subseteq \mathrm{At}(z)$. Therefore $\psi^d_{Ul}(z) \le \psi^i_{Ul}(z)$. If a decision tree $\Gamma \in \mathrm{Tree}(U)$ solves the problem z deterministically, then the decision tree Γ solves the problem z nondeterministically. Therefore $\psi^a_{Ul}(z) \le \psi^d_{Ul}(z)$. □

Let (U, ψ) be a sm-pair, $n \in \omega$ and $b, c \in \{i, d, a\}$. The notation $\mathcal{U}^{bc}_{lU\psi}(n) = \infty$ means that the set $\{\psi^b_{Ul}(z) : z \in \mathrm{Probl}(U), \psi^c_{Ul}(z) \le n\}$ is infinite. Evidently, if $\mathcal{U}^{bc}_{U\psi l}(n) = \infty$, then $\mathcal{U}^{bc}_{U\psi l}(n+1) = \infty$. It is not difficult to prove the following statement.

Lemma 3.2 *Let (U, ψ) be a sm-pair and $b, c \in \{i, d, a\}$. Then*

(a) *If there exists $n \in \omega$ such that $\mathcal{U}^{bc}_{U\psi l}(n) = \infty$, then* $\mathrm{typ}(\mathcal{U}^{bc}_{U\psi l}) = \epsilon$.

(b) *If there is no $n \in \omega$ such that $\mathcal{U}^{bc}_{U\psi l}(n) = \infty$, then* $\mathrm{Dom}(\mathcal{U}^{bc}_{U\psi l}) = \{n : n \in \omega, n \ge n_0\}$, *where* $n_0 = \min\{\psi^c_{Ul}(z) : z \in \mathrm{Probl}(U)\}$.

Let (U, ψ) be a sm-pair and $b, c, e, f \in \{i, d, a\}$. The notation $\mathcal{U}^{bc}_{U\psi l} \lhd \mathcal{U}^{ef}_{U\psi l}$ means that, for any $n \in \omega$, the following statements hold:

(a) If the value $\mathcal{U}^{bc}_{U\psi l}(n)$ is definite, then either $\mathcal{U}^{ef}_{U\psi l}(n) = \infty$ or the value $\mathcal{U}^{ef}_{U\psi l}(n)$ is definite and the inequality $\mathcal{U}^{bc}_{U\psi l}(n) \le \mathcal{U}^{ef}_{U\psi l}(n)$ holds.

(b) If $\mathcal{U}^{bc}_{U\psi l}(n) = \infty$, then $\mathcal{U}^{ef}_{U\psi l}(n) = \infty$.

Let $\preceq$ be a linear order on the set $\{\alpha, \beta, \gamma, \delta, \epsilon\}$ such that $\alpha \preceq \beta \preceq \gamma \preceq \delta \preceq \epsilon$.

Lemma 3.3 *Let (U, ψ) be a sm-pair. Then* $\mathrm{typ}(\mathcal{U}^{bi}_{U\psi l}) \preceq \mathrm{typ}(\mathcal{U}^{bd}_{U\psi l}) \preceq \mathrm{typ}(\mathcal{U}^{ba}_{U\psi l})$ *and* $\mathrm{typ}(\mathcal{U}^{ab}_{U\psi l}) \preceq \mathrm{typ}(\mathcal{U}^{db}_{U\psi l}) \preceq \mathrm{typ}(\mathcal{U}^{ib}_{U\psi l})$ *for any $b \in \{i, d, a\}$.*

Proof From the definition of the functions $\mathcal{U}^{bc}_{U\psi l}$, $b, c \in \{i, d, a\}$, and from Lemma 3.1 it follows that $\mathcal{U}^{bi}_{U\psi l} \lhd \mathcal{U}^{bd}_{U\psi l} \lhd \mathcal{U}^{ba}_{U\psi l}$ and $\mathcal{U}^{ab}_{U\psi l} \lhd \mathcal{U}^{db}_{U\psi l} \lhd \mathcal{U}^{ib}_{U\psi l}$ for any $b \in \{i, d, a\}$. Using these relations and Lemma 3.2, we obtain the statement of the lemma. □

Lemma 3.4 *Let (U, ψ) be a sm-pair and $b, c \in \{i, d, a\}$. Then*

(a) $\mathrm{typ}(\mathcal{U}^{bc}_{U\psi l}) = \alpha$ *if and only if the function ψ^b_{Ul} is bounded from above on the set* $\mathrm{Probl}(U)$.

(b) *If the function ψ^b_{Ul} is unbounded from above on* $\mathrm{Probl}(U)$, *then* $\mathrm{typ}(\mathcal{U}^{bb}_{U\psi l}) = \gamma$.

Proof The statement (a) is obvious. (b) Let the function ψ^b_{Ul} be unbounded from above on $\mathrm{Probl}(U)$. One can show that in this case the equality $\mathcal{U}^{bb}_{U\psi l}(n) = n$ holds for infinitely many $n \in \omega$. Therefore $\mathrm{typ}(\mathcal{U}^{bb}_{U\psi l}) = \gamma$. □

Corollary 3.1 *Let (U, ψ) be a sm-pair and $b \in \{i, d, a\}$. Then* $\mathrm{typ}(\mathcal{U}^{bb}_{U\psi l}) \in \{\alpha, \gamma\}$.

Lemma 3.5 *Let (U, ψ) be a sm-pair and $\mathrm{typ}(\mathcal{U}^{ii}_{U\psi l}) \ne \alpha$. Then*

$$\mathrm{typ}(\mathcal{U}^{id}_{U\psi l}) = \mathrm{typ}(\mathcal{U}^{ia}_{U\psi l}) = \epsilon\ .$$

Proof Using Lemma 3.4, we conclude that the function ψ^i_{Ul} is unbounded from above on Probl(U). Let $m \in \omega$. Then there exists a problem $z = (\nu, f_1, \ldots, f_n) \in$ Probl(U) such that $\psi^i_{Ul}(z) \geq m$. Let us consider the problem $z' = (\nu', f_1, \ldots, f_n)$, where $\nu' \equiv \{0\}$. It is clear that $\psi^i_{Ul}(z') \geq m$. Let Γ be a decision tree, which consists of the root, the terminal node labeled with 0 and the edge connecting these two nodes. One can show that the tree Γ solves the problem z' deterministically. Therefore $\psi^a_{Ul}(z') \leq \psi^d_{Ul}(z') \leq \psi(\Gamma) = \psi(\lambda)$. Taking into account that m is an arbitrary number from ω, we obtain $\mathcal{U}^{id}_{U\psi l}(\psi(\lambda)) = \infty$ and $\mathcal{U}^{ia}_{U\psi l}(\psi(\lambda)) = \infty$. Using Lemma 3.2, we conclude that typ$(\mathcal{U}^{id}_{U\psi l}) =$ typ$(\mathcal{U}^{ia}_{U\psi l}) = \epsilon$. □

Lemma 3.6 *Let (U, ψ) be a sm-pair. Then* typ$(\mathcal{U}^{ai}_{U\psi l}) \in \{\alpha, \gamma\}$.

Proof Let $U = (A, F)$ and $f : A \to E_k$ for any $f \in F$. Using Lemma 3.3 and Corollary 3.1, we obtain typ$(\mathcal{U}^{ai}_{U\psi l}) \in \{\alpha, \beta, \gamma\}$. Assume that typ$(\mathcal{U}^{ai}_{U\psi l}) = \beta$. Then there exists $m \in \omega \setminus \{0\}$ such that $\mathcal{U}^{ai}_{U\psi l}(n) < n$ for any $n \in \omega$, $n > m$. Let us prove by induction on n that, for any problem $z \in$ Probl(U), if $\psi^i_{Ul}(z) \leq n$, then $\psi^a_{Ul}(z) \leq m_0$, where $m_0 = \max\{m, \psi(\lambda)\}$. Using Lemma 3.1, we conclude that, under the condition $n \leq m$, the considered statement holds. Let it hold for some n, $n \geq m$. Let us show that this statement holds for $n + 1$ too. Let $z \in$ Probl(U) and $\psi^i_{Ul}(z) \leq n + 1$. Since $n + 1 > m$, we obtain $\psi^a_{Ul}(z) \leq n$. Let $\Gamma \in$ Tree(U), At$(\Gamma) \subseteq$ At(z), $\psi(\Gamma) = \psi^a_{Ul}(z)$ and Γ solve the problem z nondeterministically. Assume that in Γ there exists a complete path ξ in which there are no working nodes. In this case, a decision tree, which consists of the root, the terminal node labeled with $\tau(\xi)$ and the edge connecting these two nodes, solves the problem z nondeterministically. Therefore $\psi^a_{Ul}(z) \leq \psi(\lambda) \leq m_0$. Assume now that each complete path in the decision tree Γ contains a working node. Let $\xi \in$ Path(Γ), $\xi = v_0, d_0, \ldots, v_p, d_p, v_{p+1}$ and, for $i = 1, \ldots, p$, the node v_i be labeled with the attribute f_i, and the edge d_i be labeled with the number δ_i. Let us consider the problem $z_\xi = (\nu_\xi, f_1, \ldots, f_p)$, where $\nu_\xi(\delta_1, \ldots, \delta_p) = \{\tau(\xi)\}$ and $\nu_\xi(\bar{\sigma}) = \{\tau(\xi) + 1\}$ for any p-tuple $\bar{\sigma} \in E^p_k$ such that $\bar{\sigma} \neq (\delta_1, \ldots, \delta_p)$. It is clear that $\psi^i_{Ul}(z_\xi) \leq n$. Using the inductive hypothesis, we conclude that there exists a decision tree $\Gamma_\xi \in$ Tree(U), which has the following properties: Γ_ξ solves the problem z_ξ nondeterministically, At$(\Gamma_\xi) \subseteq$ At(z_ξ) and $\psi(\Gamma_\xi) \leq m_0$. Let $\mathcal{A}(\xi) \neq \emptyset$. We denote by $\tilde{\Gamma}_\xi$ a tree obtained from Γ_ξ by removal all nodes and edges that satisfy the following condition: there is no a complete path ξ' in Γ_ξ, which contains this node or edge and for which $\tau(\xi') = \tau(\xi)$. Let $\{\xi : \xi \in$ Path$(\Gamma), \mathcal{A}(\xi) \neq \emptyset\} = \{\xi_1, \ldots, \xi_r\}$. Let us identify the roots of the trees $\tilde{\Gamma}_{\xi_1}, \ldots, \tilde{\Gamma}_{\xi_r}$. We denote by G the obtained tree. It is not difficult to show that $G \in$ Tree(U), At$(G) \subseteq$ At(z), $\psi(G) \leq m_0$ and the decision tree G solves the problem z nondeterministically. Thus, the considered statement holds. Using Lemma 3.4, we conclude that typ$(\mathcal{U}^{ai}_{U\psi l}) = \alpha$. The obtained contradiction shows that typ$(\mathcal{U}^{ai}_{U\psi l}) \in \{\alpha, \gamma\}$. □

Lemma 3.7 *Let (U, ψ) be a limited sm-pair and* typ$(\mathcal{U}^{ai}_{U\psi l}) = \alpha$. *Then* typ$(\mathcal{U}^{di}_{U\psi l}) \in \{\alpha, \beta\}$.

Proof Let $U = (A, F)$ and $f : A \to E_k$ for any $f \in F$. Using Lemma 3.4, we conclude that there exists $r \in \omega$ such that the inequality $\psi^a_{Ul}(z) \leq r$ holds for any problem $z \in \mathrm{Probl}(U)$. Let $f \in F$ and $f \not\equiv const$. Let us consider a problem $z_f = (\nu, f)$, where $\nu(\delta) = \{\delta\}$ for any $\delta \in E_k$. Let $\Gamma \in \mathrm{Tree}(U)$, $\mathrm{At}(\Gamma) \subseteq \mathrm{At}(z_f)$, $\psi(\Gamma) = \psi^a_{UI}(z_f)$ and Γ solve the problem z_f nondeterministically. Since $f \not\equiv const$, we have $f \in \mathrm{At}(\Gamma)$. Using the property (b) of the complexity measure ψ, we obtain $\psi(\Gamma) \geq \psi(f)$. Consequently, $\psi(f) \leq r$.

Let $f_1, \ldots, f_n \in F$, $\delta_1, \ldots, \delta_n \in E_k$ and the system of equations

$$\{f_1(x) = \delta_1, \ldots, f_n(x) = \delta_n\} \tag{3.1}$$

be compatible on A. Let us consider a problem $z = (\nu, f_1, \ldots, f_n)$, where $\nu : E_k^n \to \{\{0\}, \{1\}\}$ and $\nu(\bar{\sigma}) = \{1\}$ if and only if $\bar{\sigma} = (\delta_1, \ldots, \delta_n)$. Taking into account that $\psi^a_{UI}(z) \leq r$ and the complexity measure ψ has the property (c), it is not difficult to show that there exists a subsystem of the system (3.1), which contains at most r equations and has the same set of solutions just as the system (3.1).

Let the system (3.1) be incompatible on A. Let us show that there exists an incompatible subsystem of the system (3.1), which contains at most $r + 1$ equations. If the system $\{f_1(x) = \delta_1\}$ is incompatible on A, then the considered statement holds. Otherwise, there exists $i \in \{1, \ldots, n-1\}$ such that the system

$$\{f_1(x) = \delta_1, \ldots, f_i(x) = \delta_i\} \tag{3.2}$$

is compatible on A and the system, which is obtained from the system (3.2) by adding the equation $f_{i+1}(x) = \delta_{i+1}$, is incompatible on A. According to proved above, there exists a subsystem S of the system (3.2), which contains at most r equations and has the same set of solutions just as the system (3.2). By adding the equation $f_{i+1}(x) = \delta_{i+1}$ to the system S, we obtain a subsystem of the system (3.1), which is incompatible on A and contains at most $r + 1$ equations.

Let $z_1 \in \mathrm{Probl}(U)$. It is clear that there exists a problem $z_2 \in \mathrm{Probl}(U)$, which has the following properties: $z_1(a) = z_2(a)$ for any $a \in A$, $\mathrm{At}(z_2) \subseteq \mathrm{At}(z_1)$, and $f \not\equiv const$ for any $f \in \mathrm{At}(z_2)$. Let $z_2 = (\nu, f_1, \ldots, f_n)$. We now consider the problem $z_3 = (\nu', f_1, \ldots, f_n)$ from $\mathrm{Probl}(U)$, where ν' satisfies the following conditions: $|\nu'(\bar{\delta})| = 1$ for any $\bar{\delta} \in E_k^n$, and $\nu'(\bar{\delta}_1) \neq \nu'(\bar{\delta}_2)$ for any $\bar{\delta}_1, \bar{\delta}_2 \in E_k^n$, $\bar{\delta}_1 \neq \bar{\delta}_2$. It is not difficult to show that $\psi^d_{UI}(z_1) \leq \psi^d_{UI}(z_2) \leq \psi^d_{UI}(z_3)$. For $\bar{\delta} = (\delta_1, \ldots, \delta_n) \in E_k^n$, let us denote by $A(\bar{\delta})$ the set of solutions on A for the following system of equations:

$$S(\bar{\delta}) = \{f_1(x) = \delta_1, \ldots, f_n(x) = \delta_n\} .$$

Let $\Delta(z_3) = \{\bar{\delta} : \bar{\delta} \in E_k^n, A(\bar{\delta}) \neq \emptyset\}$ and $N = |\Delta(z_3)|$. Taking into account that any compatible on A system $S(\bar{\delta})$ has a subsystem with at most r equations and with the same set of solutions just as the system $S(\bar{\delta})$, it is not difficult to show that

$$N \leq (n+1)^r k^r . \tag{3.3}$$

For any $i \in \{1, \ldots, n\}$ and $\delta \in E_k$, we denote by $N(i, \delta)$ the number of n-tuples from $\Delta(z_3)$ in which ith digit is equal to δ.

It is clear that $A = \bigcup_{\bar{\delta}\in\Delta(z_3)} A(\bar{\delta})$ and, for any $a_1, a_2 \in A$, the relation $z_3(a_1) \cap z_3(a_2) \neq \emptyset$ holds if and only if $a_1 \in A(\bar{\delta})$ and $a_2 \in A(\bar{\delta})$ for some $\bar{\delta} \in \Delta(z_3)$. Thus, when we begin to solve the problem z_3 for an element $a \in A$, we know only that the element a belongs to one of the sets $A(\bar{\delta}), \bar{\delta} \in \Delta(z_3)$. When the problem z_3 will be solved, we will be able to point to the concrete n-tuple $\bar{\delta} \in \Delta(z_3)$ such that $a \in A(\bar{\delta})$. If the value of the attribute f_i is computed and $f_i(a) = \delta$, then the number of sets $A(\bar{\delta})$, in which the element a can be contained, reduces from N to $N(i, \delta)$. Let $N \geq 2$. Let us show that we can choose at most $r + 1$ attributes from $\{f_1, \ldots, f_n\}$, which have the following property: the computation of these attributes, independently of obtained values, leads to reduction of the set number in twice. Let $i \in \{1, \ldots, n\}$ and σ_i be a number from E_k such that $N(i, \sigma_i) = \max\{N(i, \delta) : \delta \in E_k\}$. Evidently,

$$N(i, \delta) \leq N/2 \tag{3.4}$$

for any $\delta \in E_k, \delta \neq \sigma_i$. Let $\bar{\sigma} = (\sigma_1, \ldots, \sigma_n)$. We now consider the system of equations $S(\bar{\sigma})$. According to proved above, there exists a subsystem $\{f_{i_1}(x) = \sigma_{i_1}, \ldots, f_{i_m}(x) = \sigma_{i_m}\}$ of the system $S(\bar{\sigma})$, which has the same set of solutions just as the system $S(\bar{\sigma})$ and for which $m \leq r + 1$. Let we compute values of the attributes $f_{i_1}, \ldots, f_{i_m}$ on element $a \in A$, and let $f_{i_1}(a) = t_1, \ldots, f_{i_m}(a) = t_m$. We denote by Δ' the set $\{(\delta_1, \ldots, \delta_n) \in \Delta(z_3) : \delta_{i_1} = t_1, \ldots, \delta_{i_m} = t_m\}$. It is clear that $a \in \bigcup_{\bar{\delta}\in\Delta'} A(\bar{\delta})$. If the equality $t_j = \sigma_{i_j}$ holds for any $j \in \{1, \ldots, m\}$, then $\Delta' = \{\bar{\sigma}\}$. Let, for some $j \in \{1, \ldots, m\}$, the relation $t_j \neq \sigma_{i_j}$ hold. Using (3.4), we obtain $|\Delta'| \leq N/2$. Thus, after the computation values of $m \leq r + 1$ attributes, the number of the sets $A(\bar{\delta})$ in which the element a can be contained reduced in twice. If $|\Delta'| = 1$, then the problem z_3 is solved for the element a. If $|\Delta'| > 1$, then we in the same way look for proper attributes for the set Δ', etc. Using the inequality (3.3), we conclude that there exists a decision tree Γ, which solves the problem z_3 deterministically and for which $\mathrm{At}(\Gamma) \subseteq \mathrm{At}(z_3)$ and $h(\Gamma) \leq (r + 1)\lceil \log_2 N \rceil \leq (r + 1)^2 \log_2(k(n + 1))$. Taking into account that $\psi(f_i) \leq r$ for any attribute $f_i \in \mathrm{At}(z_3)$ and the complexity measure ψ has the property (a), we obtain

$$\psi^d_{Ul}(z_3) \leq r(r + 1)^2 \log_2(k(n + 1)) .$$

Consequently, $\psi^d_{Ul}(z_1) \leq r(r + 1)^2 \log_2(k(n + 1))$. Taking into account that the complexity measure ψ has the property (c), we obtain $\psi^i_{Ul}(z_1) \geq n$. Since z_1 is an arbitrary problem over U, $\mathrm{Dom}^+(\mathcal{U}^{di}_{U\psi l})$ is a finite set. Therefore $\mathrm{typ}(\mathcal{U}^{di}_{U\psi l}) \neq \gamma$. Using Lemma 3.3 and Corollary 3.1, we obtain $\mathrm{typ}(\mathcal{U}^{di}_{U\psi l}) \in \{\alpha, \beta\}$. □

Proof of Proposition 3.1 Let (U, ψ) be a sm-pair. Using Corollary 3.1, we conclude that $\mathrm{typ}(\mathcal{U}^{ii}_{U\psi l}) \in \{\alpha, \gamma\}$. Using Corollary 3.1 and Lemma 3.3, we obtain $\mathrm{typ}(\mathcal{U}^{di}_{U\psi l}) \in \{\alpha, \beta, \gamma\}$. From Lemma 3.6 it follows that $\mathrm{typ}(\mathcal{U}^{ai}_{U\psi l}) \in \{\alpha, \gamma\}$.

(a) Let $\mathrm{typ}(\mathcal{U}^{ii}_{U\psi l}) = \alpha$. Using Lemmas 3.3 and 3.4, we obtain $\mathrm{typ}_{lu}(U, \psi) = t_1$.

(b) Let $\mathrm{typ}(\mathcal{U}^{ii}_{U\psi l}) = \gamma$ and $\mathrm{typ}(\mathcal{U}^{di}_{U\psi l}) = \alpha$. Using Lemmas 3.3, 3.4 and 3.5, we have $\mathrm{typ}_{lu}(U, \psi) = t_2$.

(c) Let $\text{typ}(\mathscr{U}^{ii}_{U\psi l}) = \gamma$ and $\text{typ}(\mathscr{U}^{di}_{U\psi l}) = \beta$. From Lemma 3.5 it follows that $\text{typ}(\mathscr{U}^{id}_{U\psi l}) = \text{typ}(\mathscr{U}^{ia}_{U\psi l}) = \epsilon$. Using Lemmas 3.3 and 3.6, we obtain $\text{typ}(\mathscr{U}^{ai}_{U\psi l}) = \alpha$. From this equality and from Lemma 3.4 it follows that $\text{typ}(\mathscr{U}^{ad}_{U\psi l}) = \text{typ}(\mathscr{U}^{aa}_{U\psi l}) = \alpha$. Using the equality $\text{typ}(\mathscr{U}^{di}_{U\psi l}) = \beta$, Lemma 3.3 and Corollary 3.1, we obtain $\text{typ}(\mathscr{U}^{dd}_{U\psi l}) = \gamma$. From the equalities $\text{typ}(\mathscr{U}^{dd}_{U\psi l}) = \gamma$, $\text{typ}(\mathscr{U}^{aa}_{U\psi l}) = \alpha$ and from Lemmas 3.2 and 3.4 it follows that $\text{typ}(\mathscr{U}^{da}_{U\psi l}) = \epsilon$. Thus, $\text{typ}_{lu}(U, \psi) = t_3$.

(d) Let $\text{typ}(\mathscr{U}^{ii}_{U\psi l}) = \text{typ}(\mathscr{U}^{di}_{U\psi l}) = \gamma$ and $\text{typ}(\mathscr{U}^{ai}_{U\psi l}) = \alpha$. Using Lemma 3.5, we obtain $\text{typ}(\mathscr{U}^{id}_{U\psi l}) = \text{typ}(\mathscr{U}^{ia}_{U\psi l}) = \epsilon$. From Lemma 3.4 it follows that $\text{typ}(\mathscr{U}^{ad}_{U\psi l}) = \text{typ}(\mathscr{U}^{aa}_{U\psi l}) = \alpha$. Using Lemma 3.3 and Corollary 3.1, we obtain $\text{typ}(\mathscr{U}^{dd}_{U\psi l}) = \gamma$. From this equality, equality $\text{typ}(\mathscr{U}^{aa}_{U\psi l}) = \alpha$ and from Lemmas 3.2 and 3.4 it follows that $\text{typ}(\mathscr{U}^{da}_{U\psi l}) = \epsilon$. Thus, $\text{typ}_{lu}(U, \psi) = t_4$.

(e) Let $\text{typ}(\mathscr{U}^{ii}_{U\psi l}) = \text{typ}(\mathscr{U}^{di}_{U\psi l}) = \text{typ}(\mathscr{U}^{ai}_{U\psi l}) = \gamma$. Using Lemma 3.5, we conclude that $\text{typ}(\mathscr{U}^{id}_{U\psi l}) = \text{typ}(\mathscr{U}^{ia}_{U\psi l}) = \epsilon$. Using Lemma 3.3 and Corollary 3.1, we obtain $\text{typ}(\mathscr{U}^{dd}_{U\psi l}) = \text{typ}(\mathscr{U}^{ad}_{U\psi l}) = \text{typ}(\mathscr{U}^{aa}_{U\psi l}) = \gamma$. Using Lemma 3.3, we obtain $\text{typ}(\mathscr{U}^{da}_{U\psi l}) \in \{\gamma, \delta, \epsilon\}$. Therefore $\text{typ}_{lu}(U, \psi) \in \{t_5, t_6, t_7\}$.

***Proof of Proposition* 3.2** Let (U, ψ) be a limited sm-pair. Taking into account that the complexity measure ψ has the property (c) and using Lemma 3.4, we obtain $\text{typ}(\mathscr{U}^{ii}_{U\psi l}) \neq \alpha$. Therefore $\text{typ}_{lu}(U, \psi) \neq t_1$. Using Lemma 3.7, we obtain $\text{typ}_{lu}(U, \psi) \neq t_4$. From these relations and Proposition 3.1 it follows that the statement of the proposition holds.

3.4 Realizable Local Upper Types of SM-Pairs

In this section, all realizable local upper types of sm-pairs are listed.

Proposition 3.3 *For any $i \in \{1, 2, 3, 4, 5, 6, 7\}$, there exists a sm-pair (U, ψ) such that*

$$\text{typ}_{lu}(U, \psi) = t_i \ .$$

Proposition 3.4 *For any $i \in \{2, 3, 5, 6, 7\}$, there exists a limited sm-pair (U, h) such that*

$$\text{typ}_{lu}(U, h) = t_i \ .$$

We divide the proofs of the propositions into a sequence of lemmas.

Let us define a sm-pair (U_1, π) as follows: $U_1 = (\omega, F_1)$, where $F_1 = \{f\}$, $f \equiv 0$, and $\pi \equiv 0$.

Lemma 3.8 $\text{typ}_{lu}(U_1, \pi) = t_1$.

Proof Using Lemma 3.4, we conclude that $\text{typ}(\mathscr{U}^{ii}_{U_1\pi l}) = \alpha$. From this equality and from Proposition 3.1 it follows that $\text{typ}_{lu}(U_1, \pi) = t_1$. □

Let us define a sm-pair (U_2, h) as follows: $U_2 = (\omega, F_2)$, where $F_2 = F_1$.

Lemma 3.9 $\mathrm{typ}_{lu}(U_2, h) = t_2$.

Proof It is not difficult to show that $h^d_{U_2 l}(z) = 0$ for any problem $z \in \mathrm{Probl}(U_2)$. It is clear that the function $h^i_{U_2 l}$ is unbounded from above on $\mathrm{Probl}(U_2)$. Using Lemma 3.4, we conclude that $\mathrm{typ}(\mathcal{U}^{ii}_{U_2 hl}) = \gamma$ and $\mathrm{typ}(\mathcal{U}^{dd}_{U_2 hl}) = \alpha$. From these equalities and from Proposition 3.1 it follows that $\mathrm{typ}_{lu}(U_2, h) = t_2$. □

Let us define a sm-pair (U_3, h) as follows: $U_3 = (\omega, F_3)$, where $F_3 = \{l_i : i \in \omega \setminus \{0\}\}$ and, for any $i \in \omega \setminus \{0\}$, $j \in \omega$, if $j \le i$, then $l_i(j) = 0$ and if $j > i$, then $l_i(j) = 1$.

Lemma 3.10 $\mathrm{typ}_{lu}(U_3, h) = t_3$.

Proof It is not difficult to show that, for any compatible on ω system of equations

$$\{l_{i_1}(x) = \delta_1, \ldots, l_{i_m}(x) = \delta_m\} ,$$

where $l_{i_1}, \ldots, l_{i_m} \in F_3$ and $\delta_1, \ldots, \delta_m \in E_2$, there exists a subsystem, which has the same set of solutions and which contains at most two equations. Using this fact, it is not difficult to show that $h^a_{U_3 l}(z) \le 2$ for any problem $z \in \mathrm{Probl}(U_3)$. From here and from Lemma 3.4 it follows that $\mathrm{typ}(\mathcal{U}^{ai}_{U_3 hl}) = \alpha$. It is clear that (U_3, h) is a limited sm-pair. Using Lemma 3.7, we conclude that $\mathrm{typ}(\mathcal{U}^{di}_{U_3 hl}) \in \{\alpha, \beta\}$. Let us show that $\mathrm{typ}(\mathcal{U}^{di}_{U_3 hl}) = \beta$. Assume the contrary: $\mathrm{typ}(\mathcal{U}^{di}_{U_3 hl}) = \alpha$. Using Lemma 3.4, we conclude that there exists a number $m \in \omega$ such that $h^d_{U_3 l}(z) \le m$ for any problem $z \in \mathrm{Probl}(U_3)$. Therefore there exists a number $n \in \omega$ that satisfies the following condition: for any problem $z \in \mathrm{Probl}(U_3)$, there exists a decision tree $\Gamma \in \mathrm{Tree}(U_3)$, which solves the problem z deterministically and for which $|\mathrm{At}(\Gamma)| \le n$ and $\mathrm{At}(\Gamma) \subseteq \mathrm{At}(z)$. Let us consider a problem $z' = (\nu, l_1, \ldots, l_{n+1})$ from $\mathrm{Probl}(U_3)$, where $\nu : E_2^{n+1} \to \mathcal{P}(\omega)$ and $\nu(\bar{\delta}_1) \cap \nu(\bar{\delta}_2) = \emptyset$ for any $\bar{\delta}_1, \bar{\delta}_2 \in E_2^{n+1}$ such that $\bar{\delta}_1 \ne \bar{\delta}_2$. Let Γ be an arbitrary decision tree over U_3, which solves the problem z' deterministically and for which $\mathrm{At}(\Gamma) \subseteq \mathrm{At}(z')$. It is not difficult to show that $\mathrm{At}(\Gamma) = \mathrm{At}(z')$ and, hence, $|\mathrm{At}(\Gamma)| \ge n + 1$. We obtained a contradiction. Therefore $\mathrm{typ}(\mathcal{U}^{di}_{U_3 hl}) = \beta$. From this equality and from Proposition 3.1 it follows that $\mathrm{typ}_{lu}(U_3, h) = t_3$. □

Let us define a sm-pair (U_4, μ) as follows: $U_4 = (\omega, F_4)$, where $F_4 = F_3$, $\mu(\lambda) = 0$, $\mu(l_{i_1} \cdots l_{i_m}) = 1$ if $m = 1$ or $m = 2$ and $i_1 > i_2$, and $\mu(l_{i_1} \cdots l_{i_m}) = \max\{i_1, \ldots, i_m\}$ in other cases.

Lemma 3.11 $\mathrm{typ}_{lu}(U_4, \mu) = t_4$.

Proof By analogy with the proof of Lemma 3.10, one can show that $\mathrm{typ}(\mathcal{U}^{ai}_{U_4 \mu l}) = \alpha$.

Let us prove that $\mathrm{typ}(\mathcal{U}^{di}_{U_4 \mu l}) = \gamma$. Let $n \in \omega$ and $n \ge 4$. We now consider a problem $z = (\nu, l_1, \ldots, l_n)$ from $\mathrm{Probl}(U_4)$, where $\nu : E_2^n \to \mathcal{P}(\omega)$ and $\nu(\bar{\delta}_1) \cap \nu(\bar{\delta}_2) = \emptyset$ for any $\bar{\delta}_1, \bar{\delta}_2 \in E_2^n$ such that $\bar{\delta}_1 \ne \bar{\delta}_2$. It is clear that $\mu^i_{U_4 l}(z) = n$. Let Γ be a decision tree over U_4, which solves the problem z deterministically and for which $\mathrm{At}(\Gamma) \subseteq \mathrm{At}(z)$ and $\mu(\Gamma) = \mu^d_{U_4l}(z)$. It is not difficult to prove that $\mathrm{At}(\Gamma) = \mathrm{At}(z)$. Therefore $|\mathrm{At}(\Gamma)| = n \geq 4$. Using Lemma 3.1, we conclude that $\mu(\Gamma) \leq n$. Let us show that $\mu(\Gamma) = n$. Assume the contrary: $\mu(\Gamma) < n$. Since $l_n \in \mathrm{At}(\Gamma)$, there exists a complete path ξ in the decision tree Γ such that the letter l_n is in the word $\varphi(\xi)$. Since $\mu(\varphi(\xi)) < n$, we have $\varphi(\xi) = l_n$ or $\varphi(\xi) = l_n l_i$ where $i < n$. Let an edge d leave the root of the tree Γ. Then the node, which d enters, is labeled with the attribute l_n. Taking into account that $\mu(\Gamma) < n$ and using the properties of the complexity measure μ, we obtain $h(\Gamma) \leq 2$. Consequently, $|\mathrm{At}(\Gamma)| \leq 3$. We obtained a contradiction. Therefore $\mu(\Gamma) = n$ and $\mu^d_{U_4l}(z) = n$. Thus, $\mathscr{U}^{di}_{U_4\mu l}(n) = n$. Taking into account that n is an arbitrary number from ω for which $n \geq 4$, we obtain $\mathrm{typ}(\mathscr{U}^{di}_{U_4\mu l}) = \gamma$. From this equality, equality $\mathrm{typ}(\mathscr{U}^{ai}_{U_4\mu l}) = \alpha$ and Proposition 3.1 it follows that $\mathrm{typ}_{lu}(U_4, \mu) = t_4$. □

Let us define a sm-pair (U_5, h) as follows: $U_5 = (\omega, F_5)$, where $F_5 = \{f_i : i \in \omega \setminus \{0\}\}$ and, for any $i \in \omega \setminus \{0\}$, $j \in \omega$, if $i = j$, then $f_i(j) = 1$ and if $i \neq j$, then $f_i(j) = 0$.

Lemma 3.12 $\mathrm{typ}_{lu}(U_5, h) = t_5$.

Proof Let $z \in \mathrm{Probl}(U_5)$. We will show that $h^d_{U_5l}(z) = h^a_{U_5l}(z)$. Let Γ be a decision tree over U_5, which solves the problem z nondeterministically and for which $\mathrm{At}(\Gamma) \subseteq \mathrm{At}(z)$ and $h(\Gamma) = h^a_{U_5l}(z)$. Let ξ be a complete path in the tree Γ such that $0 \in \mathscr{A}(\xi)$. If in the complete path ξ there are no working nodes, then, as it is not difficult to show, $h^d_{U_5l}(z) = h^a_{U_5l}(z) = 0$. Let there be $m > 0$ working nodes in the path ξ and $\mathscr{S}(\xi) = \{f_{i_1}(x) = \delta_1, \ldots, f_{i_m}(x) = \delta_m\}$. Since $0 \in \mathscr{A}(\xi)$, we have $\delta_1 = \ldots = \delta_m = 0$. It is clear that, for any $p \in \mathscr{A}(\xi)$, the relation $\tau(\xi) \in z(p)$ holds.

Let us describe a decision tree Γ_1 over U_5, which solves the problem z deterministically and for which $\mathrm{At}(\Gamma_1) \subseteq \mathrm{At}(z)$ and $h(\Gamma_1) \leq m$. For an arbitrary $p \in \omega$, the decision tree Γ_1 computes the values $f_{i_1}(p), \ldots, f_{i_m}(p)$. If $f_{i_1}(p) = \cdots = f_{i_m}(p) = 0$, then $p \in \mathscr{A}(\xi)$ and, hence, the problem z is solved since we know that $\tau(\xi) \in z(p)$. Let, for some $j \in \{1, \ldots, m\}$, the equality $f_{i_j}(p) = 1$ hold. Then $p = i_j$ and, hence, the problem z is solved too since we know all the set $z(p)$.

It is clear that $h(\Gamma) \geq m$. Therefore $h(\Gamma_1) \leq h^a_{U_5l}(z)$ and $h^d_{U_5l}(z) \leq h^a_{U_5l}(z)$. Using Lemma 3.1, we conclude that $h^d_{U_5l}(z) \geq h^a_{U_5l}(z)$ and, hence, $h^d_{U_5l}(z) = h^a_{U_5l}(z)$. From this equality it follows that there is no $n \in \omega$ for which $\mathscr{U}^{da}_{U_5hl}(n) = \infty$. Using Lemma 3.2, we conclude that $\mathrm{Dom}(\mathscr{U}^{da}_{U_5hl}) = \{n : n \in \omega, n \geq n_0\}$ for some $n_0 \in \omega$ and $\mathscr{U}^{da}_{U_5hl}(n) \leq n$ for any $n \in \mathrm{Dom}(\mathscr{U}^{da}_{U_5hl})$.

Let $n \in \mathrm{Dom}(\mathscr{U}^{da}_{U_5hl})$ and $n \geq 1$. We now show that $\mathscr{U}^{da}_{U_5hl}(n) = n$. Let $z = (\nu, f_1, \ldots, f_n)$ be a problem from $\mathrm{Probl}(U_5)$ for which $\nu(\bar{\delta}_1) \cap \nu(\bar{\delta}_2) = \emptyset$ for any $\bar{\delta}_1, \bar{\delta}_2 \in E_2^n$ such that $\bar{\delta}_1 \neq \bar{\delta}_2$. Let Γ be a decision tree over U_5, which solves the problem z nondeterministically and for which $\mathrm{At}(\Gamma) \subseteq \mathrm{At}(z)$ and $h(\Gamma) = h^a_{U_5l}(z)$. Let ξ be a complete path in the tree Γ such that $0 \in \mathscr{A}(\xi)$. It is clear that in the complete path ξ there is at least one working node. Let $\mathscr{S}(\xi) = \{f_{i_1}(x) = \delta_1, \ldots, f_{i_m}(x) = \delta_m\}$. Since $0 \in \mathscr{A}(\xi)$, $\delta_1 = \cdots = \delta_m = 0$. It is clear that $i \notin \mathscr{A}(\xi)$ for any $i \in \{1, \ldots, n\}$.

Therefore, for any $i \in \{1, \dots, n\}$, the equation $f_i(x) = 0$ is contained in the system $\mathscr{S}(\xi)$. Consequently, $h(\Gamma) \geq n$ and $h^a_{U_5 l}(z) \geq n$. It is clear that $h^i_{U_5 l}(z) = n$. Using Lemma 3.1, we obtain $h^a_{U_5 l}(z) = n$ and $h^d_{U_5 l}(z) = n$. Therefore $\mathscr{U}^{da}_{U_5 hl}(n) \geq n$ and, as proved above, $\mathscr{U}^{da}_{U_5 hl}(n) = n$. Taking into account that n is an arbitrary number from ω such that $n \geq \max(n_0, 1)$, we conclude that $\mathrm{Dom}^-(\mathscr{U}^{da}_{U_5 hl})$ and $\mathrm{Dom}^+(\mathscr{U}^{da}_{U_5 hl})$ are infinite sets and $\mathrm{typ}(\mathscr{U}^{da}_{U_5 hl}) = \gamma$. Using Proposition 3.1, we obtain $\mathrm{typ}_{lu}(U_5, h) = t_5$. □

Let us define a sm-pair (U_6, h) as follows: $U_6 = (\omega, F_6)$, where $F_6 = F_5 \cup G_6$, $G_6 = \{g_{2i+1} : i \in \omega\}$ and, for any $i \in \omega$, $j \in \omega$, if $j \in \{2i + 1, 2i + 2\}$, then $g_{2i+1}(j) = 1$ and if $j \notin \{2i + 1, 2i + 2\}$, then $g_{2i+1}(j) = 0$.

Lemma 3.13 $\mathrm{typ}_{lu}(U_6, h) = t_6$.

Proof Let $z \in \mathrm{Probl}(U_6)$. We now show that $h^d_{U_6 l}(z) \leq h^a_{U_6 l}(z) + 1$. Let Γ be a decision tree over U_6, which solves the problem z nondeterministically and for which $\mathrm{At}(\Gamma) \subseteq \mathrm{At}(z)$ and $h(\Gamma) = h^a_{U_6 l}(z)$. Let ξ be a complete path in the tree Γ such that $0 \in \mathscr{A}(\xi)$. If in the complete path ξ there are no working nodes, then, as it is not difficult to show, $h^d_{U_6 l}(z) = h^a_{U_6 l}(z) = 0$. Let there be $m > 0$ working nodes in the path ξ and $\mathscr{S}(\xi) = \{q_1(x) = \delta_1, \dots, q_m(x) = \delta_m\}$, where $q_i \in F_6, 1 \leq i \leq m$. Since $0 \in \mathscr{A}(\xi)$, $\delta_1 = \cdots = \delta_m = 0$. It is clear that, for any $p \in \mathscr{A}(\xi)$, the relation $\tau(\xi) \in z(p)$ holds.

Let us describe a decision tree Γ_1 over U_6, which solves the problem z deterministically and for which $\mathrm{At}(\Gamma_1) \subseteq \mathrm{At}(z)$ and $h(\Gamma_1) \leq m + 1$. For an arbitrary $p \in \omega$, the decision tree Γ_1 computes the values $q_1(p), \dots, q_m(p)$. If $q_1(p) = \cdots = q_m(p) = 0$, then $p \in \mathscr{A}(\xi)$ and, hence, the problem z is solved since we know that $\tau(\xi) \in z(p)$. Let, for some $j \in \{1, \dots, m\}$, the equality $q_j(p) = 1$ hold. If $q_j = f_i$, then $p = i$ and, hence, the problem z is solved too since we know all the set $z(p)$. Let $q_j = g_{2i+1}$. Then $p \in \{2i + 1, 2i + 2\}$. Let there be no attributes f_{2i+1} and f_{2i+2} in the set $\mathrm{At}(z)$. One can show that in this case $z(2i + 1) \cap z(2i + 2) \neq \emptyset$ and, hence, the problem z is solved. Let at least one from the attributes f_{2i+1} and f_{2i+2} be in the set $\mathrm{At}(z)$. For example, let $f_{2i+1} \in \mathrm{At}(z)$. Then we compute the value $f_{2i+1}(p)$. If $f_{2i+1}(p) = 1$, then $p = 2i + 1$ and if $f_{2i+1}(p) = 0$, then $p = 2i + 2$. In these cases, the problem z is solved too.

It is clear that $h(\Gamma) \geq m$. Therefore $h(\Gamma_1) \leq h^a_{U_6 l}(z) + 1$ and $h^d_{U_6 l}(z) \leq h^a_{U_6 l}(z) + 1$. From this inequality it follows that there is no $n \in \omega$ for which $\mathscr{U}^{da}_{U_6 hl}(n) = \infty$. Using Lemma 3.2, we conclude that $\mathrm{Dom}(\mathscr{U}^{da}_{U_6 hl}) = \{n : n \in \omega, n \geq n_0\}$ for some $n_0 \in \omega$.

Let $n \in \mathrm{Dom}(\mathscr{U}^{da}_{U_6 hl})$ and $n \geq 1$. We now show that $\mathscr{U}^{da}_{U_6 hl}(n) \geq n + 1$. Let $z = (\nu, g_1, f_1, f_2, \dots, g_{2n-1}, f_{2n-1}, f_{2n})$ be a problem from $\mathrm{Probl}(U_6)$ such that $\nu(\bar{\delta}_1) \cap \nu(\bar{\delta}_2) = \emptyset$ for any $\bar{\delta}_1, \bar{\delta}_2 \in E_2^{3n}$, $\bar{\delta}_1 \neq \bar{\delta}_2$. It is not difficult to show that $h^a_{U_6 l}(z) \leq n$. Let us prove that $h^d_{U_6 l}(z) \geq n + 1$. Let Γ be a decision tree over U_6, which solves the problem z deterministically and for which $\mathrm{At}(\Gamma) \subseteq \mathrm{At}(z)$ and $h(\Gamma) = h^d_{U_6 l}(z)$. Let ξ be a complete path in the tree Γ such that $0 \in \mathscr{A}(\xi)$. It is clear that in the complete path ξ there is at least one working node. Let $\mathscr{S}(\xi) = \{q_1(x) = \delta_1, \dots, q_m(x) = \delta_m\}$, where $q_i \in F_6, 1 \le i \le m$. Since $0 \in \mathscr{A}(\xi)$, $\delta_1 = \cdots = \delta_m = 0$. It is clear that $i \notin \mathscr{A}(\xi)$ for any $i \in \{1, \ldots, 2n\}$. Therefore the equation $g_{2i-1}(x) = 0$ or both equations $f_{2i-1}(x) = 0$ and $f_{2i}(x) = 0$ must belong to the system $\mathscr{S}(\xi)$ for any $i \in \{1, \ldots, n\}$. If the number of working nodes in the path ξ is greater than n, then the considered statement holds. Let the number of working nodes in the path ξ be equal to n. Then $\mathscr{S}(\xi) = \{g_1(x) = 0, \ldots, g_{2n-1}(x) = 0\}$. Let v be the last working node in the path ξ and let v be labeled with the attribute g_{2i-1}. Let us consider the complete path ξ' in the tree Γ such that $2i - 1 \in \mathscr{A}(\xi')$. Since Γ is a deterministic decision tree, the path ξ' contains the node v. Assume that in the path ξ' there are exactly n working nodes. Then $\mathscr{S}(\xi') = \{g_{2\cdot1-1}(x) = 0, \ldots, g_{2(i-1)-1}(x) = 0, g_{2i-1}(x) = 1, g_{2(i+1)-1}(x) = 0, \ldots, g_{2n-1}(x) = 0\}$ and $\mathscr{A}(\xi') = \{2i - 1, 2i\}$, but this is impossible since $z(2i - 1) \cap z(2i) = \emptyset$. Consequently, in the path ξ', there are at least $n + 1$ working nodes and $h(\Gamma) \ge n + 1$. Therefore $h^d_{U_6l}(z) \ge n + 1$ and $\mathscr{U}^{da}_{U_6hl}(n) \ge n + 1$. Taking into account that n is an arbitrary number from ω such that $n \ge \max(n_0, 1)$, we conclude that $\mathrm{Dom}^-(\mathscr{U}^{da}_{U_6hl})$ is a finite set and $\mathrm{Dom}^+(\mathscr{U}^{da}_{U_6hl})$ is an infinite set. Consequently, $\mathrm{typ}(\mathscr{U}^{da}_{U_6hl}) = \delta$. Using Proposition 3.1, we obtain $\mathrm{typ}_{lu}(U_6, h) = t_6$. □

Let us define a sm-pair (U_7, h) as follows: $U_7 = (\omega, F_7)$, where $F_7 = F_3 \cup F_5$.

Lemma 3.14 $\mathrm{typ}_{lu}(U_7, h) = t_7$.

Proof Using Lemmas 3.2 and 3.10, we conclude that there exists $n \in \omega$ such that $\mathscr{U}^{da}_{U_3hl}(n) = \infty$. Using this equality and the relation $F_3 \subseteq F_7$, it is not difficult to show that $\mathscr{U}^{da}_{U_7hl}(n) = \infty$. From here and from Lemma 3.2 it follows that $\mathrm{typ}(\mathscr{U}^{da}_{U_7hl}) = \epsilon$.

From Lemmas 3.4 and 3.12 it follows that the function $h^a_{U_5l}$ is unbounded from above on the set $\mathrm{Probl}(U_5)$. Using the relation $F_5 \subseteq F_7$, it is not difficult to show that the function $h^a_{U_7l}$ is unbounded from above on the set $\mathrm{Probl}(U_7)$. Using Lemma 3.4, we obtain $\mathrm{typ}(\mathscr{U}^{aa}_{U_7hl}) = \gamma$. From the equalities $\mathrm{typ}(\mathscr{U}^{da}_{U_7hl}) = \epsilon$, $\mathrm{typ}(\mathscr{U}^{aa}_{U_7hl}) = \gamma$ and from Proposition 3.1 it follows that $\mathrm{typ}_{lu}(U_7, h) = t_7$. □

Proof of Proposition 3.3 The statement of the proposition follows from Lemmas 3.8-3.14.

Proof of Proposition 3.4 The statement of the proposition follows from Lemmas 3.9, 3.10, 3.12, 3.13 and 3.14.

3.5 Proofs of Theorems 3.1 and 3.2

Let us define a function $\rho : \{\alpha, \beta, \gamma, \delta, \epsilon\} \to \{\alpha, \beta, \gamma, \delta, \epsilon\}$ as follows: $\rho(\alpha) = \epsilon, \rho(\beta) = \delta, \rho(\gamma) = \gamma, \rho(\delta) = \beta, \rho(\epsilon) = \alpha$.

Proposition 3.5 *(Proposition 2.5) Let X be a nonempty set, $f : X \to \omega, g : X \to \omega$, $\mathscr{U}^{fg}(n) = \max\{f(x) : x \in X, g(x) \le n\}$ and $\mathscr{L}^{gf}(n) = \min\{g(x) : x \in X, f(x) \ge n\}$ for any $n \in \omega$. Then $\mathrm{typ}(\mathscr{L}^{gf}) = \rho(\mathrm{typ}(\mathscr{U}^{fg}))$.*

Using Proposition 3.5, we obtain the following statement.

Proposition 3.6 *Let* (U, ψ) *be a sm-pair and* $b, c \in \{i, d, a\}$. *Then*

$$\mathrm{typ}(\mathscr{L}^{cb}_{U\psi l}) = \rho(\mathrm{typ}(\mathscr{U}^{bc}_{U\psi l})) .$$

Proof of Theorem 3.1 The statement of the theorem follows from Propositions 3.1, 3.3, and 3.6.

Proof of Theorem 3.2 The statement of the theorem follows from Propositions 3.2, 3.4, and 3.6.

References

1. Moshkov, M.: Comparative analysis of complexity of deterministic and nondeterministic decision trees. Local approach. In: Actual Problems of Modern Mathematics 1, pp. 109–113. NII MIOO NGU Publishers, Novosibirsk (1995). (in Russian)
2. Moshkov, M.: Two approaches to investigation of deterministic and nondeterministic decision trees complexity. In: 2nd World Conference on the Fundamentals of Artificial Intelligence, WOCFAI 1995, pp. 275–280 (1995)
3. Moshkov, M.: Local and global approaches to comparative analysis of complexity of deterministic and nondeterministic decision trees. In: Actual Problems of Modern Mathematics 2, pp. 110–118. NII MIOO NGU Publishers, Novosibirsk (1996). (in Russian)
4. Moshkov, M.: Comparative analysis of deterministic and nondeterministic decision tree complexity. Local approach. In: Peters, J.F., Skowron, A. (eds.) Transactions on Rough Sets IV, Lecture Notes in Computer Science, vol. 3700, pp. 125–143. Springer (2005)
5. Pawlak, Z.: Information systems theoretical foundations. Inf. Syst. **6**(3), 205–218 (1981)

Chapter 4
Time and Space Complexity of Deterministic and Nondeterministic Decision Trees. Global Approach

4.1 Introduction

This chapter is devoted to the study of conventional problems over infinite binary information systems. Each such information system consists of an infinite universe and an infinite set of functions (attributes), which are defined on the universe and have values from the set {0, 1}. Any conventional problem over the information system is described by a finite number of attributes that divide the universe into domains in which these attributes have fixed values. A decision is attached to each domain. For a given object from the universe, it is required to find the decision attached to the domain containing this object. As algorithms solving these problems, deterministic and nondeterministic decision trees are considered. As time and space complexity of decision trees, we use their depth and the number of nodes. Note that nondeterministic decision trees can be considered as representations of decision rule systems.

We study relationships between time and space complexity for deterministic and nondeterministic decision trees. Decision trees are among the most interpretable models for classifying and representing knowledge [1]. In order to better understand decision trees, we should not only minimize the number of their nodes, but also the depth of decision trees to avoid the consideration of long conjunctions of conditions corresponding to long paths in these trees. When we consider decision trees as algorithms (usually sequential for deterministic decision trees and parallel for nondeterministic decision trees), we should have in mind the same bi-criteria optimization problems to minimize space and time complexity of these algorithms.

There are two approaches to decision tree investigation: the local approach, where for a problem, the decision trees are considered using only attributes from the problem description, and the global one, where for problem solving, all attributes from the considered information system can be used in decision trees. In this chapter, decision trees are investigated within the frameworks of the global approach.

K. Durdymyradov et al., *Decision Trees Versus Systems of Decision Rules*,
Studies in Big Data 160, https://doi.org/10.1007/978-3-031-71586-0_4

Based on the obtained results we can describe possible types of behavior of four functions h_U^{gd}, h_U^{ga}, L_U^{gd}, L_U^{ga} that characterize worst case time and space complexity of deterministic and nondeterministic decision trees solving problems over an infinite binary information system U (index g refers to the global approach).

The function h_U^{gd} characterizes the growth in the worst case of the minimum depth of a deterministic decision tree solving a problem with the growth of the number of attributes in the problem description. The function h_U^{gd} is either bounded from below by logarithm and bounded from above by logarithm to the power $1 + \varepsilon$, where ε is an arbitrary positive real number, or grows linearly.

The function h_U^{ga} characterizes the growth in the worst case of the minimum depth of a nondeterministic decision tree solving a problem with the growth of the number of attributes in the problem description. The function h_U^{ga} is either bounded from above by a constant or grows linearly.

The function L_U^{gd} characterizes the growth in the worst case of the minimum number of nodes in a deterministic decision tree solving a problem with the growth of the number of attributes in the problem description. The function L_U^{gd} has either polynomial or exponential growth.

The function L_U^{ga} characterizes the growth in the worst case of the minimum number of nodes in a nondeterministic decision tree solving a problem with the growth of the number of attributes in the problem description. The function L_U^{ga} has either polynomial or exponential growth.

We see that each of the functions h_U^{gd}, h_U^{ga}, L_U^{gd}, L_U^{ga} has two types of behavior. Thus, the tuple $(h_U^{gd}, h_U^{ga}, L_U^{gd}, L_U^{ga})$ can have (in general) 16 types of behavior. However (and this is one of the main results of the chapter), the tuple $(h_U^{gd}, h_U^{ga}, L_U^{gd}, L_U^{ga})$ have only five types of behavior. All these types are described in the chapter and each type is illustrated by an example. Similar result without proofs and with weaker bounds on the functions h_U^{gd}, h_U^{ga}, L_U^{gd}, L_U^{ga} was announced in [2].

There are five complexity classes of infinite binary information systems corresponding to the five possible types of the tuple $(h_U^{gd}, h_U^{ga}, L_U^{gd}, L_U^{ga})$. For each class, we study joint behavior of time and space complexity of decision trees.

A pair of functions (φ, ψ) is called a boundary gd-pair of the information system U if, for any problem over U, there exists a deterministic decision tree over U, which solves this problem and for which the depth is at most $\varphi(n)$ and the number of nodes is at most $\psi(n)$, where n is the number of attributes in the problem description. A boundary gd-pair (φ, ψ) of the information system U is called optimal if, for any boundary gd-pair (φ', ψ') of U, the inequalities $\varphi'(n) \geq \varphi(n)$ and $\psi'(n) \geq \psi(n)$ hold for any natural n. An information system U is called gd-reachable if the pair (h_U^{gd}, L_U^{gd}) is boundary (and, consequently, optimal boundary) gd-pair of the system U. For nondeterministic decision trees, the notions of a boundary ga-pair of an information system, an optimal ga-pair, and ga-reachable information system are defined in a similar way. For deterministic decision trees, the best situation is when the considered information system is gd-reachable: for any boundary gd-pair (φ, ψ) for an information system U and any natural n, $\varphi(n) \geq h_U^{gd}(n)$ and $\psi(n) \geq L_U^{gd}(n)$. For nondeterministic decision trees, the best situation is when the information system is ga-reachable.

For four out of the five complexity classes, all information systems from the class are gd-reachable. One class contains both information systems that are gd-reachable and information systems that are not gd-reachable. For each information system U that is not gd-reachable, we find a nontrivial boundary gd-pair, which is sufficiently close to the pair (h_U^{gd}, L_U^{gd}). For two out of the five complexity classes, all information systems from the class are ga-reachable. For the rest three classes, all information systems from the class are not ga-reachable. For some information systems U that are not ga-reachable, we find nontrivial boundary ga-pairs, which are sufficiently close to (h_U^{ga}, L_U^{ga}). For the rest of information systems U that are not ga-reachable, the pair $(n, L_U^{ga}(n))$ is the optimal boundary ga-pair. Note that for these information systems, the function h_U^{ga} is bounded from above by a constant.

The obtained results are related to time-space trade-off for deterministic and nondeterministic decision trees. For any information system U and for each problem over U, there exists a deterministic decision tree solving this problem, whose depth is at most $h_U^{gd}(n)$, and there exists a deterministic decision tree solving this problem for which the number of nodes is at most $L_U^{gd}(n)$, where n is the number of attributes in the problem description. If an information system U is not gd-reachable, then there exists a problem such that there is no a deterministic decision tree solving this problem, whose depth is at most $h_U^{gd}(n)$ and the number of nodes is at most $L_U^{gd}(n)$, where n is the number of attributes in the problem description. Similar situation is with nondeterministic decision trees for information systems that are not ga-reachable.

Let us consider an information system U for which the function h_U^{ga} is bounded from above by a natural number c, and $(n, L_U^{ga}(n))$ is the optimal boundary ga-pair. For each problem over U, there exists a nondeterministic decision tree solving this problem, whose depth is at most c. However, for any natural n greater than c, there is no a finite upper bound on the number of nodes in such trees for problems described by at most n attributes.

Note that this chapter will often talk about decision trees that solve a problem deterministically or nondeterministically, rather than deterministic or nondeterministic decision trees that solve a problem.

This chapter is based on the paper [5]. The rest of the chapter is organized as follows: Sect. 4.2 contains main results and Sects. 4.3–4.5—proofs of these results.

4.2 Main Results

Let A be an infinite set and F be an infinite set of functions that are defined on A and have values from the set $\{0, 1\}$. The pair $U = (A, F)$ is called an *infinite binary information system* [8], the elements of the set A are called *objects*, and the functions from F are called *attributes*. The set A is called sometimes the *universe* of the information system U.

A *problem over U* is a tuple of the form $z = (\nu, f_1, \ldots, f_n)$, where $\nu : \{0, 1\}^n \to \mathbb{N}$, $\mathbb{N}$ is the set of natural numbers $\{1, 2, \ldots\}$, and $f_1, \ldots, f_n \in F$. The problem z consists in finding the value of the function $z(x) = \nu(f_1(x), \ldots, f_n(x))$ for a given object $a \in A$. The value $\dim z = n$ is called the *dimension of the problem z*.

Various problems of combinatorial optimization, pattern recognition, fault diagnosis, probabilistic reasoning, computational geometry, etc., can be represented in this form. We will mention here some applications of decision trees considered in the research monograph [4]. Numerous combinatorial optimization problems, in particular, the traveling salesman problem with n cities, n-stone problem, and problem on 0-1-knapsack with n objects can be represented as problems over linear information systems. For example, for traveling salesman problem with $n \geq 4$ cities, there exists a linear decision tree that solves it and has depth at most n^7. Problems of diagnosis of constant faults in combinatorial circuits can be represented as problems over appropriate finite information systems. Mostly negative results relative to the depth of decision trees have been obtained for arbitrary circuits and encouraging results—for read-once circuits. The problem of recognition of words of fixed length n in a regular language can be also represented as a problem over a finite information system. For all regular languages, it was possible to understand how the minimum depth of decision trees solving the problem of recognition grows with the growth of n. For Bayesian networks in which each node has at most one entering edge and each variable has at most two values, the depth of decision trees that recognize values of all open (observable) variables was studied.

As algorithms for problem solving we consider decision trees. A *decision tree over the information system U* is a directed tree with a root in which the root and edges leaving the root are not labeled, each terminal node is labeled with a number from $\mathbb{N}$, each working node (which is neither the root nor a terminal node) is labeled with an attribute from F, and each edge leaving a working node is labeled with a number from the set $\{0, 1\}$. A decision tree is called *deterministic* if only one edge leaves the root and edges leaving an arbitrary working node are labeled with different numbers.

Let Γ be a decision tree over U and

$$\xi = v_0, d_0, v_1, d_1, \ldots, v_m, d_m, v_{m+1}$$

be a directed path from the root v_0 to a terminal node v_{m+1} of Γ (we call such path *complete*). Define a subset $A(\xi)$ of the set A as follows. If $m = 0$, then $A(\xi) = A$. Let $m > 0$ and, for $i = 1, \ldots, m$, the node v_i be labeled with the attribute f_{j_i} and the edge d_i be labeled with the number δ_i. Then

$$A(\xi) = \{a : a \in A, f_{j_1}(a) = \delta_1, \ldots, f_{j_m}(a) = \delta_m\} .$$

The decision tree Γ solves the problem z *nondeterministically* if, for any object $a \in A$, there exists a complete path ξ of Γ such that $a \in A(\xi)$ and, for each $a \in A$ and each complete path ξ such that $a \in A(\xi)$, the terminal node of ξ is labeled with the number $z(a)$ (in this case, we can say that Γ is a *nondeterministic decision tree solving the problem z*). In particular, if the decision tree Γ solves the problem z nondeterministically, then, for each complete path ξ of Γ, either the set $A(\xi)$ is empty or the function $z(x)$ is constant on the set $A(\xi)$. The decision tree Γ solves the problem z *deterministically* if Γ is a deterministic decision tree, which solves the problem z nondeterministically (in this case, we can say that Γ is a *deterministic decision tree solving the problem z*).

The *depth* of the decision tree Γ is the maximum number of working nodes in a complete path of Γ. Denote by $h(\Gamma)$ the depth of Γ and by $L(\Gamma)$—the number of nodes in Γ.

For a problem z over U, let $h_U^{gd}(z)$ be the minimum depth of a decision tree over U solving the problem z deterministically, $h_U^{ga}(z)$ be the minimum depth of a decision tree over U solving the problem z nondeterministically, $L_U^{gd}(z)$ be the minimum number of nodes in a decision tree over U solving the problem z deterministically, and $L_U^{ga}(z)$ be the minimum number of nodes in a decision tree over U solving the problem z nondeterministically.

We consider four functions defined on the set $\mathbb{N}$ in the following way: $h_U^{gd}(n) = \max h_U^{gd}(z)$, $h_U^{ga}(n) = \max h_U^{ga}(z)$, $L_U^{gd}(n) = \max L_U^{gd}(z)$, and $L_U^{ga}(n) = \max L_U^{ga}(z)$, where the maximum is taken among all problems z over U with $\dim z \le n$. These functions describe how the minimum depth and the minimum number of nodes of deterministic and nondeterministic decision trees solving problems are growing in the worst case with the growth of problem dimension. To describe possible types of behavior of these four functions, we need to define some properties of infinite binary information systems.

For $m \in \mathbb{N}$, a nonempty subset B of the set A will be called a *(m, U)-set* if B coincides with the set of solutions from A of an equation system of the form $\{f_1(x) = \delta_1, \ldots, f_m(x) = \delta_m\}$, where $f_1, \ldots, f_m$ are attributes from F (not necessary pairwise different), and $\delta_1, \ldots, \delta_m \in \{0, 1\}$. We call such system an *(m, U)-system of equations*. It is clear that an (m, U)-set is also an $(m + 1, U)$-set.

Definition 4.1 We will say that the information system U satisfies the *condition of coverage* if there exists an $m \in \mathbb{N}$ such that any $(m + 1, U)$-set is a union of a finite number of (m, U)-sets. In this case, we will say that U satisfies the *condition of coverage with parameter m*.

Definition 4.2 We will say that the information system U satisfies the *condition of restricted coverage* if there exist $m, t \in \mathbb{N}$ such that any $(m + 1, U)$-set is a union of at most t (m, U)-sets. In this case, we will say that U satisfies the *condition of restricted coverage with parameters m and t*.

If the information system U satisfies the condition of restricted coverage, then it satisfies the condition of coverage.

A subset $\{f_1, \ldots, f_m\}$ of the set F will be called *independent* if, for any $\delta_1, \ldots, \delta_m \in \{0, 1\}$, the system of equations $\{f_1(x) = \delta_1, \ldots, f_m(x) = \delta_m\}$ has a solution from the set A. The empty set of attributes is independent by definition.

Definition 4.3 We define the parameter $I(U)$, which is called the *independence dimension* or I-*dimension* of the information system U (this notion is similar to the notion of independence number of family of sets [7]) as follows. If, for each $m \in \mathbb{N}$, the set F contains an independent subset of cardinality m, then $I(U) = \infty$. Otherwise, $I(U)$ is the maximum cardinality of an independent subset of the set F.

To illustrate these definitions, we consider a slightly modified Example 8.8 from the book [6].

Example 4.1 Let P be the set of all points in the Euclidean plane and l be a straight line in the plane. This line divides the plane into two open half-planes H_1 and H_2 and the line l. We correspond one attribute to the line l. This attribute takes value 0 on points from H_1, and value 1 on points from H_2 and l. Denote by F the set of all attributes corresponding to lines in the plane. Let us consider the information system $U_0 = (P, F)$.

The information system U_0 has a finite I-dimension equal to 2: there are no three lines which divide the plane into eight domains and there are two lines that divide the plane into four domains (see Fig. 4.1). The information system U_0 satisfies the condition of restricted coverage with parameters 3 and 2: each $(4, U_0)$-set is a union of two $(3, U_0)$-sets (see Fig. 4.2).

Later we will discuss five more examples of information systems $U_1, \ldots, U_5$—see Lemmas 4.7–4.11.

Fig. 4.1 Information system U_0 has a finite I-dimension equal to 2

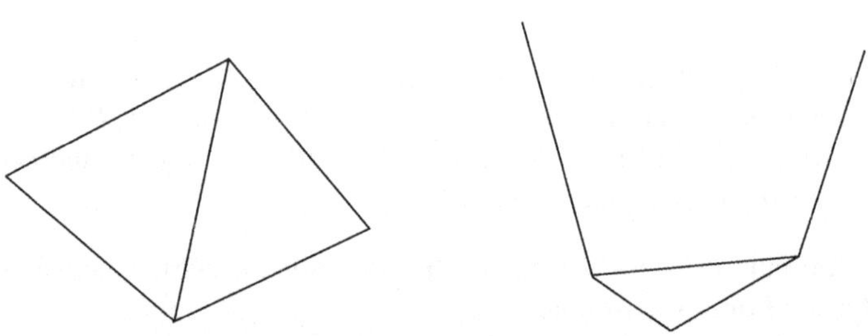

Fig. 4.2 Information system U_0 satisfies the condition of restricted coverage with parameters 3 and 2

We now consider four statements that describe possible types of behavior of functions $h_U^{gd}(n)$, $h_U^{ga}(n)$, $L_U^{gd}(n)$, and $L_U^{ga}(n)$. The next statement follows immediately from Theorem 2.1 [3] and the simple fact that $h_U^{gd}(n) \le n$ for any $n \in \mathbb{N}$.

Proposition 4.1 *For any infinite binary information system U, the function $h_U^{gd}(n)$ has one of the following two types of behavior:*

(LOG) *If the system U has finite I-dimension and satisfies the condition of restricted coverage, then for any ε, $0 < \varepsilon < 1$, there exists a positive constant c such that, for any $n \in \mathbb{N}$,*

$$\log_2(n+1) \le h_U^{gd}(n) \le c(\log_2 n)^{1+\varepsilon} + 1 .$$

(LIN) *If the system U has infinite I-dimension or does not satisfy the condition of restricted coverage, then for any $n \in \mathbb{N}$,*

$$h_U^{gd}(n) = n.$$

Proposition 4.2 *For any infinite binary information system $U = (A, F)$, the function $h_U^{ga}(n)$ has one of the following two types of behavior:*

(CON) *If the system U satisfies the condition of coverage, then there exists a positive constant c such that, for any $n \in \mathbb{N}$,*

$$h_U^{ga}(n) \le c .$$

(LIN) *If the system U does not satisfy the condition of coverage, then for any $n \in \mathbb{N}$,*

$$h_U^{ga}(n) = n .$$

Proposition 4.3 *For any infinite binary information system U, the function $L_U^{gd}(n)$ has one of the following two types of behavior:*

(POL) *If the system U has finite I-dimension, then for any $n \in \mathbb{N}$,*

$$2(n+1) \le L_U^{gd}(n) \le 2(4n)^{I(U)} .$$

(EXP) *If the system U has infinite I-dimension, then for any $n \in \mathbb{N}$,*

$$L_U^{gd}(n) = 2^{n+1} .$$

Proposition 4.4 *For any infinite binary information system U and any $n \in \mathbb{N}$,*

$$L_U^{ga}(n) = L_U^{gd}(n) .$$

Let U be an infinite binary information system. Proposition 4.1 allows us to correspond to the function $h_U^{gd}(n)$ its type of behavior from the set {LOG, LIN}. Proposition 4.2 allows us to correspond to the function $h_U^{ga}(n)$ its type of behavior

Table 4.1 Possible global types of infinite binary information systems

	$h_U^{gd}(n)$	$h_U^{ga}(n)$	$L_U^{gd}(n)$	$L_U^{ga}(n)$
1	LOG	CON	POL	POL
2	LIN	CON	POL	POL
3	LIN	LIN	POL	POL
4	LIN	CON	EXP	EXP
5	LIN	LIN	EXP	EXP

from the set {CON, LIN}. Propositions 4.3 and 4.4 allow us to correspond to each of the functions $L_U^{gd}(n)$ and $L_U^{ga}(n)$ its type of behavior from the set {POL, EXP}.

Definition 4.4 A tuple obtained from the tuple

$$(h_U^{gd}(n), h_U^{ga}(n), L_U^{gd}(n), L_U^{ga}(n))$$

by replacing functions with their types of behavior will be called the *global type* of the information system U.

We now describe all possible global types of infinite binary information systems.

Theorem 4.1 *For any infinite binary information system, its global type coincides with one of the rows of Table 4.1. Each row of Table 4.1 is the global type of some infinite binary information system.*

For $i = 1, \ldots, 5$, we denote by W_i^g the class of all infinite binary information systems, whose global type coincides with the ith row of Table 4.1. We now study for each of these complexity classes joint behavior of the depth and number of nodes in decision trees solving problems.

For a given infinite binary information system U, we will consider pairs of functions φ, ψ such that, for any problem z over U, there exists a deterministic decision tree over U solving z with the depth at most $\varphi(\dim z)$ and the number of nodes at most $\psi(\dim z)$. We will study such pairs and will call them boundary gd-pairs.

Definition 4.5 A pair of functions (φ, ψ), where $\varphi : \mathbb{N} \to \mathbb{N} \cup \{0\}$ and $\psi : \mathbb{N} \to \mathbb{N} \cup \{0\}$, will be called a *boundary gd-pair* of the information system U if, for any problem z over U, there exists a decision tree Γ over U, which solves the problem z deterministically and for which $h(\Gamma) \leq \varphi(n)$ and $L(\Gamma) \leq \psi(n)$, where $n = \dim z$.

We are interested in finding boundary gd-pairs with functions that grow as slowly as possible. For some information systems, we will be able to prove the existence of the optimal boundary gd-pair and find it.

Definition 4.6 A boundary gd-pair (φ, ψ) of the information system U will be called *optimal* if, for any boundary gd-pair (φ', ψ') of U, the inequalities $\varphi'(n) \geq \varphi(n)$ and $\psi'(n) \geq \psi(n)$ hold for any $n \in \mathbb{N}$.

It is clear that, for any boundary gd-pair (φ, ψ) of the information system U, the following inequality hold: $\varphi(n) \geq h_U^{gd}(n)$ and $\psi(n) \geq L_U^{gd}(n)$. So the best possible situation is when (h_U^{gd}, L_U^{gd}) is a boundary gd-pair of U.

Definition 4.7 An information system U will be called *gd-reachable* if the pair (h_U^{gd}, L_U^{gd}) is a boundary (and, consequently, optimal boundary) gd-pair of the system U.

We now consider similar notions for nondeterministic decision trees: the notion of boundary ga-pair, optimal boundary ga-pair, and the notion of ga-reachable information system.

Definition 4.8 A pair of functions (φ, ψ), where $\varphi : \mathbb{N} \to \mathbb{N} \cup \{0\}$ and $\psi : \mathbb{N} \to \mathbb{N} \cup \{0\}$, will be called a *boundary ga-pair* of the information system U if, for any problem z over U, there exists a decision tree Γ over U, which solves the problem z nondeterministically and for which $h(\Gamma) \leq \varphi(n)$ and $L(\Gamma) \leq \psi(n)$, where $n = \dim z$.

Definition 4.9 A boundary ga-pair (φ, ψ) of the information system U will be called *optimal* if, for any boundary ga-pair (φ', ψ') of U, the inequalities $\varphi'(n) \geq \varphi(n)$ and $\psi'(n) \geq \psi(n)$ hold for any $n \in \mathbb{N}$.

Definition 4.10 An information system U will be called *ga-reachable* if the pair (h_U^{ga}, L_U^{ga}) is a boundary (and, consequently, optimal boundary) ga-pair of the system U.

Note that for deterministic decision trees, the best situation is when the considered information system is gd-reachable, for nondeterministic decision trees—when the information system is ga-reachable.

Each information system from the classes W_2^g, W_3^g, W_4^g, W_5^g is gd-reachable. The class W_1^g contains both information systems that are gd-reachable and information systems that are not gd-reachable. For each information system U that is not gd-reachable, we find a nontrivial boundary gd-pair, which is sufficiently close to the pair (h_U^{gd}, L_U^{gd}).

Each information system from the classes W_3^g, W_5^g is ga-reachable. Each information system from the classes W_1^g, W_2^g, W_4^g is not ga-reachable. For some information systems U, which are not ga-reachable, we find nontrivial boundary ga-pairs that are sufficiently close to (h_U^{ga}, L_U^{ga}). For the rest of information systems U that are not ga-reachable, the pair $(n, L_U^{ga}(n))$ is the optimal boundary ga-pair. Note that, for such information systems, the function h_U^{ga} is bounded from above by a constant.

The obtained results are related to time-space trade-off for deterministic and nondeterministic decision trees. Details can be found in the following five theorems.

Theorem 4.2 (a) *The class W_1^g contains both information systems that are gd-reachable and information systems that are not gd-reachable. Let U be an information system from W_1^g, which is not gd-reachable. Then, for each ε, $0 < \varepsilon < 1$, there* *exists a positive constant c such that* $(c(\log_2 n)^{1+\varepsilon} + 1, 2^{c(\log_2 n)^{1+\varepsilon}+2})$ *is a boundary gd-pair of the system U.*

(b) *Let U be an information system from the class* W_1^g*. Then the system U is not ga-reachable and, for each* ε, $0 < \varepsilon < 1$, *there exist positive constants* c_1, c_2, *and* c_3 *such that* $(c_1, 2^{c_2(\log_2 n)^{1+\varepsilon}+c_3})$ *is a boundary ga-pair of the system U.*

Theorem 4.3 *Let U be an information system from the class* W_2^g*. Then*

(a) *The system U is gd-reachable.*

(b) *The system U is not ga-reachable and* $(n, L_U^{ga}(n))$ *is the optimal boundary ga-pair of the system U.*

Theorem 4.4 *Let U be an information system from the class* W_3^g*. Then*

(a) *The system U is gd-reachable.*

(b) *The system U is ga-reachable.*

Theorem 4.5 *The class* W_4^g *contains both information systems that satisfy the condition of restricted coverage and information systems that do not satisfy this condition. Let U be an information system from the class* W_4^g*. Then*

(a) *The system U is gd-reachable.*

(b) *The system U is not ga-reachable. If the system U does not satisfy the condition of restricted coverage, then* $(n, L_U^{ga}(n))$ *is the optimal boundary ga-pair of the system U. If the system U satisfies the condition of restricted coverage, then there exist positive constants* c_1 *and* c_2 *such that* (c_1, c_2^n) *is a boundary ga-pair of the system U.*

Theorem 4.6 *Let U be an information system from the class* W_5^g*. Then*

(a) *The system U is gd-reachable.*

(b) *The system U is ga-reachable.*

Table 4.2 summarizes Theorems 4.1–4.6. The first column contains the name of the complexity class. The next four columns describe the global type of information systems from this class. The last two columns "*gd*-pairs" and "*ga*-pairs" contain information about boundary *gd*-pairs and boundary *ga*-pairs for information systems from the considered class: "*gd*-reachable" means that all information systems from the class are *gd*-reachable, "*ga*-reachable" means that all information systems from the class are *ga*-reachable, Theorem 4.2(a), ..., Theorem 4.5(b) are links to corresponding statements Theorem 4.2(a), ..., Theorem 4.5(b).

Table 4.2 Summary of Theorems 4.1–4.6

	$h_U^{gd}(n)$	$h_U^{ga}(n)$	$L_U^{gd}(n)$	$L_U^{ga}(n)$	*gd*-pairs	*ga*-pairs
W_1^g	LOG	CON	POL	POL	Theorem 4.2(a)	Theorem 4.2(b)
W_2^g	LIN	CON	POL	POL	*gd*-reachable	Theorem 4.3(b)
W_3^g	LIN	LIN	POL	POL	*gd*-reachable	*ga*-reachable
W_4^g	LIN	CON	EXP	EXP	*gd*-reachable	Theorem 4.5(b)
W_5^g	LIN	LIN	EXP	EXP	*gd*-reachable	*ga*-reachable

4.3 Proofs of Propositions 4.2–4.4

In this section, we prove a number of auxiliary statements and the three mentioned propositions.

Lemma 4.1 *Let $U = (A, F)$ be an infinite binary information system. Then*

(a) *If U satisfies the condition of coverage with parameter m, then for any $n \in \mathbb{N}$, any (n, U)-set is a union of a finite number of (m, U)-sets.*

(b) *If U satisfies the condition of restricted coverage with parameters m and t, then for any $n \in \mathbb{N}$, any (n, U)-set is a union of at most t^n (m, U)-sets.*

Proof (a) Let U satisfy the condition of coverage with parameter m: any $(m + 1, U)$-set is a union of a finite number of (m, U)-sets. We now show by induction on n that, for any $n \in \mathbb{N}$, any (n, U)-set is a union of a finite number of (m, U)-sets. If $n \le m + 1$, then evidently, the considered statement holds. Let for some n, $n \ge m + 1$, the considered statement hold. Let us show that it holds for $n + 1$. We consider an arbitrary $(n + 1, U)$-set B, which is the set of solutions from A of a $(n + 1, U)$-system of equations

$$S = \{f_1(x) = \delta_1, \dots, f_{n+1}(x) = \delta_{n+1}\} .$$

Let us consider the (n, U)-system of equations

$$S' = \{f_1(x) = \delta_1, \dots, f_n(x) = \delta_n\}$$

and the set B' of solutions from A of this system. Then B' is a (n, U)-set and, according to the inductive hypothesis, B' is a union of a finite number of (m, U)-sets $B_1, \dots, B_k$, where for $i = 1, \dots, k$, the set B_i is the set of solutions of an (m, U)-system of equations S_i. One can show that the set B is equal to the union of the sets of solutions on A of the systems of equations $S_i \cup \{f_{n+1}(x) = \delta_{n+1}\}$, $i = 1, \dots, k$. Each of these sets is an $(m + 1, U)$-set and therefore is a union of a finite number of (m, U)-sets. Thus, B is a union of a finite number of (m, U)-sets.

(b) Let U satisfy the condition of restricted coverage with parameters m and t: any $(m + 1, U)$-set is a union of at most t (m, U)-sets. We now show by induction on n that, for any $n \in \mathbb{N}$, any (n, U)-set is a union of at most t^n (m, U)-sets. If $n \le m + 1$, then evidently, the considered statement holds. Let for some n, $n \ge m + 1$, the considered statement hold. Let us show that it holds for $n + 1$. We consider an arbitrary $(n + 1, U)$-set B, which is the set of solutions from A of a $(n + 1, U)$-system of equations

$$S = \{f_1(x) = \delta_1, \dots, f_{n+1}(x) = \delta_{n+1}\} .$$

Let us consider the (n, U)-system of equations

$$S' = \{f_1(x) = \delta_1, \dots, f_n(x) = \delta_n\}$$

and the set B' of solutions from A of this system. Then B' is a (n, U)-set and, according to the inductive hypothesis, B' is a union of (m, U)-sets $B_1, \dots, B_k$, where $k \le t^n$ and, for $i = 1, \dots, k$, the set B_i is the set of solutions from A of an (m, U)-system of equations S_i. One can show that the set B is equal to the union of the sets of solutions on A of the systems of equations $S_i \cup \{f_{n+1}(x) = \delta_{n+1}\}, i = 1, \dots, k$. Each of these sets is an $(m + 1, U)$-set and therefore is a union of at most t (m, U)-sets. Thus, B is a union of at most t^{n+1} (m, U)-sets. □

Proof of Proposition 4.2. (CON) Let U satisfy the condition of coverage: there exists $m \in \mathbb{N}$ such that any $(m + 1, U)$-set is a union of a finite number of (m, U)-sets. From Lemma 4.1 it follows that, for any $n \in \mathbb{N}$, any (n, U)-set is a union of a finite number of (m, U)-sets.

Let $z = (\nu, f_1, \dots, f_n)$ be a problem over U. We now show that $h_U^{ga}(z) \le m$. Let $\bar{\delta} = (\delta_1, \dots, \delta_n)$ be a tuple from $\{0, 1\}^n$ such that the equation system

$$S(\bar{\delta}) = \{f_1(x) = \delta_1, \dots, f_n(x) = \delta_n\}$$

has a solution from A. For each solution $a \in A$ of this system, we have $z(a) = \nu(\bar{\delta})$. The set of solutions of $S(\bar{\delta})$ is a union of a finite number of (m, U)-sets $D_1, \dots, D_s$. Each of these sets D_i is the set of solutions of an (m, U)-system of equations. For the considered (m, U)-system of equations, we construct a complete path ξ_i with m working nodes such that $A(\xi_i) = D_i$ and the terminal node of ξ_i is labeled with the number $\nu(\bar{\delta})$. Denote $\Sigma(\bar{\delta}) = \{\xi_1, \dots, \xi_s\}$. Let $\Sigma = \bigcup \Sigma(\bar{\delta})$, where the union is considered among all tuples $\bar{\delta} \in \{0, 1\}^n$ such that the system of equations $S(\bar{\delta})$ has a solution from A. We identify initial nodes of all paths from Σ. As a result, we obtain a decision tree over the information system U, which solves the problem z nondeterministically and whose depth is at most m. Taking into account that z is an arbitrary problem over U, we obtain $h_U^{ga}(n) \le m$ for any $n \in \mathbb{N}$.

(LIN) Let U do not satisfy the condition of coverage. Assume that there exists $m \in \mathbb{N}$ such that $h_U^{ga}(m + 1) \le m$. Let B be an arbitrary $(m + 1, U)$-set given by an $(m + 1, U)$-system of equations S. We show that the set B is a union of a finite number of (m, U)-sets. Let

$$S = \{f_1(x) = \delta_1, \dots, f_{m+1}(x) = \delta_{m+1}\}\ .$$

Consider the problem $z = (\nu, f_1, \dots, f_{m+1})$ over U such that, for any $\bar{\sigma} \in \{0, 1\}^{m+1}$, $\nu(\bar{\sigma}) \in \{1, 2\}$ and $\nu(\bar{\sigma}) = 1$ if and only if $\bar{\sigma} = (\delta_1, \dots, \delta_{m+1})$.

Let Γ be a decision tree, which solves the problem z nondeterministically and for which $h(\Gamma) \le m$. Choose all complete paths ξ in Γ in which the terminal node is labeled with the number 1 and the set $A(\xi)$ is nonempty. Each such path ξ describes a (m, U)-set $A(\xi)$. The number of these paths is finite. The union of sets described by these paths is equal to B. Since B is an arbitrary $(m + 1, U)$-set, the information system U satisfies the condition of coverage with parameter m, but this contradicts our assumption. Thus, $h_U^{ga}(n) = n$ for any $n \in \mathbb{N}$. □

We now consider some auxiliary statements related to the space complexity of decision trees. Let Γ be a decision tree over an information system $U = (A, F)$ and d be an edge of Γ entering a node w. We denote by $\Gamma(d)$ a subtree of Γ, whose root is the node w. We say that a complete path ξ of Γ is *realizable* if $A(\xi) \neq \emptyset$.

Lemma 4.2 *Let $U = (A, F)$ be an infinite binary information system, $z = (\nu, f_1, \ldots, f_n)$ be a problem over U, and Γ be a decision tree over U, which solves the problem z deterministically and for which $L(\Gamma) = L_U^{gd}(z)$. Then*

(a) *For each node of Γ, there exists a realizable complete path that passes through this node.*

(b) *Each working node of Γ has two edges leaving this node.*

Proof (a) It is clear that there exists at least one realizable complete path that passes through the root of Γ. Let us assume that w is a node of Γ different from the root and such that there is no a realizable complete path, which passes through w. Let d be an edge entering the node w. We remove from Γ the edge d and the subtree $\Gamma(d)$. As a result, we obtain a decision tree Γ', which solves z deterministically and for which $L(\Gamma') < L(\Gamma)$ but this is impossible by definition of Γ.

(b) Let us assume that in Γ there exists a working node w, which has only one leaving edge d entering a node w_1. We remove from Γ the node w and the edge d and connect the edge entering the node w to the node w_1. As a result, we obtain a decision tree Γ', which solves the problem z deterministically and for which $L(\Gamma') < L(\Gamma)$ but this is impossible by definition of Γ. □

Let U be an infinite binary information system, Γ be a decision tree over U, and d be an edge of Γ. The subtree $\Gamma(d)$ will be called *full* if there exist edges $d_1, \ldots, d_m$ in $\Gamma(d)$ such that the removal of these edges and subtrees $\Gamma(d_1), \ldots, \Gamma(d_m)$ transforms the subtree $\Gamma(d)$ into a tree G such that each terminal node of G is a terminal node of Γ, and exactly two edges labeled with the numbers 0 and 1 respectively leave each working node of G.

Lemma 4.3 *Let $U = (A, F)$ be an infinite binary information system, $z = (\nu, f_1, \ldots, f_n)$ be a problem over U, and Γ be a decision tree over U, which solves the problem z nondeterministically and for which $L(\Gamma) = L_U^{ga}(z)$. Then*

(a) *For each node of Γ, there exists a realizable complete path that passes through this node.*

(b) *If a working node w of Γ has m leaving edges $d_1, \ldots, d_m$ labeled with the same number and $m \geq 2$, then the subtrees $\Gamma(d_1), \ldots, \Gamma(d_m)$ are not full.*

(c) *If the root r of Γ has m leaving edges $d_1, \ldots, d_m$ and $m \geq 2$, then the subtrees $\Gamma(d_1), \ldots, \Gamma(d_m)$ are not full.*

Proof (a) The proof of item (a) is almost identical to the proof of item (a) of Lemma 4.2.

(b) Let w be a working node of Γ, which has m leaving edges $d_1, \ldots, d_m$ labeled with the same number, $m \geq 2$, and at least one of the subtrees $\Gamma(d_1), \ldots, \Gamma(d_m)$ is full. For the definiteness, we assume that $\Gamma(d_1)$ is full. Remove from Γ the edges

$d_2, \ldots, d_m$ and subtrees $\Gamma(d_2), \ldots, \Gamma(d_m)$. We now show that the obtained tree Γ' solves the problem z nondeterministically. Assume the contrary. Then there exists an object $a \in A$ such that, for each complete path ξ with $a \in A(\xi)$, the path ξ passes through one of the edges $d_2, \ldots, d_m$ but it is not true. Let ξ be a complete path such that $a \in A(\xi)$. Then, according to the assumption, this path passes through the node w. Let ξ' be the part of this path from the root of Γ to the node w. Since the edges $d_1, \ldots, d_m$ are labeled with the same number and $\Gamma(d_1)$ is a full subtree, we can find in $\Gamma(d_1)$ the continuation of ξ' to a terminal node of $\Gamma(d_1)$ such that the obtained complete path ξ'' of Γ satisfies the condition $a \in A(\xi'')$. Hence Γ' is a decision tree, which solves the problem z nondeterministically and for which $L(\Gamma') < L(\Gamma)$ but this is impossible by definition of Γ.

(c) Item (c) can be proven in the same way as item (b). □

We now prove a number of statements about classes of decision trees. Let Γ be a decision tree. We denote by $L_t(\Gamma)$ the number of terminal nodes in Γ and by $L_w(\Gamma)$ the number of working nodes in Γ. It is clear that $L(\Gamma) = 1 + L_t(\Gamma) + L_w(\Gamma)$.

Let U be an infinite binary information systems. We denote by $G_d(U)$ the set of all deterministic decision trees over U and by $G_d^2(U)$ the set of all decision trees from $G_d(U)$ such that each working node of the tree has two leaving edges.

Lemma 4.4 *Let U be an infinite binary information system. Then*

(a) *If $\Gamma \in G_d^2(U)$, then $L_w(\Gamma) = L_t(\Gamma) - 1$.*

(b) *If $\Gamma \in G_d(U) \setminus G_d^2(U)$, then $L_w(\Gamma) > L_t(\Gamma) - 1$.*

Proof (a) The equality $L_w(\Gamma) = L_t(\Gamma) - 1$ for trees from $G_d^2(U)$ is well-known. However, for the sake of completeness, we consider its proof by induction on $L_t(\Gamma)$. If $L_t(\Gamma) \leq 2$, then this equality holds. Let $m \geq 2$ and, for each $\Gamma \in G_d^2(U)$ with $L_t(\Gamma) \leq m$, the considered equality hold. Let $\Gamma \in G_d^2(U)$ and $L_t(\Gamma) = m + 1$. It is clear that there exists a node w of the tree Γ such that all children of w are terminal nodes. Remove children of w and edges entering these children and attach to w a number from $\mathbb{N}$. We denote the obtained tree by Γ'. It is clear that $\Gamma' \in G_d^2(U)$ and $L_t(\Gamma') = m$. By the inductive hypothesis, $L_w(\Gamma') = L_t(\Gamma') - 1$. Taking into account that $L_w(\Gamma') = L_w(\Gamma) - 1$ and $L_t(\Gamma') = L_t(\Gamma) - 1$, we obtain $L_w(\Gamma) = L_t(\Gamma) - 1$.

(b) Let $\Gamma \in G_d(U) \setminus G_d^2(U)$ and there be $m \geq 1$ working nodes in Γ each of which has exactly one leaving edge. We add m new terminal nodes to Γ and, as a result, obtain a tree $\Gamma' \in G_d^2(U)$. Then $L_w(\Gamma') = L_t(\Gamma') - 1$, $L_w(\Gamma') = L_w(\Gamma)$, and $L_t(\Gamma') = L_t(\Gamma) + m$. Therefore $L_w(\Gamma) = L_t(\Gamma) + m - 1$. Since $m \geq 1$, we obtain $L_w(\Gamma) > L_t(\Gamma) - 1$. □

We denote by $G_a^f(U)$ the set of all decision trees Γ over U that satisfy the following conditions: (i) if a working node of Γ has m leaving edges $d_1, \ldots, d_m$ labeled with the same number and $m \geq 2$, then the subtrees $\Gamma(d_1), \ldots, \Gamma(d_m)$ are not full, and (ii) if the root of Γ has m leaving edges $d_1, \ldots, d_m$ and $m \geq 2$, then the subtrees $\Gamma(d_1), \ldots, \Gamma(d_m)$ are not full. One can show that $G_d^2(U) \subseteq G_d(U) \subseteq G_a^f(U)$.

Lemma 4.5 *Let U be an infinite binary information system. If $\Gamma \in G_a^f(U) \setminus G_d^2(U)$, then $L_w(\Gamma) > L_t(\Gamma) - 1$.*

Proof We prove the considered statement by induction on $L_t(\Gamma)$. Let $\Gamma \in G_a^f(U) \setminus G_d^2(U)$ and $L_t(\Gamma) = 1$. Then Γ consists of one complete path with $k \geq 0$ working nodes. If $k = 0$, then $\Gamma \in G_d^2(U)$ but this is impossible by the choice of Γ. Therefore $L_w(\Gamma) = k \geq 1$ and $L_t(\Gamma) = 1$. Hence $L_w(\Gamma) > L_t(\Gamma) - 1$.

Let, for some $m \geq 1$ for each decision tree $\Gamma \in G_a^f(U) \setminus G_d^2(U)$ with $L_t(\Gamma) \leq m$, the inequality $L_w(\Gamma) > L_t(\Gamma) - 1$ hold. Consider a decision tree Γ such that $\Gamma \in G_a^f(U) \setminus G_d^2(U)$ and $L_t(\Gamma) = m + 1$. We now show that $L_w(\Gamma) > L_t(\Gamma) - 1$. If $\Gamma \in G_d(U) \setminus G_d^2(U)$, then by Lemma 4.4, $L_w(\Gamma) > L_t(\Gamma) - 1$. Let $\Gamma \in G_a^f(U) \setminus G_d(U)$. Then the tree Γ contains a node v (the root or a working node), which has two leaving edges labeled with the same number (if v is a working node) or do not labeled with numbers (if v is the root). We call such edges equally labeled. Since Γ is a finite tree, there is a node w of Γ, which has two leaving edges d_1 and d_2 that are equally labeled, and in the subtrees $\Gamma(d_1)$ and $\Gamma(d_2)$ there are no nodes with two leaving edges that are equally labeled.

It is clear that the subtree $\Gamma(d_1)$ is not full. We add to this tree a node and an edge leaving this node and entering the root of $\Gamma(d_1)$. As a result, we obtain a decision tree from $G_d(U) \setminus G_d^2(U)$. Using Lemma 4.4, we obtain $L_w(\Gamma(d_1)) > L_t(\Gamma(d_1)) - 1$.

Remove from Γ the edge d_1 and the subtree $\Gamma(d_1)$. Denote by Γ' the obtained tree. Since the subtree $\Gamma(d_2)$ is not full, $\Gamma' \notin G_d^2(U)$. It is clear that $\Gamma' \in G_a^f(U)$. One can show that $L_t(\Gamma') < L_t(\Gamma)$. By the inductive hypothesis, $L_w(\Gamma') > L_t(\Gamma') - 1$, and hence $L_w(\Gamma') \geq L_t(\Gamma')$. Therefore $L_w(\Gamma') + L_w(\Gamma(d_1)) > L_t(\Gamma') + L_t(\Gamma(d_1)) - 1$. Since $L_w(\Gamma) = L_w(\Gamma') + L_w(\Gamma(d_1))$ and $L_t(\Gamma) = L_t(\Gamma') + L_t(\Gamma(d_1))$, we obtain $L_w(\Gamma) > L_t(\Gamma) - 1$. □

Let $U = (A, F)$ be an infinite binary information system. For $f_1, \ldots, f_n \in F$, we denote by $N_U(f_1, \ldots, f_n)$ the number of n-tuples $(\delta_1, \ldots, \delta_n) \in \{0, 1\}^n$ for which the system of equations

$$\{f_1(x) = \delta_1, \ldots, f_n(x) = \delta_n\}$$

has a solution from A. For $n \in \mathbb{N}$, denote

$$N_U(n) = \max\{N_U(f_1, \ldots, f_n) : f_1, \ldots, f_n \in F\} .$$

It is clear that, for any $m, n \in \mathbb{N}$, if $m \leq n$ then $N_U(m) \leq N_U(n)$.

Proposition 4.5 *Let $U = (A, F)$ be an infinite binary information system. Then, for any $n \in \mathbb{N}$,*

$$L_U^{ga}(n) = L_U^{gd}(n) = 2N_U(n) .$$

Proof Let $z = (\nu, f_1, \ldots, f_m)$ be a problem over U and $m \leq n$. Let Γ be a decision tree over U, which solves the problem z deterministically, uses only attributes from the set $\{f_1, \ldots, f_m\}$, and has the minimum number of nodes among such decision trees. In the same way as in the proof of Lemma 4.2, one can prove that each working node of Γ has two edges leaving this node and, for each node of Γ, there exists a realizable complete path that passes through this node. Let ξ_1 and ξ_2 be different complete paths in Γ, $a_1 \in A(\xi_1)$, and $a_2 \in A(\xi_2)$. It is easy to show that $(f_1(a_1), \ldots, f_m(a_1)) \neq (f_1(a_2), \ldots, f_m(a_2))$. Therefore $L_t(\Gamma) \leq N_U(f_1, \ldots, f_m) \leq N_U(n)$. It is clear that $\Gamma \in G_d^2(U)$. By Lemma 4.4, $L_w(\Gamma) = L_t(\Gamma) - 1$. Hence $L(\Gamma) \leq 2N_U(n)$. Taking into account that z is an arbitrary problem over U with $\dim z \leq n$, we obtain

$$L_U^{gd}(n) \leq 2N_U(n) .$$

Since any decision tree solving the problem z deterministically solves it nondeterministically, we obtain

$$L_U^{ga}(n) \leq L_U^{gd}(n).$$

We now show that $2N_U(n) \leq L_U^{ga}(n)$. Let us consider a problem $z = (\nu, f_1, \ldots, f_n)$ over U such that

$$N_U(f_1, \ldots, f_n) = N_U(n)$$

and, for any $\bar{\delta}_1, \bar{\delta}_2 \in \{0, 1\}^n$, if $\bar{\delta}_1 \neq \bar{\delta}_2$, then $\nu(\bar{\delta}_1) \neq \nu(\bar{\delta}_2)$. Let Γ be a decision tree over U, which solves the problem z nondeterministically and for which $L(\Gamma) = L_U^{ga}(z)$. By Lemma 4.3, $\Gamma \in G_a^f(U)$. Using Lemmas 4.4 and 4.5, we obtain $L_w(\Gamma) \geq L_t(\Gamma) - 1$. It is clear that $L_t(\Gamma) \geq N_U(f_1, \ldots, f_n) = N_U(n)$. Therefore $L(\Gamma) \geq 2N_U(n)$, $L_U^{ga}(z) \geq 2N_U(n)$, and $L_U^{ga}(n) \geq 2N_U(n)$. □

The next statement follows directly from Lemmas 5.1 and 5.2 [4] and the evident inequality $N_U(n) \leq 2^n$, which is true for any infinite binary information system U. The proof of Lemma 5.1 from [4] is based on Theorems 4.6 and 4.7 from the same monograph that are similar to results obtained in [9, 10].

Proposition 4.6 *For any infinite binary information system U, the function $N_U(n)$ has one of the following two types of behavior:*

(POL) *If the system U has finite I-dimension, then for any $n \in \mathbb{N}$,*

$$n + 1 \leq N_U(n) \leq (4n)^{I(U)} .$$

(EXP) *If the system U has infinite I-dimension, then for any $n \in \mathbb{N}$,*

$$N_U(n) = 2^n .$$

We now prove Propositions 4.3 and 4.4.

Proof of Proposition 4.3. The statement of the proposition follows immediately from Propositions 4.5 and 4.6. □

Proof of Proposition 4.4. The statement of the proposition follows immediately from Proposition 4.5. □

4.4 Proof of Theorem 4.1

First, we prove six auxiliary statements.

Lemma 4.6 *For any infinite binary information system, its global type coincides with one of the rows of Table 4.1.*

Proof To prove this statement we fill Table 4.3. In the first column "Cover.", we have either "Yes" or "No": "Yes" if the considered information system satisfies the condition of coverage and "No" otherwise. In the second column "Restr. cover.", we also have either "Yes" or "No": "Yes" if the considered information system satisfies the condition of restricted coverage and "No" otherwise. In the third column "I-dim." we have either "Fin" or "Inf": "Fin" if the considered information system has finite I-dimension and "Inf" if the considered information system has infinite I-dimension.

If an information system does not satisfy the condition of coverage, then this information system does not satisfy the condition of restricted coverage. It means that there are only six possible tuples of values of the considered three parameters of information systems, which correspond to the six rows of Table 4.3. The values of the considered three parameters define the types of behavior of functions $h_U^{gd}(n)$, $h_U^{ga}(n)$, $L_U^{gd}(n)$, and $L_U^{ga}(n)$ according to Propositions 4.1–4.4. We see that the set of possible tuples of values in the last four columns coincides with the set of rows of Table 4.1. □

For each row of Table 4.1, we consider an example of infinite binary information system, whose global type coincides with this row.

For any $i \in \mathbb{N}$, we define two functions $p_i : \mathbb{N} \to \{0, 1\}$ and $l_i : \mathbb{N} \to \{0, 1\}$. Let $j \in \mathbb{N}$. Then $p_i(j) = 1$ if and only if $j = i$, and $l_i(j) = 1$ if and only if $j > i$.

Define an information system $U_1 = (A_1, F_1)$ as follows: $A_1 = \mathbb{N}$ and $F_1 = \{l_i : i \in \mathbb{N}\}$.

Lemma 4.7 *The information system U_1 belongs to the class W_1^g, $h_{U_1}^{gd}(n) = \lceil \log_2(n+1) \rceil$, $h_{U_1}^{ga}(1) = 1$ and $h_{U_1}^{ga}(n) = 2$ if $n > 1$, $L_{U_1}^{gd}(n) = 2(n+1)$, and $L_{U_1}^{ga}(n) = 2(n+1)$ for any $n \in \mathbb{N}$. This information system satisfies the condition*

Table 4.3 Parameters and global types of infinite binary information systems

Cover.	Restr. cover.	I-dim.	$h_U^{gd}(n)$	$h_U^{ga}(n)$	$L_U^{gd}(n)$	$L_U^{ga}(n)$
Yes	Yes	Fin	LOG	CON	POL	POL
Yes	No	Fin	LIN	CON	POL	POL
No	No	Fin	LIN	LIN	POL	POL
Yes	Yes	Inf	LIN	CON	EXP	EXP
Yes	No	Inf	LIN	CON	EXP	EXP
No	No	Inf	LIN	LIN	EXP	EXP

of coverage with parameter 3*, satisfies the condition of restricted coverage with parameters* 3 *and* 1*, and has finite I-dimension equal* 1*. The information system* U_1 *is* gd*-reachable.*

Proof It is easy to show that $N_{U_1}(n) = n + 1$ for any $n \in \mathbb{N}$. Using Proposition 4.5, we obtain $L^{gd}_{U_1}(n) = L^{ga}_{U_1}(n) = 2(n + 1)$ for any $n \in \mathbb{N}$. Let $n \in \mathbb{N}$. Consider a problem $z = (\nu, l_1, \ldots, l_n)$ over U_1 such that, for each $\bar{\delta}_1, \bar{\delta}_2 \in \{0, 1\}^n$ with $\bar{\delta}_1 \neq \bar{\delta}_2$, $\nu(\bar{\delta}_1) \neq \nu(\bar{\delta}_2)$. It is clear that $N_{U_1}(l_1, \ldots, l_n) = n + 1$. Therefore each decision tree Γ over U_1 that solves the problem z deterministically has at least $n + 1$ terminal nodes. One can show that the number of terminal nodes in Γ is at most $2^{h(\Gamma)}$. Hence $n + 1 \leq 2^{h(\Gamma)}$ and $\log_2(n + 1) \leq h(\Gamma)$. Since $h(\Gamma)$ is an integer, $\lceil \log_2(n + 1) \rceil \leq h(\Gamma)$. Thus, $h^{gd}_{U_1}(n) \geq \lceil \log_2(n + 1) \rceil$. Set $m = \lceil \log_2(n + 1) \rceil$. Then $n \leq 2^m - 1$. One can show that $h^{gd}_{U_1}(2^m - 1) \leq m$ (the construction of an appropriate decision tree is based on an analog of binary search, and we use only attributes from the problem description) and $h^{gd}_{U_1}(n) \leq h^{gd}_{U_1}(2^m - 1)$. Therefore $h^{gd}_{U_1}(n) \leq \lceil \log_2(n + 1) \rceil$ and $h^{gd}_{U_1}(n) = \lceil \log_2(n + 1) \rceil$. It is clear that $h^{ga}_{U_1}(1) = 1$. Let $n \geq 2$ and $z = (\nu, f_1, \ldots, f_n)$ be an arbitrary problem over U_1 and $l_{i_1}, \ldots, l_{i_m}$ be all pairwise different attributes from the set $\{f_1, \ldots, f_n\}$ ordered such that $i_1 < \cdots < i_m$. Then these attributes divide the set $\mathbb{N}$ into $m + 1$ nonempty domains that are sets of solutions on $\mathbb{N}$ of the following systems of equations: $\{l_{i_1}(x) = 0\}$, $\{l_{i_1}(x) = 1, l_{i_2}(x) = 0\}$, ..., $\{l_{i_{m-1}}(x) = 1, l_{i_m}(x) = 0\}$, $\{l_{i_m}(x) = 1\}$. The value $z(x)$ is constant in each of the considered domains. Using these facts, it is easy to show that there exists a decision tree Γ over U_1, which solves the problem z nondeterministically and for which $h(\Gamma) = 2$ if $m \geq 2$. Therefore $h^{ga}_{U_1}(n) \leq 2$. One can show that there exists a problem z over U_1 such that $\dim z = n$ and $h^{ga}_{U_1}(z) \geq 2$. Therefore $h^{ga}_{U_1}(n) = 2$.

Since the function $h^{gd}_{U_1}$ has the type of behavior LOG, the information system U_1 belongs to the class W^g_1—see Table 4.1. One can show that the information system U_1 satisfies the condition of coverage with parameter 3, satisfies the condition of restricted coverage with parameters 3 and 1, and has finite I-dimension equal 1.

Let $z = (\nu, f_1, \ldots, f_n)$ be a problem over U_1. We know that there exists a decision tree Γ over U_1, which solves this problem deterministically, uses only attributes from the set $\{f_1, \ldots, f_n\}$, and whose depth is at most $\lceil \log_2(n + 1) \rceil$. By removal of some nodes and edges from Γ, we can obtain a decision tree Γ' over U_1, which solves the problem z deterministically, and in which each working node has exactly two leaving edges and each complete path is realizable. It is clear that $N_{U_1}(f_1, \ldots, f_n) \leq n + 1$. Therefore $L_t(\Gamma') \leq n + 1$. By Lemma 4.4, $L_w(\Gamma') \leq n$. Therefore $L(\Gamma') \leq 2(n + 1) = L^{gd}_{U_1}(n)$. Taking into account that $h(\Gamma') \leq \lceil \log_2(n + 1) \rceil = h^{gd}_{U_1}(n)$ and z is an arbitrary problem over U_1 with $\dim z = n$, we obtain that U_1 is gd-reachable. □

Define an information system $U_2 = (A_2, F_2)$ as follows: $A_2 = \mathbb{N}$ and $F_2 = \{p_i : i \in \mathbb{N}\} \cup \{l_{2^i} : i \in \mathbb{N}\}$.

Lemma 4.8 *The information system* U_2 *belongs to the class* W_2^g, $h_{U_2}^{gd}(n) = n$, $h_{U_2}^{ga}(n) = 1$, $L_{U_2}^{gd}(n) = 2(n+1)$, *and* $L_{U_2}^{ga}(n) = 2(n+1)$ *for any* $n \in \mathbb{N}$. *This information system satisfies the condition of coverage with parameter* 2, *does not satisfy the condition of the restricted coverage, and has finite I-dimension equal* 1.

Proof It is easy to show that $N_{U_2}(n) = n + 1$ for any $n \in \mathbb{N}$. Using Proposition 4.5, we obtain $L_{U_2}^{gd}(n) = L_{U_2}^{ga}(n) = 2(n+1)$ for any $n \in \mathbb{N}$. Let $n \in \mathbb{N}$. Choose $t \in \mathbb{N}$ such that $2^t > n$. Consider a problem $z = (\nu, p_{2^t+1}, \ldots, p_{2^t+n})$ over U_2 such that, for each $\bar{\delta}_1, \bar{\delta}_2 \in \{0, 1\}^n$ with $\bar{\delta}_1 \neq \bar{\delta}_2$, $\nu(\bar{\delta}_1) \neq \nu(\bar{\delta}_2)$. Consider an arbitrary decision tree Γ over U_2 that solves the problem z deterministically and a complete path ξ of Γ such that $2^t + n + 1 \in A_2(\xi)$. One can show that if the number of working nodes in ξ is less than n, then the function $z(x)$ is not constant on the set $A_2(\xi)$ but this is impossible by the choice of Γ. Therefore $h(\Gamma) \geq n$ and $h_{U_2}^{gd}(n) \geq n$. It is clear that $h_{U_2}^{gd}(n) \leq n$. Hence $h_{U_2}^{gd}(n) = n$. It is easy to show that $h_{U_2}^{ga}(n) \geq 1$. We now show that $h_{U_2}^{ga}(n) \leq 1$. Let $z = (\nu, f_1, \ldots, f_n)$ be an arbitrary problem over U_2. Each attribute from the set $\{f_1, \ldots, f_n\}$ is of the form p_i or l_i. We refer to the number i as the index of the considered attribute. Let j be the maximum index of an attribute from the set $\{f_1, \ldots, f_n\}$ and t be a number from $\mathbb{N}$ such that $2^t > j$. Then the function $z(x)$ is constant on the sets of solutions of equation systems $\{p_1(x) = 1\}, \ldots, \{p_{2^t}(x) = 1\}$, $\{l_{2^t}(x) = 1\}$ on A_2, and the union of these sets of solutions is equal to A_2. Using these facts, it is easy to show that there exists a decision tree Γ over U_2, which solves the problem z nondeterministically and for which $h(\Gamma) = 1$. Therefore $h_{U_2}^{ga}(n) \leq 1$. Hence $h_{U_1}^{ga}(n) = 1$.

Since the function $h_{U_2}^{gd}$ has the type of behavior LIN, the function $h_{U_2}^{ga}$ has the type of behavior CON, and the functions $L_{U_2}^{gd}$ and $L_{U_2}^{ga}$ have the type of behavior POL, the information system U_2 belongs to the class W_2^g—see Table 4.1. One can show that the information system U_2 satisfies the condition of coverage with parameter 2 and has finite I-dimension equal 1. Using Proposition 4.1, we obtain that this information system does not satisfy the condition of restricted coverage. □

Define an information system $U_3 = (A_3, F_3)$ as follows: $A_3 = \mathbb{N}$ and $F_3 = \{p_i : i \in \mathbb{N}\}$.

Lemma 4.9 *The information system* U_3 *belongs to the class* W_3^g, $h_{U_3}^{gd}(n) = n$, $h_{U_3}^{ga}(n) = n$, $L_{U_3}^{gd}(n) = 2(n+1)$, *and* $L_{U_3}^{ga}(n) = 2(n+1)$ *for any* $n \in \mathbb{N}$. *This information system does not satisfy the conditions of coverage and restricted coverage, and has finite I-dimension equal* 1.

Proof It is easy to show that $N_{U_3}(n) = n + 1$ for any $n \in \mathbb{N}$. Using Proposition 4.5, we obtain $L_{U_3}^{gd}(n) = L_{U_3}^{ga}(n) = 2(n+1)$ for any $n \in \mathbb{N}$. Let $n \in \mathbb{N}$. Consider a problem $z = (\nu, p_1, \ldots, p_n)$ over U_3 such that, for each $\bar{\delta}_1, \bar{\delta}_2 \in \{0, 1\}^n$ with $\bar{\delta}_1 \neq \bar{\delta}_2$, $\nu(\bar{\delta}_1) \neq \nu(\bar{\delta}_2)$. Consider an arbitrary decision tree Γ_1 over U_3 that solves the problem z deterministically and a complete path ξ in Γ_1 in which each edge leaving a working node is labeled with the number 0. One can show that if the number of working nodes in ξ is less than n, then the function $z(x)$ is not constant on the set $A_3(\xi)$ but this is impossible by the choice of Γ_1. Therefore $h(\Gamma_1) \geq n$ and $h^{gd}_{U_3}(n) \geq n$. It is clear that $h^{gd}_{U_3}(n) \leq n$. Hence $h^{gd}_{U_3}(n) = n$. Consider an arbitrary decision tree Γ_2 over U_3 that solves the problem z nondeterministically. Let ξ_1 be a complete path of Γ_2 in which at least one edge leaving a working node is labeled with the number 1. Then the set $A_3(\xi_1)$ contains at most one element from A_3. The set A_3 is infinite, the number of complete paths in Γ_2 is finite, and the union of the sets $A_3(\xi)$ for all complete paths ξ in Γ is equal to A_3. Therefore there exists a complete path ξ_0 in Γ_2 in which each edge leaving a working node is labeled with the number 0. One can show that if the number of working nodes in ξ_0 is less than n, then the function $z(x)$ is not constant on the set $A_3(\xi_0)$ but this is impossible by the choice of Γ_2. Therefore $h(\Gamma_2) \geq n$ and $h^{ga}_{U_3}(n) \geq n$. It is clear that $h^{ga}_{U_3}(n) \leq n$. Hence $h^{ga}_{U_3}(n) = n$.

Since the functions $h^{gd}_{U_3}$ and $h^{ga}_{U_3}$ have the type of behavior LIN and the functions $L^{gd}_{U_3}$ and $L^{ga}_{U_3}$ have the type of behavior POL, the information system U_3 belongs to the class W^g_3—see Table 4.1. One can show that this information system has finite I-dimension equal 1. Using Propositions 4.1 and 4.2, we obtain that the information system U_3 does not satisfy the conditions of coverage and restricted coverage. □

Define an information system $U_4 = (A_4, F_4)$ as follows: $A_4 = \mathbb{N}$ and F_4 is the set of all functions from $\mathbb{N}$ to $\{0, 1\}$.

Lemma 4.10 *The information system U_4 belongs to the class W^g_4, $h^{gd}_{U_4}(n) = n$, $h^{ga}_{U_4}(n) = 1$, $L^{gd}_{U_4}(n) = 2^{n+1}$, and $L^{ga}_{U_4}(n) = 2^{n+1}$ for any $n \in \mathbb{N}$. This information system satisfies the condition of coverage with parameter 2, satisfies the condition of restricted coverage with parameters 2 and 1, and has infinite I-dimension.*

Proof It is easy to show that the information system U_4 has infinite I-dimension. By Proposition 4.6, $N_{U_4}(n) = 2^n$ for any $n \in \mathbb{N}$. Using Proposition 4.5, we obtain $L^{gd}_{U_4}(n) = L^{ga}_{U_4}(n) = 2^{n+1}$ for any $n \in \mathbb{N}$. Let $n \in \mathbb{N}$. By Proposition 4.1, $h^{gd}_{U_4}(n) = n$. It is easy to show that $h^{ga}_{U_4}(n) \geq 1$. We now show that $h^{ga}_{U_4}(n) \leq 1$. Let $z = (\nu, f_1, \ldots, f_n)$ be an arbitrary problem over U_4. Let $(\delta_1, \ldots, \delta_n) \in \{0, 1\}^n$ and the set B of solutions from A_4 of the equation system $\{f_1(x) = \delta_1, \ldots, f_n(x) = \delta_n\}$ is nonempty. Then there exists a function $f \in F_4$ such that the set B is the set of solutions from A_4 of the equation system $\{f(x) = 1\}$. Using this fact, it is easy to show that there exists a decision tree Γ over U_4, which solves the problem z nondeterministically and for which $h(\Gamma) = 1$. Therefore $h^{ga}_{U_4}(n) \leq 1$. Hence $h^{ga}_{U_4}(n) = 1$.

Since the function $h^{gd}_{U_4}$ has the type of behavior LIN, the function $h^{ga}_{U_4}$ has the type of behavior CON, and the functions $L^{gd}_{U_4}$ and $L^{ga}_{U_4}$ have the type of behavior EXP, the information system U_4 belongs to the class W^g_4—see Table 4.1. One can show that this information system satisfies the condition of coverage with parameter 2 and satisfies the condition of restricted coverage with parameters 2 and 1. □

Define an information system $U_5 = (A_5, F_5)$ as follows: A_5 is the set of all infinite sequences $a_1, a_2, \ldots$, where $a_i \in \{0, 1\}$ for any $i \in \mathbb{N}$, and $F_5 = \{f_i : i \in \mathbb{N}\}$, where $f_i(a_1, a_2, \ldots) = a_i$ for any $a_1, a_2, \ldots \in A_5$ and $i \in \mathbb{N}$.

Lemma 4.11 *The information system U_5 belongs to the class W_5^g, $h_{U_5}^{gd}(n) = n$, $h_{U_5}^{ga}(n) = n$, $L_{U_5}^{gd}(n) = 2^{n+1}$, and $L_{U_5}^{ga}(n) = 2^{n+1}$ for any $n \in \mathbb{N}$. This information system does not satisfy the condition of coverage and the condition of restricted coverage, and has infinite I-dimension.*

Proof Let $n \in \mathbb{N}$. One can show that the equation system $\{f_1(x) = \delta_1, \ldots, f_n(x) = \delta_n\}$ has a solution from the set A_5 for any $(\delta_1, \ldots, \delta_n) \in \{0, 1\}^n$. Therefore the information system U_5 has infinite I-dimension. By Proposition 4.6, $N_{U_5}(n) = 2^n$ for any $n \in \mathbb{N}$. Using Proposition 4.5, we obtain $L_{U_5}^{gd}(n) = L_{U_5}^{ga}(n) = 2^{n+1}$ for any $n \in \mathbb{N}$. Let $n \in \mathbb{N}$. By Proposition 4.1, $h_{U_5}^{gd}(n) = n$. Consider a problem $z = (\nu, f_1, \ldots, f_n)$ over U_4 such that, for each $\bar{\delta}_1, \bar{\delta}_2 \in \{0, 1\}^n$ with $\bar{\delta}_1 \neq \bar{\delta}_2$, $\nu(\bar{\delta}_1) \neq \nu(\bar{\delta}_2)$. Let Γ be an arbitrary decision tree over U_5 that solves the problem z nondeterministically and ξ be an arbitrary complete path of Γ. We denote by G the set of attributes attached to working nodes of ξ. One can show that, if $\{f_1, \ldots, f_n\}$ is not a subset of the set G, then the function $z(x)$ is not constant on the set $A_5(\xi)$ but this is impossible by the choice of Γ. Therefore $h(\Gamma) \geq n$ and $h_{U_5}^{ga}(n) \geq n$. It is clear that $h_{U_5}^{ga}(n) \leq n$. Thus, $h_{U_5}^{ga}(n) = n$.

Since the functions $h_{U_5}^{gd}$ and $h_{U_5}^{ga}$ have the type of behavior LIN and the functions $L_{U_5}^{gd}$ and $L_{U_5}^{ga}$ have the type of behavior EXP, the information system U_5 belongs to the class W_5^g—see Table 4.1. By Proposition 4.2, this information system does not satisfy the condition of coverage and therefore does not satisfy the condition of restricted coverage. □

Proof of Theorem 4.1. The statements of the theorem follow from Lemmas 4.6–4.11. □

4.5 Proofs of Theorems 4.2–4.6

First, we prove seven auxiliary statements.

Lemma 4.12 *Let U be an infinite binary information system such that $h_U^{gd}(n) = n$ for any $n \in \mathbb{N}$. Then the information system U is gd-reachable.*

Proof Let $z = (\nu, f_1, \ldots, f_n)$ be a problem over U and Γ be a decision tree that solves the problem z deterministically and satisfies the following conditions: the number of working nodes in each complete path of Γ is equal to n and these nodes in the order from the root to a terminal node are labeled with attributes $f_1, \ldots, f_n$. Remove from Γ all nodes and edges that do not belong to realizable complete paths. Let w be a working node in the obtained tree that has only one leaving edge d entering a node v. We remove the node w and edge d and connect the edge e entering w to the node v. We do the same with all working nodes with only one leaving edge. Denote by Γ' the obtained decision tree. It is clear that Γ' solves the problem z deterministically, $\Gamma' \in G_d^2(U)$, and $L_t(\Gamma') \leq N_U(f_1, \ldots, f_n) \leq N_U(n)$. By Lemma 4.4, $L_w(\Gamma') = L_t(\Gamma') - 1$. Therefore $L(\Gamma') \leq 2N_U(n)$. Using Proposition 4.5, we obtain $L(\Gamma') \leq L_U^{gd}(n)$. It is clear that $h(\Gamma') \leq n = h_U^{gd}(n)$. Therefore U is gd-reachable. □

Lemma 4.13 *Let U be an infinite binary information system such that $h_U^{ga}(n) = n$ for any $n \in \mathbb{N}$. Then the information system U is ga-reachable.*

Proof Let $z = (\nu, f_1, \ldots, f_n)$ be a problem over U. Construct for this problem the decision tree Γ' as in the proof of Lemma 4.12. It is clear that Γ' solves the problem z nondeterministically. We know that $L(\Gamma') \leq 2N_U(n)$. Using Proposition 4.5, we obtain $L(\Gamma') \leq L_U^{ga}(n)$ It is clear that $h(\Gamma') \leq n = h_U^{ga}(n)$. Therefore U is ga-reachable. □

Lemma 4.14 *Let U be an infinite binary information system, which satisfies the condition of coverage. Then the information system U is not ga-reachable.*

Proof By Proposition 4.2, the function $h_U^{ga}(n)$ is bounded from above by a positive constant c. By Proposition 4.6, the function $N_U(n)$ is not bounded from above by a constant. Choose $n \in \mathbb{N}$ such that $N_U(n) > 2^{2c}$. Let $z = (\nu, f_1, \ldots, f_n)$ be a problem over U such that $\nu(\bar{\delta}_1) \neq \nu(\bar{\delta}_2)$ for any $\bar{\delta}_1, \bar{\delta}_2 \in \{0, 1\}^n$, $\bar{\delta}_1 \neq \bar{\delta}_2$, and $N_U(f_1, \ldots, f_n) = N_U(n)$. Let Γ be a decision tree over U, which solves the problem z nondeterministically, for which $h(\Gamma) \leq h_U^{ga}(n) \leq c$, and which has the minimum number of nodes among such trees. In the same way as it was done in the proof of Lemma 4.3, we can prove that $\Gamma \in G_a^f(U)$. It is clear that $L_t(\Gamma) \geq N_U(f_1, \ldots, f_n) = N_U(n)$. Let us assume that $\Gamma \in G_d^2(U)$. Then it is easy to show that $h(\Gamma) \geq \log_2 L_t(\Gamma) \geq \log_2 N_U(n) > 2c$, which is impossible by the choice of Γ. Therefore $\Gamma \in G_a^f(U) \setminus G_d^2(U)$. By Lemma 4.5, $L_w(\Gamma) > L_t(\Gamma) - 1 \geq N_U(n) - 1$. Using Proposition 4.5, we obtain $L(\Gamma) > 2N_U(n) = L_U^{ga}(n)$. Therefore U is not ga-reachable. □

Lemma 4.15 *Let U be an infinite binary information system, which does not satisfy the condition of restricted coverage. Then $(n, L_U^{ga}(n))$ is the optimal boundary ga-pair of the information system U.*

Proof The proof of the fact that $(n, L_U^{ga}(n))$ is a boundary ga-pair of the system U is similar to the proof of Lemma 4.13.

We now prove that this is the optimal boundary ga-pair of the system U. Let (q, r) be a boundary ga-pair of the system U. It is clear that $r(n) \geq L_U^{ga}(n)$ for any $n \in \mathbb{N}$. Let us show that $q(n) \geq n$ for any $n \in \mathbb{N}$. Assume the contrary: there exists $m \in \mathbb{N}$ such that $q(m + 1) \leq m$. Let us consider an arbitrary $(m + 1, U)$-set B given by $(m + 1, U)$-system of equations

$$\{f_1(x) = \delta_1, \ldots, f_{m+1}(x) = \delta_{m+1}\} .$$

Consider the problem $z = (\nu, f_1, \ldots, f_{m+1})$ over U such that, for any $\bar{\delta} \in \{0, 1\}^{m+1}$, $\nu(\bar{\delta}) \in \{1, 2\}$ and $\nu(\bar{\delta}) = 1$ if and only if $\bar{\delta} = (\delta_1, \ldots, \delta_{m+1})$. Let Γ be a decision tree over U, which solves the problem z nondeterministically and for which $h(\Gamma) \le q(m + 1) \le m$ and $L(\Gamma) \le r(m + 1)$. Let $\xi_1, \ldots, \xi_t$ be all realizable complete paths of Γ in which terminal nodes are labeled with the number 1. It is clear that $A(\xi_1), \ldots, A(\xi_t)$ are (m, U)-sets, $B = A(\xi_1) \cup \ldots \cup A(\xi_t)$, and $t \le r(m + 1)$. Since B is an arbitrary $(m + 1)$-set, we obtain that U satisfies the condition of restricted coverage but this contradicts our assumption. Therefore $(n, L^{ga}_U(n))$ is the optimal boundary ga-pair of the system U. □

Let $\mathbb{R}_+$ be the set of nonnegative real numbers. Define an infinite binary information system $U_6 = (A_6, F_6)$ as follows: $A_6 = \mathbb{R}_+$ and $F_6 = \{p_i : i \in \mathbb{N}\} \cup \{q_i : i \in \mathbb{N}\}$, where, for any $a \in A_6$, $p_i(a) = 1$ if and only if $a = i$, and $q_i(a) = 1$ if and only if $a \ge i + \frac{1}{2}$.

Lemma 4.16 *The information system U_6 belongs to the class W^g_1. This information system is not gd-reachable.*

Proof It is easy to show that $N_{U_6}(n) = n + 1$ for any $n \in \mathbb{N}$. Using Proposition 4.5, we obtain $L^{gd}_{U_6}(n) = 2(n + 1)$ for any $n \in \mathbb{N}$. We now show that $h^{gd}_{U_6}(n) \le \lceil \log_2(n + 1)\rceil + 1$ for any $n \in \mathbb{N}$. Let $z = (\nu, f_1, \ldots, f_n)$ be an arbitrary problem over U_6. We describe a decision tree Γ over U_6 that solves the problem z deterministically. This tree will use attributes from a set G containing all attributes $f_1, \ldots, f_n$ and, possibly, some additional attributes from the set $\{q_i : i \in \mathbb{N}\}$. Initially, $G = \{f_1, \ldots, f_n\}$. Let $p_{i_1}, \ldots, p_{i_m}$ be all attributes from the set $\{p_i : i \in \mathbb{N}\} \cap \{f_1, \ldots, f_n\}$ ordered such that $i_1 < \ldots < i_m$. For $t = 1, \ldots, m - 1$, if the set G does not contain any attribute q_j such that $i_t \le j < i_{t+1}$, then we add to G the attribute q_{i_t}. As a result, we obtain a set of attributes G that contains at most n attributes from the set $\{q_i : i \in \mathbb{N}\}$. Let there be attributes $q_{j_1}, \ldots, q_{j_k}$, $k \le n$. It is easy to construct a decision tree Γ' that finds values of these attributes on a given element $a \in A_6$ and has depth at most $\lceil \log_2(n + 1)\rceil$. If we know the values of all attributes $q_{j_1}, \ldots, q_{j_k}$, then we know values of all attributes $p_{i_1}, \ldots, p_{i_m}$ on a with the exception of at most one attribute. If we compute the value of this attribute we will know the values of all attributes $f_1, \ldots, f_n$ on a and the value $z(a)$. So we can transform the decision tree Γ' into a deterministic decision tree Γ over U_6, which solves the problem z and whose depth is at most $\lceil \log_2(n + 1)\rceil + 1$. Therefore $h^{gd}_{U_6}(n) \le \lceil \log_2(n + 1)\rceil + 1$. Since the function $h^{gd}_{U_6}$ has the type of behavior LOG, the information system U_6 belongs to the class W^g_1—see Table 4.1.

We now show that the information system U_6 is not gd-reachable. Assume the contrary. Choose a natural n such that $\lceil \log_2(n + 1)\rceil + 1 < n$. Consider a problem $z = (\nu, p_1, \ldots, p_n)$ such that, for any $\bar{\delta}_1, \bar{\delta}_2 \in \{0, 1\}^n$, if $\bar{\delta}_1 \ne \bar{\delta}_2$, then $\nu(\bar{\delta}_1) \ne \nu(\bar{\delta}_2)$. Let for the definiteness, $z(i) = i$ for $i = 1, \ldots, n$ and $z(a) = n + 1$ for any $a \in A_6 \setminus \{1, \ldots, n\}$. According to the assumption, there exists a deterministic decision tree Γ over U_6, which solves the problem z and for which $h(\Gamma) \le \lceil \log_2(n + 1)\rceil + 1$ and $L(\Gamma) \le 2(n + 1)$. It is clear, that for each $i \in \{1, \ldots, n, n + 1\}$, there is at least one complete path ξ of Γ in which the terminal node is labeled with the number i. Let us assume that there exists exactly one complete path ξ of Γ in which the terminal node is labeled with the number $n+1$. In this case, $A_6(\xi) = A_6 \setminus \{1, \ldots, n\}$. It means that, for each $i \in \{1, \ldots, n\}$, there exists a working node of ξ that is labeled with the attribute p_i. Hence $h(\Gamma) \geq n$, which is impossible by the choice of Γ. Therefore, $L_t(\Gamma) \geq n+2$. Using Lemma 4.4, we obtain $L_w(\Gamma) \geq L_t(\Gamma) - 1 \geq n+1$. Thus, $L(\Gamma) \geq 2(n+2)$ but this is impossible by the choice of Γ. Therefore U_6 is not gd-reachable. □

Let $\mathbb{Z}$ be the set of integers and $\mathbb{Z}_- = \mathbb{Z} \setminus \mathbb{N}$. Define an information system $U_7 = (A_7, F_7)$ as follows: $A_7 = \mathbb{Z}$ and $F_7 = \{p_i : i \in \mathbb{N}\} \cup \{l_{2^i} : i \in \mathbb{N}\} \cup F_0$. The functions p_i and l_{2^i} have value 0 on the set Z_-, and F_0 is the set of all functions from $\mathbb{Z}$ to $\{0, 1\}$ that are equal to 0 on the set $\mathbb{N}$.

Lemma 4.17 *The information system U_7 belongs to the class W_4^g and does not satisfy the condition of restricted coverage.*

Proof Let $n \in \mathbb{N}$. One can show that there are attributes $g_1, \ldots, g_n \in F_0$ such that $N_{U_7}(g_1, \ldots, g_n) = 2^n$. Therefore the information system U_7 has infinite I-dimension. Using Proposition 4.3, we obtain that the function $L_{U_7}^{gd}$ has the type of behavior EXP. We now show that $h_{U_7}^{ga}(n) \leq 1$. Let $z = (\nu, f_1, \ldots, f_n)$ be an arbitrary problem over U_7. The attributes $f_1, \ldots, f_n$ divide the set A_7 into finite number of nonempty domains in each of which these attributes have fixed values. One can show that each domain can be represented as a union of finite number of subdomains such that each subdomain is the set of solutions on A_7 of equation system of the form $\{f(x) = 1\}$, where $f \in F_0$, or $\{p_i(x) = 1\}$, or $\{l_{2^i}(x) = 1\}$. Using these facts, it is easy to show that there exists a decision tree Γ over U_7, which solves the problem z nondeterministically and for which $h(\Gamma) = 1$. Therefore $h_{U_7}^{ga}(n) \leq 1$. Hence the function $h_{U_7}^{ga}$ has the type of behavior CON. Taking into account that the function L_U^{gd} has the type of behavior EXP, we obtain that the information system U_7 belongs to the class W_4^g—see Table 4.1.

Let us assume that the information system U_7 satisfies the condition of restricted coverage with parameters m and t. Let r be a natural number such that $t < 2^r - m + 1$. Let B be the set of solutions on A_7 of the equation system

$$\{l_{2^r}(x) = 1, p_{2^r+1}(x) = 0, \ldots, p_{2^r+m-1}(x) = 0, l_{2^{r+1}}(x) = 0\} .$$

It is clear that $B = \{2^r + m, 2^r + m + 1, \ldots, 2^{r+1}\}$ and $|B| = 2^r - m + 1$. Let us assume that B is a union of at most t (U_7, m)-sets. Let C be a (U_7, m)-set such that $C \subseteq B$. One can show that $|C| = 1$. Therefore $t \geq 2^r - m + 1$ but this is impossible by the choice of r. Therefore the information system U_7 does not satisfy the condition of restricted coverage. □

Lemma 4.18 *Let $U = (A, F)$ be an information system from the class W_4^g, which satisfies the condition of restricted coverage. Then there exist positive constants c_1 and c_2 such that $(c_1, c_2^{\,n})$ is a boundary a-pair of the system U.*

Proof Let U satisfy the condition of restricted coverage with parameters m and t: any $(m+1, U)$-set is a union of at most t (m, U)-sets. From Lemma 4.1 it follows that, for any $n \in \mathbb{N}$, any (n, U)-set is a union of at most t^n (m, U)-sets.

Let $z = (\nu, f_1, \ldots, f_n)$ be a problem over U. We now show that there exists a decision tree Γ over U, which solves the problem z nondeterministically and for which $h(\Gamma) \leq m$ and $L(\Gamma) \leq (m+2)2^n t^n$. Let $\bar{\delta} = (\delta_1, \ldots, \delta_n)$ be a tuple from $\{0, 1\}^n$ such that the equation system

$$S(\bar{\delta}) = \{f_1(x) = \delta_1, \ldots, f_n(x) = \delta_n\}$$

has a solution from A. For each solution $a \in A$ of this system, we have $z(a) = \nu(\bar{\delta})$. The set of solutions of $S(\bar{\delta})$ is a union of (m, U)-sets $D_1, \ldots, D_s$, where $s \leq t^n$. Each of these sets D_i is the set of solutions of an (m, U)-system of equations. For the considered (m, U)-system of equations, we construct a complete path ξ_i with m working nodes such that $A(\xi_i) = D_i$ and the terminal node of ξ_i is labeled with the number $\nu(\bar{\delta})$. Denote $\Sigma(\bar{\delta}) = \{\xi_1, \ldots, \xi_s\}$. Let $\Sigma = \bigcup \Sigma(\bar{\delta})$, where the union is considered among all tuples $\bar{\delta} \in \{0, 1\}^n$ such that the set of equations $S(\bar{\delta})$ has a solution from A (the number of such tuples is at most 2^n). We identify initial nodes of all paths from Σ. As a result, we obtain a decision tree Γ over the information system U, which solves the problem z nondeterministically and for which $h(\Gamma) \leq m$ and $L(\Gamma) \leq (m+2)2^n t^n$. Denote $c_1 = m$ and $c_2 = (m+2)2t$. Then $h(\Gamma) \leq c_1$ and $L(\Gamma) \leq c_2^n$. Taking into account that z is an arbitrary problem over U with $\dim z = n$, we obtain that (c_1, c_2^n) is a boundary *ga*-pair of the system U. □

Proof of Theorem 4.2. Each information system from the class W_1^g satisfies the condition of coverage and the condition of restricted coverage, and has finite I-dimension (see Table 4.3).

(a) The existence of both *gd*-reachable and not *gd*-reachable information systems in the class W_1^g follows from Lemmas 4.7 and 4.16. Let U be an information system from W_1^g, which is not *gd*-reachable. Then U satisfies the condition of restricted coverage and has finite I-dimension. From here and from Theorem 2.1 [3] it follows that, for each ε, $0 < \varepsilon < 1$, there exists a positive constant c such that $(c(\log_2 n)^{1+\varepsilon} + 1, 2^{c(\log_2 n)^{1+\varepsilon}+2})$ is a boundary *gd*-pair of the system U.

(b) Let $U = (A, F)$ be an information system from the class W_1^g. Then U satisfies the condition of coverage. Using Lemma 4.14, we obtain that the system U is not *ga*-reachable. We know that the system U satisfies the condition of restricted coverage and has finite I-dimension. In [3] it is proven (see item (a) of Theorem 2.2) that, for each ε, $0 < \varepsilon < 1$, there exist positive constants c_1, c_2 such that, for each problem z over U, there exists a system Δ of decision rules of the form

$$(f_1(x) = \delta_1) \wedge \cdots \wedge (f_m(x) = \delta_m) \rightarrow z(x) = \sigma ,$$

where $f_1, \ldots, f_m \in F$, $\delta_1, \ldots, \delta_m \in \{0, 1\}$, and $\sigma \in \mathbb{N}$, that satisfies the following conditions: (i) each rule is true for the problem z, (ii) for each $a \in A$, there exists a rule from Δ that accepts a, (iii) the number of conditions in the left-hand side of each rule is at most c_1, and (iv) the number of riles in Δ is at most $2^{c_2(\log_2 n)^{1+\varepsilon}+1}$, where $n = \dim z$. One can transform the system of decision rules Δ into a decision tree Γ over U, which solves the problem z nondeterministically and for which $h(\Gamma) \leq c_1$ and $L(\Gamma) \leq (c_1 + 2)2^{c_2(\log_2 n)^{1+\varepsilon}+1} = 2^{c_2(\log_2 n)^{1+\varepsilon}+c_3}$, where $c_3 = \log_2(c_1 + 2) + 1$. Note that complete paths in Γ correspond to rules from the system Δ. Thus, $(c_1, 2^{c_2(\log_2 n)^{1+\varepsilon}+c_3})$ is a boundary ga-pair of the system U. □

Proof of Theorem 4.3. Each information system from the class W_2^g satisfies the condition of coverage, does not satisfy the condition of restricted coverage, and has finite I-dimension (see Table 4.3).

(a) Let U be an information system from the class W_2^g. Then U does not satisfy the condition of restricted coverage. By Proposition 4.1, $h_U^{gd}(n) = n$ for any $n \in \mathbb{N}$. Using Lemma 4.12, we obtain that the system U is gd-reachable.

(b) Let U be an information system from the class W_2^g. Then U satisfies the condition of coverage. Using Lemma 4.14, we obtain that the system U is not ga-reachable. We know that U does not satisfy the condition of restricted coverage. Using Lemma 4.15, we obtain that $(n, L_U^{ga}(n))$ is the optimal boundary ga-pair of the system U. □

Proof of Theorem 4.4. Each information system from the class W_3^g does not satisfy the condition of coverage and the condition of restricted coverage, and has finite I-dimension (see Table 4.3).

(a) Let U be an information system from the class W_3^g. Then U does not satisfy the condition of restricted coverage. By Proposition 4.1, $h_U^{gd}(n) = n$ for any $n \in \mathbb{N}$. Using Lemma 4.12, we obtain that the system U is gd-reachable.

(b) Let U be an information system from the class W_3^g. Then U does not satisfy the condition of coverage. By Proposition 4.2, $h_U^{ga}(n) = n$ for any $n \in \mathbb{N}$. Using Lemma 4.13, we obtain that the system U is ga-reachable. □

Proof of Theorem 4.5. Each information system from the class W_4^g satisfies the condition of coverage and has infinite I-dimension (see Table 4.3). From Lemmas 4.10 and 4.17 it follows that the class W_4^g contains both information systems that satisfy the condition of restricted coverage and information systems that do not satisfy this condition.

(a) Let U be an information system from the class W_4^g. Then U has infinite I-dimension. By Proposition 4.1, $h_U^{gd}(n) = n$ for any $n \in \mathbb{N}$. Using Lemma 4.12, we obtain that the system U is gd-reachable.

(b) Let U be an information system from the class W_4^g. Then U satisfies the condition of coverage. Using Lemma 4.14, we obtain that the system U is not ga-reachable. Let U do not satisfy the condition of restricted coverage. Using Lemma 4.15, we obtain that $(n, L_U^{ga}(n))$ is the optimal boundary ga-pair of the system U. Let U satisfy the condition of restricted coverage. From Lemma 4.18 it follows that there exist positive constants c_1 and c_2 such that (c_1, c_2^n) is a boundary ga-pair of the system U. □

Proof of Theorem 4.6. Each information system from the class W_5^g does not satisfy the condition of coverage and the condition of restricted coverage, and has infinite I-dimension (see Table 4.3).

(a) Let U be an information system from the class W_5^g. Then U does not satisfy the condition of restricted coverage. By Proposition 4.1, $h_U^{gd}(n) = n$ for any $n \in \mathbb{N}$. Using Lemma 4.12, we obtain that the system U is gd-reachable.

(b) Let U be an information system from the class W_5^g. Then U does not satisfy the condition of coverage. By Proposition 4.2, $h_U^{ga}(n) = n$ for any $n \in \mathbb{N}$. Using Lemma 4.13, we obtain that the system U is ga-reachable. □

References

1. Molnar, C.: Interpretable Machine Learning: A Guide for Making Black Box Models Explainable, 2nd edn (2022). https://christophm.github.io/interpretable-ml-book/
2. Moshkov, M.: On time and space complexity of deterministic and nondeterministic decision trees. In: 8th International Conference Information Processing and Management of Uncertainty in Knowledge-Based Systems, IPMU 2000, vol. 3, pp. 1932–1936 (2000)
3. Moshkov, M.: Classification of infinite information systems depending on complexity of decision trees and decision rule systems. Fundam. Inform. **54**(4), 345–368 (2003)
4. Moshkov, M.: Time complexity of decision trees. In: Peters, J.F., Skowron, A. (eds.) Transactions on Rough Sets III. Lecture Notes in Computer Science, vol. 3400, pp. 244–459. Springer, Berlin (2005)
5. Moshkov, M.: Time and space complexity of deterministic and nondeterministic decision trees. Ann. Math. Artif. Intell. **91**(1), 45–74 (2023)
6. Moshkov, M., Zielosko, B.: Combinatorial Machine Learning—A Rough Set Approach. Studies in Computational Intelligence, vol. 360. Springer, Berlin (2011)
7. Naiman, D.Q., Wynn, H.P.: Independence number and the complexity of families of sets. Discr. Math. **154**, 203–216 (1996)
8. Pawlak, Z.: Information systems theoretical foundations. Inf. Syst. **6**(3), 205–218 (1981)
9. Sauer, N.: On the density of families of sets. J. Comb. Theory (A) **13**, 145–147 (1972)
10. Shelah, S.: A combinatorial problem; stability and order for models and theories in infinitary languages. Pac. J. Math. **41**, 241–261 (1972)

Chapter 5
Time and Space Complexity of Deterministic and Nondeterministic Decision Trees. Local Approach

5.1 Introduction

In this chapter, we study conventional problems over infinite binary information systems. Each such information system consists of an infinite universe and an infinite set of functions (attributes), which are defined on the universe and have values from the set {0, 1}. Any conventional problem over the information system is described by a finite number of attributes that divide the universe into domains in which these attributes have fixed values. A decision is attached to each domain. For a given object from the universe, it is required to find the decision attached to the domain containing this object. As algorithms solving these problems, deterministic and nondeterministic decision trees are considered. As time and space complexity of decision trees, we use their depth and the number of nodes. Note that nondeterministic decision trees can be considered as representations of decision rule systems.

We study relationships between time and space complexity for deterministic and nondeterministic decision trees. Interest in such a study is due to the following considerations. Decision trees are among the most interpretable models for classifying and representing knowledge [5]. In order to better understand decision trees, we should not only minimize the number of their nodes, but also the depth of decision trees to avoid the consideration of long conjunctions of conditions corresponding to long paths in these trees. When we consider decision trees as algorithms (usually sequential for deterministic decision trees and parallel for nondeterministic decision trees), we should have in mind the same bi-criteria optimization problems to minimize space and time complexity of these algorithms.

There are two approaches to decision tree investigation: the local approach, where for a problem, the decision trees are considered using only attributes from the problem description, and the global one, where for problem solving, all attributes from the considered information system can be used in decision trees. In this chapter, decision trees are investigated within the frameworks of the local approach.

K. Durdymyradov et al., *Decision Trees Versus Systems of Decision Rules*, Studies in Big Data 160, https://doi.org/10.1007/978-3-031-71586-0_5

Based on the results obtained in the present chapter and in [6, 8], we describe possible types of behavior of four functions $h_U^{ld}, h_U^{la}, L_U^{ld}, L_U^{la}$ that characterize worst case time and space complexity of deterministic and nondeterministic decision trees solving problems over an infinite binary information system U (index l refers to the local approach). Decision trees solving a problem can use only attributes from the problem description (as we say, they should be decision trees over the problem).

The function h_U^{ld} characterizes the growth in the worst case of the minimum depth of a deterministic decision tree solving a problem with the growth of the number of attributes in the problem description. The function h_U^{ld} is either grows as a logarithm or linearly.

The function h_U^{la} characterizes the growth in the worst case of the minimum depth of a nondeterministic decision tree solving a problem with the growth of the number of attributes in the problem description. The function h_U^{la} is either bounded from above by a constant or grows linearly.

The function L_U^{ld} characterizes the growth in the worst case of the minimum number of nodes in a deterministic decision tree solving a problem with the growth of the number of attributes in the problem description. The function L_U^{ld} has either polynomial or exponential growth.

The function L_U^{la} characterizes the growth in the worst case of the minimum number of nodes in a nondeterministic decision tree solving a problem with the growth of the number of attributes in the problem description. The function L_U^{la} has either polynomial or exponential growth.

Each of the functions $h_U^{ld}, h_U^{la}, L_U^{ld}, L_U^{la}$ has two types of behavior. The tuple $(h_U^{ld}, h_U^{la}, L_U^{ld}, L_U^{la})$ has three types of behavior. All these types are described in the chapter and each type is illustrated by an example.

There are three complexity classes of infinite binary information systems corresponding to the three possible types of the tuple $(h_U^{ld}, h_U^{la}, L_U^{ld}, L_U^{la})$. For each class, we study joint behavior of time and space complexity of decision trees. The obtained results are related to time-space trade-off for deterministic and nondeterministic decision trees.

A pair of functions (φ, ψ) is called a boundary ld-pair of the information system U if, for any problem over U, there exists a deterministic decision tree over this problem, which solves the problem and for which the depth is at most $\varphi(n)$ and the number of nodes is at most $\psi(n)$, where n is the number of attributes in the problem description. An information system U is called ld-reachable if the pair (h_U^{ld}, L_U^{ld}) is a boundary ld-pair of the system U. For nondeterministic decision trees, the notions of a boundary la-pair of an information system and la-reachable information system are defined in a similar way. For deterministic decision trees, the best situation is when the considered information system is ld-reachable: for any boundary ld-pair (φ, ψ) for an information system U and any natural n, $\varphi(n) \geq h_U^{ld}(n)$ and $\psi(n) \geq L_U^{ld}(n)$. For nondeterministic decision trees, the best situation is when the information system is la-reachable.

For all complexity classes, information systems from the class are ld-reachable. For two out of the three complexity classes, all information systems from the class are la-reachable. For the remaining class, all information systems from the class are not la-reachable. For all information systems U that are not la-reachable, we find nontrivial boundary la-pairs, which are sufficiently close to (h_U^{la}, L_U^{la}).

Note that this chapter will often talk about decision trees that solve a problem deterministically or nondeterministically, rather than deterministic or nondeterministic decision trees that solve a problem.

This chapter is based on the papers [3, 4]. The rest of the chapter is organized as follows: Sect. 5.2 contains main results and Sects. 5.3–5.5—proofs of these results.

5.2 Main Results

Let A be an infinite set and F be an infinite set of functions that are defined on A and have values from the set $\{0, 1\}$. The pair $U = (A, F)$ is called an *infinite binary information system* [10], the elements of the set A are called *objects*, and the functions from F are called *attributes*. The set A is called sometimes the *universe* of the information system U.

A *problem over* U is a tuple of the form $z = (\nu, f_1, \ldots, f_n)$, where $\nu : \{0, 1\}^n \to \mathbb{N}$, $\mathbb{N}$ is the set of natural numbers $\{1, 2, \ldots\}$, and $f_1, \ldots, f_n \in F$. We do not require attributes $f_1, \ldots, f_n$ to be pairwise distinct. The problem z consists in finding the value of the function $z(x) = \nu(f_1(x), \ldots, f_n(x))$ for a given object $a \in A$. The value $\dim z = n$ is called the *dimension* of the problem z.

Various problems of combinatorial optimization, pattern recognition, fault diagnosis, probabilistic reasoning, computational geometry, etc., can be represented in this form.

As algorithms for problem solving, we consider decision trees. A *decision tree over the information system* U is a directed tree with a root in which the root and edges leaving the root are not labeled, each terminal node is labeled with a number from $\mathbb{N}$, each *working* node (which is neither the root nor a terminal node) is labeled with an attribute from F, and each edge leaving a working node is labeled with a number from the set $\{0, 1\}$. A decision tree is called *deterministic* if only one edge leaves the root and edges leaving an arbitrary working node are labeled with different numbers.

Let Γ be a decision tree over U and

$$\xi = v_0, d_0, v_1, d_1, \ldots, v_m, d_m, v_{m+1}$$

be a directed path from the root v_0 to a terminal node v_{m+1} of Γ (we call such path *complete*). Define a subset $A(\xi)$ of the set A as follows. If $m = 0$, then $A(\xi) = A$. Let $m > 0$ and, for $i = 1, \ldots, m$, the node v_i be labeled with the attribute f_{j_i} and the edge d_i be labeled with the number δ_i. Then

$$A(\xi) = \{a : a \in A, f_{j_1}(a) = \delta_1, \ldots, f_{j_m}(a) = \delta_m\}.$$

The *depth* of the decision tree Γ is the maximum number of working nodes in a complete path of Γ. Denote by $h(\Gamma)$ the depth of Γ and by $L(\Gamma)$—the number of nodes in Γ.

A decision tree over the information system U is called a *decision tree over the problem* $z = (\nu, f_1, \ldots, f_n)$ if each working node of Γ is labeled with an attribute from the set $\{f_1, \ldots, f_n\}$.

The decision tree Γ over z solves the problem z *nondeterministically* if, for any object $a \in A$, there exists a complete path ξ of Γ such that $a \in A(\xi)$ and, for each $a \in A$ and each complete path ξ such that $a \in A(\xi)$, the terminal node of ξ is labeled with the number $z(a)$ (in this case, we can say that Γ is a *nondeterministic decision tree solving the problem* z).

In particular, if the decision tree Γ solves the problem z nondeterministically, then, for each complete path ξ of Γ, either the set $A(\xi)$ is empty or the function $z(x)$ is constant on the set $A(\xi)$. The decision tree Γ over z solves the problem z *deterministically* if Γ is a deterministic decision tree, which solves the problem z nondeterministically (in this case, we can say that Γ is a *deterministic decision tree solving the problem* z).

For a problem z over U, let $h_U^{ld}(z)$ be the minimum depth of a decision tree over z solving the problem z deterministically, $h_U^{la}(z)$ be the minimum depth of a decision tree over z solving the problem z nondeterministically, $L_U^{ld}(z)$ be the minimum number of nodes in a decision tree over z solving the problem z deterministically, and $L_U^{la}(z)$ be the minimum number of nodes in a decision tree over z solving the problem z nondeterministically.

We consider four functions defined on the set $\mathbb{N}$ in the following way: $h_U^{ld}(n) = \max h_U^{ld}(z)$, $h_U^{la}(n) = \max h_U^{la}(z)$, $L_U^{ld}(n) = \max L_U^{ld}(z)$, and $L_U^{la}(n) = \max L_U^{la}(z)$, where the maximum is taken among all problems z over U with $\dim z \leq n$. These functions describe how the minimum depth and the minimum number of nodes of deterministic and nondeterministic decision trees solving problems are growing in the worst case with the growth of problem dimension. To describe possible types of behavior of these four functions, we need to define some properties of infinite binary information systems.

Definition 5.1 We will say that the information system $U = (A, F)$ satisfies the *condition of reduction* if there exists $m \in \mathbb{N}$ such that, for each compatible on A system of equations $\{f_1(x) = \delta_1, \ldots, f_r(x) = \delta_r\}$, where $r \in \mathbb{N}$, $f_1, \ldots, f_r \in F$ and $\delta_1, \ldots, \delta_r \in \{0, 1\}$, there exists a subsystem of this system, which has the same set of solutions from A and contains at most m equations. In this case, we will say that U satisfies the *condition of reduction with parameter* m.

We now consider two examples of infinite binary information systems that satisfy the condition of reduction. These examples are close to ones considered in Sect. 3.4 of the book [1].

Example 5.1 Let $d, t \in \mathbb{N}$, $f_1, \ldots, f_t$ be functions from $\mathbb{R}^d$ to $\mathbb{R}$, where $\mathbb{R}$ is the set of real numbers, and s be a function from $\mathbb{R}$ to $\{0, 1\}$ such that $s(x) = 0$ if $x < 0$ and $s(x) = 1$ if $x \geq 0$. Then the infinite binary information system $(\mathbb{R}^d, F)$, where $F =$

$\{s(f_i + c) : i = 1, \ldots, t, c \in \mathbb{R}\}$, satisfies the condition of reduction with parameter $2t$. If $f_1, \ldots, f_t$ are linear functions, then we deal with attributes corresponding to t families of parallel hyperplanes in $\mathbb{R}^d$ what is common for decision trees for datasets with t numerical attributes only [2].

Example 5.2 Let P be the Euclidean plane and l be a straight line (line in short) in the plane. This line divides the plane into two open half-planes H_1 and H_2, and the line l. Two attributes correspond to the line l. The first attribute takes value 0 on points from H_1, and value 1 on points from H_2 and l. The second one takes value 0 on points from H_2, and value 1 on points from H_1 and l. We denote by $\mathscr{L}$ the set of all attributes corresponding to lines in the plane. Infinite binary information systems of the form (P, L), where $L \subseteq \mathscr{L}$, are called *linear* information systems.

Let l be a line in the plane. Let us denote by $\mathscr{L}(l)$ the set of all attributes corresponding to lines, which are parallel to l. Let p be a point in the plane. We denote by $\mathscr{L}(p)$ the set of all attributes corresponding to lines, which pass through p. A set C of attributes from $\mathscr{L}$ is called a *clone* if $C \subseteq \mathscr{L}(l)$ for some line l or $C \subseteq \mathscr{L}(p)$ for some point p. In [7], it was proved that a linear information system (P, L) satisfies the condition of reduction if and only if L is the union of a finite number of clones.

Definition 5.2 Let $U = (A, F)$ be an infinite binary information system. A subset $\{f_1, \ldots, f_m\}$ of the set F will be called *independent* if, for any $\delta_1, \ldots, \delta_m \in \{0, 1\}$, the system of equations $\{f_1(x) = \delta_1, \ldots, f_m(x) = \delta_m\}$ has a solution from the set A. The empty set of attributes is independent by definition.

Definition 5.3 We define the parameter $I(U)$, which is called the *independence dimension* or I-*dimension* of the information system U (this notion is similar to the notion of independence number of family of sets [9]) as follows. If, for each $m \in \mathbb{N}$, the set F contains an independent subset of cardinality m, then $I(U) = \infty$. Otherwise, $I(U)$ is the maximum cardinality of an independent subset of the set F.

We now consider examples of infinite binary information systems with finite I-dimension and with infinite I-dimension. More examples can be found in Lemmas 5.7–5.9.

Example 5.3 Let $m, t \in \mathbb{N}$. We denote by $Pol(m)$ the set of all polynomials, which have integer coefficients and depend on variables $x_1, \ldots, x_m$. We denote by $Pol(m, t)$ the set of all polynomials from $Pol(m)$ such that the degree of each polynomial is at most t. We define infinite binary information systems $U(m)$ and $U(m, t)$ as follows: $U(m) = (\mathbb{R}^m, F(m))$ and $U(m, t) = (\mathbb{R}^m, F(m, t))$, where $F(m) = \{s(p) : p \in Pol(m)\}$, $F(m, t) = \{s(p) : p \in Pol(m, t)\}$, and $s(x) = 0$ if $x < 0$ and $s(x) = 1$ if $x \geq 0$. One can show that the system $U(m)$ has infinite I-dimension and the system $U(m, t)$ has finite I-dimension.

We now consider four statements that describe possible types of behavior of functions $h_U^{ld}(n)$, $h_U^{la}(n)$, $L_U^{ld}(n)$, and $L_U^{la}(n)$. The next statement follows immediately from Theorem 4.3 [6].

Proposition 5.1 *For any infinite binary information system U, the function $h_U^{ld}(n)$ has one of the following two types of behavior:*

(LOG) *If the system U satisfies the condition of reduction, then $h_U^{ld}(n) = \Theta(\log n)$.*

(LIN) *If the system U does not satisfy the condition of reduction, then $h_U^{ld}(n) = n$ for any $n \in \mathbb{N}$.*

The next statement follows immediately from Theorem 8.2 [8].

Proposition 5.2 *For any infinite binary information system U, the function $h_U^{la}(n)$ has one of the following two types of behavior:*

(CON) *If the system U satisfies the condition of reduction, then $h_U^{la}(n) = O(1)$.*

(LIN) *If the system U does not satisfy the condition of reduction, then $h_U^{la}(n) = n$ for any $n \in \mathbb{N}$.*

Proposition 5.3 *For any infinite binary information system U, the function $L_U^{ld}(n)$ has one of the following two types of behavior:*

(POL) *If the system U has finite I-dimension, then for any $n \in \mathbb{N}$,*

$$2(n+1) \leq L_U^{ld}(n) \leq 2(4n)^{I(U)} .$$

(EXP) *If the system U has infinite I-dimension, then for any $n \in \mathbb{N}$,*

$$L_U^{ld}(n) = 2^{n+1} .$$

Proposition 5.4 *For any infinite binary information system U and any $n \in \mathbb{N}$,*

$$L_U^{la}(n) = L_U^{ld}(n) .$$

Let U be an infinite binary information system. Proposition 5.1 allows us to correspond to the function $h_U^{ld}(n)$ its type of behavior from the set $\{\text{LOG}, \text{LIN}\}$. Proposition 5.2 allows us to correspond to the function $h_U^{la}(n)$ its type of behavior from the set $\{\text{CON}, \text{LIN}\}$. Propositions 5.3 and 5.4 allow us to correspond to each of the functions $L_U^{ld}(n)$ and $L_U^{la}(n)$ its type of behavior from the set $\{\text{POL}, \text{EXP}\}$.

Definition 5.4 A tuple obtained from the tuple

$$(h_U^{ld}(n), h_U^{la}(n), L_U^{ld}(n), L_U^{la}(n))$$

by replacing functions with their types of behavior will be called the *local type* of the information system U.

We now describe all possible local types of infinite binary information systems.

Theorem 5.1 *For any infinite binary information system, its local type coincides with one of the rows of Table 5.1. Each row of Table 5.1 is the local type of some infinite binary information system.*

Table 5.1 Possible local types of infinite binary information systems

	$h_U^{ld}(n)$	$h_U^{la}(n)$	$L_U^{ld}(n)$	$L_U^{la}(n)$
1	LOG	CON	POL	POL
2	LIN	LIN	POL	POL
3	LIN	LIN	EXP	EXP

For $i = 1, 2, 3$, we denote by W_i^l the class of all infinite binary information systems, whose local type coincides with the ith row of Table 5.1. We now study for each of these complexity classes joint behavior of the depth and number of nodes in decision trees solving problems.

For a given infinite binary information system U, we will consider pairs of functions (φ, ψ) such that, for any problem z over U, there exists a deterministic decision tree over z solving z with the depth at most $\varphi(\dim z)$ and the number of nodes at most $\psi(\dim z)$. We will study such pairs and will call them boundary ld-pairs.

Definition 5.5 A pair of functions (φ, ψ), where $\varphi : \mathbb{N} \to \mathbb{N} \cup \{0\}$ and $\psi : \mathbb{N} \to \mathbb{N} \cup \{0\}$, will be called a *boundary ld-pair* of the information system U if, for any problem z over U, there exists a decision tree Γ over z, which solves the problem z deterministically and for which $h(\Gamma) \le \varphi(n)$ and $L(\Gamma) \le \psi(n)$, where $n = \dim z$.

We are interested in finding boundary ld-pairs with functions that grow as slowly as possible. It is clear that, for any boundary ld-pair (φ, ψ) of the information system U, the following inequalities hold: $\varphi(n) \ge h_U^{ld}(n)$ and $\psi(n) \ge L_U^{ld}(n)$. So the best possible situation is when (h_U^{ld}, L_U^{ld}) is a boundary ld-pair of U.

Definition 5.6 An information system U will be called *ld-reachable* if the pair (h_U^{ld}, L_U^{ld}) is a boundary ld-pair of the system U.

We now consider similar notions for nondeterministic decision trees: the notion of boundary la-pair and the notion of la-reachable information system.

Definition 5.7 A pair of functions (φ, ψ), where $\varphi : \mathbb{N} \to \mathbb{N} \cup \{0\}$ and $\psi : \mathbb{N} \to \mathbb{N} \cup \{0\}$, will be called a *boundary la-pair* of the information system U if, for any problem z over U, there exists a decision tree Γ over z, which solves the problem z nondeterministically and for which $h(\Gamma) \le \varphi(n)$ and $L(\Gamma) \le \psi(n)$, where $n = \dim z$.

Definition 5.8 An information system U will be called *la-reachable* if the pair (h_U^{la}, L_U^{la}) is a boundary la-pair of the system U.

Note that for deterministic decision trees, the best situation is when the considered information system is ld-reachable and for nondeterministic decision trees—when the information system is la-reachable.

Each information system from the classes W_1^l, W_2^l, and W_3^l is ld-reachable. Each information system from the classes W_2^l and W_3^l is la-reachable. Each information

Table 5.2 Summary of Theorems 5.1–5.4

	$h_U^{ld}(n)$	$h_U^{la}(n)$	$L_U^{ld}(n)$	$L_U^{la}(n)$	ld-pairs	la-pairs
W_1^l	LOG	CON	POL	POL	ld-reachable	Theorem 5.2(b)
W_2^l	LIN	LIN	POL	POL	ld-reachable	la-reachable
W_3^l	LIN	LIN	EXP	EXP	ld-reachable	la-reachable

system from the class W_1^l is not la-reachable. For all information systems U, which are not la-reachable, we find nontrivial boundary la-pairs that are sufficiently close to (h_U^{la}, L_U^{la}).

The obtained results are related to time-space trade-off for deterministic and nondeterministic decision trees. Details can be found in the following three theorems.

Theorem 5.2 *Let U be an information system from the class W_1^l. Then*
(a) *The system U is ld-reachable.*
(b) *The system U is not la-reachable and there exists $m \in \mathbb{N}$ such that*

$$(m, (m + 1)L_U^{la}(n)/2 + 1)$$

is a boundary la-pair of the system U.

Theorem 5.3 *Let U be an information system from the class W_2^l. Then*
(a) *The system U is ld-reachable.*
(b) *The system U is la-reachable.*

Theorem 5.4 *Let U be an information system from the class W_3^l. Then*
(a) *The system U is ld-reachable.*
(b) *The system U is la-reachable.*

Table 5.2 summarizes Theorems 5.1–5.4. The first column contains the name of the complexity class. The next four columns describe the local type of information systems from this class. The last two columns "ld-pairs" and "la-pairs" contain information about boundary ld-pairs and boundary la-pairs for information systems from the considered class: "ld-reachable" means that all information systems from the class are ld-reachable, "la-reachable" means that all information systems from the class are la-reachable, Theorem 5.2(b) is a link to the corresponding statement Theorem 5.2(b).

5.3 Proofs of Propositions 5.3 and 5.4

In this section, we consider a number of auxiliary statements and prove the two mentioned propositions.

Let Γ be a decision tree over an infinite binary information system $U = (A, F)$ and d be an edge of Γ entering a node w. We denote by $\Gamma(d)$ a subtree of Γ, whose root is the node w. We say that a complete path ξ of Γ is *realizable* if $A(\xi) \neq \emptyset$.

Lemma 5.1 *Let $U = (A, F)$ be an infinite binary information system, $z = (\nu, f_1, \ldots, f_n)$ be a problem over U, and Γ be a decision tree over z, which solves the problem z deterministically and for which $L(\Gamma) = L_U^{ld}(z)$. Then*

(a) *For each node of Γ, there exists a realizable complete path that passes through this node.*

(b) *Each working node of Γ has two edges leaving this node.*

Proof (a) It is clear that there exists at least one realizable complete path that passes through the root of Γ. Let us assume that w is a node of Γ different from the root and such that there is no a realizable complete path, which passes through w. Let d be an edge entering the node w. We remove from Γ the edge d and the subtree $\Gamma(d)$. As a result, we obtain a decision tree Γ' over z, which solves z deterministically and for which $L(\Gamma') < L(\Gamma)$ but this is impossible by definition of Γ.

(b) Let us assume that in Γ there exists a working node w, which has only one leaving edge d entering a node w_1. We remove from Γ the node w and the edge d and connect the edge entering the node w to the node w_1. As a result, we obtain a decision tree Γ' over z, which solves the problem z deterministically and for which $L(\Gamma') < L(\Gamma)$ but this is impossible by definition of Γ. □

Let U be an infinite binary information system, Γ be a decision tree over U, and d be an edge of Γ. The subtree $\Gamma(d)$ will be called *full* if there exist edges $d_1, \ldots, d_m$ in $\Gamma(d)$ such that the removal of these edges and subtrees $\Gamma(d_1), \ldots, \Gamma(d_m)$ transforms the subtree $\Gamma(d)$ into a tree G such that each terminal node of G is a terminal node of Γ, and exactly two edges labeled with the numbers 0 and 1 respectively leave each working node of G.

Lemma 5.2 *Let $U = (A, F)$ be an infinite binary information system, $z = (\nu, f_1, \ldots, f_n)$ be a problem over U, and Γ be a decision tree over z, which solves the problem z nondeterministically and for which $L(\Gamma) = L_U^{la}(z)$. Then*

(a) *For each node of Γ, there exists a realizable complete path that passes through this node.*

(b) *If a working node w of Γ has m leaving edges $d_1, \ldots, d_m$ labeled with the same number and $m \geq 2$, then the subtrees $\Gamma(d_1), \ldots, \Gamma(d_m)$ are not full.*

(c) *If the root r of Γ has m leaving edges $d_1, \ldots, d_m$ and $m \geq 2$, then the subtrees $\Gamma(d_1), \ldots, \Gamma(d_m)$ are not full.*

Proof (a) The proof of item (a) is almost identical to the proof of item (a) of Lemma 5.1.

(b) Let w be a working node of Γ, which has m leaving edges $d_1, \ldots, d_m$ labeled with the same number, $m \geq 2$, and at least one of the subtrees $\Gamma(d_1), \ldots, \Gamma(d_m)$ is full. For the definiteness, we assume that $\Gamma(d_1)$ is full. Remove from Γ the edges $d_2, \ldots, d_m$ and subtrees $\Gamma(d_2), \ldots, \Gamma(d_m)$. We now show that the obtained tree Γ' solves the problem z nondeterministically. Assume the contrary. Then there exists an object $a \in A$ such that, for each complete path ξ with $a \in A(\xi)$, the path ξ passes through one of the edges $d_2, \ldots, d_m$ but it is not true. Let ξ be a complete path such that $a \in A(\xi)$. Then, according to the assumption, this path passes through the node w. Let ξ' be the part of this path from the root of Γ to the node w. Since the edges $d_1, \ldots, d_m$ are labeled with the same number and $\Gamma(d_1)$ is a full subtree, we can find in $\Gamma(d_1)$ the continuation of ξ' to a terminal node of $\Gamma(d_1)$ such that the obtained complete path ξ'' of Γ satisfies the condition $a \in A(\xi'')$. Hence Γ' is a decision tree over z, which solves the problem z nondeterministically and for which $L(\Gamma') < L(\Gamma)$, but this is impossible by definition of Γ.

(c) Item (c) can be proven in the same way as item (b). □

Let Γ be a decision tree. We denote by $L_t(\Gamma)$ the number of terminal nodes in Γ and by $L_w(\Gamma)$ the number of working nodes in Γ. It is clear that $L(\Gamma) = 1 + L_t(\Gamma) + L_w(\Gamma)$.

Let U be an infinite binary information systems. We denote by $G_d(U)$ the set of all deterministic decision trees over U and by $G_d^2(U)$ the set of all decision trees from $G_d(U)$ such that each working node of the tree has two leaving edges.

Lemma 5.3 *(Lemma 4.4) Let U be an infinite binary information system. Then*
(a) If $\Gamma \in G_d^2(U)$, then $L_w(\Gamma) = L_t(\Gamma) - 1$.
(b) If $\Gamma \in G_d(U) \setminus G_d^2(U)$, then $L_w(\Gamma) > L_t(\Gamma) - 1$.

We denote by $G_a^f(U)$ the set of all decision trees Γ over U that satisfy the following conditions: (i) if a working node of Γ has m leaving edges $d_1, \ldots, d_m$ labeled with the same number and $m \geq 2$, then the subtrees $\Gamma(d_1), \ldots, \Gamma(d_m)$ are not full, and (ii) if the root of Γ has m leaving edges $d_1, \ldots, d_m$ and $m \geq 2$, then the subtrees $\Gamma(d_1), \ldots, \Gamma(d_m)$ are not full. One can show that $G_d^2(U) \subseteq G_d(U) \subseteq G_a^f(U)$.

Lemma 5.4 *(Lemma 4.5) Let U be an infinite binary information system. If $\Gamma \in G_a^f(U) \setminus G_d^2(U)$, then $L_w(\Gamma) > L_t(\Gamma) - 1$.*

Let $U = (A, F)$ be an infinite binary information system. For $f_1, \ldots, f_n \in F$, we denote by $N_U(f_1, \ldots, f_n)$ the number of n-tuples $(\delta_1, \ldots, \delta_n) \in \{0, 1\}^n$ for which the system of equations

$$\{f_1(x) = \delta_1, \ldots, f_n(x) = \delta_n\}$$

has a solution from A. For $n \in \mathbb{N}$, denote

$$N_U(n) = \max\{N_U(f_1, \ldots, f_n) : f_1, \ldots, f_n \in F\} .$$

It is clear that, for any $m, n \in \mathbb{N}$, if $m \leq n$, then $N_U(m) \leq N_U(n)$.

Proposition 5.5 *Let $U = (A, F)$ be an infinite binary information system. Then, for any $n \in \mathbb{N}$,*

$$L_U^{la}(n) = L_U^{ld}(n) = 2N_U(n) .$$

Proof Let $z = (\nu, f_1, \ldots, f_m)$ be a problem over U and $m \leq n$. Let Γ be a decision tree over z, which solves the problem z deterministically and for which $L(\Gamma) = L_U^{ld}(z)$. From Lemma 5.1 it follows that each working node of Γ has two edges leaving this node and, for each node of Γ, there exists a realizable complete path that passes through this node. Let ξ_1 and ξ_2 be different complete paths in Γ, $a_1 \in A(\xi_1)$, and $a_2 \in A(\xi_2)$. It is easy to show that $(f_1(a_1), \ldots, f_m(a_1)) \neq (f_1(a_2), \ldots, f_m(a_2))$. Therefore $L_t(\Gamma) \leq N_U(f_1, \ldots, f_m) \leq N_U(n)$. It is clear that $\Gamma \in G_d^2(U)$. By Lemma 5.3, $L_w(\Gamma) = L_t(\Gamma) - 1$. Hence $L(\Gamma) \leq 2N_U(n)$. Taking into account that z is an arbitrary problem over U with $\dim z \leq n$, we obtain

$$L_U^{ld}(n) \leq 2N_U(n) .$$

Since any decision tree solving the problem z deterministically solves it nondeterministically, we obtain

$$L_U^{la}(n) \leq L_U^{ld}(n) .$$

We now show that $2N_U(n) \leq L_U^{la}(n)$. Let us consider a problem $z = (\nu, f_1, \ldots, f_n)$ over U such that

$$N_U(f_1, \ldots, f_n) = N_U(n)$$

and, for any $\bar{\delta}_1, \bar{\delta}_2 \in \{0, 1\}^n$, if $\bar{\delta}_1 \neq \bar{\delta}_2$, then $\nu(\bar{\delta}_1) \neq \nu(\bar{\delta}_2)$. Let Γ be a decision tree over z, which solves the problem z nondeterministically and for which $L(\Gamma) = L_U^{la}(z)$. By Lemma 5.2, $\Gamma \in G_a^f(U)$. Using Lemmas 5.3 and 5.4, we obtain $L_w(\Gamma) \geq L_t(\Gamma) - 1$. It is clear that $L_t(\Gamma) \geq N_U(f_1, \ldots, f_n) = N_U(n)$. Therefore $L(\Gamma) \geq 2N_U(n)$, $L_U^{la}(z) \geq 2N_U(n)$, and $L_U^{la}(n) \geq 2N_U(n)$. □

The next statement follows directly from Lemmas 5.1 and 5.2 [6] and the evident inequality $N_U(n) \leq 2^n$, which is true for any infinite binary information system U. The proof of Lemma 5.1 from [6] is based on Theorems 4.6 and 4.7 from the same monograph that are similar to results obtained in [11, 12].

Proposition 5.6 *For any infinite binary information system U, the function $N_U(n)$ has one of the following two types of behavior:*

(POL) *If the system U has finite I-dimension, then for any $n \in \mathbb{N}$,*

$$n + 1 \leq N_U(n) \leq (4n)^{I(U)} .$$

(EXP) *If the system U has infinite I-dimension, then for any $n \in \mathbb{N}$,*

$$N_U(n) = 2^n .$$

We now prove Propositions 5.3 and 5.4.

Proof of Proposition 5.3. The statement of the proposition follows immediately from Propositions 5.5 and 5.6. □

Proof of Proposition 5.4. The statement of the proposition follows immediately from Proposition 5.5. □

5.4 Proof of Theorem 5.1

First, we prove five auxiliary statements.

Lemma 5.5 *Let $U = (A, F)$ be an infinite binary information system, which has infinite I-dimension. Then U does not satisfy the condition of reduction.*

Proof Let us assume the contrary: U satisfies the condition of reduction. Then U satisfies the condition of reduction with parameter m for some $m \in \mathbb{N}$. Since $I(U) = \infty$, there exists an independent subset $\{f_1, \ldots, f_{m+1}\}$ of the set F. It is clear that the system of equations

$$S = \{f_1(x) = 0, \ldots, f_{m+1}(x) = 0\}$$

is compatible on A and each proper subsystem of the system S has the set of solutions different from the set of solutions of S. Therefore U does not satisfy the condition of reduction with parameter m. □

Lemma 5.6 *For any infinite binary information system, its local type coincides with one of the rows of Table 5.1.*

Proof To prove this statement we fill Table 5.3. In the first column "I-dim." we have either "Fin" or "Inf": "Fin" if the considered information system has finite I-dimension and "Inf" if the considered information system has infinite I-dimension. In the second column "Reduct.", we have either "Yes" or "No": "Yes" if the considered information system satisfies the condition of reduction and "No" otherwise.

Table 5.3 Parameters and local types of infinite binary information systems

I-dim.	Reduct.	$h_U^{ld}(n)$	$h_U^{la}(n)$	$L_U^{ld}(n)$	$L_U^{la}(n)$
Fin	Yes	LOG	CON	POL	POL
Fin	No	LIN	LIN	POL	POL
Inf	No	LIN	LIN	EXP	EXP

By Lemma 5.5, if an information system has infinite I-dimension, then this information system does not satisfy the condition of reduction. It means that there are only three possible tuples of values of the considered two parameters of information systems, which correspond to the three rows of Table 5.3. The values of the considered two parameters define the types of behavior of functions $h_U^{ld}(n)$, $h_U^{la}(n)$, $L_U^{ld}(n)$, and $L_U^{la}(n)$ according to Propositions 5.1–5.4. We see that the set of possible tuples of values in the last four columns coincides with the set of rows of Table 5.1. □

For each row of Table 5.1, we consider an example of infinite binary information system, whose local type coincides with this row.

For any $i \in \mathbb{N}$, we define two functions $p_i : \mathbb{N} \to \{0, 1\}$ and $l_i : \mathbb{N} \to \{0, 1\}$. Let $j \in \mathbb{N}$. Then $p_i(j) = 1$ if and only if $j = i$, and $l_i(j) = 1$ if and only if $j > i$.

Define an information system $U_1 = (A_1, F_1)$ as follows: $A_1 = \mathbb{N}$ and $F_1 = \{l_i : i \in \mathbb{N}\}$.

Lemma 5.7 *The information system U_1 belongs to the class W_1^l, $h_{U_1}^{ld}(n) = \lceil \log_2(n+1) \rceil$, $h_{U_1}^{la}(1) = 1$ and $h_{U_1}^{la}(n) = 2$ if $n > 1$, $L_{U_1}^{ld}(n) = 2(n+1)$, and $L_{U_1}^{la}(n) = 2(n+1)$ for any $n \in \mathbb{N}$. This information system satisfies the condition of reduction with parameter 2 and has finite I-dimension equal 1.*

Proof It is easy to show that $N_{U_1}(n) = n + 1$ for any $n \in \mathbb{N}$. Using Proposition 5.5, we obtain $L_{U_1}^{ld}(n) = L_{U_1}^{la}(n) = 2(n+1)$ for any $n \in \mathbb{N}$. Let $n \in \mathbb{N}$. Consider a problem $z = (\nu, l_1, \dots, l_n)$ over U_1 such that, for each $\bar{\delta}_1, \bar{\delta}_2 \in \{0, 1\}^n$ with $\bar{\delta}_1 \neq \bar{\delta}_2$, $\nu(\bar{\delta}_1) \neq \nu(\bar{\delta}_2)$. It is clear that $N_{U_1}(l_1, \dots, l_n) = n + 1$. Therefore each decision tree Γ over z that solves the problem z deterministically has at least $n + 1$ terminal nodes. One can show that the number of terminal nodes in Γ is at most $2^{h(\Gamma)}$. Hence $n + 1 \leq 2^{h(\Gamma)}$ and $\log_2(n+1) \leq h(\Gamma)$. Since $h(\Gamma)$ is an integer, $\lceil \log_2(n+1) \rceil \leq h(\Gamma)$. Thus, $h_{U_1}^{ld}(n) \geq \lceil \log_2(n+1) \rceil$. Set $m = \lceil \log_2(n+1) \rceil$. Then $n \leq 2^m - 1$. One can show that $h_{U_1}^{ld}(2^m - 1) \leq m$ (the construction of an appropriate decision tree is based on an analog of binary search, and we use only attributes from the problem description) and $h_{U_1}^{ld}(n) \leq h_{U_1}^{ld}(2^m - 1)$. Therefore $h_{U_1}^{ld}(n) \leq \lceil \log_2(n+1) \rceil$ and $h_{U_1}^{ld}(n) = \lceil \log_2(n+1) \rceil$. It is clear that $h_{U_1}^{la}(1) = 1$. Let $n \geq 2$, $z = (\nu, f_1, \dots, f_n)$ be an arbitrary problem over U_1 and $l_{i_1}, \dots, l_{i_m}$ be all pairwise different attributes from the set $\{f_1, \dots, f_n\}$ ordered such that $i_1 < \cdots < i_m$. Then these attributes divide the set $\mathbb{N}$ into $m + 1$ nonempty domains that are sets of solutions on $\mathbb{N}$ of the following systems of equations: $\{l_{i_1}(x) = 0\}$, $\{l_{i_1}(x) = 1, l_{i_2}(x) = 0\}$, ..., $\{l_{i_{m-1}}(x) = 1, l_{i_m}(x) = 0\}$, $\{l_{i_m}(x) = 1\}$. The value $z(x)$ is constant in each of the considered domains. Using these facts, it is easy to show that there exists a decision tree Γ over z, which solves the problem z nondeterministically and for which $h(\Gamma) = 2$ if $m \geq 2$. Therefore $h_{U_1}^{la}(n) \leq 2$. One can show that there exists a problem z over U_1 such that $\dim z = n$ and $h_{U_1}^{la}(z) \geq 2$. Therefore $h_{U_1}^{la}(n) = 2$.

Since the function $h_{U_1}^{ld}$ has the type of behavior LOG, the information system U_1 belongs to the class W_1^l—see Lemma 5.6 and Table 5.1. One can show that the information system U_1 satisfies the condition of reduction with parameter 2 and has finite I-dimension equal 1. □

Define an information system $U_2 = (A_2, F_2)$ as follows: $A_2 = \mathbb{N}$ and $F_2 = \{p_i : i \in \mathbb{N}\}$.

Lemma 5.8 *The information system U_2 belongs to the class W_2^l, $h_{U_2}^{ld}(n) = n$, $h_{U_2}^{la}(n) = n$, $L_{U_2}^{ld}(n) = 2(n+1)$, and $L_{U_2}^{la}(n) = 2(n+1)$ for any $n \in \mathbb{N}$. This information system does not satisfy the condition of reduction and has finite I-dimension equal* 1.

Proof It is easy to show that $N_{U_2}(n) = n + 1$ for any $n \in \mathbb{N}$. Using Proposition 5.5, we obtain $L_{U_2}^{ld}(n) = L_{U_2}^{la}(n) = 2(n+1)$ for any $n \in \mathbb{N}$.

Let $n \in \mathbb{N}$. It is clear that the system of equations

$$S_n = \{p_1(x) = 0, \ldots, p_n(x) = 0\}$$

is compatible on A_2 and each proper subsystem of the system S_n has the set of solutions different from the set of solutions of S_n. Therefore U_2 does not satisfy the condition of reduction. Using Propositions 5.1 and 5.2, we obtain $h_{U_2}^{ld}(n) = h_{U_2}^{la}(n) = n$ for any $n \in \mathbb{N}$.

Since the function $h_{U_2}^{ld}$ has the type of behavior LIN and the function $L_{U_2}^{ld}$ has the type of behavior POL, the information system U_2 belongs to the class W_2^l—see Lemma 5.6 and Table 5.1. One can show that this information system has finite I-dimension equal 1. □

Define an information system $U_3 = (A_3, F_3)$ as follows: $A_3 = \mathbb{N}$ and F_3 is the set of all functions from $\mathbb{N}$ to $\{0, 1\}$.

Lemma 5.9 *The information system U_3 belongs to the class W_3^l, $h_{U_3}^{ld}(n) = n$, $h_{U_3}^{la}(n) = n$, $L_{U_3}^{ld}(n) = 2^{n+1}$, and $L_{U_3}^{la}(n) = 2^{n+1}$ for any $n \in \mathbb{N}$. This information system does not satisfy the condition of reduction and has infinite I-dimension.*

Proof It is easy to show that the information system U_3 has infinite I-dimension. Using Lemma 5.5, we obtain that the system U_3 does not satisfy the condition of reduction. Let $n \in \mathbb{N}$. By Propositions 5.1 and 5.2, $h_{U_3}^{ld}(n) = h_{U_3}^{la}(n) = n$. By Propositions 5.3 and 5.4, $L_{U_3}^{ld}(n) = L_{U_3}^{la}(n) = 2^{n+1}$.

Since the function $L_{U_3}^{ld}$ has the type of behavior EXP, the information system U_3 belongs to the class W_3^l—see Lemma 5.6 and Table 5.1. □

Proof of Theorem 5.1. The statements of the theorem follow from Lemmas 5.6–5.9. □

5.5 Proofs of Theorems 5.2–5.4

First, we prove a number of auxiliary statements.

Lemma 5.10 *Let $U = (A, F)$ be an infinite binary information system. Then the information system U is ld-reachable.*

Proof Let $z = (\nu, f_1, \ldots, f_n)$ be a problem over U. Then there exists a decision tree Γ over z, which solves this problem deterministically and whose depth is at most $h_U^{ld}(n)$. By removal of some nodes and edges from Γ, we can obtain a decision tree Γ' over z, which solves the problem z deterministically and in which each working node has exactly two leaving edges and each complete path is realizable. Let ξ_1 and ξ_2 be different complete paths in Γ', $a_1 \in A(\xi_1)$, and $a_2 \in A(\xi_2)$. It is easy to show that $(f_1(a_1), \ldots, f_n(a_1)) \neq (f_1(a_2), \ldots, f_n(a_2))$. Therefore $L_t(\Gamma') \leq N_U(f_1, \ldots, f_n) \leq N_U(n)$. It is clear that $\Gamma' \in G_d^2(U)$. By Lemma 5.3, $L_w(\Gamma') = L_t(\Gamma') - 1$. Therefore $L(\Gamma') \leq 2N_U(n)$. By Proposition 5.5, $2N_U(n) = L_U^{ld}(n)$. Taking into account that $h(\Gamma') \leq h_U^{ld}(n)$ and z is an arbitrary problem over U with $\dim z = n$, we obtain that U is ld-reachable. □

Lemma 5.11 *Let U be an infinite binary information system such that $h_U^{la}(n) = n$ for any $n \in \mathbb{N}$. Then the information system U is la-reachable.*

Proof Let $z = (\nu, f_1, \ldots, f_n)$ be a problem over U and Γ be a decision tree over z that solves the problem z deterministically and satisfies the following conditions: the number of working nodes in each complete path of Γ is equal to n and these nodes in the order from the root to a terminal node are labeled with attributes $f_1, \ldots, f_n$. Remove from Γ all nodes and edges that do not belong to realizable complete paths. Let w be a working node in the obtained tree that has only one leaving edge d entering a node v. We remove the node w and edge d and connect the edge e entering w to the node v. We do the same with all working nodes with only one leaving edge. Denote by Γ' the obtained decision tree. It is clear that Γ' solves the problem z deterministically and hence nondeterministically, $\Gamma' \in G_d^2(U)$, and $L_t(\Gamma') \leq N_U(f_1, \ldots, f_n) \leq N_U(n)$. By Lemma 5.3, $L_w(\Gamma') = L_t(\Gamma') - 1$. Therefore $L(\Gamma') \leq 2N_U(n)$. Using Proposition 5.5, we obtain $L(\Gamma') \leq L_U^{la}(n)$. It is clear that $h(\Gamma') \leq n = h_U^{la}(n)$. Therefore U is la-reachable. □

Lemma 5.12 *Let U be an infinite binary information system, which satisfies the condition of reduction. Then the information system U is not la-reachable.*

Proof By Proposition 5.2, the function $h_U^{la}(n)$ is bounded from above by a positive constant c. By Proposition 5.6, the function $N_U(n)$ is not bounded from above by a constant. Choose $n \in \mathbb{N}$ such that $N_U(n) > 2^{2c}$. Let $z = (\nu, f_1, \ldots, f_n)$ be a problem over U such that $\nu(\bar{\delta}_1) \neq \nu(\bar{\delta}_2)$ for any $\bar{\delta}_1, \bar{\delta}_2 \in \{0, 1\}^n$, $\bar{\delta}_1 \neq \bar{\delta}_2$, and $N_U(f_1, \ldots, f_n) = N_U(n)$. Let Γ be a decision tree over z, which solves the problem z nondeterministically, for which $h(\Gamma) \leq h_U^{la}(n) \leq c$, and which has the minimum number of nodes among such trees. In the same way as it was done in the proof of Lemma 5.2, we can prove that $\Gamma \in G_a^f(U)$. It is clear that $L_t(\Gamma) \geq N_U(f_1, \ldots, f_n) = N_U(n)$. Let us assume that $\Gamma \in G_d^2(U)$. Then it is easy to show that $h(\Gamma) \geq \log_2 L_t(\Gamma) \geq \log_2 N_U(n) > 2c$, which is impossible by the choice of Γ. Therefore $\Gamma \in G_a^f(U) \setminus G_d^2(U)$. By Lemma 5.4, $L_w(\Gamma) > L_t(\Gamma) - 1 \geq N_U(n) - 1$. Using Proposition 5.5, we obtain $L(\Gamma) > 2N_U(n) = L_U^{la}(n)$. Therefore U is not la-reachable. □

Lemma 5.13 *Let $U = (A, F)$ be an infinite binary information system, which satisfies the condition of reduction with parameter m. Then $(m, (m+1)L_U^{la}(n)/2 + 1)$ is a boundary la-pair of the system U.*

Proof Let $z = (\nu, f_1, \ldots, f_n)$ be a problem over U. We now describe a decision tree Γ over z, which solves the problem z nondeterministically and for which $h(\Gamma) \leq m$ and $L(\Gamma) \leq (m+1)L_U^{la}(n)/2 + 1$. For each tuple $\bar{\delta} = (\delta_1, \ldots, \delta_n) \in \{0, 1\}^n$ for which the system of equations

$$S_{\bar{\delta}} = \{f_1(x) = \delta_1, \ldots, f_n(x) = \delta_n\}$$

has a solution from A, we describe a complete path $\xi_{\bar{\delta}}$. Since the information system U satisfies the condition of reduction with parameter m, there exists a subsystem

$$S'_{\bar{\delta}} = \{f_{i_1}(x) = \delta_{i_1}, \ldots, f_{i_t}(x) = \delta_{i_t}\}$$

of the system $S_{\bar{\delta}}$, which has the same set of solutions and for which $t \leq m$. Then

$$\xi_{\bar{\delta}} = v_0, d_0, v_1, d_1, \ldots, v_t, d_t, v_{t+1} ,$$

where the node v_0 and the edge d_0 are not labeled, for $j = 1, \ldots, t$, the node v_j is labeled with the attribute f_{i_j} and the edge d_j is labeled with the number δ_{i_j}, and the node v_{t+1} is labeled with the number $\nu(\bar{\delta})$. We merge initial nodes of all such complete paths and denote by Γ the obtained tree. One can show that Γ is a decision tree over z, which solves the problem z nondeterministically and for which $h(\Gamma) \leq m$. The number of the considered complete paths is equal to $N_U(f_1, \ldots, f_n) \leq N_U(n)$. The number of nodes in each complete paths is at most $m + 2$. Therefore $L(\Gamma) \leq (m+1)N_U(n) + 1$. By Proposition 5.5, $N_U(n) = L_U^{la}(n)/2$. Hence $L(\Gamma) \leq (m+1)L_U^{la}(n)/2 + 1$. Thus, $(m, (m+1)L_U^{la}(n)/2 + 1)$ is a boundary la-pair of the system U. □

Proof of Theorem 5.2. Each information system from the class W_1^l satisfies the condition of reduction (see Table 5.3).

(a) Let U be an information system from the class W_1^l. Using Lemma 5.10, we obtain that the system U is ld-reachable.

(b) Let U be an information system from the class W_1^l. Then, for some $m \in \mathbb{N}$, the system U satisfies the condition of reduction with parameter m. Using Lemma 5.12, we obtain that the system U is not la-reachable. Using Lemma 5.13, we obtain that $(m, (m+1)L_U^{la}(n)/2 + 1)$ is a boundary la-pair of the system U. □

Proof of Theorem 5.3. Each information system from the class W_2^l does not satisfy the condition of reduction (see Table 5.3).

(a) Let U be an information system from the class W_2^l. Using Lemma 5.10, we obtain that the system U is ld-reachable.

(b) Let U be an information system from the class W_2^l. By Proposition 5.2, $h_U^{la}(n) = n$ for any $n \in \mathbb{N}$. Using Lemma 5.11, we obtain that the system U is la-reachable. □

Proof of Theorem 5.4. Each information system from the class W_3^l does not satisfy the condition of reduction (see Table 5.3).

(a) Let U be an information system from the class W_3^l. Using Lemma 5.10, we obtain that the system U is ld-reachable.

(b) Let U be an information system from the class W_3^l. By Proposition 5.2, $h_U^{la}(n) = n$ for any $n \in \mathbb{N}$. Using Lemma 5.11, we obtain that the system U is la-reachable. □

References

1. AbouEisha, H., Amin, T., Chikalov, I., Hussain, S., Moshkov, M.: Extensions of Dynamic Programming for Combinatorial Optimization and Data Mining. In: Intelligent Systems Reference Library, vol. 146. Springer, Berlin (2019)
2. Breiman, L., Friedman, J.H., Olshen, R.A., Stone, C.J.: Classification and Regression Trees. Wadsworth and Brooks (1984)
3. Durdymyradov, K., Moshkov, M.: A local approach to studying the time and space complexity of deterministic and nondeterministic decision trees. CoRR (2023). arXiv:2311.17306
4. Durdymyradov, K., Moshkov, M.: Time and space complexity of deterministic and nondeterministic decision trees: local approach. In: IEEE International Conference on Big Data, BigData 2023, Sorrento, Italy, December 15–18, 2023, pp. 6040–6045. IEEE (2023)
5. Molnar, C.: Interpretable Machine Learning: A Guide for Making Black Box Models Explainable, 2nd edn (2022). https://christophm.github.io/interpretable-ml-book/
6. Moshkov, M.: Time complexity of decision trees. In: Peters, J.F., Skowron, A. (eds.) Transactions on Rough Sets III. Lecture Notes in Computer Science, vol. 3400, pp. 244–459. Springer, Berlin (2005)
7. Moshkov, M.: On the class of restricted linear information systems. Discret. Math. **307**(22), 2837–2844 (2007)
8. Moshkov, M., Zielosko, B.: Combinatorial Machine Learning—A Rough Set Approach. Studies in Computational Intelligence, vol. 360. Springer, Berlin (2011)
9. Naiman, D.Q., Wynn, H.P.: Independence number and the complexity of families of sets. Discr. Math. **154**, 203–216 (1996)
10. Pawlak, Z.: Information systems theoretical foundations. Inf. Syst. **6**(3), 205–218 (1981)
11. Sauer, N.: On the density of families of sets. J. Comb. Theory (A) **13**, 145–147 (1972)
12. Shelah, S.: A combinatorial problem; stability and order for models and theories in infinitary languages. Pac. J. Math. **41**, 241–261 (1972)

Part II
Decision Tables from Closed Classes

In this part, we consider decision tables from classes of tables closed under the operations of changing of decisions and removal of attributes (columns), and sometimes some other operations. For decision tables with many-valued decisions and conventional decision tables, we study relationships among three parameters: the minimum complexity of a deterministic decision tree, the minimum complexity of a nondeterministic decision tree, and the complexity of the set of attributes attached to columns of the table (in one of the chapters, instead of this parameter, we consider close to it parameter called the complexity of the decision table). For decision tables with 0-1-decisions, we study relationships among four parameters: the minimum complexity of a deterministic decision tree, the minimum complexity of a strongly nondeterministic decision tree, the minimum complexity of a test, and the complexity of the set of attributes attached to columns of the table. This part consists of four chapters.

In Chap. 6, we consider classes of decision tables with many-valued decisions closed under operations of removal of columns, changing of decisions, permutation of columns, and duplication of columns. We study rough classification of relationships for each ordered pair of the following three parameters of decision tables: the minimum complexity of a deterministic decision tree, the minimum complexity of a nondeterministic decision tree, and the complexity of the table.

In Chap. 7, we consider classes of decision tables with many-valued decisions closed under operations of removal of columns and changing of decisions. We study the dependence of the minimum complexity of deterministic and nondeterministic decision trees on the complexity of the set of attributes attached to columns. We also study the dependence of the minimum complexity of deterministic decision trees on the minimum complexity of nondeterministic decision trees.

In Chap. 8, we consider classes of conventional decision tables closed under operations of removal of columns and changing of decisions. We study the dependence of the minimum complexity of deterministic and nondeterministic decision trees on the complexity of the set of attributes attached to columns. We also study the dependence of the minimum complexity of deterministic decision trees on the minimum complexity of nondeterministic decision trees. In addition, we consider the dependence of the complexity of deterministic and nondeterministic decision trees constructed by some algorithms on the complexity of the set of attributes attached to columns.

In Chap. 9, we consider classes of decision tables with 0-1-decisions closed under operations of removal of columns and changing of decisions. We study the dependence of the minimum complexity of deterministic decision trees on various parameters of the tables: the minimum complexity of a test, the complexity of the set of attributes attached to columns, and the minimum complexity of a strongly nondeterministic decision tree. We also study the dependence of the minimum complexity of strongly nondeterministic decision trees on the complexity of the set of attributes attached to columns.

Chapter 6
Comparative Analysis of Deterministic and Nondeterministic Decision Trees for Decision Tables from Closed Classes

6.1 Introduction

In this chapter, we consider closed classes of decision tables with many-valued decisions and study relationships among three parameters of these tables: the complexity of a decision table (if we consider the depth of decision trees, then the complexity of a decision table is the number of columns in it), the minimum complexity of a deterministic decision tree, and the minimum complexity of a nondeterministic decision tree for this table.

A decision table with many-valued decisions is a rectangular table in which columns are labeled with attributes, rows are pairwise different and each row is labeled with a nonempty finite set of decisions. Rows are interpreted as tuples of values of the attributes. For a given row, it is required to find a decision from the set of decisions attached to the row. To this end, we can use the following queries: we can choose an attribute and ask what is the value of this attribute in the considered row. We study two types of algorithms based on these queries: deterministic and nondeterministic decision trees. One can interpret nondeterministic decision trees for a decision table as a way to represent an arbitrary system of true decision rules for this table that cover all rows. To evaluate time complexity of decision trees, we consider both general complexity measures and limited complexity measures including depth and weighted depth of decision trees.

Decision tables with many-valued decisions often appear in data analysis, where they are known as multi-label decision tables [2, 9, 10]. Moreover, decision tables with many-valued decisions are common in such areas as combinatorial optimization, computational geometry, and fault diagnosis, where they are used to represent and explore problems [1, 5].

We study classes of decision tables with many-valued decisions closed under four operations: removal of columns, changing of decisions, permutation of columns, and duplication of columns. The most natural examples of such classes are closed classes of decision tables generated by information systems [8]. An information system

K. Durdymyradov et al., *Decision Trees Versus Systems of Decision Rules*,
Studies in Big Data 160, https://doi.org/10.1007/978-3-031-71586-0_6

consists of a set of objects (universe) and a set of attributes (functions) defined on the universe and with values from a finite set. A problem with many-valued decisions over an information system is specified by a finite number of attributes that divide the universe into nonempty domains in which these attributes have fixed values. A nonempty finite set of decisions is attached to each domain. For a given object from the universe, it is required to find a decision from the set attached to the domain containing this object.

A decision table with many-valued decisions corresponds to this problem in a natural way: columns of this table are labeled with the considered attributes, rows correspond to domains and are labeled with sets of decisions attached to domains. The set of decision tables corresponding to problems with many-valued decisions over an information system forms a closed class generated by this system. Note that the family of all closed classes is essentially wider than the family of closed classes generated by information systems. In particular, the union of two closed classes generated by two information systems is a closed class. However, generally, there is no an information system that generates this class.

In the present chapter, we study so-called cm-pairs $(\mathscr{C}, \psi)$, where $\mathscr{C}$ is a class of decision tables closed under the considered four operations and ψ is a complexity measure for this class. The cm-pair is called limited if ψ is a limited complexity measure. For any decision table $T \in \mathscr{C}$, we have three parameters:

- $\psi^i(T)$–the complexity of the decision table T. This parameter is equal to the complexity of a deterministic decision tree for the table T, which sequentially computes values of all attributes attached to columns of T.
- $\psi^d(T)$–the minimum complexity of a deterministic decision tree for the table T.
- $\psi^a(T)$–the minimum complexity of a nondeterministic decision tree for the table T.

We investigate the relationships between any two of these parameters for decision tables from $\mathscr{C}$. Let us consider, for example, the parameters $\psi^i(T)$ and $\psi^d(T)$. Let $n \in \omega = \{0, 1, 2, \ldots\}$. We will study relations of the kind $\psi^i(T) \le n \Rightarrow \psi^d(T) \le u$, which are true for any table $T \in \mathscr{C}$. The minimum value of u is the most interesting for us. This value (if exists) is equal to

$$\mathscr{U}^{di}_{\mathscr{C}\psi}(n) = \max\left\{\psi^d(T) : T \in \mathscr{C}, \psi^i(T) \le n\right\} .$$

We will also study relations of the kind $\psi^i(T) \ge n \Rightarrow \psi^d(T) \ge l$. In this case, the maximum value of l is the most interesting for us. This value (if exists) is equal to

$$\mathscr{L}^{di}_{\mathscr{C}\psi}(n) = \min\left\{\psi^d(T) : T \in \mathscr{C}, \psi^i(T) \ge n\right\} .$$

The two functions $\mathscr{U}^{di}_{\mathscr{C}\psi}$ and $\mathscr{L}^{di}_{\mathscr{C}\psi}$ describe how the behavior of the parameter $\psi^d(T)$ depends on the behavior of the parameter $\psi^i(T)$ for tables from $\mathscr{C}$.

There are 18 such functions for all ordered pairs of parameters $\psi^i(T)$, $\psi^d(T)$, and $\psi^a(T)$. These 18 functions well describe the relationships among the considered parameters. It would be very interesting to list 18-tuples of these functions for all cm-pairs and all limited cm-pairs, but it is most likely impossible.

In this chapter, instead of functions, we will study types of functions. With any partial function $f : \omega \to \omega$, we will associate its type from the set $\{\alpha, \beta, \gamma, \delta, \epsilon\}$. For example, if the function f has an infinite domain, and it is bounded from above, then its type is equal to α. If the function f has an infinite domain, is not bounded from above, and the inequality $f(n) \geq n$ holds for a finite number of $n \in \omega$, then its type is equal to β, etc. Thus, we will list 18-tuples of types of functions. These tuples will be represented in tables called the types of cm-pairs. We will prove that there are only seven realizable types of cm-pairs and only five realizable types of limited cm-pairs.

First, we will study 9-tuples of types of functions $\mathcal{U}^{bc}_{\mathcal{C}\psi}$, $b, c \in \{i, d, a\}$. These tuples will be represented in tables called upper types of cm-pairs. We will list all realizable upper types of cm-pairs and limited cm-pairs. After that, we will extend the results obtained for upper types of cm-pairs to the case of types of cm-pairs. We will also define the notion of a union of two cm-pairs and study the upper type of the resulting cm-pair depending on the upper types of the initial cm-pairs.

This chapter is based on Chap. 3 in which similar results are obtained for classes of problems over information systems and which is based on the paper [3]. We generalize proofs from Chap. 3 to the case of decision tables from closed classes and use some results from this chapter to prove the existence of cm-pairs and limited cm-pairs with given upper types.

The results considered in this chapter are published in [6, 7]. The chapter consists of seven sections. In Sect. 6.2, basic definitions are considered. In Sect. 6.3, we provide the main results related to types of cm-pairs and limited cm-pairs. In Sects. 6.4–6.6, we study upper types of cm-pairs and limited cm-pairs. Section 6.7 contains proofs of the main results.

6.2 Basic Definitions

6.2.1 Decision Tables and Closed Classes

Let $\omega = \{0, 1, 2, \ldots\}$ be the set of nonnegative integers. For any $k \in \omega \setminus \{0, 1\}$, let $E_k = \{0, 1, \ldots, k-1\}$. The set of nonempty finite subsets of the set ω will be denoted by $\mathcal{P}(\omega)$. Let F be a nonempty set of *attributes* (really, names of attributes).

Definition 6.1 We now define the set of decision tables $\mathcal{M}^{\infty}_{k}(F)$ (index ∞ refers to decision tables with many-valued decisions). An arbitrary *decision table* T from this set is a rectangular table with $n \in \omega \setminus \{0\}$ columns labeled with attributes $f_1, \ldots, f_n \in F$, where any two columns labeled with the same attribute are equal. The rows of this table are pairwise different and are filled with numbers from E_k. Each row is interpreted as a tuple of values of attributes $f_1, \ldots, f_n$. For each row in the table, a set from $\mathscr{P}(\omega)$ is attached, which is interpreted as a set of decisions for this row.

Example 6.1 Three decision tables T_1, T_2, and T_3 from the set $\mathscr{M}_2^\infty(F_0)$, where $F_0 = \{f_1, f_2, f_3\}$, are shown in Fig. 6.1.

We correspond to the table T the following *problem*: for a given row of T, we should recognize a decision from the set of decisions attached to this row. To this end, we can use queries about the values of attributes for this row.

We denote by $\mathrm{At}(T)$ the set $\{f_1, \ldots, f_n\}$ of attributes attached to the columns of T. By $\Pi(T)$, we denote the intersection of the sets of decisions attached to the rows of T, and by $\Delta(T)$, we denote the set of rows of the table T. Decisions from $\Pi(T)$ are called *common decisions* for T. The table T will be called *degenerate* if $\Delta(T) = \varnothing$ or $\Pi(T) \neq \varnothing$. We denote by $\mathscr{M}_k^{\infty c}(F)$ the set of degenerate decision tables from $\mathscr{M}_k^\infty(F)$.

Example 6.2 Two degenerate decision tables D_1 and D_2 are shown in Fig. 6.2.

Definition 6.2 A *subtable* of the table T is a table obtained from T by removal of some of its rows. Let $\Omega(T) = \{(f, \delta) : f \in \mathrm{At}(T), \delta \in E_k\}$ and $\Omega^*(T)$ be the set of all finite words in the alphabet $\Omega(T)$ including the empty word λ. Let $\alpha \in \Omega^*(T)$. We now define a subtable $T\alpha$ of the table T. If $\alpha = \lambda$, then $T\alpha = T$. Let $\alpha = (f_{i_1}, \delta_1) \cdots (f_{i_m}, \delta_m)$. Then $T\alpha$ consists of all rows of T that in the intersection with columns $f_{i_1}, \ldots, f_{i_m}$ have values $\delta_1, \ldots, \delta_m$, respectively.

Example 6.3 Two subtables of the tables T_1 and T_2 (depicted in Fig. 6.1) are shown in Fig. 6.3.

We now define four operations on the set $\mathscr{M}_k^\infty(F)$ of decision tables.

$T_1 =$

f_1	f_2	
0	0	{1}
1	0	{2,3}
0	1	{2}
1	1	{4}

$T_2 =$

f_1	f_2	f_3	
1	0	0	{1,2}
0	1	0	{1,3}
0	0	1	{4}
0	0	0	{1,2,3}

$T_3 =$

f_1	f_1	f_3	
0	0	0	{1,3}
1	1	0	{1}
0	0	1	{2}
1	1	1	{1,2}

Fig. 6.1 Decision tables T_1, T_2, and T_3

$D_1 =$

f_1	f_2	

$D_2 =$

f_1	f_2	f_3	
1	0	0	{1,2}
0	1	0	{1,3}
0	0	0	{1,2,3}

Fig. 6.2 Degenerate decision tables D_1 and D_2

$T_1(f_1,1) =$

f_1	f_2	
1	0	{2,3}
1	1	{4}

$T_2(f_1,0)(f_2,0)(f_3,0) =$

f_1	f_2	f_3	
0	0	0	{1,2,3}

Fig. 6.3 Subtables $T_1(f_1, 1)$ and $T_2(f_1, 0)(f_2, 0)(f_3, 0)$ of tables T_1 and T_2 shown in Fig. 6.1

Definition 6.3 (*Removal of columns*) We can remove an arbitrary column in a table T with at least two columns. As a result, the obtained table can have groups of equal rows. We keep only the first row in each such group.

Definition 6.4 (*Changing of decisions*) In a given table T, we can change in an arbitrary way sets of decisions attached to rows.

Definition 6.5 (*Permutation of columns*) We can swap any two columns in a table T, including the attached attribute names.

Definition 6.6 (*Duplication of columns*) For any column in a table T, we can add its duplicate next to that column.

Example 6.4 Decision tables T_1', T_2', T_1'', and T_2'' depicted in Fig. 6.4 are obtained from decision tables T_1 and T_2 shown in Fig. 6.1 by operations of changing the decisions, removal of columns, permutation of columns, and duplication of columns, respectively.

Definition 6.7 Let $T \in \mathcal{M}_k^\infty(F)$. The closure of the table T is a set, which contains all tables that can be obtained from T by the operations of removal of columns, changing of decisions, permutation of columns, and duplication of columns and only such tables. We denote the closure of the table T by $[T]$. It is clear that $T \in [T]$.

$T_1' =$

f_1	f_2	
0	0	{1,4}
1	0	{2,3}
0	1	{3}
1	1	{4}

$T_2' =$

f_1	
1	{1,2}
0	{1,3}

$T_1'' =$

f_2	f_1	
0	0	{1}
0	1	{2,3}
1	0	{2}
1	1	{4}

$T_2'' =$

f_1	f_2	f_2	f_3	
1	0	0	0	{1,2}
0	1	1	0	{1,3}
0	0	0	1	{4}
0	0	0	0	{1,2,3}

Fig. 6.4 Decision tables T_1', T_2', T_1'', and T_2'' obtained from tables T_1 and T_2 shown in Fig. 6.1 by operations of changing the decisions, removal of columns, permutation of columns and duplication of columns, respectively

Definition 6.8 Let $\mathscr{C} \subseteq \mathscr{M}_k^\infty(F)$. The closure $[\mathscr{C}]$ of the set $\mathscr{C}$ is defined in the following way: $[\mathscr{C}] = \bigcup_{T \in \mathscr{C}} [T]$. We will say that $\mathscr{C}$ is a closed class if $\mathscr{C} = [\mathscr{C}]$. In particular, the empty set of tables is a closed class.

Example 6.5 We now consider a closed class $\mathscr{C}_0$ of decision tables from the set $\mathscr{M}_2^\infty(\{f_1, f_2\})$, which is equal to $[Q]$, where the decision table Q is depicted in Fig. 6.5. The closed class $\mathscr{C}_0$ contains all tables depicted in Fig. 6.6 and all tables that can be obtained from them by operations of duplication of columns and permutation of columns.

If $\mathscr{C}_1$ and $\mathscr{C}_2$ are closed classes belonging to $\mathscr{M}_k^\infty(F)$, then $\mathscr{C}_1 \cup \mathscr{C}_2$ is also a closed class. We can consider closed classes $\mathscr{C}_1$ and $\mathscr{C}_2$ belonging to different sets of decision tables. Let $\mathscr{C}_1 \subseteq \mathscr{M}_{k_1}^\infty(F_1)$ and $\mathscr{C}_2 \subseteq \mathscr{M}_{k_2}^\infty(F_2)$. Then $\mathscr{C}_1 \cup \mathscr{C}_2$ is a closed class and $\mathscr{C}_1 \cup \mathscr{C}_2 \subseteq \mathscr{M}_{\max(k_1,k_2)}^\infty(F_1 \cup F_2)$.

6.2.2 Deterministic and Nondeterministic Decision Trees

A *finite directed tree with the root* is a finite directed tree in which exactly one node has no entering edges. This node is called the *root*. Nodes of the tree, which have no outgoing edges are called *terminal* nodes. Nodes that are neither the root nor the terminal are called *working* nodes. A *complete path* in a finite directed tree with the root is any sequence of nodes and edges starting from the root node and ending with a terminal node $\xi = v_0, d_0, \ldots, v_m, d_m, v_{m+1}$, where d_i is the edge outgoing from the node v_i and entering the node v_{i+1}, $i = 0, \ldots, m$.

Definition 6.9 A *decision tree over the set of decision tables* $\mathscr{M}_k^\infty(F)$ is a labeled finite directed tree with the root with at least two nodes (the root and a terminal node) possessing the following properties:

$Q =$

f_1	f_2	
1	0	$\{1\}$
0	1	$\{2\}$
0	0	$\{3\}$

Fig. 6.5 Decision table Q

$Q_1 =$

f_1	f_2	
1	0	d_1
0	1	d_2
0	0	d_3

$Q_2 =$

f_1	
1	d_4
0	d_5

$Q_3 =$

f_2	
0	d_6
1	d_7

Fig. 6.6 Decision tables from closed class $\mathscr{C}_0$, where $d_1, \ldots, d_7 \in \mathscr{P}(\omega)$

- The root and the edges outgoing from the root are not labeled.
- Each working node is labeled with an attribute from the set F.
- Each edge outgoing from a working node is labeled with a number from E_k.
- Each terminal node is labeled with a number from ω.

We denote by $\mathcal{T}_k(F)$ the set of decision trees over the set of decision tables $\mathcal{M}_k^\infty(F)$.

Definition 6.10 A decision tree from $\mathcal{T}_k(F)$ is called *deterministic* if it satisfies the following conditions:

- Exactly one edge leaves the root.
- Edges outgoing from each working node are labeled with pairwise different numbers.

Let Γ be a decision tree from $\mathcal{T}_k(F)$. Denote by $\mathrm{At}(\Gamma)$ the set of attributes attached to working nodes of Γ. Set $\Omega(\Gamma) = \{(f, \delta) : f \in \mathrm{At}(\Gamma), \delta \in E_k\}$. Denote by $\Omega^*(\Gamma)$ the set of all finite words in the alphabet $\Omega(\Gamma)$ including the empty word λ. We correspond to an arbitrary complete path $\xi = v_0, d_0, \ldots, v_m, d_m, v_{m+1}$ in Γ, a word $\pi(\xi)$. If $m = 0$, then $\pi(\xi) = \lambda$. Let $m > 0$ and, for $i = 1, \ldots, m$, the node v_i be labeled with an attribute f_{j_i} and the edge d_i be labeled with the number δ_i. Then $\pi(\xi) = (f_{j_1}, \delta_1) \cdots (f_{j_m}, \delta_m)$. We denote by $\tau(\xi)$ the number attached to the terminal node of the path ξ. We denote by $\mathrm{Path}(\Gamma)$ the set of complete paths in the tree Γ.

Definition 6.11 Let $T \in \mathcal{M}_k^\infty(F)$. A *nondeterministic decision tree* for the table T is a decision tree Γ over $\mathcal{M}_k^\infty(F)$ satisfying the following conditions:

- $\mathrm{At}(\Gamma) \subseteq \mathrm{At}(T)$.
- $\bigcup_{\xi \in \mathrm{Path}(\Gamma)} \Delta(T\pi(\xi)) = \Delta(T)$.
- For any row $r \in \Delta(T)$ and any complete path $\xi \in \mathrm{Path}(\Gamma)$, if $r \in \Delta(T\pi(\xi))$, then $\tau(\xi)$ belongs to the set of decisions attached to the row r.

Example 6.6 Nondeterministic decision trees Γ_1 and Γ_2 for decision tables T_1 and T_2 shown in Fig. 6.1 are depicted in Fig. 6.7.

Definition 6.12 A *deterministic decision tree* for the table T is a deterministic decision tree over $\mathcal{M}_k^\infty(F)$, which is a nondeterministic decision tree for the table T.

Example 6.7 Deterministic decision trees Γ_1' and Γ_2' for decision tables T_1 and T_2 shown in Fig. 6.1 are depicted in Fig. 6.8.

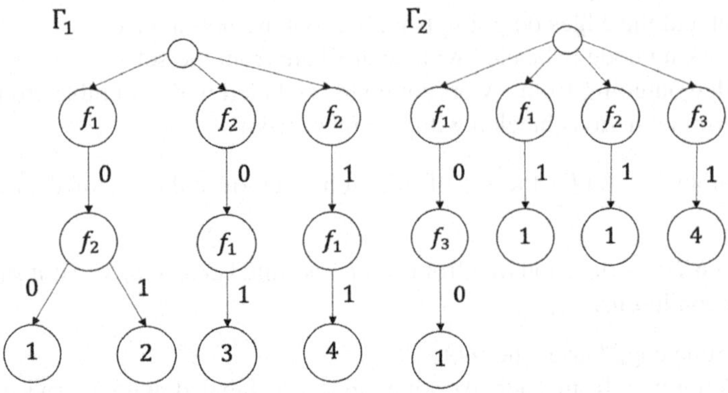

Fig. 6.7 Nondeterministic decision trees Γ_1 and Γ_2 for decision tables T_1 and T_2 depicted in Fig. 6.1

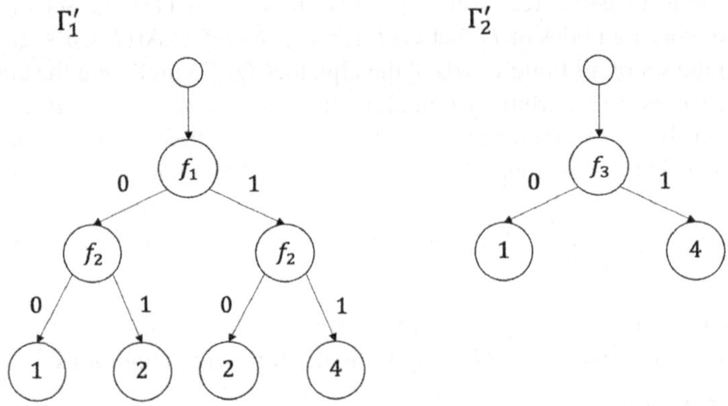

Fig. 6.8 Deterministic decision trees Γ_1' and Γ_2' for decision tables T_1 and T_2 depicted in Fig. 6.1

6.2.3 *Complexity Measures*

Denote by F^* the set of all finite words over the alphabet F including the empty word λ.

Definition 6.13 A *complexity measure* over the set of decision tables $\mathscr{M}_k^\infty(F)$ is any mapping $\psi : F^* \to \omega$.

Definition 6.14 The complexity measure ψ will be called *limited* if it possesses the following properties:

(a) $\psi(\alpha_1\alpha_2) \leq \psi(\alpha_1) + \psi(\alpha_2)$ for any $\alpha_1, \alpha_2 \in F^*$.
(b) $\psi(\alpha_1\alpha_2\alpha_3) \geq \psi(\alpha_1\alpha_3)$ for any $\alpha_1, \alpha_2, \alpha_3 \in F^*$.
(c) For any $\alpha \in F^*$, the inequality $\psi(\alpha) \geq |\alpha|$ holds, where $|\alpha|$ is the length of α.

We extend an arbitrary complexity measure ψ onto the set $\mathscr{T}_k(F)$ in the following way. Let $\Gamma \in \mathscr{T}_k(F)$. Then $\psi(\Gamma) = \max\{\psi(\varphi(\xi)) : \xi \in \text{Path}(\Gamma)\}$, where $\varphi(\xi) = \lambda$

if $\pi(\xi) = \lambda$ and $\varphi(\xi) = f_1 \cdots f_m$ if $\pi(\xi) = (f_1, \delta_1) \cdots (f_m, \delta_m)$. The value $\psi(\Gamma)$ will be called the *complexity of the decision tree* Γ.

We now consider an example of a complexity measure. Let $w : F \to \omega \setminus \{0\}$. We define the function $\psi^w : F^* \to \omega$ in the following way: $\psi^w(\alpha) = 0$ if $\alpha = \lambda$ and $\psi^w(\alpha) = \sum_{i=1}^m w(f_i)$ if $\alpha = f_1 \cdots f_m$. The function ψ^w is a limited complexity measure over $\mathscr{M}_k^\infty(F)$ and it is called a *weighted depth*. If $w \equiv 1$, then the function ψ^w is called the *depth* and is denoted by h.

Let ψ be a complexity measure over $\mathscr{M}_k^\infty(F)$ and T be a decision table from $\mathscr{M}_k^\infty(F)$ in which rows are labeled with attributes $f_1, \dots, f_n$. The value $\psi^i(T) = \psi(f_1 \cdots f_n)$ will be called the *complexity of the decision table* T. We denote by $\psi^d(T)$ the minimum complexity of a deterministic decision tree for the table T. We denote by $\psi^a(T)$ the minimum complexity of a nondeterministic decision tree for the table T.

6.2.4 Information Systems

Let A be a nonempty set and F be a nonempty set of functions from A to E_k.

Definition 6.15 Functions from F are called *attributes* and the pair $U = (A, F)$ is called an *information system*.

Definition 6.16 A *problem over* U is any $(n+1)$-tuple $z = (\nu, f_1, \dots, f_n)$, where $n \in \omega \setminus \{0\}$, $\nu : E_k^n \to \mathscr{P}(\omega)$, and $f_1, \dots, f_n \in F$.

The problem z can be interpreted as a problem of searching for at least one number from the set $z(a) = \nu(f_1(a), \dots, f_n(a))$ for a given $a \in A$. We denote by $\mathrm{Probl}^\infty(U)$ the set of problems over the information system U.

We correspond to the problem z a decision table $T(z) \in \mathscr{M}_k^\infty(F)$. This table has n columns labeled with attributes $f_1, \dots, f_n$. A tuple $\bar{\delta} = (\delta_1, \dots, \delta_n) \in E_k^n$ is a row of the table $T(z)$ if and only if the system of equations

$$\{f_1(x) = \delta_1, \dots, f_n(x) = \delta_n\}$$

has a solution from the set A. This row is labeled with the set of decisions $\nu(\bar{\delta})$. Let $\mathrm{Tab}^\infty(U) = \{T(z) : z \in \mathrm{Probl}^\infty(U)\}$. One can show the set $\mathrm{Tab}^\infty(U)$ is a closed class of decision tables.

Closed classes of decision tables based on information systems are the most natural examples of closed classes. However, the notion of a closed class is essentially wider. In particular, the union $\mathrm{Tab}^\infty(U_1) \cup \mathrm{Tab}^\infty(U_2)$, where U_1 and U_2 are information systems, is a closed class, but generally, we cannot find an information system U such that $\mathrm{Tab}^\infty(U) = \mathrm{Tab}^\infty(U_1) \cup \mathrm{Tab}^\infty(U_2)$.

6.2.5 Types of CM-Pairs

First, we define the notion of cm-pair.

Definition 6.17 A pair $(\mathscr{C}, \psi)$ where $\mathscr{C}$ is a closed class of decision tables from $\mathscr{M}_k^\infty(F)$ and ψ is a complexity measure over $\mathscr{M}_k^\infty(F)$ will be called a *class-measure-pair* (or, *cm-pair*, in short). If ψ is a limited complexity measure then cm-pair $(\mathscr{C}, \psi)$ will be called a *limited cm-pair*.

Let $(\mathscr{C}, \psi)$ be a cm-pair. We have three parameters $\psi^i(T)$, $\psi^d(T)$ and $\psi^a(T)$ for any decision table $T \in \mathscr{C}$. We now define functions that describe relationships among these parameters. Let $b, c \in \{i, d, a\}$.

Definition 6.18 We define partial functions $\mathscr{U}^{bc}_{\mathscr{C}\psi} : \omega \to \omega$ and $\mathscr{L}^{bc}_{\mathscr{C}\psi} : \omega \to \omega$ by

$$\mathscr{U}^{bc}_{\mathscr{C}\psi}(n) = \max\left\{\psi^b(T) : T \in \mathscr{C}, \psi^c(T) \le n\right\} ,$$
$$\mathscr{L}^{bc}_{\mathscr{C}\psi}(n) = \min\left\{\psi^b(T) : T \in \mathscr{C}, \psi^c(T) \ge n\right\} .$$

If the value $\mathscr{U}^{bc}_{\mathscr{C}\psi}(n)$ is definite, then it is the unimprovable upper bound on the values $\psi^b(T)$ for tables $T \in \mathscr{C}$ satisfying $\psi^c(T) \le n$. If the value $\mathscr{L}^{bc}_{\mathscr{C}\psi}(n)$ is definite, then it is the unimprovable lower bound on the values $\psi^b(T)$ for tables $T \in \mathscr{C}$ satisfying $\psi^c(T) \ge n$.

Let g be a partial function from ω to ω. We denote by $\text{Dom}(g)$ the domain of g. Denote $\text{Dom}^+(g) = \{n : n \in \text{Dom}(g), g(n) \ge n\}$ and $\text{Dom}^-(g) = \{n : n \in \text{Dom}(g), g(n) \le n\}$.

Definition 6.19 We now define the value $\text{typ}(g) \in \{\alpha, \beta, \gamma, \delta, \epsilon\}$ called the *type* of g.

- If $\text{Dom}(g)$ is an infinite set and g is a bounded from above function, then $\text{typ}(g) = \alpha$.
- If $\text{Dom}(g)$ is an infinite set, $\text{Dom}^+(g)$ is a finite set, and g is an unbounded from above function, then $\text{typ}(g) = \beta$.
- If both sets $\text{Dom}^+(g)$ and $\text{Dom}^-(g)$ are infinite, then $\text{typ}(g) = \gamma$.
- If $\text{Dom}(g)$ is an infinite set and $\text{Dom}^-(g)$ is a finite set, then $\text{typ}(g) = \delta$.
- If $\text{Dom}(g)$ is a finite set, then $\text{typ}(g) = \epsilon$.

Example 6.8 One can show that $\text{typ}(1) = \alpha$, $\text{typ}(\lceil \log_2 n \rceil) = \beta$, $\text{typ}(n) = \gamma$, $\text{typ}(n^2) = \delta$, and $\text{typ}(\frac{1}{\lfloor 1/n \rfloor}) = \epsilon$.

Definition 6.20 We now define the table $\text{typ}(\mathscr{C}, \psi)$, which will be called the *type of cm-pair* $(\mathscr{C}, \psi)$. This is a table with three rows and three columns in which rows from the top to the bottom and columns from the left to the right are labeled with indices i, d, a. The pair $\text{typ}(\mathscr{L}^{bc}_{\mathscr{C}\psi})\,\text{typ}(\mathscr{U}^{bc}_{\mathscr{C}\psi})$ is in the intersection of the row with index $b \in \{i, d, a\}$ and the column with index $c \in \{i, d, a\}$.

6.3 Main Results

The main problem investigated in this chapter is finding all types of cm-pairs and limited cm-pairs. The solution to this problem describes all possible (in terms of functions $\mathscr{U}^{bc}_{\mathscr{C}\psi}$, $\mathscr{L}^{bc}_{\mathscr{C}\psi}$ types, $b, c \in \{i, d, a\}$) relationships among the complexity of decision tables, the minimum complexity of nondeterministic decision trees for them, and the minimum complexity of deterministic decision trees for these tables. We now define seven tables:

$T_1 =$

	i	d	a
i	$\epsilon\alpha$	$\epsilon\alpha$	$\epsilon\alpha$
d	$\epsilon\alpha$	$\epsilon\alpha$	$\epsilon\alpha$
a	$\epsilon\alpha$	$\epsilon\alpha$	$\epsilon\alpha$

$T_2 =$

	i	d	a
i	$\gamma\gamma$	$\epsilon\epsilon$	$\epsilon\epsilon$
d	$\alpha\alpha$	$\epsilon\alpha$	$\epsilon\alpha$
a	$\alpha\alpha$	$\epsilon\alpha$	$\epsilon\alpha$

$T_3 =$

	i	d	a
i	$\gamma\gamma$	$\delta\epsilon$	$\epsilon\epsilon$
d	$\alpha\beta$	$\gamma\gamma$	$\epsilon\epsilon$
a	$\alpha\alpha$	$\alpha\alpha$	$\epsilon\alpha$

$T_4 =$

	i	d	a
i	$\gamma\gamma$	$\gamma\epsilon$	$\epsilon\epsilon$
d	$\alpha\gamma$	$\gamma\gamma$	$\epsilon\epsilon$
a	$\alpha\alpha$	$\alpha\alpha$	$\epsilon\alpha$

$T_5 =$

	i	d	a
i	$\gamma\gamma$	$\gamma\epsilon$	$\gamma\epsilon$
d	$\alpha\gamma$	$\gamma\gamma$	$\gamma\gamma$
a	$\alpha\gamma$	$\gamma\gamma$	$\gamma\gamma$

$T_6 =$

	i	d	a
i	$\gamma\gamma$	$\gamma\epsilon$	$\gamma\epsilon$
d	$\alpha\gamma$	$\gamma\gamma$	$\gamma\delta$
a	$\alpha\gamma$	$\beta\gamma$	$\gamma\gamma$

$T_7 =$

	i	d	a
i	$\gamma\gamma$	$\gamma\epsilon$	$\gamma\epsilon$
d	$\alpha\gamma$	$\gamma\gamma$	$\gamma\epsilon$
a	$\alpha\gamma$	$\alpha\gamma$	$\gamma\gamma$

Theorem 6.1 *For any cm-pair* $(\mathscr{C}, \psi)$, *the relation* $\text{typ}(\mathscr{C}, \psi) \in \{T_1, T_2, T_3, T_4, T_5, T_6, T_7\}$ *holds. For any* $i \in \{1, 2, 3, 4, 5, 6, 7\}$, *there exists a cm-pair* $(\mathscr{C}, \psi)$ *such that* $\text{typ}(\mathscr{C}, \psi) = T_i$.

Theorem 6.2 *For any limited cm-pair* $(\mathscr{C}, \psi)$, *the relation* $\text{typ}(\mathscr{C}, \psi) \in \{T_2, T_3, T_5, T_6, T_7\}$ *holds. For any* $i \in \{2, 3, 5, 6, 7\}$, *there exists a limited cm-pair* $(\mathscr{C}, h)$ *such that* $\text{typ}(\mathscr{C}, h) = T_i$.

6.4 Possible Upper Types of CM-Pairs

We begin our study by considering the upper type of cm-pair, which is a simpler object than the type of cm-pair.

Definition 6.21 Let $(\mathscr{C}, \psi)$ be a cm-pair. We now define table $\text{typ}_u(\mathscr{C}, \psi)$, which will be called the upper type of cm-pair $(\mathscr{C}, \psi)$. This is a table with three rows and three columns in which rows from the top to the bottom and columns from the left to the right are labeled with indices i, d, a. The value $\text{typ}(\mathscr{U}^{bc}_{\mathscr{C}\psi})$ is in the intersection of the row with index $b \in \{i, d, a\}$ and the column with index $c \in \{i, d, a\}$. The table $\text{typ}_u(\mathscr{C}, \psi)$ will be called the *upper type of cm-pair* $(\mathscr{C}, \psi)$.

In this section, all possible upper types of cm-pairs are listed. We now define seven tables:

$t_1 =$

	i	d	a
i	α	α	α
d	α	α	α
a	α	α	α

$t_2 =$

	i	d	a
i	γ	ϵ	ϵ
d	α	α	α
a	α	α	α

$t_3 =$

	i	d	a
i	γ	ϵ	ϵ
d	β	γ	ϵ
a	α	α	α

$t_4 =$

	i	d	a
i	γ	ϵ	ϵ
d	γ	γ	ϵ
a	α	α	α

$t_5 =$

	i	d	a
i	γ	ϵ	ϵ
d	γ	γ	γ
a	γ	γ	γ

$t_6 =$

	i	d	a
i	γ	ϵ	ϵ
d	γ	γ	δ
a	γ	γ	γ

$t_7 =$

	i	d	a
i	γ	ϵ	ϵ
d	γ	γ	ϵ
a	γ	γ	γ

Proposition 6.1 *For any cm-pair* $(\mathscr{C}, \psi)$, *the relation* $\mathrm{typ}_u(\mathscr{C}, \psi) \in \{t_1, t_2, t_3, t_4, t_5, t_6, t_7\}$ *holds.*

Proposition 6.2 *For any limited cm-pair* $(\mathscr{C}, \psi)$, *the relation* $\mathrm{typ}_u(\mathscr{C}, \psi) \in \{t_2, t_3, t_5, t_6, t_7\}$ *holds.*

We divide the proofs of the propositions into a sequence of lemmas.

Lemma 6.1 *Let* T *be a decision table from a set of decision tables* $\mathscr{M}_k^\infty(F)$ *and* ψ *be a complexity measure over* $\mathscr{M}_k^\infty(F)$. *Then the inequalities* $\psi^a(T) \le \psi^d(T) \le \psi^i(T)$ *hold.*

Proof Let columns of the table T be labeled with attributes $f_1, \dots, f_n$. It is not difficult to construct a deterministic decision tree Γ_0 for the table T, which sequentially computes values of attributes $f_1, \dots, f_n$. Evidently, $\psi(\Gamma_0) = \psi^i(T)$. Therefore $\psi^d(T) \le \psi^i(T)$. If a decision tree Γ is a deterministic decision tree for T, then Γ is a nondeterministic decision tree for T. Therefore $\psi^a(T) \le \psi^d(T)$. □

Let $(\mathscr{C}, \psi)$ be a cm-pair, $n \in \omega$ and $b, c \in \{i, d, a\}$. The notation $\mathscr{U}^{bc}_{\mathscr{C}\psi}(n) = \infty$ means that the set $X = \{\psi^b(T) : T \in \mathscr{C}, \psi^c(T) \le n\}$ is infinite. The notation $\mathscr{U}^{bc}_{\mathscr{C}\psi}(n) = \varnothing$ means that the set X is empty. Evidently, if $\mathscr{U}^{bc}_{\mathscr{C}\psi}(n) = \infty$, then $\mathscr{U}^{bc}_{\mathscr{C}\psi}(n+1) = \infty$. It is not difficult to prove the following statement.

Lemma 6.2 *Let* $(\mathscr{C}, \psi)$ *be a cm-pair and* $b, c \in \{i, d, a\}$. *Then*

(a) *If there exists* $n \in \omega$ *such that* $\mathscr{U}^{bc}_{\mathscr{C}\psi}(n) = \infty$, *then* $\mathrm{typ}(\mathscr{U}^{bc}_{\mathscr{C}\psi}) = \epsilon$.

(b) *If there is no* $n \in \omega$ *such that* $\mathscr{U}^{bc}_{\mathscr{C}\psi}(n) = \infty$, *then* $\mathrm{Dom}(\mathscr{U}^{bc}_{\mathscr{C}\psi}) = \{n : n \in \omega, n \ge n_0\}$, *where* $n_0 = \min\{\psi^c(T) : T \in \mathscr{C}\}$.

Let $(\mathscr{C}, \psi)$ be a cm-pair and $b, c, e, f \in \{i, d, a\}$. The notation $\mathscr{U}^{bc}_{\mathscr{C}\psi} \lhd \mathscr{U}^{ef}_{\mathscr{C}\psi}$ means that, for any $n \in \omega$, the following statements hold:

(a) If the value $\mathscr{U}^{bc}_{\mathscr{C}\psi}(n)$ is definite, then either $\mathscr{U}^{ef}_{\mathscr{C}\psi}(n) = \infty$ or the value $\mathscr{U}^{ef}_{\mathscr{C}\psi}(n)$ is definite and the inequality $\mathscr{U}^{bc}_{\mathscr{C}\psi}(n) \le \mathscr{U}^{ef}_{\mathscr{C}\psi}(n)$ holds.

(b) If $\mathscr{U}^{bc}_{\mathscr{C}\psi}(n) = \infty$, then $\mathscr{U}^{ef}_{\mathscr{C}\psi}(n) = \infty$.

Let $\preceq$ be a linear order on the set $\{\alpha, \beta, \gamma, \delta, \epsilon\}$ such that $\alpha \preceq \beta \preceq \gamma \preceq \delta \preceq \epsilon$.

Lemma 6.3 *Let* $(\mathcal{C}, \psi)$ *be a cm-pair. Then* $\text{typ}(\mathcal{U}^{bi}_{\mathcal{C}\psi}) \preceq \text{typ}(\mathcal{U}^{bd}_{\mathcal{C}\psi}) \preceq \text{typ}(\mathcal{U}^{ba}_{\mathcal{C}\psi})$ *and* $\text{typ}(\mathcal{U}^{ab}_{\mathcal{C}\psi}) \preceq \text{typ}(\mathcal{U}^{db}_{\mathcal{C}\psi}) \preceq \text{typ}(\mathcal{U}^{ib}_{\mathcal{C}\psi})$ *for any* $b \in \{i, d, a\}$.

Proof From the definition of the functions $\mathcal{U}^{bc}_{\mathcal{C}\psi}$, $b, c \in \{i, d, a\}$, and from Lemma 6.1 it follows that $\mathcal{U}^{bi}_{\mathcal{C}\psi} \lhd \mathcal{U}^{bd}_{\mathcal{C}\psi} \lhd \mathcal{U}^{ba}_{\mathcal{C}\psi}$ and $\mathcal{U}^{ab}_{\mathcal{C}\psi} \lhd \mathcal{U}^{db}_{\mathcal{C}\psi} \lhd \mathcal{U}^{ib}_{\mathcal{C}\psi}$ for any $b \in \{i, d, a\}$. Using these relations and Lemma 6.2 we obtain the statement of the lemma. □

Lemma 6.4 *Let* $(\mathcal{C}, \psi)$ *be a cm-pair and* $b, c \in \{i, d, a\}$. *Then*

(a) $\text{typ}(\mathcal{U}^{bc}_{\mathcal{C}\psi}) = \alpha$ *if and only if the function* ψ^b *is bounded from above on the closed class* $\mathcal{C}$.

(b) *If the function* ψ^b *is unbounded from above on* $\mathcal{C}$, *then* $\text{typ}(\mathcal{U}^{bb}_{\mathcal{C}\psi}) = \gamma$.

Proof The statement (a) is obvious. (b) Let the function ψ^b be unbounded from above on $\mathcal{C}$. One can show that in this case the equality $\mathcal{U}^{bb}_{\mathcal{C}\psi}(n) = n$ holds for infinitely many $n \in \omega$. Therefore $\text{typ}(\mathcal{U}^{bb}_{\mathcal{C}\psi}) = \gamma$. □

Corollary 6.1 *Let* $(\mathcal{C}, \psi)$ *be a cm-pair and* $b \in \{i, d, a\}$. *Then* $\text{typ}(\mathcal{U}^{bb}_{\mathcal{C}\psi}) \in \{\alpha, \gamma\}$.

Lemma 6.5 *Let* $(\mathcal{C}, \psi)$ *be a cm-pair and* $\text{typ}(\mathcal{U}^{ii}_{\mathcal{C}\psi}) \neq \alpha$. *Then*

$$\text{typ}(\mathcal{U}^{id}_{\mathcal{C}\psi}) = \text{typ}(\mathcal{U}^{ia}_{\mathcal{C}\psi}) = \epsilon .$$

Proof Using Lemma 6.4, we conclude that the function ψ^i is unbounded from above on $\mathcal{C}$. Let $m \in \omega$. Then there exists a decision table $T \in \mathcal{C}$ for which the inequality $\psi^i(T) \geq m$ holds. Let us consider a degenerate decision table $T' \in \mathcal{C}$ obtained from T by replacing the sets of decisions attached to rows by the set $\{0\}$. It is clear that $\psi^i(T') \geq m$. Let Γ be a decision tree, which consists of the root, the terminal node labeled with 0, and the edge connecting these two nodes. One can show that Γ is a deterministic decision tree for the table T'. Therefore $\psi^a(T') \leq \psi^d(T') \leq \psi(\Gamma) = \psi(\lambda)$. Taking into account that m is an arbitrary number from ω, we obtain $\mathcal{U}^{id}_{\mathcal{C}\psi}(\psi(\lambda)) = \infty$ and $\mathcal{U}^{ia}_{\mathcal{C}\psi}(\psi(\lambda)) = \infty$. Using Lemma 6.2, we conclude that $\text{typ}(\mathcal{U}^{id}_{\mathcal{C}\psi}) = \text{typ}(\mathcal{U}^{ia}_{\mathcal{C}\psi}) = \epsilon$. □

Example 6.9 Let us consider a cm-pair $(\mathcal{C}_0, h)$, where $\mathcal{C}_0$ is closed class described in Example 6.5. It is clear that the function h^i is unbounded from above on $\mathcal{C}_0$ and the functions h^a and h^d are bounded from above on $\mathcal{C}_0$. Using Lemma 6.4, we obtain that $\text{typ}(\mathcal{U}^{ab}_{\mathcal{C}_0 h}) = \text{typ}(\mathcal{U}^{db}_{\mathcal{C}_0 h}) = \alpha$ for any $b \in \{i, d, a\}$ and $\text{typ}(\mathcal{U}^{ii}_{\mathcal{C}_0 h}) = \gamma$. By Lemma 6.5, $\text{typ}(\mathcal{U}^{id}_{\mathcal{C}_0 h}) = \text{typ}(\mathcal{U}^{ia}_{\mathcal{C}_0 h}) = \epsilon$. Therefore, $\text{typ}_u(\mathcal{C}_0, h) = t_2$.

Lemma 6.6 *Let* $(\mathcal{C}, \psi)$ *be a cm-pair. Then* $\text{typ}(\mathcal{U}^{ai}_{\mathcal{C}\psi}) \in \{\alpha, \gamma\}$.

Proof Using Lemma 6.3 and Corollary 6.1, we obtain $\text{typ}(\mathcal{U}^{ai}_{\mathcal{C}\psi}) \in \{\alpha, \beta, \gamma\}$. By Lemma 6.2, $\text{Dom}(\mathcal{U}^{ai}_{\mathcal{C}\psi}) = \{n : n \in \omega, n \geq n_0\}$ for some $n_0 \in \omega$. Set $D = \text{Dom}(\mathcal{U}^{ai}_{\mathcal{C}\psi})$. Assume that $\text{typ}(\mathcal{U}^{ai}_{\mathcal{C}\psi}) = \beta$. Then there exists $m \in D$ such that

$\mathscr{U}^{ai}_{\mathscr{C}\psi}(n) < n$ for any $n \in D$, $n > m$. Let us prove by induction on n that, for any decision table T from $\mathscr{C}$, if $\psi^i(T) \leq n$, then $\psi^a(T) \leq m_0$, where $m_0 = \max\{m, \psi(\lambda)\}$. Using Lemma 6.1, we conclude that under the condition $n \leq m$ the considered statement holds. Let it hold for some $n, n \geq m$. Let us show that this statement holds for $n+1$ too. Let $T \in \mathscr{C}$, $\psi^i(T) \leq n+1$ and columns of the table T be labeled with attributes $f_{i_1}, \ldots, f_{i_k}$. Since $n+1 > m$, we obtain $\psi^a(T) \leq n$. Let Γ be a nondeterministic decision tree for the table T and $\psi(\Gamma) = \psi^a(T)$. Assume that in Γ there exists a complete path ξ in which there are no working nodes. In this case, a decision tree, which consists of the root, the terminal node labeled with $\tau(\xi)$ and the edge connecting these two nodes is a nondeterministic decision tree for the table T. Therefore $\psi^a(T) \leq \psi(\lambda) \leq m_0$. Assume now that each complete path in the decision tree Γ contains a working node. Let $\xi \in \mathrm{Path}(\Gamma)$, $\Delta(T\pi(\xi)) \neq \varnothing$, $\xi = v_0, d_0, \ldots, v_p, d_p, v_{p+1}$ and, for $i = 1, \ldots, p$, the node v_i be labeled with the attribute f_i, and the edge d_i be labeled with the number δ_i. Let the decision table T' be obtained from the decision table T by operations of permutation of columns and duplication of columns so that its columns are labeled with attributes $f_1, \ldots, f_p, f_{i_1}, \ldots, f_{i_k}$. We obtain the decision table T'' from T' by removal the last k columns. Let us denote by T_ξ the decision table obtained from T'' by changing the set of decisions corresponding to the row $(\delta_1, \ldots, \delta_p)$ with $\{\tau(\xi)\}$, and for the remaining rows with $\{\tau(\xi)+1\}$. It is clear that $\psi^i\left(T_\xi\right) \leq n$. Using the inductive hypothesis, we conclude that there exists a nondeterministic decision tree Γ_ξ for the table T_ξ such that $\psi(\Gamma_\xi) \leq m_0$. We denote by $\tilde{\Gamma}_\xi$ a tree obtained from Γ_ξ by removal of all nodes and edges that satisfy the following condition: there is no a complete path ξ' in Γ_ξ, which contains this node or edge and for which $\tau\left(\xi'\right) = \tau(\xi)$. Let $\{\xi : \xi \in \mathrm{Path}(\Gamma), \Delta(T\pi(\xi)) \neq \varnothing\} = \{\xi_1, \ldots, \xi_r\}$. Let us identify the roots of the trees $\tilde{\Gamma}_{\xi_1}, \ldots, \tilde{\Gamma}_{\xi_r}$. We denote by G the obtained tree. It is not difficult to show that G is a nondeterministic decision tree for the table T and $\psi(G) \leq m_0$. Thus, the considered statement holds. Using Lemma 6.4, we conclude that $\mathrm{typ}(\mathscr{U}^{ai}_{\mathscr{C}\psi}) = \alpha$. The obtained contradiction shows that $\mathrm{typ}(\mathscr{U}^{ai}_{\mathscr{C}\psi}) \in \{\alpha, \gamma\}$. □

Let T be a decision table from $\mathscr{M}_k^\infty(F)$. We now give definitions of parameters $N(T)$ and $M(T)$ of the table T.

Definition 6.22 We denote by $N(T)$ the number of rows in the table T.

Definition 6.23 Let columns of table T be labeled with attributes $f_1, \ldots, f_n \in F$. We now define the parameter $M(T)$. If table T is degenerate, then $M(T) = 0$. Let now T be a nondegenerate table and $\bar{\delta} = (\delta_1, \ldots, \delta_n) \in E_k^n$. Then $M(T, \bar{\delta})$ is the minimum natural m such that there exist attributes $f_{i_1}, \ldots, f_{i_m} \in \mathrm{At}(T)$ for which $T(f_{i_1}, \delta_{i_1}) \cdots (f_{i_m}, \delta_{i_m})$ is a degenerate table. We denote $M(T) = \max\{M(T, \bar{\delta}) : \bar{\delta} \in E_k^n\}$.

The next statement follows immediately from Theorem 3.5 [4].

Lemma 6.7 *Let T be a nonempty decision table from $\mathscr{M}_k^\infty(F)$ in which each row is labeled with a set containing only one decision. Then*

$$h^d(T) \le M(T) \log_2 N(T) .$$

Lemma 6.8 *Let* $(\mathscr{C}, \psi)$ *be a limited cm-pair and* $\mathrm{typ}(\mathscr{U}^{ai}_{\mathscr{C}\psi}) = \alpha$. *Then* $\mathrm{typ}(\mathscr{U}^{di}_{\mathscr{C}\psi}) \in \{\alpha, \beta\}$.

Proof Using Lemma 6.4, we conclude that there exists $r \in \omega$ such that the inequality $\psi^a(T) \le r$ holds for any table $T \in \mathscr{C}$.

Let T be a nonempty table from $\mathscr{C}$ in which columns are labeled with attributes $f_1, \dots, f_n$ and $\bar{\delta} = (\delta_1, \dots, \delta_n) \in E_k^n$. We now show that there exist attributes $f_{i_1}, \dots, f_{i_m} \in \mathrm{At}(T)$ such that the subtable $T(\bar{\delta}) = T(f_1, \delta_1) \cdots (f_n, \delta_n)$ is equal to the subtable $T(f_{i_1}, \delta_{i_1}) \cdots (f_{i_m}, \delta_{i_m})$ and $m \le r$ if $\bar{\delta}$ is a row of T, and $m \le r + 1$ if $\bar{\delta}$ is not a row of T.

Let $\bar{\delta}$ be a row of T. Let us change the set of decisions attached to the row $\bar{\delta}$ with the set $\{1\}$ and for the remaining rows of T with the set $\{0\}$. We denote the obtained table by T'. It is clear that $T' \in \mathscr{C}$. Taking into account that $\psi^a(T') \le r$ and the complexity measure ψ has the property (c), it is not difficult to show that there exist attributes $f_{i_1}, \dots, f_{i_m} \in \mathrm{At}(T') = \mathrm{At}(T)$ such that $m \le r$ and $T'(f_{i_1}, \delta_{i_1}) \cdots (f_{i_m}, \delta_{i_m})$ contains only the row $\bar{\delta}$. From here it follows that $T(\bar{\delta}) = T(f_{i_1}, \delta_{i_1}) \cdots (f_{i_m}, \delta_{i_m})$.

Let $\bar{\delta}$ be not a row of T. Let us show that there exist attributes $f_{i_1}, \dots, f_{i_m} \in \mathrm{At}(T)$ such that $m \le r + 1$ and the subtable $T(f_{i_1}, \delta_{i_1}) \cdots (f_{i_m}, \delta_{i_m})$ is empty. If $T(f_1, \delta_1)$ is empty, then the considered statement holds. Otherwise, there exists $q \in \{1, \dots, n-1\}$ such that the subtable $T(f_1, \delta_1) \cdots (f_q, \delta_q)$ is nonempty but the subtable $T(f_1, \delta_1) \cdots (f_{q+1}, \delta_{q+1})$ is empty. We denote by T' the table obtained from T by removal of attributes $f_{q+1}, \dots, f_n$. It is clear that $T' \in \mathscr{C}$ and $(\delta_1, \dots, \delta_q)$ is a row of T'. According to proven above, there exist attributes $f_{i_1}, \dots, f_{i_p} \in \{f_1, \dots, f_q\}$ such that

$$T'(f_{i_1}, \delta_{i_1}) \cdots (f_{i_p}, \delta_{i_p}) = T'(f_1, \delta_1) \cdots (f_q, \delta_q)$$

and $p \le r$. Using this fact one can show that $T(f_{i_1}, \delta_{i_1}) \cdots (f_{i_p}, \delta_{i_p})(f_{q+1}, \delta_{q+1})$ is empty and is equal to $T(\bar{\delta})$.

Let $T_1 \in \mathscr{C}$. We denote by T_2 the decision table obtained from T_1 by removal of all columns in which all numbers are equal. Let columns of T_2 be labeled with attributes $f_1, \dots, f_n$. We now consider the decision table T_3, which is obtained from T_2 by changing decisions so that the decision set attached to each row of table T_3 contains only one decision and, for any two non-equal rows, corresponding decisions are different. It is clear that $T_3 \in \mathscr{C}$. It is not difficult to show that $\psi^d(T_1) \le \psi^d(T_2) \le \psi^d(T_3)$.

We now show that the inequality $\psi(f) \le r$ holds for any attribute $f \in \mathrm{At}(T_3)$. Let us denote by T' the decision table obtained from T_3 by removal of all columns except the column labeled with the attribute f. If there is more than one column in T_3, which is labeled with the attribute f, then we keep only one of them. Let the decision table T_f be obtained from T' by changing the set of decisions for each row (δ) with the set of decisions $\{\delta\}$. It is clear that $T_f \in \mathscr{C}$. Let Γ be a nondeterministic decision tree for the table T_f and $\psi(\Gamma) = \psi^a(T_f) \le r$. Since the column f contains different numbers, we have $f \in \mathrm{At}(\Gamma)$. Using the property (b) of the complexity measure ψ, we obtain $\psi(\Gamma) \geq \psi(f)$. Consequently, $\psi(f) \leq r$.

Taking into account that, for any $\bar{\delta} \in \Delta(T_3)$, there exist attributes $f_{i_1}, \ldots, f_{i_m} \in \{f_1, \ldots, f_n\}$ such that $m \leq r$, and $T_3(f_{i_1}, \delta_{i_1}) \cdots (f_{i_m}, \delta_{i_m})$ contains only the row $\bar{\delta}$, it is not difficult to show that

$$N(T_3) \leq n^r \cdot k^r \ . \tag{6.1}$$

According to the proven above, for any $\bar{\delta} \in E_k^n$, there exist attributes $f_{i_1}, \ldots, f_{i_m} \in \{f_1, \ldots, f_n\}$ such that $m \leq r+1$, and $T_3(f_{i_1}, \delta_{i_1}) \cdots (f_{i_m}, \delta_{i_m}) = T_3(f_1, \delta_1) \cdots (f_n, \delta_n)$. Taking into account this equality one can show that

$$M(T_3) \leq r + 1 \ . \tag{6.2}$$

Using Lemma 6.7, and inequalities (6.1) and (6.2), we conclude that there exists a deterministic decision tree Γ for the table T_3 with $h(\Gamma) \leq M(T_3) \log_2 N(T_3) \leq (r+1)^2 \log_2(kn)$. Taking into account that $\psi(f) \leq r$ for any attribute $f \in \mathrm{At}(T_3)$ and the complexity measure ψ has the property (a), we obtain

$$\psi^d(T_3) \leq (r+1)^3 \log_2(kn) \ .$$

Consequently, $\psi^d(T_1) \leq (r+1)^3 \log_2(kn)$. Taking into account that the complexity measure ψ has the property (c), we obtain $\psi^i(T_1) \geq n$. Since T_1 is an arbitrary decision table from $\mathscr{C}$, we have $\mathrm{Dom}^+(\mathscr{U}^{di}_{\mathscr{C}\psi})$ is a finite set. Therefore $\mathrm{typ}(\mathscr{U}^{di}_{U\psi}) \neq \gamma$. Using Lemma 6.3 and Corollary 6.1, we obtain $\mathrm{typ}(\mathscr{U}^{di}_{\mathscr{C}\psi}) \in \{\alpha, \beta\}$. □

Proof of Proposition 6.1. Let $(\mathscr{C}, \psi)$ be a cm-pair. Using Corollary 6.1, we conclude that $\mathrm{typ}(\mathscr{U}^{ii}_{\mathscr{C}\psi}) \in \{\alpha, \gamma\}$. Using Corollary 6.1 and Lemma 6.3, we obtain $\mathrm{typ}(\mathscr{U}^{di}_{\mathscr{C}\psi}) \in \{\alpha, \beta, \gamma\}$. From Lemma 6.6 it follows that $\mathrm{typ}(\mathscr{U}^{ai}_{\mathscr{C}\psi}) \in \{\alpha, \gamma\}$.

(a) Let $\mathrm{typ}(\mathscr{U}^{ii}_{\mathscr{C}\psi}) = \alpha$. Using Lemmas 6.3 and 6.4, we obtain $\mathrm{typ}_u(\mathscr{C}, \psi) = t_1$.

(b) Let $\mathrm{typ}(\mathscr{U}^{ii}_{\mathscr{C}\psi}) = \gamma$ and $\mathrm{typ}(\mathscr{U}^{di}_{\mathscr{C}\psi}) = \alpha$. Using Lemmas 6.3, 6.4, and 6.5, we obtain $\mathrm{typ}_u(\mathscr{C}, \psi) = t_2$.

(c) Let $\mathrm{typ}(\mathscr{U}^{ii}_{\mathscr{C}\psi}) = \gamma$ and $\mathrm{typ}(\mathscr{U}^{di}_{\mathscr{C}\psi}) = \beta$. From Lemma 6.5 it follows that $\mathrm{typ}(\mathscr{U}^{id}_{\mathscr{C}\psi}) = \mathrm{typ}(\mathscr{U}^{ia}_{\mathscr{C}\psi}) = \epsilon$. Using Lemmas 6.3 and 6.6, we obtain $\mathrm{typ}(\mathscr{U}^{ai}_{\mathscr{C}\psi}) = \alpha$. From this equality and from Lemma 6.4 it follows that $\mathrm{typ}(\mathscr{U}^{ad}_{\mathscr{C}\psi}) = \mathrm{typ}(\mathscr{U}^{aa}_{\mathscr{C}\psi}) = \alpha$. Using the equality $\mathrm{typ}(\mathscr{U}^{di}_{\mathscr{C}\psi}) = \beta$, Lemma 6.3, and Corollary 6.1, we obtain $\mathrm{typ}(\mathscr{U}^{dd}_{\mathscr{C}\psi}) = \gamma$. From the equalities $\mathrm{typ}(\mathscr{U}^{dd}_{\mathscr{C}\psi}) = \gamma$, $\mathrm{typ}(\mathscr{U}^{aa}_{\mathscr{C}\psi}) = \alpha$ and from Lemmas 6.2 and 6.4 it follows that $\mathrm{typ}(\mathscr{U}^{da}_{\mathscr{C}\psi}) = \epsilon$. Thus, $\mathrm{typ}_u(\mathscr{C}, \psi) = t_3$.

(d) Let $\mathrm{typ}(\mathscr{U}^{ii}_{\mathscr{C}\psi}) = \mathrm{typ}(\mathscr{U}^{di}_{\mathscr{C}\psi}) = \gamma$ and $\mathrm{typ}(\mathscr{U}^{ai}_{\mathscr{C}\psi}) = \alpha$. Using Lemma 6.5, we obtain $\mathrm{typ}(\mathscr{U}^{id}_{\mathscr{C}\psi}) = \mathrm{typ}(\mathscr{U}^{ia}_{\mathscr{C}\psi}) = \epsilon$. From Lemma 6.4 it follows that $\mathrm{typ}(\mathscr{U}^{ad}_{\mathscr{C}\psi}) = \mathrm{typ}(\mathscr{U}^{aa}_{\mathscr{C}\psi}) = \alpha$. Using Lemma 6.3 and Corollary 6.1, we obtain $\mathrm{typ}(\mathscr{U}^{dd}_{\mathscr{C}\psi}) = \gamma$. From this equality, equality $\mathrm{typ}(\mathscr{U}^{aa}_{\mathscr{C}\psi}) = \alpha$ and from Lemmas 6.2 and 6.4 it follows that $\mathrm{typ}(\mathscr{U}^{da}_{\mathscr{C}\psi}) = \epsilon$. Thus, $\mathrm{typ}_u(\mathscr{C}, \psi) = t_4$.

(e) Let $\mathrm{typ}(\mathcal{U}^{ii}_{\mathscr{C}\psi}) = \mathrm{typ}(\mathcal{U}^{di}_{\mathscr{C}\psi}) = \mathrm{typ}(\mathcal{U}^{ai}_{\mathscr{C}\psi}) = \gamma$. Using Lemma 6.5 we conclude that $\mathrm{typ}(\mathcal{U}^{id}_{\mathscr{C}\psi}) = \mathrm{typ}(\mathcal{U}^{ia}_{\mathscr{C}\psi}) = \epsilon$. Using Lemma 6.3 and Corollary 6.1, we obtain $\mathrm{typ}(\mathcal{U}^{dd}_{\mathscr{C}\psi}) = \mathrm{typ}(\mathcal{U}^{ad}_{\mathscr{C}\psi}) = \mathrm{typ}(\mathcal{U}^{aa}_{\mathscr{C}\psi}) = \gamma$. Using Lemma 6.3, we obtain $\mathrm{typ}(\mathcal{U}^{da}_{\mathscr{C}\psi}) \in \{\gamma, \delta, \epsilon\}$. Therefore $\mathrm{typ}_u(\mathscr{C}, \psi) \in \{t_5, t_6, t_7\}$. □

Proof of Proposition 6.2. Let $(\mathscr{C}, \psi)$ be a limited cm-pair. Taking into account that the complexity measure ψ has the property (c), and using Lemma 6.4, we obtain $\mathrm{typ}(\mathcal{U}^{ii}_{\mathscr{C}\psi}) \neq \alpha$. Therefore $\mathrm{typ}_u(\mathscr{C}, \psi) \neq t_1$. Using Lemma 6.8, we obtain $\mathrm{typ}_u(\mathscr{C}, \psi) \neq t_4$. From these relations and Proposition 6.1 it follows that the statement of the proposition holds. □

6.5 Realizable Upper Types of CM-Pairs

In this section, all realizable upper types of cm-pairs are listed.

Proposition 6.3 *For any* $i \in \{1, 2, 3, 4, 5, 6, 7\}$, *there exists a cm-pair* $(\mathscr{C}, \psi)$ *such that*

$$\mathrm{typ}_u(\mathscr{C}, \psi) = t_i .$$

Proposition 6.4 *For any* $i \in \{2, 3, 5, 6, 7\}$, *there exists a limited cm-pair* $(\mathscr{C}, h)$ *such that*

$$\mathrm{typ}_u(\mathscr{C}, h) = t_i .$$

Proofs of these propositions are based on results obtained for information systems in Chap. 3.

Let $U = (A, F)$ be an information system, where attributes from F have values from E_k, and ψ be a complexity measure over U–see Chap. 3. Note that ψ is also a complexity measure over the set of decision tables $\mathscr{M}_k^{\infty}(F)$. Let $z = (\nu, f_1, \ldots, f_n)$ be a problem over U. In Chap. 3, three parameters of the problem z were defined: $\psi^i_U(z) = \psi(f_1 \cdots f_n)$ called the complexity of the problem z description, $\psi^d_U(z)$–the minimum complexity of a decision tree with attributes from the set $\{f_1, \ldots, f_n\}$, which solves the problem z deterministically, and $\psi^a_U(z)$–the minimum complexity of a decision tree with attributes from the set $\{f_1, \ldots, f_n\}$, which solves the problem z nondeterministically.

Let $b, c \in \{i, d, a\}$. In Chap. 3, the partial function $\mathcal{U}^{bc}_{U\psi} : \omega \to \omega$ was defined:

$$\mathcal{U}^{bc}_{U\psi}(n) = \max\{\psi^b_U(z) : z \in \mathrm{Probl}^{\infty}(U), \psi^c_U(z) \leq n\} .$$

The table $\mathrm{typ}_{lu}(U, \psi)$ for the pair (U, ψ) was defined in Chap. 3 as follows: this is a table with three rows and three columns in which rows from the top to the bottom and columns from the left to the right are labeled with indices i, d, a. The value $\mathrm{typ}(\mathcal{U}^{bc}_{U\psi})$ is in the intersection of the row with the index $b \in \{i, d, a\}$ and the column with the index $c \in \{i, d, a\}$.

We now prove the following proposition:

Proposition 6.5 *Let U be an information system and ψ be a complexity measure over U. Then*

$$\mathrm{typ}_{lu}(U, \psi) = \mathrm{typ}_u(\mathrm{Tab}^{\infty}(U), \psi) .$$

Proof Let $z = (\nu, f_1, \ldots, f_n)$ be a problem over U and $T(z)$ be the decision table corresponding to this problem. It is easy to see that $\psi_U^i(z) = \psi^i(T(z))$. One can show the set of decision trees solving the problem z nondeterministically and using only attributes from the set $\{f_1, \ldots, f_n\}$ (see corresponding definitions in Chap. 3) is equal to the set of nondeterministic decision trees for the table $T(z)$. From here it follows that $\psi_U^a(z) = \psi^a(T(z))$ and $\psi_U^d(z) = \psi^d(T(z))$. Using these equalities, we can show that $\mathrm{typ}_{lu}(U, \psi) = \mathrm{typ}_u(\mathrm{Tab}^{\infty}(U), \psi)$. □

This proposition allows us to transfer results obtained for information systems in Chap. 3 to the case of closed classes of decision tables. Before each of the following seven lemmas, we define a pair (U, ψ), where U is an information system and ψ is a complexity measure over U.

Let us define a pair (U_1, π) as follows: $U_1 = (\omega, F_1)$, where $F_1 = \{f\}$, $f \equiv 0$, and $\pi \equiv 0$.

Lemma 6.9 $\mathrm{typ}_u(\mathrm{Tab}^{\infty}(U_1), \pi) = t_1$.

Proof From Lemma 3.8 it follows that $\mathrm{typ}_{lu}(U_1, \pi) = t_1$. Using Proposition 6.5, we obtain $\mathrm{typ}_u(\mathrm{Tab}^{\infty}(U_1), \pi) = t_1$. □

Let us define a pair (U_2, h) as follows: $U_2 = (\omega, F_2)$, where $F_2 = F_1$.

Lemma 6.10 $\mathrm{typ}_u(\mathrm{Tab}^{\infty}(U_2), h) = t_2$.

Proof From Lemma 3.9 it follows that $\mathrm{typ}_{lu}(U_2, h) = t_2$. Using Proposition 6.5, we obtain $\mathrm{typ}_u(\mathrm{Tab}^{\infty}(U_2), h) = t_2$. □

Let us define a pair (U_3, h) as follows: $U_3 = (\omega, F_3)$, where $F_3 = \{l_i : i \in \omega \setminus \{0\}\}$ and, for any $i \in \omega \setminus \{0\}$, $j \in \omega$, if $j \leq i$, then $l_i(j) = 0$, and if $j > i$, then $l_i(j) = 1$.

Lemma 6.11 $\mathrm{typ}_u(\mathrm{Tab}^{\infty}(U_3), h) = t_3$.

Proof From Lemma 3.10 it follows that $\mathrm{typ}_{lu}(U_3, h) = t_3$. Using Proposition 6.5, we obtain $\mathrm{typ}_u(\mathrm{Tab}^{\infty}(U_3), h) = t_3$. □

Let us define a pair (U_4, μ) as follows: $U_4 = (\omega, F_4)$, where $F_4 = F_3$, $\mu(\lambda) = 0$, $\mu(l_{i_1} \cdots l_{i_m}) = 1$ if $m = 1$ or $m = 2$ and $i_1 > i_2$, and $\mu(l_{i_1} \cdots l_{i_m}) = \max\{i_1, \ldots, i_m\}$ in other cases.

Lemma 6.12 $\mathrm{typ}_u(\mathrm{Tab}^{\infty}(U_4), \mu) = t_4$.

Proof From Lemma 3.11 it follows that $\mathrm{typ}_{lu}(U_4, \mu) = t_4$. Using Proposition 6.5, we obtain $\mathrm{typ}_u(\mathrm{Tab}^\infty(U_4), \mu) = t_4$. □

Let us define a pair (U_5, h) as follows: $U_5 = (\omega, F_5)$, where $F_5 = \{f_i : i \in \omega \setminus \{0\}\}$ and, for any $i \in \omega \setminus \{0\}$, $j \in \omega$, if $i = j$, then $f_i(j) = 1$, and if $i \neq j$, then $f_i(j) = 0$.

Lemma 6.13 $\mathrm{typ}_u(\mathrm{Tab}^\infty(U_5), h) = t_5$.

Proof From Lemma 3.12 it follows that $\mathrm{typ}_{lu}(U_5, h) = t_5$. Using Proposition 6.5, we obtain $\mathrm{typ}_u(\mathrm{Tab}^\infty(U_5), h) = t_5$. □

Let us define a pair (U_6, h) as follows: $U_6 = (\omega, F_6)$, where $F_6 = F_5 \cup G$, $G = \{g_{2i+1} : i \in \omega\}$ and, for any $i \in \omega$, $j \in \omega$, if $j \in \{2i+1, 2i+2\}$, then $g_{2i+1}(j) = 1$, and if $j \notin \{2i+1, 2i+2\}$, then $g_{2i+1}(j) = 0$.

Lemma 6.14 $\mathrm{typ}_u(\mathrm{Tab}^\infty(U_6), h) = t_6$.

Proof From Lemma 3.13 it follows that $\mathrm{typ}_{lu}(U_6, h) = t_6$. Using Proposition 6.5, we obtain $\mathrm{typ}_u(\mathrm{Tab}^\infty(U_6), h) = t_6$. □

Let us define a pair (U_7, h) as follows: $U_7 = (\omega, F_7)$, where $F_7 = F_3 \cup F_5$.

Lemma 6.15 $\mathrm{typ}_u(\mathrm{Tab}^\infty(U_7), h) = t_7$.

Proof From Lemma 3.14 it follows that $\mathrm{typ}_{lu}(U_7, h) = t_7$. Using Proposition 6.5, we obtain $\mathrm{typ}_u(\mathrm{Tab}^\infty(U_7), h) = t_7$. □

Proof of Proposition 6.3. The statement of the proposition follows from Lemmas 6.9–6.15. □

Proof of Proposition 6.4. The statement of the proposition follows from Lemmas 6.10, 6.11, 6.13, 6.14 and 6.15. □

6.6 Union of CM-Pairs

In this section, we define a union of two cm-pairs, which is also a cm-pair, and study its upper type. Let $\tau_1 = (\mathscr{C}_1, \psi_1)$ and $\tau_2 = (\mathscr{C}_2, \psi_2)$ be cm-pairs, where $\mathscr{C}_1 \subseteq \mathscr{M}_{k_1}^\infty(F_1)$ and $\mathscr{C}_2 \subseteq \mathscr{M}_{k_2}^\infty(F_2)$. These two cm-pairs will be called *compatible* if $F_1 \cap F_2 = \varnothing$ and $\psi_1(\lambda) = \psi_2(\lambda)$. We now define a cm-pair $\tau = (\mathscr{C}, \psi)$, which is called a *union* of compatible cm-pairs τ_1 and τ_2.

Definition 6.24 The closed class $\mathscr{C}$ in τ is defined as follows: $\mathscr{C} = \mathscr{C}_1 \cup \mathscr{C}_2 \subseteq \mathscr{M}_{\max(k_1,k_2)}^\infty(F_1 \cup F_2)$. The complexity measure ψ in τ is defined for any word $\alpha \in (F_1 \cup F_2)^*$ in the following way: if $\alpha \in F_1^*$, then $\psi(\alpha) = \psi_1(\alpha)$, if $\alpha \in F_2^*$, then $\psi(\alpha) = \psi_2(\alpha)$, if α contains letters from both F_1 and F_2, then $\psi(\alpha)$ can have arbitrary value from ω. In particular, if $\psi_1 = \psi_2 = h$, then as ψ we can use the depth h.

We now consider the upper type of cm-pair $\tau = (\mathscr{C}, \psi)$. We denote by $\widetilde{\max}$ the function maximum for the linear order $\alpha \preceq \beta \preceq \gamma \preceq \delta \preceq \epsilon$.

Theorem 6.3 *The equality* $\mathrm{typ}(\mathscr{U}^{bc}_{\mathscr{C}\psi}) = \widetilde{\max}(\mathrm{typ}(\mathscr{U}^{bc}_{\mathscr{C}_1\psi_1}), \mathrm{typ}(\mathscr{U}^{bc}_{\mathscr{C}_2\psi_2}))$ *holds for any* $b, c \in \{i, d, a\}$ *except for the case* $bc = da$ *and* $\mathrm{typ}(\mathscr{U}^{da}_{\mathscr{C}_1\psi_1}) = \mathrm{typ}(\mathscr{U}^{da}_{\mathscr{C}_2\psi_2}) = \gamma$. *In the last case,* $\mathrm{typ}(\mathscr{U}^{da}_{\mathscr{C}\psi}) \in \{\gamma, \delta\}$.

Proof Let $n \in \omega$ and $b, c \in \{i, d, a\}$. We now define the value $M = \underline{\max}(\mathscr{U}_1, \mathscr{U}_2)$, where $\mathscr{U}_1 = \mathscr{U}^{bc}_{\mathscr{C}_1\psi_1}(n)$ and $\mathscr{U}_2 = \mathscr{U}^{bc}_{\mathscr{C}_2\psi_2}(n)$. Both $\mathscr{U}_1$ and $\mathscr{U}_2$ have values from the set $\{\varnothing, \infty\} \cup \omega$ (see definitions before Lemma 6.2). If $\mathscr{U}_1 = \mathscr{U}_2 = \varnothing$, then $M = \varnothing$. If one of $\mathscr{U}_1, \mathscr{U}_2$ is equal to $\varnothing$ and another one is equal to a number $m \in \omega$, then $M = m$. If $\mathscr{U}_1, \mathscr{U}_2 \in \omega$, then $M = \max(\mathscr{U}_1, \mathscr{U}_2)$. If at least one of $\mathscr{U}_1, \mathscr{U}_2$ is equal to ∞, then $M = \infty$.

The following equality follows from the definition of partial function $\mathscr{U}^{bc}_{\mathscr{C}\psi}(n)$, where $n \in \omega$ and $b, c \in \{i, d, a\}$: $\mathscr{U}^{bc}_{\mathscr{C}\psi}(n) = \underline{\max}(\mathscr{U}^{bc}_{\mathscr{C}_1\psi_1}(n), \mathscr{U}^{bc}_{\mathscr{C}_2\psi_2}(n))$. Later in the proof, we will use this equality without special mention. From this equality we obtain $\mathrm{typ}(\mathscr{U}^{bc}_{\mathscr{C}_1\psi_1}) \preceq \mathrm{typ}(\mathscr{U}^{bc}_{\mathscr{C}\psi})$ and $\mathrm{typ}(\mathscr{U}^{bc}_{\mathscr{C}_2\psi_2}) \preceq \mathrm{typ}(\mathscr{U}^{bc}_{\mathscr{C}\psi})$. We now consider two different cases separately: (1) $\mathrm{typ}(\mathscr{U}^{bc}_{\mathscr{C}_1\psi_1}) = \mathrm{typ}(\mathscr{U}^{bc}_{\mathscr{C}_2\psi_2})$ and (2) $\mathrm{typ}(\mathscr{U}^{bc}_{\mathscr{C}_1\psi_1}) \neq \mathrm{typ}(\mathscr{U}^{bc}_{\mathscr{C}_2\psi_2})$.

(1) Let $\mathrm{typ}(\mathscr{U}^{bc}_{\mathscr{C}_1\psi_1}) = \mathrm{typ}(\mathscr{U}^{bc}_{\mathscr{C}_2\psi_2})$.

(a) Let $\mathrm{typ}(\mathscr{U}^{bc}_{\mathscr{C}_1\psi_1}) = \mathrm{typ}(\mathscr{U}^{bc}_{\mathscr{C}_2\psi_2}) = \alpha$. Since the functions $\mathscr{U}^{bc}_{\mathscr{C}_1\psi_1}$ and $\mathscr{U}^{bc}_{\mathscr{C}_2\psi_2}$ are both bounded from above, we obtain the function $\mathscr{U}^{bc}_{\mathscr{C}\psi} = \underline{\max}(\mathscr{U}^{bc}_{\mathscr{C}_1\psi_1}, \mathscr{U}^{bc}_{\mathscr{C}_2\psi_2})$ is also bounded from above. From this, it follows that $\mathrm{typ}(\mathscr{U}^{bc}_{\mathscr{C}\psi}) = \widetilde{\max}(\mathrm{typ}(\mathscr{U}^{bc}_{\mathscr{C}_1\psi_1}), \mathrm{typ}(\mathscr{U}^{bc}_{\mathscr{C}_2\psi_2})) = \alpha$.

(b) Let $\mathrm{typ}(\mathscr{U}^{bc}_{\mathscr{C}_1\psi_1}) = \mathrm{typ}(\mathscr{U}^{bc}_{\mathscr{C}_2\psi_2}) = \beta$. From the fact that $\mathrm{Dom}^+(\mathscr{U}^{bc}_{\mathscr{C}_1\psi_1})$ and $\mathrm{Dom}^+(\mathscr{U}^{bc}_{\mathscr{C}_2\psi_2})$ are both finite, we obtain $\mathrm{Dom}^+(\mathscr{U}^{bc}_{\mathscr{C}\psi})$ is also finite. Similarly, one can show that $\mathscr{U}^{bc}_{\mathscr{C}\psi}$ is unbounded from above on $\mathscr{C}$. From here it follows that $\mathrm{typ}(\mathscr{U}^{bc}_{\mathscr{C}\psi}) = \widetilde{\max}(\mathrm{typ}(\mathscr{U}^{bc}_{\mathscr{C}_1\psi_1}), \mathrm{typ}(\mathscr{U}^{bc}_{\mathscr{C}_2\psi_2})) = \beta$.

(c) Let $\mathrm{typ}(\mathscr{U}^{bc}_{\mathscr{C}_1\psi_1}) = \mathrm{typ}(\mathscr{U}^{bc}_{\mathscr{C}_2\psi_2}) = \gamma$. From here, it follows that the function ψ^b is unbounded from above on $\mathscr{C}$. From Proposition 6.1 it follows that bc belongs to the set $\{ii, di, dd, da, ai, ad, aa\}$. Let $c = b$. Using Lemma 6.4, we obtain $\mathrm{typ}(\mathscr{U}^{bb}_{\mathscr{C}\psi}) = \gamma$. Let $bc \in \{di, ai, ad\}$. Using Lemma 6.3 and inequalities $\mathrm{typ}(\mathscr{U}^{bc}_{\mathscr{C}_1\psi_1}) \preceq \mathrm{typ}(\mathscr{U}^{bc}_{\mathscr{C}\psi})$ and $\mathrm{typ}(\mathscr{U}^{bc}_{\mathscr{C}_2\psi_2}) \preceq \mathrm{typ}(\mathscr{U}^{bc}_{\mathscr{C}\psi})$, we obtain $\mathrm{typ}(\mathscr{U}^{bc}_{\mathscr{C}\psi}) = \gamma$. The only case left is when $bc = da$. Since, there is no $n \in \omega$ for which $\mathscr{U}^{bc}_{\mathscr{C}_1\psi_1}(n) = \infty$ or $\mathscr{U}^{bc}_{\mathscr{C}_2\psi_2}(n) = \infty$, according to Lemma 6.2, we obtain $\mathrm{Dom}(\mathscr{U}^{bc}_{\mathscr{C}\psi})$ is an infinite set. Therefore, $\mathrm{typ}(\mathscr{U}^{bc}_{\mathscr{C}\psi}) \neq \epsilon$ and hence $\mathrm{typ}(\mathscr{U}^{bc}_{\mathscr{C}\psi}) \in \{\gamma, \delta\}$. From Proposition 6.6 it follows that both cases are possible.

(d) Let $\mathrm{typ}(\mathscr{U}^{bc}_{\mathscr{C}_1\psi_1}) = \mathrm{typ}(\mathscr{U}^{bc}_{\mathscr{C}_2\psi_2}) = \delta$. From here, it follows that there is no $n \in \omega$ for which $\mathscr{U}^{bc}_{\mathscr{C}_1\psi_1}(n) = \infty$ or $\mathscr{U}^{bc}_{\mathscr{C}_2\psi_2}(n) = \infty$. Using Lemma 6.2, we conclude that $\mathrm{Dom}(\mathscr{U}^{bc}_{\mathscr{C}\psi})$ is an infinite set. From the fact that $\mathrm{Dom}^-(\mathscr{U}^{bc}_{\mathscr{C}_1\psi_1})$ and $\mathrm{Dom}^-(\mathscr{U}^{bc}_{\mathscr{C}_2\psi_2})$ are both finite, we obtain $\mathrm{Dom}^-(\mathscr{U}^{bc}_{\mathscr{C}\psi})$ is also finite. Therefore, $\mathrm{typ}(\mathscr{U}^{bc}_{\mathscr{C}\psi}) = \widetilde{\max}(\mathrm{typ}(\mathscr{U}^{bc}_{\mathscr{C}_1\psi_1}), \mathrm{typ}(\mathscr{U}^{bc}_{\mathscr{C}_2\psi_2})) = \delta$.

(e) Let $\text{typ}(\mathcal{U}^{bc}_{\mathcal{C}_1\psi_1}) = \text{typ}(\mathcal{U}^{bc}_{\mathcal{C}_2\psi_2}) = \epsilon$. Since both $\text{Dom}(\mathcal{U}^{bc}_{\mathcal{C}_1\psi_1})$ and $\text{Dom}(\mathcal{U}^{bc}_{\mathcal{C}_2\psi_2})$ are finite sets, we obtain $\text{Dom}(\mathcal{U}^{bc}_{\mathcal{C}\psi})$ is also a finite set. Therefore, $\text{typ}(\mathcal{U}^{bc}_{\mathcal{C}\psi}) = \widetilde{\max}(\text{typ}(\mathcal{U}^{bc}_{\mathcal{C}_1\psi_1}), \text{typ}(\mathcal{U}^{bc}_{\mathcal{C}_2\psi_2})) = \epsilon$.

(2) Let $\text{typ}(\mathcal{U}^{bc}_{\mathcal{C}_1\psi_1}) \neq \text{typ}(\mathcal{U}^{bc}_{\mathcal{C}_2\psi_2})$. Denote $f = \mathcal{U}^{bc}_{\mathcal{C}_1\psi_1}$ and $g = \mathcal{U}^{bc}_{\mathcal{C}_2\psi_2}$. Let $\text{typ}(f) \preceq \text{typ}(g)$. We now consider a number of cases.

(a) Let $\text{typ}(g) = \epsilon$. From here, it follows that $\text{Dom}(g)$ is a finite set. Taking into account this fact, we obtain $\text{Dom}(\underline{\max}(f, g))$ is also a finite set. Therefore, $\text{typ}(\underline{\max}(f, g)) = \widetilde{\max}(\text{typ}(f), \text{typ}(g)) = \epsilon$. Later we will assume that $\text{typ}(g) \neq \epsilon$.

(b) Let $\text{typ}(f) = \alpha$. Then both f and g are nondecreasing functions, f is bounded from above and g is unbounded from above. From here, it follows that there exists $n_0 \in \omega$ such that $f(n) < g(n)$ for any $n \in \omega, n \geq n_0$. Using this fact, we conclude that $\underline{\max}(f(n), g(n)) = g(n)$ for $n \geq n_0$. Therefore, $\text{typ}(\underline{\max}(f, g)) = \widetilde{\max}(\text{typ}(f), \text{typ}(g)) = \text{typ}(g)$. Later we will assume that $\text{typ}(f) \neq \alpha$. It means we should consider only pairs $(\text{typ}(f), \text{typ}(g)) \in \{(\beta, \delta), (\beta, \gamma), (\gamma, \delta)\}$.

(c) Let $\text{typ}(f) = \beta, \text{typ}(g) = \delta$. From here, it follows that $\text{Dom}^-(f), \text{Dom}^+(g)$ are both infinite sets and $\text{Dom}^+(f), \text{Dom}^-(g)$ are both finite sets. Taking into account that both f and g are nondecreasing functions, we obtain that there exists $n_0 \in \omega$ such that $f(n) < g(n)$ for any $n \in \omega, n \geq n_0$. Therefore, $\text{typ}(\underline{\max}(f, g)) = \widetilde{\max}(\text{typ}(f), \text{typ}(g)) = \text{typ}(g) = \delta$.

(d) Let $\text{typ}(f) = \beta, \text{typ}(g) = \gamma$. Then $\text{Dom}^+(\underline{\max}(f, g))$ is an infinite set. Taking into account that $\text{Dom}^-(g)$ is an infinite set and $\text{Dom}^+(f)$ is a finite set, we obtain $\text{Dom}^-(\underline{\max}(f, g))$ is also an infinite set. Therefore, $\text{typ}(\underline{\max}(f, g)) = \widetilde{\max}(\text{typ}(f), \text{typ}(g)) = \text{typ}(g) = \gamma$.

(e) Let $\text{typ}(f) = \gamma, \text{typ}(g) = \delta$. From here, it follows that $\text{Dom}^+(\underline{\max}(f, g))$ is an infinite set and $\text{Dom}^-(\underline{\max}(f, g))$ is a finite set. Therefore, $\text{typ}(\underline{\max}(f, g)) = \widetilde{\max}(\text{typ}(f), \text{typ}(g)) = \text{typ}(g) = \delta$. □

The next statement follows immediately from Proposition 6.1 and Theorem 6.3.

Corollary 6.2 *Let τ_1 and τ_2 be compatible cm-pairs and τ be a union of these cm-pairs. Then the possible values of $\text{typ}_u(\tau)$ are in the table shown in Fig. 6.9, in the intersection of the row labeled with $\text{typ}_u(\tau_1)$ and the column labeled with $\text{typ}_u(\tau_2)$.*

To finalize the study of unions of cm-pairs, we prove the following statement.

Proposition 6.6 *There exist compatible cm-pairs τ_1^1 and τ_2^1 and their union τ^1 such that $\text{typ}_u(\tau_1^1) = \text{typ}_u(\tau_2^1) = \text{typ}_u(\tau^1) = t_5$.*

There exist compatible cm-pairs τ_1^2 and τ_2^2 and their union τ^2 such that $\text{typ}_u(\tau_1^2) = \text{typ}_u(\tau_2^2) = t_5$ and $\text{typ}_u(\tau^2) = t_6$.

Proof For $i \in \omega$, we denote $F_i = \{a_i, b_i, c_i\}$ and G_i the decision table depicted in Fig. 6.10. We study the cm-pair $(\mathcal{T}_i, \psi_i)$, where $\mathcal{T}_i$ is the closed class of decision tables from $\mathcal{M}_2^\infty(F_i)$, which is equal to $[G_i]$, and ψ_i is a complexity measure over $\mathcal{M}_2^\infty(F_i)$ defined in the following way: $\psi_i(\lambda) = 0$, $\psi_i(a_i) = \psi_i(b_i) = \psi_i(c_i) = i$ and $\psi_i(\alpha) = i + 1$ if $\alpha \in F_i^*$ and $|\alpha| \geq 2$.

We now study the function $\mathcal{U}^{da}_{\mathcal{T}_i\psi_i}$. Since the operations of duplication of columns and permutation of columns do not change the minimum complexity of deterministic

	t_1	t_2	t_3	t_4	t_5	t_6	t_7
t_1	t_1	t_2	t_3	t_4	t_5	t_6	t_7
t_2	t_2	t_2	t_3	t_4	t_5	t_6	t_7
t_3	t_3	t_3	t_3	t_4	t_7	t_7	t_7
t_4	t_4	t_4	t_4	t_4	t_7	t_7	t_7
t_5	t_5	t_5	t_7	t_7	t_5,t_6	t_6	t_7
t_6	t_6	t_6	t_7	t_7	t_6	t_6	t_7
t_7	t_7	t_7	t_7	t_7	t_7	t_7	t_7

Fig. 6.9 Possible upper types of a union of two compatible cm-pairs

$G_i =$

a_i	b_i	c_i	
1	0	0	{1}
0	1	0	{2}
0	0	1	{3}

Fig. 6.10 Decision table G_i

and nondeterministic decision trees, we only consider the operations of changing of decisions and removal of columns.

By these operations, decision tables from $\mathcal{T}_i$ can be obtained from G_i in three ways: (1) only changing of decisions, (2) removing one column and changing of decisions, and (3) removing two columns and changing of decisions. Figure 6.11 demonstrates examples of decision tables from $\mathcal{T}_i$ for each case. Without loss of generality, we can restrict ourselves to considering these three tables H_1, H_2, and H_3.

(a) There are three different cases for the table H_1: (i) the sets of decisions d_1, d_2, d_3 are pairwise disjoint, (ii) there are $l, t \in \{1, 2, 3\}$ such that $l \neq t$, $d_l \cap d_t \neq \varnothing$ and $d_1 \cap d_2 \cap d_3 = \varnothing$, and (iii) $d_1 \cap d_2 \cap d_3 \neq \varnothing$. In the first case, $\psi_i^a(H_1) = i$ and $\psi_i^d(H_1) = i + 1$. In the second case, $\psi_i^a(H_1) = i$ and $\psi_i^d(H_1) = i$. In the third case, $\psi_i^a(H_1) = 0$ and $\psi_i^d(H_1) = 0$.

(b) There are three different cases for the table H_2: (i) the sets of decisions d_4, d_5, d_6 are pairwise disjoint, (ii) there are $l, t \in \{4, 5, 6\}$ such that $l \neq t$, $d_l \cap d_t \neq \varnothing$ and $d_4 \cap d_5 \cap d_6 = \varnothing$, and (iii) $d_4 \cap d_5 \cap d_6 \neq \varnothing$. In the first case, $\psi_i^a(H_2) = i + 1$ and $\psi_i^d(H_2) = i + 1$. In the second case, we have either $\psi_i^a(H_2) = \psi_i^d(H_2) = i + 1$ or

1) $H_1 =$

a_i	b_i	c_i	
1	0	0	d_1
0	1	0	d_2
0	0	1	d_3

2) $H_2 =$

a_i	b_i	
0	0	d_4
1	0	d_5
0	1	d_6

3) $H_3 =$

c_i	
0	d_7
1	d_8

Fig. 6.11 Decision tables from closed class $\mathcal{T}_i$, where $d_1, \ldots, d_8 \in \mathcal{P}(\omega)$

$\psi_i^a(H_2) = \psi_i^d(H_2) = i$ depending on the intersecting decision sets. In the third case, $\psi_i^a(H_2) = 0$ and $\psi_i^d(H_2) = 0$.

(c) There are two different cases for the table H_3: (i) $d_7 \cap d_8 = \varnothing$ and (ii) $d_7 \cap d_8 \neq \varnothing$. In the first case, $\psi_i^a(H_3) = i$ and $\psi_i^d(H_3) = i$. In the second case, $\psi_i^a(H_3) = 0$ and $\psi_i^d(H_3) = 0$.

As a result, we obtain that, for any $n \in \omega$,

$$\mathcal{U}^{da}_{\mathcal{T}_i \psi_i}(n) = \begin{cases} 0, & n < i\,, \\ i+1, & n \geq i\,. \end{cases} \tag{6.3}$$

Let K be an infinite subset of the set ω. Denote $F_K = \bigcup_{i \in K} F_i$ and $\mathcal{T}_K = \bigcup_{i \in K}[G_i]$. It is clear that $\mathcal{T}_K$ is a closed class of decision tables from $\mathcal{M}_2^\infty(F_K)$. We now define a complexity measure ψ_K over $\mathcal{M}_2^\infty(F_K)$. Let $\alpha \in F_K^*$. If $\alpha \in F_i^*$ for some $i \in K$, then $\psi_K(\alpha) = \psi_i(\alpha)$. If α contains letters both from F_i and F_j, $i \neq j$, then $\psi_K(\alpha) = 0$.

Let $K = \{n_j : j \in \omega\}$ and $n_j < n_{j+1}$ for any $j \in \omega$. We define a function $\varphi_K : \omega \to \omega$ as follows. Let $n \in \omega$. If $n < n_0$, then $\varphi_K(n) = 0$. Let, for some $j \in \omega$, $n_j \leq n < n_{j+1}$. Then $\varphi_K(n) = n_j$. Using (6.3), one can show that, for any $n \in \omega$,

$$\mathcal{U}^{da}_{\mathcal{T}_K \psi_K}(n) = \varphi_K(n)\,.$$

Using this equality, one can prove that $\mathrm{typ}(\mathcal{U}^{da}_{\mathcal{T}_K \psi_K}) = \gamma$ if the set $\omega \setminus K$ is infinite and $\mathrm{typ}(\mathcal{U}^{da}_{\mathcal{T}_K \psi_K}) = \delta$ if the set $\omega \setminus K$ is finite.

Denote $K_1^1 = \{3j : j \in \omega\}$, $K_2^1 = \{3j+1 : j \in \omega\}$ and $K^1 = K_1^1 \cup K_2^1$. Denote $\tau_1^1 = (\mathcal{T}_{K_1^1}, \psi_{K_1^1})$, $\tau_2^1 = (\mathcal{T}_{K_2^1}, \psi_{K_2^1})$ and $\tau^1 = (\mathcal{T}_{K^1}, \psi_{K^1})$. One can show that cm-pairs τ_1^1 and τ_2^1 are compatible and τ^1 is a union of τ_1^1 and τ_2^1. It is easy to prove that $\mathrm{typ}(\mathcal{U}^{da}_{\mathcal{T}_{K_1^1} \psi_{K_1^1}}) = \mathrm{typ}(\mathcal{U}^{da}_{\mathcal{T}_{K_2^1} \psi_{K_2^1}}) = \mathrm{typ}(\mathcal{U}^{da}_{\mathcal{T}_{K^1} \psi_{K^1}}) = \gamma$. Using Proposition 6.2, we obtain $\mathrm{typ}_u(\tau_1^1) = \mathrm{typ}_u(\tau_2^1) = \mathrm{typ}_u(\tau^1) = t_5$.

Denote $K_1^2 = \{2j : j \in \omega\}$, $K_2^2 = \{2j+1 : j \in \omega\}$ and $K^2 = K_1^2 \cup K_2^2 = \omega$. Denote $\tau_1^2 = (\mathcal{T}_{K_1^2}, \psi_{K_1^2})$, $\tau_2^2 = (\mathcal{T}_{K_2^2}, \psi_{K_2^2})$ and $\tau^2 = (\mathcal{T}_{K^2}, \psi_{K^2})$. One can show that cm-pairs τ_1^2 and τ_2^2 are compatible and τ^2 is a union of τ_1^2 and τ_2^2. It is easy to prove that $\mathrm{typ}(\mathcal{U}^{da}_{\mathcal{T}_{K_1^2} \psi_{K_1^2}}) = \mathrm{typ}(\mathcal{U}^{da}_{\mathcal{T}_{K_2^2} \psi_{K_2^2}}) = \gamma$ and $\mathrm{typ}(\mathcal{U}^{da}_{\mathcal{T}_{K^2} \psi_{K^2}}) = \delta$. Using Proposition 6.2, we obtain $\mathrm{typ}_u(\tau_1^2) = \mathrm{typ}_u(\tau_2^2) = t_5$ and $\mathrm{typ}_u(\tau^2) = t_6$. □

6.7 Proofs of Theorems 6.1 and 6.2

First, we consider some auxiliary statements.

Definition 6.25 Let us define a function $\rho : \{\alpha, \beta, \gamma, \delta, \epsilon\} \to \{\alpha, \beta, \gamma, \delta, \epsilon\}$, as follows: $\rho(\alpha) = \epsilon$, $\rho(\beta) = \delta$, $\rho(\gamma) = \gamma$, $\rho(\delta) = \beta$, $\rho(\epsilon) = \alpha$.

Proposition 6.7 (Proposition 2.5) *Let X be a nonempty set, $f : X \to \omega, g : X \to \omega$, $\mathcal{U}^{fg}(n) = \max\{f(x) : x \in X, g(x) \leq n\}$ and $\mathcal{L}^{gf}(n) = \min\{g(x) : x \in X, f(x) \geq n\}$ for any $n \in \omega$. Then* $\text{typ}(\mathcal{L}^{gf}) = \rho(\text{typ}\left(\mathcal{U}^{fg}\right))$.

Using Proposition 6.7, we obtain the following statement.

Proposition 6.8 *Let $(\mathcal{C}, \psi)$ be a cm-pair and $b, c \in \{i, d, a\}$. Then* $\text{typ}(\mathcal{L}^{cb}_{\mathcal{C}\psi}) = \rho(\text{typ}(\mathcal{U}^{bc}_{\mathcal{C}\psi}))$.

Corollary 6.3 *Let $(\mathcal{C}, \psi)$ be a cm-pair and $i \in \{1, \ldots, 7\}$. Then* $\text{typ}_u(\mathcal{C}, \psi) = t_i$ *if and only if* $\text{typ}(\mathcal{C}, \psi) = T_i$.

Proof of Theorem 6.1. The statement of the theorem follows from Propositions 6.1 and 6.3 and Corollary 6.3. □

Proof of Theorem 6.2. The statement of the theorem follows from Propositions 6.2 and 6.4 and Corollary 6.3. □

References

1. Alsolami, F., Azad, M., Chikalov, I., Moshkov, M.: Decision and Inhibitory Trees and Rules for Decision Tables with Many-valued Decisions. In: Intelligent Systems Reference Library, vol. 156. Springer (2020)
2. Boutell, M.R., Luo, J., Shen, X., Brown, C.M.: Learning multi-label scene classification. Pattern Recognit. **37**(9), 1757–1771 (2004)
3. Moshkov, M.: Comparative analysis of deterministic and nondeterministic decision tree complexity local approach. In: Peters, J.F., Skowron, A. (eds.) Transactions on Rough Sets IV. Lecture Notes in Computer Science, vol. 3700, pp. 125–143. Springer (2005)
4. Moshkov, M.: Time complexity of decision trees. In: Peters, J.F., Skowron, A. (eds.) Transactions on Rough Sets III. Lecture Notes in Computer Science, vol. 3400, pp. 244–459. Springer (2005)
5. Moshkov, M., Zielosko, B.: Combinatorial Machine Learning—A Rough Set Approach. Studies in Computational Intelligence, vol. 360. Springer (2011)
6. Ostonov, A., Moshkov, M.: Comparative analysis of deterministic and nondeterministic decision trees for decision tables from closed classes (2023). CoRR arXiv:2304.10594
7. Ostonov, A., Moshkov, M.: Comparative analysis of deterministic and nondeterministic decision trees for decision tables from closed classes. Entropy **26**(6), 519 (2024)
8. Pawlak, Z.: Information systems theoretical foundations. Inf. Syst. **6**(3), 205–218 (1981)
9. Vens, C., Struyf, J., Schietgat, L., Dzeroski, S., Blockeel, H.: Decision trees for hierarchical multi-label classification. Mach. Learn. **73**(2), 185–214 (2008)
10. Zhou, Z., Zhang, M., Huang, S., Li, Y.: Multi-instance multi-label learning. Artif. Intell. **176**(1), 2291–2320 (2012)

Chapter 7
Complexity of Deterministic and Nondeterministic Decision Trees for Decision Tables with Many-Valued Decisions from Closed Classes

7.1 Introduction

In this chapter, we consider closed classes of decision tables with many-valued decisions. For tables from an arbitrary closed class, we study functions that characterize the dependence in the worst case of the minimum complexity of deterministic and nondeterministic decision trees on the complexity of the set of attributes attached to columns (in particular, for the depth, the complexity of a set of attributes is equal to its cardinality). We also study the dependence in the worst case of the minimum complexity of deterministic decision trees on the minimum complexity of nondeterministic decision trees.

A decision table with many-valued decisions is a rectangular table in which columns are labeled with attributes, rows are pairwise different and each row is labeled with a nonempty finite set of decisions. Rows are interpreted as tuples of values of the attributes. For a given row, it is required to find a decision from the set of decisions attached to the row. To this end, we can use the following queries: we can choose an attribute and ask what is the value of this attribute in the considered row. We study two types of algorithms based on these queries: deterministic and nondeterministic decision trees. One can interpret nondeterministic decision trees for a decision table as a way to represent an arbitrary system of true decision rules for this table that cover all rows. We consider so-called bounded complexity measures that characterize the time complexity of decision trees, for example, the depth of decision trees.

Decision tables with many-valued decisions often appear in data analysis, where they are known as multi-label decision tables [2, 11, 12]. Moreover, decision tables with many-valued decisions are common in such areas as combinatorial optimization, computational geometry, and fault diagnosis, where they are used to represent and explore problems [1, 5].

K. Durdymyradov et al., *Decision Trees Versus Systems of Decision Rules*, Studies in Big Data 160, https://doi.org/10.1007/978-3-031-71586-0_7

We study classes of decision tables with many-valued decisions closed under removal of columns (attributes) and changing the decisions (really, sets of decisions). The most natural examples of such classes are closed classes of decision tables generated by information systems [10]. An information system consists of a set of objects (universe) and a set of attributes (functions) defined on the universe and with values from a finite set. A problem with many-valued decisions over an information system is specified by a finite number of attributes that divide the universe into nonempty domains in which these attributes have fixed values. A nonempty finite set of decisions is attached to each domain. For a given object from the universe, it is required to find a decision from the set attached to the domain containing this object.

A decision table with many-valued decisions corresponds to this problem in a natural way: columns of this table are labeled with the considered attributes, rows correspond to domains and are labeled with sets of decisions attached to domains. The set of decision tables corresponding to problems with many-valued decisions over an information system forms a closed class generated by this system. Note that the family of all closed classes is essentially wider than the family of closed classes generated by information systems. In particular, the union of two closed classes generated by two information systems is a closed class. However, generally, there is no an information system that generates this class.

Let A be a class of decision tables with many-valued decisions closed under removal of columns and changing of decisions, and ψ be a bounded complexity measure. In this chapter, we study three functions: $\mathcal{F}^{\infty}_{\psi,A}(n)$, $\mathcal{G}^{\infty}_{\psi,A}(n)$, and $\mathcal{H}^{\infty}_{\psi,A}(n)$.

The function $\mathcal{F}^{\infty}_{\psi,A}(n)$ characterizes the growth in the worst case of the minimum complexity of a deterministic decision tree for a decision table from A with the growth of the complexity of the set of attributes attached to columns of the table. We prove that the function $\mathcal{F}^{\infty}_{\psi,A}(n)$ is either bounded from above by a constant, or grows as a logarithm of n, or grows almost linearly depending on n (it is bounded from above by n and is equal to n for infinitely many n).

The function $\mathcal{G}^{\infty}_{\psi,A}(n)$ characterizes the growth in the worst case of the minimum complexity of a nondeterministic decision tree for a decision table from A with the growth of the complexity of the set of attributes attached to columns of the table. We prove that the function $\mathcal{G}^{\infty}_{\psi,A}(n)$ is either bounded from above by a constant or grows almost linearly depending on n (it is bounded from above by n and is equal to n for infinitely many n).

The function $\mathcal{H}^{\infty}_{\psi,A}(n)$ characterizes the growth in the worst case of the minimum complexity of a deterministic decision tree for a decision table from A with the growth of the minimum complexity of a nondeterministic decision tree for the table. We indicated the condition for the function $\mathcal{H}^{\infty}_{\psi,A}(n)$ to be defined everywhere. Let $\mathcal{H}^{\infty}_{\psi,A}(n)$ be everywhere defined. We proved that this function is either bounded from above by a constant, or is greater than or equal to n for infinitely many n. In particular, for any nondecreasing function φ such that $\varphi(n) \geq n$ and $\varphi(0) = 0$, the function $\mathcal{H}^{\infty}_{\psi,A}(n)$ can grow between $\varphi(n)$ and $\varphi(n) + n$. We indicated also conditions for the function $\mathcal{H}^{\infty}_{\psi,A}(n)$ to be bounded from above by a polynomial on n.

In this chapter, we also study the joint behavior of the functions $\mathcal{F}^{\infty}_{\psi,A}$ and $\mathcal{G}^{\infty}_{\psi,A}$, and list all three possible types of this behavior.

Note that there is a similarity between some results obtained in this chapter for closed classes of decision tables and results from the book [4] obtained for problems over information systems. However, the results of the present chapter are more general since the family of all closed classes is essentially wider than the family of closed classes generated by information systems.

The present chapter is based mainly on the papers [6–9]. The chapter consists of six sections. In Sect. 7.2, main definitions and notation are considered. In Sect. 7.3, we provide the main results. In Sects. 7.4–7.6, we prove the main results.

7.2 Main Definitions and Notation

Denote $\omega = \{0, 1, 2, \ldots\}$, $\mathscr{P}(\omega)$ the set of nonempty finite subsets of the set ω and, for any $k \in \omega \setminus \{0, 1\}$, denote $E_k = \{0, 1, \ldots, k-1\}$. Let $P = \{f_i : i \in \omega\}$ be the set of *attributes* (really, names of attributes). Two attributes $f_i, f_j \in P$ are considered *different* if $i \neq j$.

7.2.1 Decision Tables

First, we define the notion of a decision table.

Definition 7.1 Let $k \in \omega \setminus \{0, 1\}$. Denote by $\mathscr{M}_k^\infty$ the set of rectangular tables filled with numbers from E_k in each of which rows are pairwise different, each row is labeled with a set from $\mathscr{P}(\omega)$ (set of decisions), and columns are labeled with pairwise different attributes from P. Rows are interpreted as tuples of values of these attributes. Empty tables without rows belong also to the set $\mathscr{M}_k^\infty$. We will use the same notation Λ for these tables. Tables from $\mathscr{M}_k^\infty$ will be called *decision tables with many-valued decisions* (*decision tables*).

For a table $T \in \mathscr{M}_k^\infty$, we denote by $\Delta(T)$ the set of rows of the table T and by $\Pi(T)$ we denote the intersection of sets of decisions attached to rows of T. Decisions from $\Pi(T)$ are called *common decisions* for the table T.

Example 7.1 Figure 7.1 shows a decision table from $\mathscr{M}_2^\infty$.

Denote by $\mathscr{M}_k^{\infty c}$ the set of tables from $\mathscr{M}_k^\infty$ in each of which there exists a common decision. Let $\Lambda \in \mathscr{M}_k^{\infty c}$.

Let T be a nonempty table from $\mathscr{M}_k^\infty$. Denote by $\mathrm{At}(T)$ the set of attributes attached to columns of the table T. We denote by $\Omega_k(T)$ the set of finite words over the alphabet $\{(f_i, \delta) : f_i \in \mathrm{At}(T), \delta \in E_k\}$ including the empty word λ. For any $\alpha \in \Omega_k(T)$, we now define a subtable $T\alpha$ of the table T. If $\alpha = \lambda$, then $T\alpha = T$. If $\alpha \neq \lambda$ and $\alpha = (f_{i_1}, \delta_1) \cdots (f_{i_m}, \delta_m)$, then $T\alpha$ is the table obtained from T by

Fig. 7.1 Decision table from $\mathcal{M}_2^\infty$

f_2	f_4	f_3	
1	1	1	{1}
0	1	1	{0,1,2}
1	1	0	{1,3}
0	0	1	{2}
1	0	0	{3}
0	0	0	{2,3}

Fig. 7.2 Decision table obtained from the decision table shown in Fig. 7.1 by removal of a column and changing of decisions

f_2	f_3	
1	1	{1}
0	1	{0,1}
1	0	{0,1}
0	0	{0}

removal of all rows that do not satisfy the following condition: in columns labeled with attributes $f_{i_1}, \ldots, f_{i_m}$ the row has numbers $\delta_1, \ldots, \delta_m$, respectively.

We now define two operations on decision tables: removal of columns and changing of decisions. Let $T \in \mathcal{M}_k^\infty$.

Definition 7.2 (*Removal of columns*) Let $D \subseteq \mathrm{At}(T)$. We remove from T all columns labeled with the attributes from the set D. In each group of rows equal on the remaining columns, we keep the first one. Denote the obtained table by $I(D, T)$. In particular, $I(\emptyset, T) = T$ and $I(\mathrm{At}(T), T) = \Lambda$. It is obvious that $I(D, T) \in \mathcal{M}_k^\infty$.

Definition 7.3 (*Changing of decisions*) Let $\nu : E_k^{|\mathrm{At}(T)|} \to \mathcal{P}(\omega)$ (by definition, $E_k^0 = \emptyset$). For each row $\bar{\delta}$ of the table T, we replace the set of decisions attached to this row with $\nu(\bar{\delta})$. We denote the obtained table by $J(\nu, T)$. It is obvious that $J(\nu, T) \in \mathcal{M}_k^\infty$.

Definition 7.4 Set $[T] = \{J(\nu, I(D, T)) : D \subseteq \mathrm{At}(T), \nu : E_k^{|\mathrm{At}(T)\setminus D|} \to \mathcal{P}(\omega)\}$. The set $[T]$ is the *closure of the table* T under the operations of removal of columns and changing of decisions.

Example 7.2 Figure 7.2 shows the table $J(\nu, I(D, T_0))$, where T_0 is the table shown in Fig. 7.1, $D = \{f_4\}$ and $\nu(x_1, x_2) = \{\min(x_1, x_2), \max(x_1, x_2)\}$.

Definition 7.5 Let $A \subseteq \mathcal{M}_k^\infty$ and $A \neq \emptyset$. Denote $[A] = \bigcup_{T \in A} [T]$. The set $[A]$ is the *closure* of the set A under the considered two operations. The class (the set) of decision tables A will be called a *closed class* if $[A] = A$.

A closed class of decision tables will be called *nontrivial* if it contains nonempty decision tables.

Let A_1 and A_2 be closed classes of decision tables from $\mathcal{M}_k^\infty$. Then $A_1 \cup A_2$ is a closed class of decision tables from $\mathcal{M}_k^\infty$.

7.2.2 Deterministic and Nondeterministic Decision Trees

A *finite tree with root* is a finite directed tree in which exactly one node called the *root* has no entering edges. The nodes without leaving edges are called *terminal* nodes.

Definition 7.6 A *k-decision tree* is a finite tree with root, which has at least two nodes and in which

- The root and edges leaving the root are not labeled.
- Each terminal node is labeled with a decision from the set ω.
- Each node, which is neither the root nor a terminal node, is labeled with an attribute from the set P. Each edge leaving such node is labeled with a number from the set E_k.

Example 7.3 Figures 7.3 and 7.4 show 2-decision trees.

We denote by $\mathscr{T}_k$ the set of all k-decision trees. Let $\Gamma \in \mathscr{T}_k$. We denote by $\mathrm{At}(\Gamma)$ the set of attributes attached to nodes of Γ that are neither the root nor terminal nodes. A *complete path* of Γ is a sequence $\tau = v_1, d_1, \dots, v_m, d_m, v_{m+1}$ of nodes and edges of Γ in which v_1 is the root of Γ, v_{m+1} is a terminal node of Γ and, for $j = 1, \dots, m$, the edge d_j leaves the node v_j and enters the node v_{j+1}. Let $T \in \mathscr{M}_k^\infty$. If $\mathrm{At}(\Gamma) \subseteq \mathrm{At}(T)$, then we correspond to the table T and the complete path τ a word $\pi(\tau) \in \Omega_k(T)$. If $m = 1$, then $\pi(\tau) = \lambda$. If $m > 1$ and, for $j = 2, \dots, m$, the node v_j is labeled with the attribute f_{i_j} and the edge d_j is labeled with the number δ_j, then $\pi(\tau) = (f_{i_2}, \delta_2) \cdots (f_{i_m}, \delta_m)$. Denote $T(\tau) = T\pi(\tau)$.

Definition 7.7 Let $T \in \mathscr{M}_k^\infty \setminus \{\Lambda\}$. A *deterministic decision tree for the table T* is a k-decision tree Γ satisfying the following conditions:

- Only one edge leaves the root of Γ.

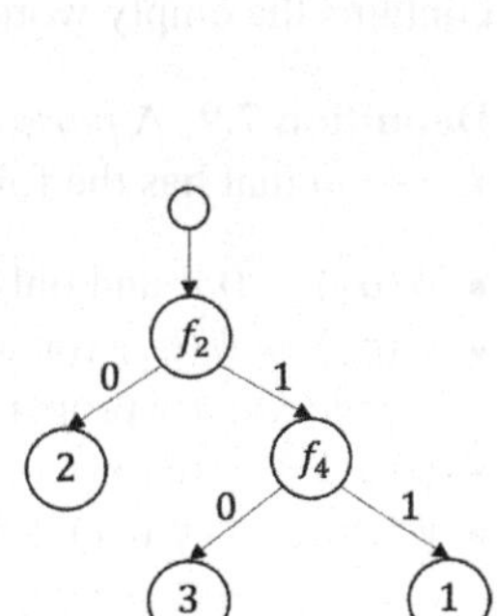

Fig. 7.3 A deterministic decision tree for the decision table shown in Fig. 7.1

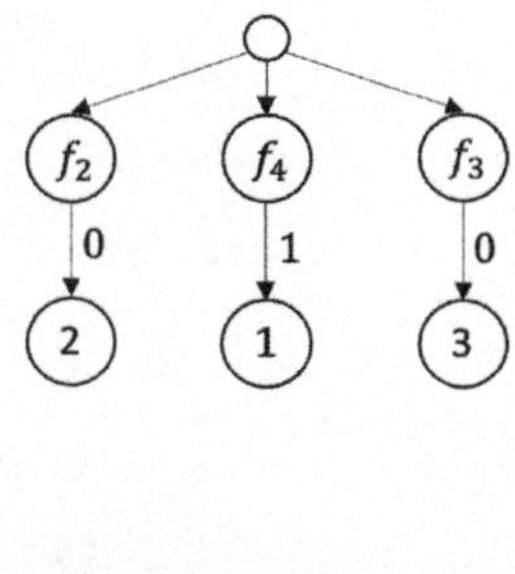

Fig. 7.4 A nondeterministic decision tree for the decision table shown in Fig. 7.1

- For any node, which is neither the root nor a terminal node, edges leaving this node are labeled with pairwise different numbers.
- $\mathrm{At}(\Gamma) \subseteq \mathrm{At}(T)$.
- For any row of T, there exists a complete path τ of Γ such that the considered row belongs to the table $T(\tau)$.
- For any complete path τ of Γ, either $T(\tau) = \Lambda$ or the decision attached to the terminal node of τ is a common decision for the table $T(\tau)$.

Example 7.4 The 2-decision tree shown in Fig. 7.3 is a deterministic decision tree for the decision table shown in Fig. 7.1.

Definition 7.8 Let $T \in \mathscr{M}_k^\infty \setminus \{\Lambda\}$. A *nondeterministic decision tree for the table* T is a k-decision tree Γ satisfying the following conditions:

- $\mathrm{At}(\Gamma) \subseteq \mathrm{At}(T)$.
- For any row of T, there exists a complete path τ of Γ such that the considered row belongs to the table $T(\tau)$.
- For any complete path τ of Γ, either $T(\tau) = \Lambda$ or the decision attached to the terminal node of τ is a common decision for the table $T(\tau)$.

Example 7.5 The 2-decision tree shown in Fig. 7.4 is a nondeterministic decision tree for the decision table shown in Fig. 7.1.

7.2.3 Complexity Measures

Denote by P^* the set of all finite words over the alphabet $P = \{f_i : i \in \omega\}$, which contains the empty word λ.

Definition 7.9 A *partially bounded complexity measure* is an arbitrary function $\psi : P^* \to \omega$ that has the following properties: for any words $\alpha_1, \alpha_2 \in P^*$,

- $\psi(\alpha_1) = 0$ if and only if $\alpha_1 = \lambda$ – *positivity* property.
- $\psi(\alpha_1) = \psi(\alpha_1')$ for any word α_1' obtained from α_1 by permutation of letters – *commutativity* property.
- $\psi(\alpha_1) \leq \psi(\alpha_1\alpha_2)$ – *nondecreasing* property.
- $\psi(\alpha_1\alpha_2) \leq \psi(\alpha_1) + \psi(\alpha_2)$ – *boundedness from above* property.

The following functions are partially bounded complexity measures:

- Function h for which, for any word $\alpha \in P^*$, $h(\alpha) = |\alpha|$, where $|\alpha|$ is the length of the word α. This function is called the *depth*.
- An arbitrary function $\varphi : P^* \to \omega$ such that $\varphi(\lambda) = 0$, for any $f_i \in P$, $\varphi(f_i) > 0$ and, for any nonempty word $f_{i_1} \cdots f_{i_m} \in P^*$,

$$\varphi(f_{i_1} \cdots f_{i_m}) = \sum_{j=1}^{m} \varphi(f_{i_j}) . \tag{7.1}$$

  This function is called the *weighted depth*.
- An arbitrary function $\rho : P^* \to \omega$ such that $\rho(\lambda) = 0$, for any $f_i \in P$, $\rho(f_i) > 0$, and, for any nonempty word $f_{i_1} \cdots f_{i_m} \in P^*$, $\rho(f_{i_1} \cdots f_{i_m}) = \max\{\rho(f_{i_j}) : j = 1, \ldots, m\}$.

Definition 7.10 A *bounded* complexity measure is a partially bounded complexity measure ψ, which has the *boundedness from below* property: for any word $\alpha \in P^*$, $\psi(\alpha) \geq |\alpha|$.

Any partially bounded complexity measure satisfying the equality (7.1), in particular the function h, is a bounded complexity measure. One can show that if functions ψ_1 and ψ_2 are partially bounded complexity measures, then the functions ψ_3 and ψ_4 are partially bounded complexity measures, where for any $\alpha \in P^*$, $\psi_3(\alpha) = \psi_1(\alpha) + \psi_2(\alpha)$ and $\psi_4(\alpha) = \max(\psi_1(\alpha), \psi_2(\alpha))$. If the function ψ_1 is a bounded complexity measure, then the functions ψ_3 and ψ_4 are bounded complexity measures.

Definition 7.11 Let ψ be a partially bounded complexity measure. We extend it to the set of all finite subsets of the set P. Let D be a finite subset of the set P. If $D = \emptyset$, then $\psi(D) = 0$. Let $D = \{f_{i_1}, \ldots, f_{i_m}\}$ and $m \geq 1$. Then $\psi(D) = \psi(f_{i_1} \cdots f_{i_m})$.

Definition 7.12 Let ψ be a partially bounded complexity measure. We extend it to the set of finite words Ω over the alphabet $\{(f_i, \delta) : f_i \in P, \delta \in \omega\}$ including the empty word λ. Let $\alpha \in \Omega$. If $\alpha = \lambda$, then $\psi(\alpha) = 0$. Let $\alpha = (f_{i_1}, \delta_1) \cdots (f_{i_m}, \delta_m)$ and $m \geq 1$. Then $\psi(\alpha) = \psi(f_{i_1} \cdots f_{i_m})$.

7.2.4 Parameters of Decision Trees and Tables

Definition 7.13 Let ψ be a partially bounded complexity measure. We extend the function ψ to the set $\mathscr{T}_k$. Let $\Gamma \in \mathscr{T}_k$. Then $\psi(\Gamma) = \max\{\psi(\pi(\tau))\}$, where the maximum is taken over all complete paths τ of the decision tree Γ. For a given partially bounded complexity measure ψ, the value $\psi(\Gamma)$ will be called the *complexity of the decision tree* Γ. The value $h(\Gamma)$ will be called the *depth of the decision tree* Γ.

Let ψ be a partially bounded complexity measure. We now describe the functions ψ^d, ψ^a, m_ψ, W_ψ, S_ψ, $\hat{S}_\psi$, N, Sep, M_ψ defined on the set $\mathscr{M}_k^\infty$ and functions Z, G defined on the set $\mathscr{M}_2^\infty$ and taking values from the set ω. By definition, the value of each of these functions for Λ is equal 0. We also describe the function l_ψ defined on the set $\mathscr{M}_k^\infty \times \omega$. By definition, the value of this function for tuple (Λ, n), $n \in \omega$, is equal 0. Let $T \in \mathscr{M}_k^\infty \setminus \{\Lambda\}$ and $n \in \omega$.

- $\psi^d(T) = \min\{\psi(\Gamma)\}$, where the minimum is taken over all deterministic decision trees Γ for the table T.
- $\psi^a(T) = \min\{\psi(\Gamma)\}$, where the minimum is taken over all nondeterministic decision trees Γ for the table T.
- $m_\psi(T) = \max\{\psi(f_i) : f_i \in \mathrm{At}(T)\}$.
- $W_\psi(T) = \psi(\mathrm{At}(T))$. This is *the complexity of the set of attributes attached to columns of the table T*.
- Let $\bar{\delta}$ be a row of the table T. Denote $S_\psi(T, \bar{\delta}) = \min\{\psi(D)\}$, where the minimum is taken over all subsets D of the set $\mathrm{At}(T)$ such that in the set of columns of T labeled with attributes from D the row $\bar{\delta}$ is different from all other rows of the table T. Then $S_\psi(T) = \max\{S_\psi(T, \bar{\delta})\}$, where the maximum is taken over all rows $\bar{\delta}$ of the table T.
- $\hat{S}_\psi(T) = \max\{S_\psi(T^*) : T^* \in [T]\}$.
- $N(T)$ is the number of rows in the table T.
- A set $D \subseteq \mathrm{At}(T)$ is called a *separating set* for the table T if in the set of columns labeled with attributes from D rows of the table T are pairwise different. Then $\mathrm{Sep}(T)$ is the minimum cardinality of a separating sets for the table T.
- If $T \in \mathscr{M}_k^{\infty c}$, then $M_\psi(T) = 0$. Let $T \notin \mathscr{M}_k^{\infty c}$, $|\mathrm{At}(T)| = n$, and columns of the table T be labeled with the attributes $f_{t_1}, \ldots, f_{t_n}$. Let $\bar{\delta} = (\delta_1, \ldots, \delta_n) \in E_k^n$. Denote by $M_\psi(T, \bar{\delta})$ the minimum number $p \in \omega$ for which there exist attributes $f_{t_{i_1}}, \ldots, f_{t_{im}} \in \mathrm{At}(T)$ such that $T(f_{t_{i_1}}, \delta_{i_1}) \cdots (f_{t_{im}}, \delta_{i_m}) \in \mathscr{M}_k^{\infty c}$ and $\psi(f_{t_{i_1}} \cdots f_{t_{im}}) = p$. Then $M_\psi(T) = \max\{M_\psi(T, \bar{\delta}) : \bar{\delta} \in E_k^n\}$.
- Let $T \in \mathscr{M}_2^\infty$. A decision table $Q \in \mathscr{M}_2^\infty$ will be called *complete* if $N(Q) = 2^{|\mathrm{At}(Q)|}$. Then $Z(T)$ is the maximum number of columns in a complete table from $[T]$ if such tables exist and 0 otherwise.
- Let $T \in \mathscr{M}_2^\infty$. A word $\alpha \in \Omega_2(T)$ will be called *annihilating* word for the table T if $T\alpha = \Lambda$ and α does not contain letters (f_i, δ) and (f_i, σ) such that $\delta \neq \sigma$. An annihilating word α for the table T will be called *irreducible* if any subword of α obtained from α by removal of some letters and different from α is not annihilating. Then $G(T)$ is the maximum length of an irreducible annihilating word for the table T if such words exist and 0 otherwise.
- Denote $\Omega_k^n(T) = \{\alpha : \alpha \in \Omega_k(T), \psi(\alpha) \leq n\}$. A finite set $U \subseteq \Omega_k^n(T)$ will be called *(ψ, n)-cover* of the table T if $\bigcup_{\alpha \in U} \Delta(T\alpha) = \Delta(T)$. The (ψ, n)-cover U will be called *irreducible* if each proper subset of U is not a (ψ, n)-cover of the table T. Then $l_\psi(T, n)$ is the maximum cardinality of an irreducible (ψ, n)-cover of the table T. It is clear that $\{\lambda\}$ is an irreducible (ψ, n)-cover of the table T. Therefore $l_\psi(T, n) \geq 1$.

For the bounded complexity measure h, we denote $W(T) = W_h(T)$, $S(T) = S_h(T)$, $\hat{S}(T) = \hat{S}_h(T)$, $M(T, \bar{\delta}) = M_h(T, \bar{\delta})$ and $M(T) = M_h(T)$. Note that $W(T)$ is the number of columns in the table T.

Example 7.6 We denote by T_0 the decision table shown in Fig. 7.1. One can show that $h^d(T_0) = 2$, $h^a(T_0) = 1$, $m_h(T_0) = 1$, $W(T_0) = 3$, $S(T_0) = 2$, $\hat{S}(T_0) = 2$, $N(T_0) = 6$, $\mathrm{Sep}(T_0) = 3$, $M(T_0) = 2$, $Z(T_0) = 2$, $G(T_0) = 3$, $l_h(T_0, 0) = 1$, $l_h(T_0, 1) = 3$ and $l_h(T_0, n) = 6$ for any $n \in \omega \setminus \{0, 1\}$.

7.3 Main Results

In this section, we consider results obtained for the functions $\mathscr{F}^{\infty}_{\psi,A}$, $\mathscr{G}^{\infty}_{\psi,A}$, and $\mathscr{H}^{\infty}_{\psi,A}$ and discuss closed classes of decision tables generated by information systems.

7.3.1 Function $\mathscr{F}^{\infty}_{\psi,A}$

Let ψ be a bounded complexity measure and A be a nontrivial closed class of decision tables from $\mathscr{M}^{\infty}_k$. We now define a function $\mathscr{F}^{\infty}_{\psi,A} : \omega \to \omega$. Let $n \in \omega$. Then

$$\mathscr{F}^{\infty}_{\psi,A}(n) = \max\{\psi^d(T) : T \in A, W_{\psi}(T) \leq n\}.$$

The function $\mathscr{F}^{\infty}_{\psi,A}$ characterizes the growth in the worst case of the minimum complexity of deterministic decision trees for decision tables from A with the growth of the complexity of the sets of attributes attached to columns of these tables.

Let $D = \{n_i : i \in \omega\}$ be an infinite subset of the set ω in which, for any $i \in \omega$, $n_i < n_{i+1}$. We now define a function $H_D : \omega \to \omega$. Let $n \in \omega$. If $n < n_0$, then $H_D(n) = 0$. If, for some $i \in \omega$, $n_i \leq n < n_{i+1}$, then $H_D(n) = n_i$.

Theorem 7.1 *Let ψ be a bounded complexity measure and A be a nontrivial closed class of decision tables from $\mathscr{M}^{\infty}_k$. Then $\mathscr{F}^{\infty}_{\psi,A}$ is an everywhere defined nondecreasing function such that $\mathscr{F}^{\infty}_{\psi,A}(n) \leq n$ for any $n \in \omega$ and $\mathscr{F}^{\infty}_{\psi,A}(0) = 0$. For this function, one of the following statements holds:*

(a) *If the functions S_{ψ} and N are bounded from above on the class A, then there exists a positive constant c_0 such that $\mathscr{F}^{\infty}_{\psi,A}(n) \leq c_0$ for any $n \in \omega$.*

(b) *If the function S_{ψ} is bounded from above on the class A and the function N is not bounded from above on the class A, then there exist positive constants c_1, c_2, c_3, c_4 such that $c_1 \log_2 n - c_2 \leq \mathscr{F}^{\infty}_{\psi,A}(n) \leq c_3 \log_2 n + c_4$ for any $n \in \omega \setminus \{0\}$.*

(c) *If the function S_{ψ} is not bounded from above on the class A, then there exists an infinite subset D of the set ω such that $H_D(n) \leq \mathscr{F}^{\infty}_{\psi,A}(n)$ for any $n \in \omega$.*

For the bounded complexity measure h, the bound from the statement (c) of Theorem 7.1 can be improved.

Theorem 7.2 *Let A be a nontrivial closed class of decision tables from $\mathcal{M}_k^\infty$ for which the function S is not bounded from above. Then $\mathcal{F}_{h,A}^\infty(n) = n$ for any $n \in \omega$.*

7.3.2 Function $\mathcal{G}_{\psi,A}^\infty$

Let ψ be a bounded complexity measure and A be a nontrivial closed class of decision tables from $\mathcal{M}_k^\infty$. We now define a function $\mathcal{G}_{\psi,A}^\infty$. Let $n \in \omega$. Then

$$\mathcal{G}_{\psi,A}^\infty(n) = \max\{\psi^a(T) : T \in A, W_\psi(T) \leq n\} .$$

The function $\mathcal{G}_{\psi,A}^\infty$ characterizes the growth in the worst case of the minimum complexity of nondeterministic decision trees for decision tables from A with the growth of the complexity of the sets of attributes attached to columns of these table.

Theorem 7.3 *Let ψ be a bounded complexity measure and A be a nontrivial closed class of decision tables from $\mathcal{M}_k^\infty$. Then $\mathcal{G}_{\psi,A}^\infty$ is an everywhere defined nondecreasing function such that $\mathcal{G}_{\psi,A}^\infty(n) \leq n$ for any $n \in \omega$ and $\mathcal{G}_{\psi,A}^\infty(0) = 0$. For this function, one of the following statements holds:*

(a) *If the function S_ψ is bounded from above on the class A, then there exists a positive constant c such that $\mathcal{G}_{\psi,A}^\infty(n) \leq c$ for any $n \in \omega$.*

(b) *If the function S_ψ is not bounded from above on the class A, then there exists an infinite subset D of the set ω such that $H_D(n) \leq \mathcal{G}_{\psi,A}^\infty(n)$ for any $n \in \omega$.*

For the bounded complexity measure h, the bound from the statement (b) of Theorem 7.3 can be improved.

Theorem 7.4 *Let A be a nontrivial closed class of decision tables from $\mathcal{M}_k^\infty$ for which the function S is not bounded from above. Then $\mathcal{G}_{h,A}^\infty(n) = n$ for any $n \in \omega$.*

7.3.3 Function $\mathcal{H}_{\psi,A}^\infty$

Let ψ be a bounded complexity measure and A be a nontrivial closed class of decision tables from $\mathcal{M}_k^\infty$. We now define possibly partial function $\mathcal{H}_{\psi,A}^\infty : \omega \to \omega$. Let $n \in \omega$. If the set $\{\psi^d(T) : T \in A, \psi^a(T) \leq n\}$ is infinite, then the value $\mathcal{H}_{\psi,A}^\infty(n)$ is undefined. Otherwise, $\mathcal{H}_{\psi,A}^\infty(n) = \max\{\psi^d(T) : T \in A, \psi^a(T) \leq n\}$.

The function $\mathcal{H}_{\psi,A}^\infty$ characterizes the growth in the worst case of the minimum complexity of deterministic decision trees for decision tables from A with the growth of the minimum complexity of nondeterministic decision trees for these tables.

We now define possibly partial function $L_{\psi,A} : \omega \to \omega$. Let $n \in \omega$. If the set $\{l_\psi(T, n) : T \in A\}$ is infinite, then the value $L_{\psi,A}(n)$ is not defined. Otherwise, $L_{\psi,A}(n) = \max\{l_\psi(T, n) : T \in A\}$. One can show that in this case $L_{\psi,A}(n) \geq 1$.

The following statement describes a criterion for the function $\mathcal{H}^{\infty}_{\psi,A}$ to be everywhere defined.

Theorem 7.5 *Let ψ be a bounded complexity measure and A be a nontrivial closed class of decision tables from $\mathcal{M}^{\infty}_{k}$. The function $\mathcal{H}^{\infty}_{\psi,A}$ is everywhere defined if and only if the function $L_{\psi,A}$ is everywhere defined.*

We now describe two possible types of behavior for everywhere defined function $\mathcal{H}_{\psi,A}$.

Theorem 7.6 *Let ψ be a bounded complexity measure, A be a nontrivial closed class of decision tables from $\mathcal{M}^{\infty}_{k}$, and the function $\mathcal{H}^{\infty}_{\psi,A}$ be everywhere defined. Then $\mathcal{H}^{\infty}_{\psi,A}$ is a nondecreasing function and $\mathcal{H}^{\infty}_{\psi,A}(0) = 0$. For this function, one of the following statements holds:*

(a) *If the function ψ^d is bounded from above on the class A, then there is a nonnegative constant c such that $\mathcal{H}^{\infty}_{\psi,A}(n) \leq c$ for any $n \in \omega$.*

(b) *If the function ψ^d is not bounded from above on the class A, then there exists an infinite subset D of the set ω such that $\mathcal{H}^{\infty}_{\psi,A}(n) \geq H_D(n)$ for any $n \in \omega$.*

Remark 7.1 From Theorem 7.1 it follows that the function ψ^d is bounded from above on the class A if and only if the functions S_ψ and N are bounded from above on the class A.

The following statement shows a wide spectrum of behavior of the everywhere defined function $\mathcal{H}^{\infty}_{\psi,A}$ in the case when the function ψ^d is not bounded from above on the class A.

Theorem 7.7 *Let $\varphi : \omega \to \omega$ be a nondecreasing function such that $\varphi(n) \geq n$ for any $n \in \omega$ and $\varphi(0) = 0$. Then there exist a closed class A of decision tables from $\mathcal{M}^{\infty}_{k}$ and a bounded complexity measure ψ such that the function $\mathcal{H}^{\infty}_{\psi,A}$ is everywhere defined and $\varphi(n) \leq \mathcal{H}^{\infty}_{\psi,A}(n) \leq \varphi(n) + n$ for any $n \in \omega$.*

Let A be a nontrivial closed class of decision tables from $\mathcal{M}^{\infty}_{2}$ and ψ be a bounded complexity measure. Let $n \in \omega$. Denote $A_\psi(n) = \{T : T \in A, m_\psi(T) \leq n\}$. Since $\Lambda \in A$, both sets $A_\psi(n)$ and $\{Z(T) : T \in A_\psi(n)\}$ are nonempty sets. We now define probably partial function $Z_{\psi,A} : \omega \to \omega$. If $\{Z(T) : T \in A_\psi(n)\}$ is an infinite set, then the value $Z_{\psi,A}(n)$ is not defined. Otherwise, $Z_{\psi,A}(n) = \max\{Z(T) : T \in A_\psi(n)\}$.

Let $n \in \omega$. Since $\Lambda \in A$, the set $\{G(T) : T \in A_\psi(n)\}$ is nonempty. We now define probably partial function $G_{\psi,A} : \omega \to \omega$. If $\{G(T) : T \in A_\psi(n)\}$ is an infinite set, then the value $G_{\psi,A}(n)$ is not defined. Otherwise, $G_{\psi,A}(n) = \max\{G(T) : T \in A_\psi(n)\}$.

The following statement describes the criterion for the everywhere defined function $\mathcal{H}^{\infty}_{\psi,A}$ to be bounded from above by a polynomial. For simplicity, we consider here only decision tables from the set $\mathcal{M}^{\infty}_{2}$.

Theorem 7.8 *Let A be a nontrivial closed class of decision tables from $\mathscr{M}_2^\infty$, ψ be a bounded complexity measure and the function $\mathscr{H}_{\psi,A}^\infty$ be everywhere defined. Then a polynomial p_0 such that $\mathscr{H}_{\psi,A}^\infty(n) \le p_0(n)$ for any $n \in \omega$ exists if and only if there exist polynomials p_1, p_2, and p_3 such that $Z_{\psi,A}(n) \le p_1(n)$, $G_{\psi,A}(n) \le p_2(n)$ and $L_{\psi,A}(n) \le 2^{p_3(n)}$ for any $n \in \omega$.*

7.3.4 Family of Closed Classes of Decision Tables

Let U be a set and $\Phi = \{f_0, f_1, \ldots\}$ be a finite or countable set of functions (attributes) defined on U and taking values from E_k. The pair (U, Φ) is called a *k-information system.*

A *problem* over (U, Φ) is an arbitrary tuple $z = (U, \nu, f_{i_1}, \ldots, f_{i_n})$, where $n \in \omega$, $\nu : E_k^n \to \mathscr{P}(\omega)$ and $f_{i_1}, \ldots, f_{i_n}$ are functions from Φ with pairwise different indices $i_1, \ldots, i_n$.

Let $n > 0$. The problem z is to determine a value from the set $\nu(f_{i_1}(u), \ldots, f_{i_n}(u))$ for a given $u \in U$. Various examples of k-information systems can be found in [3].

We denote by $T(z)$ a decision table from $\mathscr{M}_k^\infty$ with n columns labeled with attributes $f_{i_1}, \ldots, f_{i_n}$. A row $(\delta_1, \ldots, \delta_n) \in E_k^n$ belongs to the table $T(z)$ if and only if the system of equations $\{f_{i_1}(x) = \delta_1, \ldots, f_{i_n}(x) = \delta_n\}$ has a solution from the set U. This row is labeled with the set of decisions $\nu(\delta_1, \ldots, \delta_n)$.

Let the algorithms for the problem z solving be algorithms in which each elementary operation consists in calculating the value of some attribute from the set $\{f_{i_1}, \ldots, f_{i_n}\}$ on a given element $u \in U$. Then, as a model of the problem z, we can use the decision table $T(z)$, and as models of algorithms for the problem z solving–deterministic and nondeterministic decision trees for the table $T(z)$.

If $n = 0$, then we set $T(z) = \Lambda$.

Denote by $\mathrm{Probl}^\infty(U, \Phi)$ the set of problems over (U, Φ) and $\mathrm{Tab}^\infty(U, \Phi) = \{T(z) : z \in \mathrm{Probl}^\infty(U, \Phi)\}$. One can show that $\mathrm{Tab}^\infty(U, \Phi) = [\mathrm{Tab}^\infty(U, \Phi)]$, i.e., $\mathrm{Tab}^\infty(U, \Phi)$ is a closed class of decision tables from $\mathscr{M}_k^\infty$ *generated* by the information system (U, Φ).

Closed classes of decision tables generated by k-information systems are the most natural examples of closed classes. However, the notion of a closed class is essentially wider. In particular, the union $\mathrm{Tab}^\infty(U_1, \Phi_1) \cup \mathrm{Tab}^\infty(U_2, \Phi_2)$, where (U_1, Φ_1) and (U_2, Φ_2) are k-information systems, is a closed class, but generally, we cannot find an information system (U, Φ) such that $\mathrm{Tab}^\infty(U, \Phi) = \mathrm{Tab}^\infty(U_1, \Phi_1) \cup \mathrm{Tab}^\infty(U_2, \Phi_2)$.

7.3.5 Example of Information System

Let $\mathbb{R}$ be the set of real numbers and $F = \{f_i : i \in \omega\}$ be the set of functions defined on $\mathbb{R}$ and taking values from the set E_2 such that, for any $i \in \omega$ and $a \in \mathbb{R}$,

$$f_i(a) = \begin{cases} 0, & a < i\,, \\ 1, & a \geq i\,. \end{cases}$$

Let ψ be a bounded complexity measure and $A = \mathrm{Tab}^\infty(\mathbb{R}, F)$. One can prove the following statements:

- The function N is not bounded from above on the set A.
- The function S_ψ is bounded from above on the set A if and only if there exists a constant $c_0 > 0$ such that $\psi(f_i) \leq c_0$ for any $i \in \omega$.
- The function $L_{\psi,A}$ is everywhere defined if and only if, for any $n \in \omega$, the set $\{i : i \in \omega, \psi(f_i) \leq n\}$ is finite.
- A polynomial p such that $L_{\psi,A}(n) \leq 2^{p(n)}$ for any $n \in \omega$ exists if and only if there exists a polynomial q such that $|\{i : i \in \omega, \psi(f_i) \leq n\}| \leq 2^{q(n)}$ for any $n \in \omega$.
- For any $n \in \omega$, $Z_{\psi,A}(n) \leq 1$.
- For any $n \in \omega$, $G_{\psi,A}(n) \leq 2$.

7.3.6 Joint Behavior of Functions $\mathscr{F}^\infty_{\psi,A}$ and $\mathscr{G}^\infty_{\psi,A}$

In this section, we study joint behavior of functions $\mathscr{F}^\infty_{\psi,A}$ and $\mathscr{G}^\infty_{\psi,A}$. Let ψ be a bounded complexity measure and A be a nontrivial closed class of decision tables from $\mathscr{M}_k^\infty$. From Theorems 7.1–7.4 it follows that there are three possible types of the join behavior of functions $\mathscr{F}^\infty_{\psi,A}$ and $\mathscr{G}^\infty_{\psi,A}$:

- If the functions S_ψ and N are bounded from above on the class A, then $\mathscr{F}^\infty_{\psi,A}(n) = O(1)$ and $\mathscr{G}^\infty_{\psi,A}(n) = O(1)$.
- If the function S_ψ is bounded from above on the class A and the function N is not bounded from above on the class A, then $\mathscr{F}^\infty_{\psi,A}(n) = \Theta(\log n)$ and $\mathscr{G}^\infty_{\psi,A}(n) = O(1)$.
- If the function S_ψ is not bounded from above on the class A, then there exist infinite subsets D_1 and D_2 of the set ω such that $H_{D_1}(n) \leq \mathscr{F}^\infty_{\psi,A}(n) \leq n$ and $H_{D_2}(n) \leq \mathscr{G}^\infty_{\psi,A}(n) \leq n$ for any $n \in \omega$. If $\psi = h$, then $\mathscr{F}^\infty_{h,A}(n) = \mathscr{G}^\infty_{h,A}(n) = n$ for any $n \in \omega$.

All these three types are realizable for the bounded complexity measure h. Let us consider three 2-information systems (ω, F_1), (ω, F_2), and (ω, F_3), where

- $F_1 = \{q\}$ and, for any $j \in \omega$, $q(j) = 0$ if $j = 0$ and $q(j) = 1$ if $j > 0$.
- $F_2 = \{l_i : i \in \omega\}$ and, for any $i, j \in \omega$, $l_i(j) = 0$ if $j \leq i$ and $l_i(j) = 1$ if $j > i$.
- $F_3 = \{p_i : i \in \omega\}$ and, for any $i, j \in \omega$, $p_i(j) = 1$ if $i = j$ and $p_i(j) = 0$ if $i \neq j$.

One can show, that

- The functions S and N are bounded from above on the closed class $\mathrm{Tab}^{\infty}(\omega, F_1)$.
- The function S is bounded from above on the closed class $\mathrm{Tab}^{\infty}(\omega, F_2)$ and the function N is not bounded from above on this class.
- The function S is not bounded from above on the closed class $\mathrm{Tab}^{\infty}(\omega, F_3)$.

7.4 Proofs of Theorems 7.1, 7.2, 7.3 and 7.4

First, we consider a number of auxiliary statements.

It is not difficult to prove the following upper bound on the minimum complexity of deterministic decision trees for a decision table.

Lemma 7.1 *For any partially bounded complexity measure ψ and any table T from $\mathscr{M}_k^{\infty}$,*

$$\psi^d(T) \le W_\psi(T) .$$

The next upper bound on the minimum complexity of deterministic decision trees for a decision table follows from Corollary 5.2 from [4].

Lemma 7.2 *For any partially bounded complexity measure ψ and any table T from $\mathscr{M}_k^{\infty}$,*

$$\psi^d(T) \le \begin{cases} 0, & T \in \mathscr{M}_k^{\infty c} , \\ M_\psi(T) \log_2 N(T), & T \notin \mathscr{M}_k^{\infty c} . \end{cases}$$

Lemma 7.3 *For any partially bounded complexity measure ψ and any table T from $\mathscr{M}_k^{\infty}$,*

$$M_\psi(T) \le 2\hat{S}_\psi(T) .$$

Proof Let $T \in \mathscr{M}_k^{\infty c}$. Then $M_\psi(T) = 0$. Therefore $M_\psi(T) \le 2\hat{S}_\psi(T)$.

Let $T \notin \mathscr{M}_k^{\infty c}$, $W(T) = n$ and $f_{t_1}, \ldots, f_{t_n}$ be attributes attached to columns of the table T. Denote $D = \{f_i : f_i \in \mathrm{At}(T), \psi(f_i) \le S_\psi(T)\}$ and $T^* = I(\mathrm{At}(T) \setminus D, T)$. Evidently, $T^* \in [T]$. Taking into account that the function ψ has the non-decreasing and commutativity properties, we obtain that any two rows of T are different in columns labeled with attributes from the set D. Let for the definiteness, $D = \{f_{t_1}, \ldots, f_{t_m}\}$.

Let $\bar{\delta} = (\delta_1, \ldots, \delta_n) \in E_k^n$. Let $(\delta_1, \ldots, \delta_m)$ be a row of T^*. Since $T^* \in [T]$, there exist attributes $f_{t_{j_1}}, \ldots, f_{t_{j_s}}$ of the table T^* such that the row $(\delta_1, \ldots, \delta_m)$ is different from all other rows of T^* in columns labeled with these attributes and $\psi(f_{t_{j_1}} \cdots f_{t_{j_s}}) \le \hat{S}_\psi(T)$. It is clear that $T(f_{t_{j_1}}, \delta_{j_1}) \cdots (f_{t_{j_s}}, \delta_{j_s}) \in \mathscr{M}_k^{\infty c}$. Therefore $M_\psi(T, \bar{\delta}) \le \hat{S}_\psi(T)$.

Let $(\delta_1, \ldots, \delta_m)$ be not a row of T^*. We consider m tables

$$T^*(f_{t_1}, \delta_1), T^*(f_{t_1}, \delta_1)(f_{t_2}, \delta_2), \ldots, T^*(f_{t_1}, \delta_1) \cdots (f_{t_m}, \delta_m) .$$

If $T^*(f_{t_1}, \delta_1) = \Lambda$, then $T(f_{t_1}, \delta_1) = \Lambda$. Since $f_{t_1} \in D$, $\psi(f_{t_1}) \le S_\psi(T)$. Taking into account that $S_\psi(T) \le \hat{S}_\psi(T)$, we obtain $M_\psi(T, \bar{\delta}) \le \hat{S}_\psi(T)$. Let $T^*(f_{t_1}, \delta_1) \ne \Lambda$. Then there exists $p \in \{1, \dots, m-1\}$ such that

$$T^*(f_{t_1}, \delta_1) \cdots (f_{t_p}, \delta_p) \ne \Lambda$$

and $T^*(f_{t_1}, \delta_1) \cdots (f_{t_{p+1}}, \delta_{p+1}) = \Lambda$. Denote $C = \{f_{t_{p+1}}, f_{t_{p+2}}, \dots, f_{t_n}\}$ and $T^0 = I(C, T)$. Evidently, $T^0 \in [T]$. Therefore in T^0 there are attributes $f_{t_{i_1}}, \dots, f_{t_{i_l}}$ such that the row $(\delta_1, \dots, \delta_p)$ is different from all other rows of the table T^0 in columns labeled with these attributes and $\psi(f_{t_{i_1}} \cdots f_{t_{i_l}}) \le \hat{S}_\psi(T)$. One can show that

$$T^*(f_{t_{i_1}}, \delta_{i_1}) \cdots (f_{t_{i_l}}, \delta_{i_l}) = T^*(f_{t_1}, \delta_1) \cdots (f_{t_p}, \delta_p) .$$

Therefore $T^*(f_{t_{i_1}}, \delta_{i_1}) \cdots (f_{t_{i_l}}, \delta_{i_l})(f_{t_{p+1}}, \delta_{p+1}) = \Lambda$. Hence

$$T(f_{t_{i_1}}, \delta_{i_1}) \cdots (f_{t_{i_l}}, \delta_{i_l})(f_{t_{p+1}}, \delta_{p+1}) = \Lambda .$$

Since $f_{t_{p+1}} \in D$, $\psi(f_{t_{p+1}}) \le S_\psi(T) \le \hat{S}_\psi(T)$. Using boundedness from above property of the function ψ, we obtain $\psi(f_{t_{i_1}} \cdots f_{t_{i_l}} f_{t_{p+1}}) \le 2\hat{S}_\psi(T)$. Therefore

$$M_\psi(T, \bar{\delta}) \le 2\hat{S}_\psi(T) .$$

Thus, for any $\bar{\delta} \in E_k^n$, $M_\psi(T, \bar{\delta}) \le 2\hat{S}_\psi(T)$. As a result, we obtain $M_\psi(T) \le 2\hat{S}_\psi(T)$. □

Lemma 7.4 *For any table T from $\mathcal{M}_k^\infty \setminus \{\Lambda\}$,*

$$N(T) \le (kW(T))^{S(T)} .$$

Proof If $N(T) = 1$, then $S(T) = 0$ and the considered inequality holds. Let $N(T) > 1$. Then $S(T) > 0$. Denote $m = S(T)$. Evidently, for any row $\bar{\delta}$ of the table T, there exist attributes $f_{i_1}, \dots, f_{i_m} \in \mathrm{At}(T)$ and numbers $\sigma_1, \dots, \sigma_m \in E_k$ such that the table $T(f_{i_1}, \sigma_1) \cdots (f_{i_m}, \sigma_m)$ contains only the row $\bar{\delta}$. Therefore there is a one-to-one mapping of rows of the table T onto some set G of pairs of tuples of the kind $((f_{i_1}, \dots, f_{i_m}), (\sigma_1, \dots, \sigma_m))$, where $f_{i_1}, \dots, f_{i_m} \in \mathrm{At}(T)$ and $\sigma_1, \dots, \sigma_m \in E_k$. Evidently, $|G| \le W(T)^m k^m$. Therefore $N(T) \le (kW(T))^{S(T)}$. □

Lemma 7.5 *For any table T from $\mathcal{M}_k^\infty \setminus \{\Lambda\}$, there exists a mapping $\nu : E_k^{W(T)} \to \mathcal{P}(\omega)$ such that*

$$h^d(J(\nu, T)) \ge \log_k N(T) .$$

Proof Let $T \in \mathcal{M}_k^\infty \setminus \{\Lambda\}$ and $\nu : E_k^{W(T)} \to \mathcal{P}(\omega)$ be a mapping for which $\nu(\bar{\delta}) \cap \nu(\bar{\sigma}) = \emptyset$ for any $\bar{\delta}, \bar{\sigma} \in E_k^{W(T)}$ such that $\bar{\delta} \ne \bar{\sigma}$. Denote $T^* = J(\nu, T)$. Let Γ be a deterministic decision tree for the table T^* such that $h(\Gamma) = h^d(T^*)$. Denote by $L_t(\Gamma)$ the number of terminal nodes of Γ. Evidently, $N(T) \leq L_t(\Gamma)$. One can show that $L_t(\Gamma) \leq k^{h(\Gamma)}$. Therefore $N(T) \leq k^{h(\Gamma)}$. Since $T \neq \Lambda, N(T) > 0$. Hence $h(\Gamma) \geq \log_k N(T)$. Taking into account that $h(\Gamma) = h^d(T^*)$, we obtain $h^d(T^*) \geq \log_k N(T)$. □

Lemma 7.6 *For any partially bounded complexity measure* ψ *and any table* T *from* $\mathcal{M}_k^\infty$,

$$\psi^a(T) \leq \psi^d(T) .$$

Proof Let $T \in \mathcal{M}_k^\infty$. If $T = \Lambda$, then $\psi^a(T) = \psi^d(T) = 0$. Let $T \in \mathcal{M}_k^\infty \setminus \{\Lambda\}$. It is clear that each deterministic decision tree for the table T is a nondeterministic decision tree for the table T. Therefore $\psi^a(T) \leq \psi^d(T)$. □

Lemma 7.7 *For any partially bounded complexity measure* ψ *and any table* T *from* $\mathcal{M}_k^\infty$, *which contains at least two rows, there exists a table* $T^* \in [T]$ *such that*

$$\psi^a(T^*) = \psi^d(T^*) = W_\psi(T^*) = S_\psi(T^*) = S_\psi(T) .$$

Proof Let $\bar{\delta}$ be a row of the table T such that $S_\psi(T, \bar{\delta}) = S_\psi(T)$. Let D be a subset of the set $\mathrm{At}(T)$ with the minimum cardinality such that $\psi(D) = S_\psi(T, \bar{\delta})$ and in the set of columns labeled with attributes from D the row $\bar{\delta}$ is different from all other rows of the table T. Let $\bar{\sigma}$ be the tuple obtained from the row $\bar{\delta}$ by the removal of all numbers that are in the intersection with columns labeled with attributes from the set $\mathrm{At}(T) \setminus D$. Let $\nu : E_k^{|D|} \to \mathcal{P}(\omega)$ and, for any $\bar{\gamma} \in E_k^{|D|}$, if $\bar{\gamma} = \bar{\sigma}$, then $\nu(\bar{\gamma}) = \{1\}$ and if $\bar{\gamma} \neq \bar{\sigma}$, then $\nu(\bar{\gamma}) = \{0\}$. Denote $T^* = J(\nu, I(\mathrm{At}(T) \setminus D, T))$.

From the fact that D has the minimum cardinality and from the properties of the function ψ it follows that, for any attribute from the set D, there exists a row of T, which is different from the row $\bar{\delta}$ only in the column labeled with the considered attribute among attributes from D. Therefore, for any attribute of the table T^*, there exists a row of T^*, which is different from the row $\bar{\sigma}$ only in the column labeled with this attribute. Thus,

$$S_\psi(T^*, \bar{\sigma}) = W_\psi(T^*) . \tag{7.2}$$

Using properties of the function ψ, we obtain $S_\psi(T^*) \leq W_\psi(T^*)$. From this inequality and from (7.2) it follows that

$$S_\psi(T^*) = W_\psi(T^*) . \tag{7.3}$$

Let Γ be a nondeterministic decision tree for the table T^* such that $\psi(\Gamma) = \psi^a(T^*)$, τ be a complete path of Γ such that the row $\bar{\sigma}$ belongs to the table $T(\tau)$, and $\pi(\tau) = (f_{i_1}, \beta_1) \cdots (f_{i_t}, \beta_t)$. It is clear that $\bar{\sigma}$ is the only row of the table $T(\tau)$. Therefore in columns labeled with attributes $f_{i_1}, \ldots, f_{i_t}$ the row $\bar{\sigma}$ is different from all other rows of the table T^*. Thus, $\psi(\pi(\tau)) \geq S_\psi(T^*, \bar{\sigma}) = W_\psi(T^*)$. Therefore $\psi(\Gamma) \geq W_\psi(T^*)$ and $\psi^a(T^*) \geq W_\psi(T^*)$. Using Lemmas 7.1 and 7.6, we obtain $\psi^a(T^*) \leq \psi^d(T^*) \leq W_\psi(T^*)$. Therefore

$$\psi^a(T^*) = \psi^d(T^*) = W_\psi(T^*) . \tag{7.4}$$

By the choice of the set D, $W_\psi(T^*) = S_\psi(T)$. From this equality and from (7.3) and (7.4) it follows that $\psi^a(T^*) = \psi^d(T^*) = W_\psi(T^*) = S_\psi(T^*) = S_\psi(T)$. □

It is not difficult to prove the following upper bound on the minimum cardinality of a separating set for a table by the induction on the number of rows in the table.

Lemma 7.8 *For any table T from $\mathscr{M}_k^\infty \setminus \{\Lambda\}$,*

$$\mathrm{Sep}(T) \le N(T) - 1 .$$

It is not difficult to prove the following upper bounds on the minimum complexity of nondeterministic decision trees for a table.

Lemma 7.9 *For any partially bounded complexity measure ψ and any table T from $\mathscr{M}_k^\infty$,*

$$\psi^a(T) \le S_\psi(T) \le W_\psi(T) .$$

We now prove Theorems 7.1, 7.2, 7.3 and 7.4.

Proof of Theorem 7.1. Since A is a closed class, $\Lambda \in A$. By definition, $W_\psi(\Lambda) = 0$. Using this fact and Lemma 7.1, we obtain that $\mathscr{F}^\infty_{\psi,A}$ is an everywhere defined function and $\mathscr{F}^\infty_{\psi,A}(n) \le n$ for any $n \in \omega$. Evidently, $\mathscr{F}^\infty_{\psi,A}$ is a nondecreasing function. Let $T \in A$ and $W_\psi(T) \le 0$. Using positivity property of the function ψ, we obtain $T = \Lambda$. Therefore $\mathscr{F}^\infty_{\psi,A}(0) = 0$.

(a) Let the functions S_ψ and N be bounded from above on the class A. Then there are constants $a \ge 1$ and $b \ge 2$ such that $S_\psi(T) \le a$ and $N(T) \le b$ for any table $T \in A$. Let $T \in A$. Taking into account that A is a closed class, we obtain $\hat{S}_\psi(T) \le a$. By Lemma 7.3, $M_\psi(T) \le 2a$. From this inequality, inequality $N(T) \le b$ and from Lemma 7.2 it follows that $\psi^d(T) \le 2a \log_2 b$. Denote $c_0 = 2a \log_2 b$. Taking into account that T is an arbitrary table from the class A, we obtain that $\mathscr{F}^\infty_{\psi,A}(n) \le c_0$ for any $n \in \omega$.

(b) Let the function S_ψ be bounded from above on the class A and the function N be not bounded from above on A. Then there exists a constant $a \ge 2$ such that, for any table $Q \in A$,

$$S_\psi(Q) \le a . \tag{7.5}$$

Let $n \in \omega \setminus \{0\}$ and T be an arbitrary table from A such that $W_\psi(T) \le n$. If $T \in \mathscr{M}_k^{\infty c}$, then, evidently, $\psi^d(T) = 0$. Let $T \notin \mathscr{M}_k^{\infty c}$. Using the boundedness from below property of the function ψ, we obtain $W(T) \le n$ and $S(T) \le a$. From these inequalities and Lemma 7.4 it follows that $N(T) \le (kn)^a$. From (7.5) and Lemma 7.3 it follows that $M_\psi(T) \le 2a$. Using the last two inequalities and Lemma 7.2, we obtain

$$\psi^d(T) \le 2a^2 \log_2 n + 2a^2 \log_2 k . \tag{7.6}$$

Denote $c_3 = 2a^2$ and $c_4 = 2a^2 \log_2 k$. Then $\psi^d(T) \le c_3 \log_2 n + c_4$. Taking into account that n is an arbitrary number from $\omega \setminus \{0\}$ and T is an arbitrary table from A such that $W_\psi(T) \le n$, we obtain that, for any $n \in \omega \setminus \{0\}$,

$$\mathscr{F}^\infty_{\psi,A}(n) \le c_3 \log_2 n + c_4 . \tag{7.7}$$

Let $n \in \omega \setminus \{0\}$. Denote $c_1 = 1/\log_2 k$ and $c_2 = \log_k a$. We now show that

$$\mathscr{F}^\infty_{\psi,A}(n) \ge c_1 \log_2 n - c_2 . \tag{7.8}$$

Denote $m = \lfloor n/a \rfloor$. Taking into account that the function N is not bounded from above on the set A, we obtain that there exists a table $T \in A$ such that $N(T) \ge k^m$. Let C be a separating set for the table T with the minimum cardinality such that $\psi(f_i) \le a$ for any $f_i \in C$. The existence of such set follows from the inequality (7.5) and properties of commutativity and nondecreasing of the function ψ. Evidently, $|C| \ge m$. Let D be a subset of the set C such that $|D| = m$. Denote $T^0 = I(\mathrm{At}(T) \setminus D, T)$. One can show that, for any attribute of T^0, there are two rows of the table T^0 that differ only in the column labeled with this attribute. Therefore D is a separating set for the table T^0 with the minimum cardinality. By Lemma 7.8, $N(T^0) \ge m + 1 \ge n/a$.

Using Lemma 7.5, we obtain that there exists a mapping $\nu : E_k^m \to \omega$ such that

$$h^d(J(\nu, T^0)) \ge \log_k(n/a) .$$

Denote $T^* = J(\nu, T^0)$. Since ψ is a bounded complexity measure, $\psi^d(T^*) \ge \log_k(n/a)$. Using boundedness from above property of the function ψ, we obtain $W_\psi(T^*) \le \lfloor n/a \rfloor a \le n$. Hence the inequality (7.8) holds. From (7.7) and (7.8) it follows that $c_1 \log_2 n - c_2 \le \mathscr{F}^\infty_{\psi,A}(n) \le c_3 \log_2 n + c_4$ for any $n \in \omega \setminus \{0\}$.

(c) Let the function S_ψ be not bounded from above on the class A. Using Lemma 7.7, we obtain that the set $D = \{W_\psi(T) : T \in A, \psi^d(T) = W_\psi(T)\}$ is infinite. Since the class A is closed, $\Lambda \in A$ and, since $\psi^d(\Lambda) = W_\psi(\Lambda) = 0$, $0 \in D$. Evidently, for any $n \in D$, $\mathscr{F}^\infty_{\psi,A}(n) \ge n$. Taking into account that $\mathscr{F}^\infty_{\psi,A}$ is a nondecreasing function, we obtain that $\mathscr{F}^\infty_{\psi,A}(n) \ge H_D(n)$ for any $n \in \omega \setminus \{0\}$. □

Proof of Theorem 7.2. From Theorem 7.1 it follows that $\mathscr{F}^\infty_{h,A}(n) \le n$ for any $n \in \omega$. In particular, $\mathscr{F}^\infty_{h,A}(0) = 0$. Let $n \in \omega \setminus \{0\}$. Since the function S is not bounded from above on the class A, there exists a table $T \in A$ such that $S(T) \ge n$. Let $\bar{\delta}$ be a row of the table T for which $S(T, \bar{\delta}) = S(T)$. Let C be a set of attributes of T with the minimum cardinality such that in columns labeled with attributes from C the row $\bar{\delta}$ is different from all other rows of the table T. Evidently, $|C| \ge n$. Let C^* be a subset of the set C with $|C^*| = n$. Denote $T^0 = I(\mathrm{At}(T) \setminus C^*, T)$. One can show that $S(T^0) = n$. From Lemma 7.7 it follows that there exists a table $T^* \in [T^0]$ for which $h^d(T^*) = W(T^*) = n$. Therefore $\mathscr{F}^\infty_{h,A}(n) \ge n$ and $\mathscr{F}^\infty_{h,A}(n) = n$. □

Proof of Theorem 7.3. Since A is a closed class, $\Lambda \in A$. By definition, $W_\psi(\Lambda) = 0$. Using this fact and Lemma 7.9, we obtain that $\mathscr{G}_{\psi,A}$ is an everywhere defined function and $\mathcal{G}^{\infty}_{\psi,A}(n) \le n$ for any $n \in \omega$. Evidently, $\mathcal{G}^{\infty}_{\psi,A}$ is a nondecreasing function. Let $T \in A$ and $W_\psi(T) \le 0$. Using positivity property of the function ψ, we obtain $T = \Lambda$. Therefore $\mathcal{G}^{\infty}_{\psi,A}(0) = 0$.

(a) Let the function S_ψ be bounded from above on the class A. Then there is a constant $c > 0$ such that $S_\psi(T) \le c$ for any $T \in A$. By Lemma 7.9, $\psi^a(T) \le S_\psi(T) \le c$ for any $T \in A$. Therefore, for any $n \in \omega$, $\mathcal{G}^{\infty}_{\psi,A}(n) \le c$.

(b) Let the function S_ψ be not bounded from above on the class A. Using Lemma 7.7, we obtain that the set $D = \{W_\psi(T) : T \in A,\ \psi^a(T) = W_\psi(T)\}$ is infinite. Since the class A is closed, $\Lambda \in A$ and, since $\psi^a(\Lambda) = W_\psi(\Lambda) = 0$, $0 \in D$. Evidently, for any $n \in D$, $\mathcal{G}^{\infty}_{\psi,A}(n) \ge n$. Taking into account that $\mathcal{G}^{\infty}_{\psi,A}$ is a nondecreasing function, we obtain that $\mathcal{G}^{\infty}_{\psi,A}(n) \ge H_D(n)$ for any $n \in \omega \setminus \{0\}$. □

Proof of Theorem 7.4. From Theorem 7.3 it follows that $\mathcal{G}^{\infty}_{h,A}(n) \le n$ for any $n \in \omega$. In particular, $\mathcal{G}^{\infty}_{h,A}(0) = 0$. Let $n \in \omega \setminus \{0\}$. Since the function S is not bounded from above on the class A, there exists a table $T \in A$ such that $S(T) \ge n$. Let $\bar{\delta}$ be a row of the table T such that $S(T, \bar{\delta}) = S(T)$. Let C be a set of attributes of T with the minimum cardinality such that in columns labeled with the attributes from C the row $\bar{\delta}$ is different from all other rows of the table T. Evidently, $|C| \ge n$. Let C^* be a subset of the set C with $|C^*| = n$. Denote $T^0 = I(\mathrm{At}(T) \setminus C^*, T)$. One can show that $S(T^0) = n$. From Lemma 7.7 it follows that there exists a table $T^* \in [T^0]$ for which $h^a(T^*) = W(T^*) = n$. Therefore $\mathcal{G}^{\infty}_{h,A}(n) \ge n$ and $\mathcal{G}^{\infty}_{h,A}(n) = n$. □

7.5 Proofs of Theorems 7.5, 7.6 and 7.7

First, we prove the following auxiliary statement.

Lemma 7.10 *Let A be a nontrivial closed class of decision tables from $\mathcal{M}^{\infty}_k$, ψ be a bounded complexity measure, $T \in A$, $n \in \omega$ and $l_\psi(T, n) > 0$. Then there exists a mapping $\nu : E_k^{W(T)} \to \mathcal{P}(\omega)$ such that, for the table $T^* = J(\nu, T)$, $\psi^a(T^*) \le n$ and $\psi^d(T^*) \ge \log_k l_\psi(T, n)$.*

Proof Let U be an irreducible (ψ, n)-cover of the table T such that $|U| = l_\psi(T, n)$. Let $U = \{\alpha_1, \dots, \alpha_l\}$, where $l = l_\psi(T, n)$. We now define a mapping $\nu : E_k^{W(T)} \to \mathcal{P}(\omega)$. Let $\bar{\delta} \in E_k^{W(T)}$. If $\bar{\delta} \notin \Delta(T)$, then $\nu(\bar{\delta}) = \{0\}$. Let $\bar{\delta} \in \Delta(T)$. Then $\nu(\bar{\delta}) \subseteq \{1, \dots, l\}$. For any $j \in \{1, \dots, l\}$, $j \in \nu(\bar{\delta})$ if and only if $\bar{\delta} \in \Delta(T\alpha_j)$. Since the cover U is an irreducible (ψ, n)-cover of the table T, for each $j \in \{1, \dots, l\}$, there is a row $\bar{\delta}_j$ of the table T such that $\nu(\bar{\delta}_j) = \{j\}$. Set $T^* = J(\nu, T)$. Let Γ be a deterministic decision tree for the table T^* such that $h(\Gamma) = h^d(T^*)$. Denote by $L_t(\Gamma)$ the number of terminal nodes of Γ. Evidently, $l \le L_t(\Gamma)$. One can show that $L_t(\Gamma) \le k^{h(\Gamma)}$. Therefore $0 < l \le k^{h(\Gamma)}$. Hence $h(\Gamma) \ge \log_k l$. Taking into account that $h(\Gamma) = h^d(T^*)$, we obtain $h^d(T^*) \ge \log_k l_\psi(T, n)$. Using boundedness from below property of the function ψ, we obtain $\psi^d(T^*) \ge \log_k l_\psi(T, n)$. It is easy to construct a nondeterministic decision tree G for the table T^* with $l = l_\psi(T, n)$ complete paths $\tau_1, \dots, \tau_l$ such that $\pi(\tau_1) = \alpha_1, \dots, \pi(\tau_l) = \alpha_l$. It is clear that $\psi(G) \le n$. Therefore $\psi^a(T^*) \le n$. □

Proof of Theorem 7.5. Let the function $L_{\psi,A}$ be everywhere defined. Let $T \in A$ and $\psi^a(T) \le n$. Let Γ be a nondeterministic decision tree for the table T such that $\psi(\Gamma) = \psi^a(T)$ and U be the set of words $\pi(\tau)$ corresponding to the complete paths τ in Γ. Then U is a (ψ, n)-cover of the table T. Let U' be a subset of the set U, which is an irreducible (ψ, n)-cover of the table T. Then $|U'| \le L_{\psi,A}(n)$. Let $U' = \{\alpha_1, \dots, \alpha_t\}$. We now consider a deterministic decision tree G for the table T, which, for a given $\bar{\delta} \in \Delta(T)$, sequentially iterates through the words α_i from U' and check if $\bar{\delta} \in \Delta(T\alpha_i)$ by computing values of attributes from α_i. For the first i such that $\bar{\delta} \in \Delta(T\alpha_i)$, the decision tree G returns a common decision for the table $T\alpha_i$. Using boundedness from above property of the function ψ, we obtain that $\psi(G) \le nL_{\psi,A}(n)$. Therefore $\mathscr{H}^{\infty}_{\psi,A}(n) \le nL_{\psi,A}(n)$.

Let the function $L_{\psi,A}$ be not everywhere defined. Then there exist a number $n \in \omega$ and an infinite sequence of tables $T_0, T_1, \dots$ from A such that $l_\psi(T_0, n) < l_\psi(T_1, n) < \cdots$. Let $i \in \omega$. Using Lemma 7.10, we obtain that there exists a mapping $\nu : E_k^{W(T_i)} \to \mathscr{P}(\omega)$ such that, for the table $T_i^* = J(\nu, T_i)$, $\psi^a(T_i^*) \le n$ and $\psi^d(T_i^*) \ge \log_k l_\psi(T_i, n)$. Evidently, $T_i^* \in A$. Therefore, the value $\mathscr{H}^{\infty}_{\psi,A}(n)$ is not defined. □

Proof of Theorem 7.6. Immediately from the definition of the function $\mathscr{H}^{\infty}_{\psi,A}$ it follows that $\mathscr{H}^{\infty}_{\psi,A}(n) \le \mathscr{H}^{\infty}_{\psi,A}(n+1)$ for any $n \in \omega$. Let $T \in A$ and $\psi^a(T) \le 0$. Using positivity property of the function ψ, one can show that $T \in \mathscr{M}_k^{\infty c}$. From Lemma 7.2 it follows that $\psi^d(T) = 0$. Therefore $\mathscr{H}^{\infty}_{\psi,A}(0) = 0$.

(a) Let there exist a nonnegative constant c such that $\psi^d(T) \le c$ for any table $T \in A$. Then, evidently, $\mathscr{H}^{\infty}_{\psi,A}(n) \le c$ for any $n \in \omega$.

(b) Let there be no a nonnegative constant c such that $\psi^d(T) \le c$ for any table $T \in A$. Let us assume that there exists a nonnegative constant d such that $\psi^a(T) \le d$ for any table $T \in A$. Then the value $\mathscr{H}^{\infty}_{\psi,A}(d)$ is not defined but this is impossible. Therefore the set $D = \{\psi^a(T) : T \in A\}$ is infinite. Since $\Lambda \in A$, $0 \in D$. By Lemma 7.6, $\psi^a(T) \le \psi^d(T)$ for any table $T \in A$. Hence $\mathscr{H}^{\infty}_{\psi,A}(n) \ge n$ for any $n \in D$. Taking into account that $\mathscr{H}^{\infty}_{\psi,A}$ is a nondecreasing function, we obtain that $\mathscr{H}^{\infty}_{\psi,A}(n) \ge H_D(n)$ for any $n \in \omega$. □

Proof of Theorem 7.7. For each $n \in \omega \setminus \{0\}$, we define a decision table Q_n. The union of closures of these tables forms a closed class A. In the same time, we define a weighted depth ψ.

Let $n \in \omega \setminus \{0\}$ and $\varphi(n) = nm + j$, where $0 \le j \le n - 1$. Denote by Q_n a decision table from $\mathscr{M}_2^{\infty}$ with $m + 2$ columns labeled with pairwise different attributes $f_{i(1,n)}, \dots, f_{i(m+2,n)}$, respectively, and $m + 2$ rows labeled with the sets of decisions $\{1\}, \dots, \{m + 2\}$, respectively. The table Q_n is filled with zeros with the exception of the main diagonal that is filled with ones. If $j = 0$, then we should remove from the table Q_n the first row and the first column. If $j > 0$, then

$\psi(f_{i(1,n)}) = j$. For $t = 2, \ldots, m+2$, $\psi(f_{i(t,n)}) = n$. We choose attributes such that $\mathrm{At}(Q_l) \cap \mathrm{At}(Q_t) = \emptyset$ if $l \neq t$. If $f_i \in P \setminus \bigcup_{n \in \omega \setminus \{0\}} \mathrm{At}(Q_n)$, then $\psi(f_i) = 1$. Denote $A = [\{Q_1, Q_2, \ldots\}]$.

One can show that, for any $n \in \omega$, $L_{\psi,A}(n) \leq \varphi(n) + 2$, i.e., the function $L_{\psi,A}$ is everywhere defined. By Theorem 7.5, the function $\mathscr{H}^{\infty}_{\psi,A}$ is everywhere defined. By Theorem 7.6, $\mathscr{H}^{\infty}_{\psi,A}(0) = 0$. Let $n \in \omega \setminus \{0\}$. One can show that $\psi^a(Q_n) = n$ and $\psi^d(Q_n) = nm + j = \varphi(n)$. Hence $H_{\psi,A}(n) \geq \varphi(n)$.

Let us show that, for any $n \in \omega$, $\mathscr{H}^{\infty}_{\psi,A}(n) \leq \varphi(n) + n$. Evidently, this inequality holds if $n = 0$. Let $n > 0$, $T \in A$ and $\psi^a(T) \leq n$. Let $T \in [\{Q_1, \ldots, Q_n\}]$. By Lemma 7.1, $\psi^d(T) \leq W_\psi(T) \leq \varphi(n) + n$. Let $T \in [\{Q_{n+1}, Q_{n+2}, \ldots\}]$. Let Γ be a nondeterministic decision tree for the table T such that $\psi(\Gamma) \leq n$. Using commutativity and nondecreasing properties of the function ψ, we obtain that $\psi(f_i) \leq n$ for any $f_i \in \mathrm{At}(\Gamma)$. It is clear that $|\{f_i : f_i \in \mathrm{At}(T), \psi(f_i) \leq n\}| \leq 1$. Using these facts, one can show that $\psi^d(T) \leq n$. □

7.6 Proof of Theorem 7.8

We preface the proof of the theorem with several auxiliary statements.

Lemma 7.11 *Let A be a nontrivial closed class of decision tables from $\mathscr{M}^{\infty}_k$, ψ be a bounded complexity measure, and the function $L_{\psi,A}$ be everywhere defined. Then, for any $n \in \omega$ such that $L_{\psi,A}(n) > 0$, $\mathscr{H}^{\infty}_{\psi,A}(n) \geq \log_k L_{\psi,A}(n)$.*

Proof By Theorem 7.5, the function $\mathscr{H}^{\infty}_{\psi,A}$ is everywhere defined. Let $n \in \omega$ and $L_{\psi,A}(n) > 0$. Let $T \in A$ and $l_\psi(T, n) = L_{\psi,A}(n)$. Using Lemma 7.10, we obtain that there exists a mapping $\nu : E_k^{W(T)} \rightarrow \mathscr{P}(\omega)$ such that, for the table $T^* = J(\nu, T)$, $\psi^a(T^*) \leq n$ and $\psi^d(T^*) \geq \log_k l_\psi(T, n) = \log_k L_{\psi,A}(n)$. Therefore $\mathscr{H}^{\infty}_{\psi,A}(n) \geq \log_k L_{\psi,A}(n)$. □

It is not difficult to prove the following statement.

Lemma 7.12 *Let ψ be a bounded complexity measure and $T \in \mathscr{M}^{\infty}_k \setminus \mathscr{M}^{\infty c}_k$. Then $\psi^a(T) = \max\{M_\psi(T, \bar{\delta}) : \bar{\delta} \in \Delta(T)\}$.*

Let $k \in \omega \setminus \{0\}$. Let us consider the graph G_k depicted in Fig. 7.5. Nodes of this graph are divided into k layers numbered from top to bottom. The number of nodes in this graph is equal to $m(k) = \frac{k(k+1)}{2}$. Nodes of the graph G_k are numbered as it is shown in Fig. 7.5. We denote by $l(i)$ the number of the left-hand child of the node i and by $p(i)$ the number of the right-hand child of the node i. We now define mapping $\nu_k : E_2^{m(k)} \rightarrow \mathscr{P}(\omega)$. Let $\bar{\delta} = (\delta_1, \ldots, \delta_{m(k)}) \in E_2^{m(k)}$. Then $\nu_k(\bar{\delta}) \subseteq \{0, 1, \ldots, m(k)\}$. The number 0 belongs to the set $\nu_k(\bar{\delta})$ if and only if $\delta_1 = 0$. Let $i \in \{1, \ldots, m(k)\}$. If i is not a node of the kth layer, then $i \in \nu_k(\bar{\delta})$ if and only if $\delta_i = 1$ and $\delta_{l(i)} = \delta_{p(i)} = 0$. If i is a node of the kth layer, then $i \in \nu_k(\bar{\delta})$ if and only if $\delta_i = 1$.

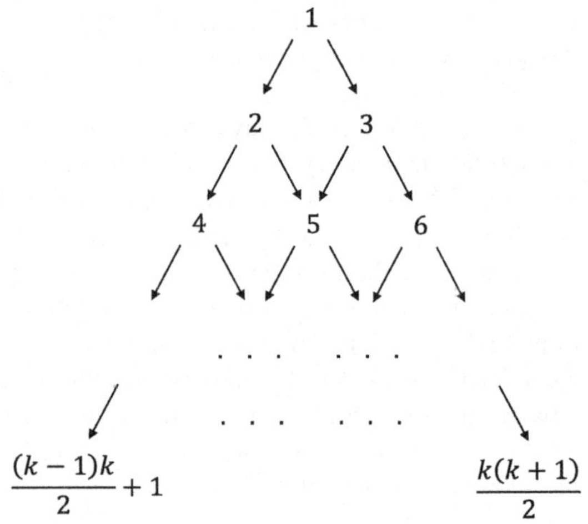

Fig. 7.5 Graph G_k

Let T_k be a decision table from $\mathscr{M}_2^\infty$ with $m(k)$ columns labeled with the attributes $f_1, \ldots, f_{m(k)}$ and $2^{m(k)}$ pairwise different rows filled with numbers from E_2. Any row $(\delta_1, \ldots, \delta_{m(k)})$ of the table T_k is labeled with the set of decisions $\nu_k(\delta_1, \ldots, \delta_{m(k)})$. It is clear that T_k is a complete decision table.

Lemma 7.13 *For any* $k \in \omega \setminus \{0\}$, $h^a(T_k) \le 3$.

Proof Let $\bar{\delta} = (\delta_1, \ldots, \delta_{m(k)}) \in E_2^{m(k)}$. If $\delta_1 = 0$, then $0 \in \Pi(T_k(f_1, 0))$, $T_k(f_1, 0) \in \mathscr{M}_2^{\infty c}$, and $M(T_k, \bar{\delta}) \le 1$. Let $\delta_1 = 1$. If the node 1 belongs to the kth layer, then $1 \in \Pi(T_k(f_1, 1))$, $T_k(f_1, 1) \in \mathscr{M}_2^{\infty c}$, and $M(T_k, \bar{\delta}) \le 1$. Let the node 1 do not belong to the kth layer. If $\delta_{l(1)} = 0$ and $\delta_{p(1)} = 0$, then $1 \in \Pi(T_k(f_1, 1)(f_{l(1)}, 0)(f_{p(1)}, 0))$, $T_k(f_1, 1)(f_{l(1)}, 0)(f_{p(1)}, 0) \in \mathscr{M}_2^{\infty c}$, and $M(T_k, \bar{\delta}) \le 3$. Let $1 \in \{\delta_{l(1)}, \delta_{p(1)}\}$ and, for definiteness, $\delta_{l(1)} = 1$. Then, instead of the number 1, we consider the number $l(1)$, etc. As a result, we either find $i \in \{1, \ldots, m(k)\}$ such that the node i does not belong to the kth layer of the graph G_k and for which $\delta_i = 1$ and $\delta_{l(i)} = \delta_{p(i)} = 0$, or we find $i \in \{1, \ldots, m(k)\}$ such that the node i belongs to the kth layer of the graph G_k and for which $\delta_i = 1$. In both cases, $M(T_k, \bar{\delta}) \le 3$. Thus, $M(T_k, \bar{\delta}) \le 3$ for any $\bar{\delta} \in E_2^{m(k)}$ and $M(T) \le 3$. Using Lemma 7.12, we obtain that $h^a(T_k) \le 3$. □

A *complete path* in the graph G_k is a directed path from the node of the 1st layer to a node of the kth layer. We correspond to a complete path π in G_k its *characteristic tuple* $\delta(\pi) = (\delta_1, \ldots, \delta_{m(k)}) \in E_2^{m(k)}$, where, for $i = 1, \ldots, m(k)$, $\delta_i = 1$ if and only if π passes through the node i of G_k. Denote by $Path_k$ the set of complete paths in the graph G_k. Evidently $|Path_k| = 2^{k-1}$. Denote by T_k^* a subtable of the table T_k, which contains only rows that are characteristic tuples of paths from $Path_k$. It is clear that each row from T_k^* is labeled with a singleton set in which the only element is the number of the last node in the path corresponding to the considered row. Let T be a subtable of the table T_k^* and $\bar{\delta}_1, \ldots, \bar{\delta}_t$ be all rows of T. Denote $r(T) = \left|\{\nu_k(\bar{\delta}_1), \ldots, \nu_k(\bar{\delta}_t)\}\right|$. Evidently, $T \in \mathscr{M}_2^{\infty c}$ if and only if $r(T) \le 1$.

Lemma 7.14 *Let $k \in \omega \setminus \{0\}$, $f_{i_1}, \ldots, f_{i_n} \in \{f_1, \ldots, f_{m(k)}\}$ and $T = T_k^*(f_{i_1}, 0) \cdots (f_{i_n}, 0)$. If $T \neq \Lambda$, then $r(T) \geq \max(1, k - n)$.*

Proof We will prove the considered statement by induction on k. Let $k = 1$. Then, evidently, the considered statement holds. Let it hold for $k - 1$ for some $k \geq 2$. Let $f_{i_1}, \ldots, f_{i_n} \in \{f_1, \ldots, f_{m(k)}\}$, $T = T_k^*(f_{i_1}, 0) \cdots (f_{i_n}, 0)$ and $T \neq \Lambda$. Without loss of generality we assume that $i_1, \ldots, i_t \leq m(k-1)$ and $i_{t+1}, \ldots, i_n > m(k-1)$. Since $T \neq \Lambda$, $T_{k-1}^*(f_{i_1}, 0) \cdots (f_{i_t}, 0) \neq \Lambda$. By induction hypothesis, $r(T_{k-1}^*(f_{i_1}, 0) \cdots (f_{i_t}, 0)) \geq \max(1, k - 1 - t))$. Let $j_1 < \cdots < j_s$ be all elements of sets attached to rows of the table $T_{k-1}^*(f_{i_1}, 0) \cdots (f_{i_t}, 0)$. Then, evidently, $l(j_1) < p(j_1) < \cdots < p(j_s)$ are elements of sets attached to rows of the table $T_k^*(f_{i_1}, 0) \cdots (f_{i_t}, 0)$. Let us show that $r(T) \geq \max(1, k - n)$. If $n \geq k$, then this inequality follows from the fact that $T \neq \Lambda$. Let $n < k$. Then the following two cases are possible: (i) $t = k - 1$ and (ii) $t < k - 1$. In the first case, the considered inequality follows from the fact that $T \neq \Lambda$. In the second case, $r(T_{k-1}^*(f_{i_1}, 0) \cdots (f_{i_t}, 0)) \geq k - t - 1$ and, as it was shown earlier, $r(T_k^*(f_{i_1}, 0) \cdots (f_{i_t}, 0)) \geq k - t$. One can show that $r(T) \geq r(T_k^*(f_{i_1}, 0) \cdots (f_{i_t}, 0)) - (n - t) \geq k - t - n + t = k - n$. □

Lemma 7.15 *For any $k \in \omega \setminus \{0\}$, $h^d(T_k) \geq k - 1$.*

Proof If $k = 1$, then the considered inequality holds. Let $k > 1$. We now show that $h^d(T_k^*) \geq k - 1$. One can prove that there exists a deterministic decision tree for the table T_k^* such that

- $h(\Gamma) = h^d(T_k^*)$.
- For each node of Γ, which is neither the root nor a terminal node, there are exactly two edges leaving this node.
- For each complete path τ of Γ, $T_k^*(\tau) \neq \Lambda$.

Let τ_0 be a complete path in Γ such that, for each node of τ_0, which is neither the root nor a terminal node, the edge leaving this node is labeled with the number 0. Evidently, $r(T_k^*(\tau_0)) = 1$. By Lemma 7.14, the number of nodes in τ_0 labeled with attributes is at least $k - 1$. Therefore $h(\Gamma) \geq k - 1$ and $h^d(T_k^*) \geq k - 1$. It is clear that each deterministic decision tree for the table T_k is a deterministic decision tree for the table T_k^*. Therefore $h^d(T_k) \geq k - 1$. □

The next statement follows immediately from Theorem 4.6 from [3].

Lemma 7.16 *For any table $T \in \mathscr{M}_2^\infty$, if $T \neq \Lambda$, then $N(T) \leq (4W(T))^{Z(T)}$.*

Lemma 7.17 *Let A be a nontrivial closed class of decision tables from $\mathscr{M}_2^\infty$ and ψ be a bounded complexity measure. If the function $Z_{\psi,A}$ is not everywhere defined, then the function $\mathscr{H}_{\psi,A}^\infty$ is not everywhere defined.*

Proof Let, for $n \in \omega$, the value $Z_{\psi,A}(n)$ be not defined and $k \in \omega \setminus \{0\}$. Then there exists a table $T \in A_\psi(n)$ such that $Z(T) \geq m(k)$. It is easy to see that the set $[T]$ contains a complete table T' with $m(k)$ columns. It is clear that $J(\nu_k, T') = T_k$, i.e., $T_k \in A_\psi(n)$. Using Lemma 7.13, and boundedness from above property of the function ψ, we obtain that $\psi^a(T_k) \le 3n$. Using Lemma 7.15, and boundedness from below property of the function ψ, we obtain that $\psi^d(T_k) \ge k-1$. Since k is an arbitrary number from $\omega \setminus \{0\}$, we obtain that the value $\mathscr{H}^{\infty}_{\psi,A}(3n)$ is not defined. □

Lemma 7.18 *Let A be a nontrivial closed class of decision tables from $\mathscr{M}_2^{\infty}$ and ψ be a bounded complexity measure. If the function $\mathscr{H}^{\infty}_{\psi,A}$ is everywhere defined, then, for any $n \in \omega$, $\mathscr{H}^{\infty}_{\psi,A}(3n) \ge \sqrt{2Z_{\psi,A}(n)} - 3$.*

Proof If $Z_{\psi,A}(n) = 0$, then, evidently, $\mathscr{H}^{\infty}_{\psi,A}(3n) \ge \sqrt{Z_{\psi,A}(n)} - 3$. Let $Z_{\psi,A}(n) > 0$ and k be the maximum number from ω such that $\frac{k(k+1)}{2} \le Z_{\psi,A}(n)$. Then, $T_k \in A_{\psi}(n)$ and, by Lemmas 7.13 and 7.15, $\psi^a(T_k) \le 3n$ and $\psi^d(T_k) \ge k-1$. Evidently, $\frac{(k+1)(k+2)}{2} > Z_{\psi,A}(n)$. Therefore $(k+2)^2 > 2Z_{\psi,A}(n)$ and $k > \sqrt{2Z_{\psi,A}(n)} - 2$. Thus, $\mathscr{H}^{\infty}_{\psi,A}(3n) \ge \sqrt{2Z_{\psi,A}(n)} - 3$. □

Lemma 7.19 *Let A be a nontrivial closed class of decision tables from $\mathscr{M}_2^{\infty}$, ψ be a bounded complexity measure, $n \in \omega$, $T \in A_{\psi}(n)$ and $G(T) > 0$. Then there exists a decision table $T^* \in [T]$ such that $\psi^a(T^*) \le n$ and $\psi^d(T^*) \ge G(T) - 1$.*

Proof Let $n \in \omega$, $T \in A_{\psi}(n)$ and $G(T) > 0$. Let $\alpha = (f_{j_1}, \sigma_1) \cdots (f_{j_m}, \sigma_m)$ be an irreducible annihilating word for the table T, which length is equal to $G(T)$ and in which the attributes $f_{j_1}, \dots, f_{j_m}$ are arranged in the same order as in the table T. Denote $Q = \mathrm{At}(T) \setminus \{f_{j_1}, \dots, f_{j_m}\}$ and $T' = I(Q, T)$. Let us define a mapping $\nu : E_2^m \to \mathscr{P}(\omega)$. Let $\bar{\delta} = (\delta_1, \dots, \delta_m) \in E_2^m$. Denote $\bar{\sigma} = (\sigma_1, \dots, \sigma_m)$. If $\bar{\delta} = \bar{\sigma}$, then $\nu(\bar{\delta}) = \{0\}$. Let $\bar{\delta} \ne \bar{\sigma}$. Then $\nu(\bar{\delta}) = \{i : i \in \{1, \dots, m\}, \delta_i \ne \sigma_i\}$. Denote $T^* = J(\nu, T')$.

Let $\bar{\delta} = (\delta_1, \dots, \delta_m) \in \Delta(T^*)$. Then there is $i \in \{1, \dots, m\}$ such that $\delta_i \ne \sigma_i$. Evidently, i is a common decision for the table $T^*(f_{j_i}, \delta_i)$. Therefore $M(T^*, \bar{\delta}) \le 1$. Using Lemma 7.12, we obtain $h^a(T^*) \le 1$. Since the word α is irreducible, for each $i \in \{1, \dots, m\}$, the set $\Delta(T^*)$ contains a row (tuple) that is different from $\bar{\sigma}$ only in the ith digit and is labeled with the set of decisions $\{i\}$. Using this fact, it is easy to show that $M(T^*, \bar{\sigma}) \ge m-1$. Therefore $M(T^*) \ge m-1$. According to Theorem 3.1 from [3], $h^d(T^*) \ge M(T^*)$. Thus, $h^d(T^*) \ge m-1 = G(T)-1$. Using boundedness from above and boundedness from below properties of the function ψ, we obtain $\psi^a(T^*) \le n$ and $\psi^d(T^*) \ge G(T)-1$. □

Lemma 7.20 *Let A be a nontrivial closed class of decision tables from $\mathscr{M}_2^{\infty}$ and ψ be a bounded complexity measure. If the function $G_{\psi,A}$ is not everywhere defined, then the function $\mathscr{H}^{\infty}_{\psi,A}$ is not everywhere defined.*

Proof Let $n \in \omega$ and the value $G_{\psi,A}(n)$ be not defined. Then there exists an infinite sequence $D_0, D_1, \dots$ of decision tables from $A_{\psi}(n)$ such that $0 < G(D_0) < G(D_1) < \cdots$. Let $i \in \omega$. From Lemma 7.19 it follows that there exists a decision table $D_i^* \in [D_i]$ such that $\psi^a(D_i^*) \le n$ and $\psi^d(D_i^*) \ge G(D_i) - 1$. Therefore the value $\mathscr{H}^{\infty}_{\psi,A}(n)$ is not defined. □

Lemma 7.21 *Let A be a nontrivial closed class of decision tables from $\mathscr{M}_2^\infty$ and ψ be a bounded complexity measure. If the function $\mathscr{H}_{\psi,A}^\infty$ is everywhere defined, then, for any $n \in \omega$, $\mathscr{H}_{\psi,A}^\infty(n) \geq G_{\psi,A}(n) - 1$.*

Proof Let $n \in \omega$. If $G_{\psi,A}(n) = 0$, then the considered inequality holds. Let $G_{\psi,A}(n) > 0$, $T \in A_\psi(n)$ and $G(T) = G_{\psi,A}(n)$. From Lemma 7.19 it follows that there exists a decision table $T^* \in [T]$ such that $\psi^a(T^*) \leq n$ and $\psi^d(T^*) \geq G(T) - 1$. Therefore $\mathscr{H}_{\psi,A}^\infty(n) \geq G_{\psi,A}(n) - 1$. □

Lemma 7.22 *Let A be a nontrivial closed class of decision tables from $\mathscr{M}_2^\infty$, ψ be a bounded complexity measure, the function $\mathscr{H}_{\psi,A}^\infty$ be everywhere defined and $n \in \omega \setminus \{0\}$. Then $\mathscr{H}_{\psi,A}^\infty(n) \leq \max(n, nG_{\psi,A}(n))Z_{\psi,A}(n)\log_2(4nL_{\psi,A}(n))$.*

Proof Let $T \in A$ and $\psi^a(T) \leq n$. We now show that

$$\psi^d(T) \leq \max(n, nG_{\psi,A}(n))Z_{\psi,A}(n)\log_2(4nL_{\psi,A}(n)) .$$

If $T \in \mathscr{M}_2^{\infty c}$, then $\psi^d(T) = 0$ and the considered inequality holds. Let $T \notin \mathscr{M}_2^{\infty c}$. One can show that there exists a nondeterministic decision tree Γ for the table T such that $\psi(\Gamma) \leq n$ and the set U of words corresponding to complete path of Γ is an irreducible (ψ, n)-cover of the table T. Let $U = \{\alpha_1, \ldots, \alpha_t\}$ and $d_1, \ldots, d_t$ be decisions attached to terminal nodes of complete paths with corresponding words $\alpha_1, \ldots, \alpha_t$, respectively. Let, for the definiteness, $\mathrm{At}(\Gamma) = \{f_1, \ldots, f_m\}$ and attributes $f_1, \ldots, f_m$ are arranged in the same order as in the table T. It is clear that $t \leq L_{\psi,A}(n)$. Using boundedness from below property of the function ψ, we obtain $m \leq nL_{\psi,A}(n)$. Denote $T' = I(\mathrm{At}(T) \setminus \mathrm{At}(\Gamma), T)$. One can show that U is an irreducible (ψ, n)-cover of the table T'. We now define a mapping $\nu : E_2^m \to \mathscr{P}(\omega)$. Let $d_0 \in \omega \setminus \{d_1, \ldots, d_t\}$ and $\bar{\delta} = (\delta_1, \ldots, \delta_m) \in E_2^m$. If $\bar{\delta} \notin \Delta(T')$, then $\nu(\bar{\delta}) = \{d_0\}$. Otherwise, $\nu(\bar{\delta}) = \{d_i : \bar{\delta} \in \Delta(T'\alpha_i), i \in \{1, \ldots, t\}\}$.

Denote $T^* = J(\nu, T')$. One can show that any deterministic decision tree for the table T^* is a deterministic decision tree for the table T. Therefore $\psi^d(T) \leq \psi^d(T^*)$. It is clear that U is an irreducible (ψ, n)-cover of the table T^*. Therefore $\psi^a(T^*) \leq n$. Using commutativity and nondecreasing properties of the function ψ, we obtain that $\psi(f_j) \leq n$ for any $f_j \in \mathrm{At}(\Gamma)$. Therefore $T^* \in A_\psi(n)$. We shown that $W(T^*) = m \leq nL_{\psi,A}(n)$. Using Lemma 7.16, we obtain $N(T^*) \leq (4nL_{\psi,A}(n))^{Z(T^*)} \leq (4nL_{\psi,A}(n))^{Z_{\psi,A}(n)}$. We now show that

$$M_\psi(T^*) \leq \max(n, nG_{\psi,A}(n)) .$$

Let $\bar{\delta} = (\delta_1, \ldots, \delta_m) \in E_2^m$. First, we consider the case when $\bar{\delta} \in \Delta(T^*)$. Using Lemma 7.12 and inequality $\psi^a(T^*) \leq n$, we obtain $M_\psi(T^*, \bar{\delta}) \leq n$. Let $\bar{\delta} \notin \Delta(T^*)$. Then the word $\alpha = (f_1, \delta_1) \cdots (f_m, \delta_m)$ is an annihilating word for the table T^*. By removal of some letters from the word α, we obtain an irreducible annihilating word β for the table T^*. It is clear that the length of β is at most $G(T^*)$ and $G(T^*) \leq G_{\psi,A}(n)$. Let $\beta = (f_{j_1}, \delta_{j_1}) \cdots (f_{j_s}, \delta_{j_s})$. Using the fact that $T^* \in A_\psi(n)$ and boundedness from above property of the function ψ, we obtain that $\psi(f_{j_1} \cdots f_{j_s}) \leq nG_{\psi,A}(n)$.

Therefore $M_{\psi}(T^*, \bar{\delta}) \leq nG_{\psi,A}(n)$. Hence $M_{\psi}(T^*) \leq \max(n, nG_{\psi,A}(n))$. Using Lemma 7.2, we obtain

$$\psi^d(T) \leq \psi^d(T^*) \leq \max(n, nG_{\psi,A}(n))Z_{\psi,A}(n) \log_2(4nL_{\psi,A}(n)) .$$

Thus, the statement of the lemma holds. □

Proof of Theorem 7.8. Using Theorem 7.5 and Lemmas 7.17 and 7.20, we obtain that the functions $Z_{\psi,A}$, $G_{\psi,A}$ and $L_{\psi,A}$ are everywhere defined. Let there exist polynomials p_1, p_2, and p_3 such that $Z_{\psi,A}(n) \leq p_1(n)$, $G_{\psi,A}(n) \leq p_2(n)$ and $L_{\psi,A}(n) \leq 2^{p_3(n)}$ for any $n \in \omega$. Using Lemma 7.22, we obtain that there exists a polynomial p_0 such that $\mathscr{H}^{\infty}_{\psi,A}(n) \leq p_0(n)$ for any $n \in \omega$. Let there be no a polynomial p_1 such that $Z_{\psi,A}(n) \leq p_1(n)$ for any $n \in \omega$ or there be no a polynomial p_2 such that $G_{\psi,A}(n) \leq p_2(n)$ for any $n \in \omega$, or there be no a polynomial p_3 such that $L_{\psi,A}(n) \leq 2^{p_3(n)}$ for any $n \in \omega$. Using Lemmas 7.11, 7.18, and 7.21, we obtain that there is no a polynomial p_0 such that $\mathscr{H}^{\infty}_{\psi,A}(n) \leq p_0(n)$ for any $n \in \omega$. □

References

1. Alsolami, F., Azad, M., Chikalov, I., Moshkov, M.: Decision and Inhibitory Trees and Rules for Decision Tables with Many-valued Decisions. In: Intelligent Systems Reference Library, vol. 156. Springer (2020)
2. Boutell, M.R., Luo, J., Shen, X., Brown, C.M.: Learning multi-label scene classification. Pattern Recognit. **37**(9), 1757–1771 (2004)
3. Moshkov, M.: Time complexity of decision trees. In: Peters, J.F., Skowron, A. (eds.) Transactions on Rough Sets III. Lecture Notes in Computer Science, vol. 3400, pp. 244–459. Springer (2005)
4. Moshkov, M.: Comparative Analysis of Deterministic and Nondeterministic Decision Trees. In: Intelligent Systems Reference Library, vol. 179. Springer (2020)
5. Moshkov, M., Zielosko, B.: Combinatorial Machine Learning—A Rough Set Approach. Studies in Computational Intelligence, vol. 360. Springer (2011)
6. Ostonov, A., Moshkov, M.: Depth of deterministic and nondeterministic decision trees for decision tables with many-valued decisions from closed classes. In: Nguyen, N.T., Chbeir, R., Manolopoulos, Y., Fujita, H., Hong, T.-P., Le Nguyen, M., Wojtkiewicz, K. (eds.) Recent Challenges in Intelligent Information and Database Systems – 16th Asian Conference on Intelligent Information and Database Systems, ACIIDS 2024, Ras Al Khaimah, UAE, April 15-18, 2024, Proceedings, Part I. Communications in Computer and Information Science. vol. 2144, pp 164–174. Springer (2024)
7. Ostonov, A., Moshkov, M.: On Complexity of Deterministic and Nondeterministic Decision Trees for Decision Tables with Many-Valued Decisions from Closed Classes. In: Rough Sets – International Joint Conference, IJCRS 2024, Halifax, NS, Canada, May 17-20, 2024, Proceedings, Part I. Lecture Notes in Computer Science. vol. 14839, pp 173–187, Springer (2024)
8. Ostonov, A., Moshkov, M.: Comparison of deterministic and nondeterministic decision trees for decision tables with many-valued decisions from closed classes (2023). CoRR arXiv:2312.01116

9. Ostonov, A., Moshkov, M.: Deterministic and nondeterministic decision trees for decision tables with many-valued decisions from closed classes. In: Rough Sets—International Joint Conference, IJCRS 2023, Krakow, Poland, October 5–8, 2023, Proceedings. Lecture Notes in Computer Science, vol. 14481, pp. 89–104. Springer (2023)
10. Pawlak, Z.: Information systems theoretical foundations. Inf. Syst. **6**(3), 205–218 (1981)
11. Vens, C., Struyf, J., Schietgat, L., Dzeroski, S., Blockeel, H.: Decision trees for hierarchical multi-label classification. Mach. Learn. **73**(2), 185–214 (2008)
12. Zhou, Z., Zhang, M., Huang, S., Li, Y.: Multi-instance multi-label learning. Artif. Intell. **176**(1), 2291–2320 (2012)

Chapter 8
Complexity of Deterministic and Nondeterministic Decision Trees for Conventional Decision Tables from Closed Classes

8.1 Introduction

In this chapter, we consider closed classes of conventional decision tables and study the dependence of the minimum complexity of deterministic and nondeterministic decision trees and the complexity of deterministic and nondeterministic decision trees constructed by some algorithms on the complexity of the set of attributes attached to columns (in particular, for the depth, the complexity of a set of attributes is equal to its cardinality). We also study the dependence of the minimum complexity of deterministic decision trees on the minimum complexity of nondeterministic decision trees.

A conventional decision table is a rectangular table in which columns are labeled with attributes, rows are pairwise different and each row is labeled with a decision. Rows are interpreted as tuples of values of the attributes. For a given row, it is required to find the decision attached to the row. To this end, we can use the following queries: we can choose an attribute and ask what is the value of this attribute in the considered row. We study two types of algorithms based on these queries: deterministic and nondeterministic decision trees. One can interpret nondeterministic decision trees for a decision table as a way to represent an arbitrary system of true decision rules for this table that cover all rows of the table. We consider so-called bounded complexity measures, for example, the depth of decision trees.

Conventional decision tables are widely used in data analysis and appear in areas such as combinatorial optimization, computational geometry, and fault diagnosis, where they are used to represent and explore problems [3, 4].

We study classes of conventional decision tables closed under two operations: removal of columns and changing of decisions. The most natural examples of such classes are closed classes of decision tables generated by information systems [7]. An information system consists of a set of objects (universe) and a set of attributes (functions) defined on the universe and with values from a finite set. A conventional problem over an information system is specified by a finite number of attributes

K. Durdymyradov et al., *Decision Trees Versus Systems of Decision Rules*,
Studies in Big Data 160, https://doi.org/10.1007/978-3-031-71586-0_8

that divide the universe into nonempty domains in which these attributes have fixed values. A decision is attached to each domain. For a given object from the universe, it is required to find the decision attached to the domain containing this object.

A conventional decision table corresponds to this problem in a natural way: columns of this table are labeled with the considered attributes, rows correspond to domains and are labeled with decisions attached to domains. The set of conventional decision tables corresponding to conventional problems over an information system forms a closed class generated by this system. Note that the family of all closed classes is essentially wider than the family of closed classes generated by information systems. In particular, the union of two closed classes generated by two information systems is a closed class. However, generally, there is no an information system that generates this class.

Let A be a class of conventional decision tables closed under removal of columns and changing of decisions, and ψ be a bounded complexity measure. In this chapter, we study five functions: $\mathcal{F}_{\psi,A}(n)$, $\mathcal{G}_{\psi,A}(n)$, $\mathcal{H}_{\psi,A}(n)$, $\mathcal{U}_{\psi,A}(n)$, and $\mathcal{V}_{\psi,A}(n)$.

The function $\mathcal{F}_{\psi,A}(n)$ characterizes the growth in the worst case of the minimum complexity of a deterministic decision tree for a decision table from A with the growth of the complexity of the set of attributes attached to columns of the table. We prove that the function $\mathcal{F}_{\psi,A}(n)$ is either bounded from above by a constant, or grows as a logarithm of n, or grows almost linearly depending on n (it is bounded from above by n and is equal to n for infinitely many n).

The function $\mathcal{G}_{\psi,A}(n)$ characterizes the growth in the worst case of the minimum complexity of a nondeterministic decision tree for a decision table from A with the growth of the complexity of the set of attributes attached to columns of the table. We prove that the function $\mathcal{G}_{\psi,A}(n)$ is either bounded from above by a constant or grows almost linearly depending on n (it is bounded from above by n and is equal to n for infinitely many n).

The function $\mathcal{H}_{\psi,A}(n)$ characterizes the growth in the worst case of the minimum complexity of a deterministic decision tree for a decision table from A with the growth of the minimum complexity of a nondeterministic decision tree for the table. This function is either not everywhere defined or is everywhere defined. Let $\mathcal{H}_{\psi,A}(n)$ be everywhere defined. We prove that this function is either bounded from above by a constant, or is greater than or equal to n for infinitely many n.

The function $\mathcal{U}_{\psi,A}(n)$ characterizes the growth in the worst case of the complexity of the deterministic decision tree for a decision table from A constructed by a greedy algorithm $U_{R,\psi}$ with the growth of the complexity of the set of attributes attached to columns of the table. We prove that the function $\mathcal{U}_{\psi,A}(n)$ is either bounded from above by a constant, or grows as a logarithm of n, or grows almost linearly depending on n (it is bounded from above by n and is equal to n for infinitely many n). We find lower and upper bounds for the function $\mathcal{U}_{\psi,A}(n)$ depending on the function $\mathcal{F}_{\psi,A}(n)$. The considered results are the essential generalization of the results obtained in [2] for the depth of decision trees.

The function $\mathcal{V}_{\psi,A}(n)$ characterizes the growth in the worst case of the complexity of the nondeterministic decision tree for a decision table from A constructed by an algorithm V^a with the growth of the complexity of the set of attributes attached to columns of the table. We prove that $\mathscr{V}_{\psi,A}(n) = \mathscr{G}_{\psi,A}(n)$ for any n.

In this chapter, we also study the joint behavior of the functions $\mathscr{F}_{\psi,A}$ and $\mathscr{G}_{\psi,A}$, and list all three possible types of this behavior.

This chapter is based mainly on the papers [5, 6]. The chapter consists of five sections. In Sect. 8.2, main definitions and notation are considered. In Sect. 8.3, we provide the main results. Section 8.4 contains auxiliary statements. In Sect. 8.5, we prove the main results.

8.2 Main Definitions and Notation

Denote $\omega = \{0, 1, 2, \ldots\}$ and, for any $k \in \omega \setminus \{0, 1\}$, denote $E_k = \{0, 1, \ldots, k-1\}$. Let $P = \{f_i : i \in \omega\}$ be the set of *attributes* (really names of attributes). Two attributes $f_i, f_j \in P$ are considered different if $i \neq j$.

8.2.1 Decision Tables

First, we define the notion of a conventional decision table.

Definition 8.1 Let $k \in \omega \setminus \{0, 1\}$. Denote by $\mathscr{M}_k$ the set of rectangular tables filled with numbers from E_k in each of which rows are pairwise different, each row is labeled with a number from ω (decision), and columns are labeled with pairwise different attributes from P. Rows are interpreted as tuples of values of these attributes. Empty tables without rows belong also to the set $\mathscr{M}_k$. We will use the same notation Λ for these tables. Tables from $\mathscr{M}_k$ will be called *conventional decision tables* (*decision tables*).

Example 8.1 Figure 8.1 shows a decision table from $\mathscr{M}_2$.

Denote by $\mathscr{M}_k^c$ the set of tables from $\mathscr{M}_k$ in each of which all rows are labeled with the same decision. Let $\Lambda \in \mathscr{M}_k^c$.

Fig. 8.1 Decision table from $\mathscr{M}_2$

f_2	f_4	f_3	
1	1	1	1
0	1	1	3
1	1	0	1
1	0	1	2
0	0	1	2
1	0	0	2
0	0	0	2

Let T be a nonempty table from $\mathcal{M}_k$. Denote by $\text{At}(T)$ the set of attributes attached to columns of the table T. Let $f_{i_1}, \ldots, f_{i_m} \in \text{At}(T)$ and $\delta_1, \ldots, \delta_m \in E_k$. We denote by $T(f_{i_1}, \delta_1) \cdots (f_{i_m}, \delta_m)$ the table obtained from T by removal of all rows that do not satisfy the following condition: in columns labeled with attributes $f_{i_1}, \ldots, f_{i_m}$, the row has numbers $\delta_1, \ldots, \delta_m$, respectively.

We now define two operations on decision tables: removal of columns and changing of decisions. Let $T \in \mathcal{M}_k$.

Definition 8.2 *Removal of columns*. Let $D \subseteq \text{At}(T)$. We remove from T all columns labeled with the attributes from the set D. In each group of rows equal on the remaining columns, we keep one with the minimum decision. Denote the obtained table by $I(D, T)$. In particular, $I(\emptyset, T) = T$ and $I(\text{At}(T), T) = \Lambda$. It is obvious that $I(D, T) \in \mathcal{M}_k$.

Definition 8.3 *Changing of decisions*. Let $\nu : E_k^{|\text{At}(T)|} \to \omega$ (by definition, $E_k^0 = \emptyset$). For each row $\bar{\delta}$ of the table T, we replace the decision attached to this row with $\nu(\bar{\delta})$. We denote the obtained table by $J(\nu, T)$. It is obvious that $J(\nu, T) \in \mathcal{M}_k$.

Definition 8.4 Denote $[T] = \{J(\nu, I(D, T)) : D \subseteq \text{At}(T), \nu : E_k^{|\text{At}(T) \setminus D|} \to \omega\}$. The set $[T]$ is the *closure of the table* T under the operations of removal of columns and changing of decisions.

Example 8.2 Figure 8.2 shows the table $J(\nu, I(D, T_0))$, where T_0 is the table shown in Fig. 8.1, $D = \{f_4\}$ and $\nu(x_1, x_2) = x_1 + x_2$.

Definition 8.5 Let $A \subseteq \mathcal{M}_k$ and $A \neq \emptyset$. Denote $[A] = \bigcup_{T \in A} [T]$. The set $[A]$ is the *closure of the set* A under the considered two operations. The class (the set) of decision tables A will be called a *closed class* if $[A] = A$.

Let A_1 and A_2 be closed classes of decision tables from $\mathcal{M}_k$. Then $A_1 \cup A_2$ is a closed class of decision tables from $\mathcal{M}_k$.

8.2.2 Deterministic and Nondeterministic Decision Trees

A *finite tree with root* is a finite directed tree in which exactly one node called the *root* has no entering edges. The nodes without leaving edges are called *terminal* nodes.

Fig. 8.2 Decision table obtained from the decision table shown in Fig. 8.1 by removal of a column and changing of decisions

f_2	f_3	
1	1	2
0	1	1
1	0	1
0	0	0

Definition 8.6 A *k-decision tree* is a finite tree with root, which has at least two nodes and in which

- The root and edges leaving the root are not labeled.
- Each terminal node is labeled with a decision from the set ω.
- Each node, which is neither the root nor a terminal node, is labeled with an attribute from the set P. Each edge leaving such node is labeled with a number from the set E_k.

Example 8.3 Figures 8.3 and 8.4 show 2-decision trees.

We denote by $\mathscr{T}_k$ the set of all k-decision trees. Let $\Gamma \in \mathscr{T}_k$. We denote by $\text{At}(\Gamma)$ the set of attributes attached to nodes of Γ that are neither the root nor terminal nodes. A *complete path* of Γ is a sequence $\tau = v_1, d_1, \ldots, v_m, d_m, v_{m+1}$ of nodes and edges of Γ in which v_1 is the root of Γ, v_{m+1} is a terminal node of Γ and, for $j = 1, \ldots, m$, the edge d_j leaves the node v_j and enters the node v_{j+1}. Let $T \in \mathscr{M}_k$. If $\text{At}(\Gamma) \subseteq \text{At}(T)$, then we correspond to the table T and the complete path τ a decision table $T(\tau)$. If $m = 1$, then $T(\tau) = T$. If $m > 1$ and, for $j = 2, \ldots, m$, the node v_j is labeled with the attribute f_{i_j} and the edge d_j is labeled with the number δ_j, then $T(\tau) = T(f_{i_2}, \delta_2) \cdots (f_{i_m}, \delta_m)$.

Definition 8.7 Let $T \in \mathscr{M}_k \setminus \{\Lambda\}$. A *deterministic decision tree* for the table T is a k-decision tree Γ satisfying the following conditions:

- Only one edge leaves the root of Γ.

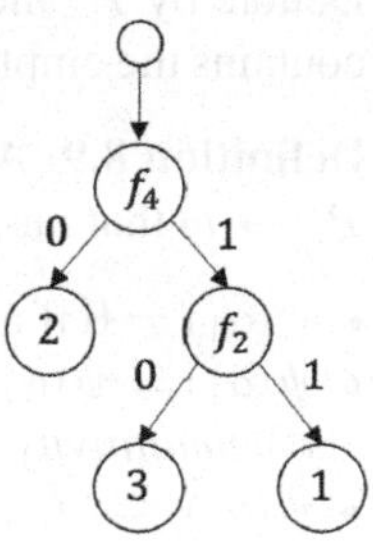

Fig. 8.3 A deterministic decision tree for the decision table shown in Fig. 8.1

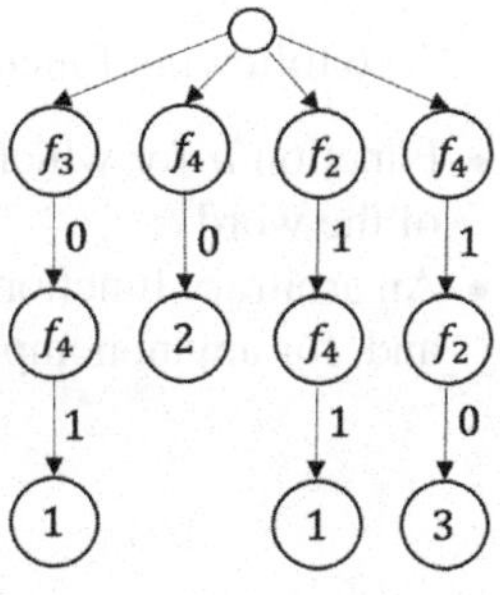

Fig. 8.4 A nondeterministic decision tree for the decision table shown in Fig. 8.1

- For any node, which is neither the root nor a terminal node, edges leaving this node are labeled with pairwise different numbers.
- $\mathrm{At}(\Gamma) \subseteq \mathrm{At}(T)$.
- For any row of T, there exists a complete path τ of Γ such that the considered row belongs to the table $T(\tau)$.
- For any complete path τ of Γ, either $T(\tau) = \Lambda$ or all rows of $T(\tau)$ are labeled with the decision attached to the terminal node of τ.

Example 8.4 The 2-decision tree shown in Fig. 8.3 is a deterministic decision tree for the decision table shown in Fig. 8.1.

Definition 8.8 Let $T \in \mathscr{M}_k \setminus \{\Lambda\}$. A *nondeterministic decision tree* for the table T is a k-decision tree Γ satisfying the following conditions:

- $\mathrm{At}(\Gamma) \subseteq \mathrm{At}(T)$.
- For any row of T, there exists a complete path τ of Γ such that the considered row belongs to the table $T(\tau)$.
- For any complete path τ of Γ, either $T(\tau) = \Lambda$ or all rows of $T(\tau)$ are labeled with the decision attached to the terminal node of τ.

Example 8.5 The 2-decision tree shown in Fig. 8.4 is a nondeterministic decision tree for the decision table shown in Fig. 8.1.

8.2.3 Complexity Measures

Denote by P^* the set of all finite words over the alphabet $P = \{f_i : i \in \omega\}$, which contains the empty word λ and on which the word concatenation operation is defined.

Definition 8.9 A *partially bounded complexity measure* is an arbitrary function $\psi : P^* \to \omega$ that has the following properties: for any words $\alpha_1, \alpha_2 \in P^*$,

- $\psi(\alpha_1) = 0$ if and only if $\alpha_1 = \lambda$–*positivity* property.
- $\psi(\alpha_1) = \psi(\alpha_1')$ for any word α_1' obtained from α_1 by permutation of letters–*commutativity* property.
- $\psi(\alpha_1) \leq \psi(\alpha_1\alpha_2)$–*nondecreasing* property.
- $\psi(\alpha_1\alpha_2) \leq \psi(\alpha_1) + \psi(\alpha_2)$–*boundedness from above* property.

The following functions are partially bounded complexity measures:

- Function h for which, for any word $\alpha \in P^*$, $h(\alpha) = |\alpha|$, where $|\alpha|$ is the length of the word α.
- An arbitrary function $\varphi : P^* \to \omega$ such that $\varphi(\lambda) = 0$, for any $f_i \in P$, $\varphi(f_i) > 0$ and, for any nonempty word $f_{i_1} \cdots f_{i_m} \in P^*$,

$$\varphi(f_{i_1} \cdots f_{i_m}) = \sum_{j=1}^{m} \varphi(f_{i_j}) . \tag{8.1}$$

- An arbitrary function $\rho : P^* \to \omega$ such that $\rho(\lambda) = 0$, for any $f_i \in P$, $\rho(f_i) > 0$, and, for any nonempty word $f_{i_1} \cdots f_{i_m} \in P^*$, $\rho(f_{i_1} \cdots f_{i_m}) = \max\{\rho(f_{i_j}) : j = 1, \ldots, m\}$.

Definition 8.10 A *bounded* complexity measure is a partially bounded complexity measure ψ, which has the *boundedness from below* property: for any word $\alpha \in P^*$, $\psi(\alpha) \geq |\alpha|$.

Any partially bounded complexity measure satisfying the equality (8.1), in particular the function h, is a bounded complexity measure. One can show that if functions ψ_1 and ψ_2 are partially bounded complexity measures, then the functions ψ_3 and ψ_4 are partially bounded complexity measures, where for any $\alpha \in P^*$, $\psi_3(\alpha) = \psi_1(\alpha) + \psi_2(\alpha)$ and $\psi_4(\alpha) = \max(\psi_1(\alpha), \psi_2(\alpha))$. If the function ψ_1 is a bounded complexity measure, then the functions ψ_3 and ψ_4 are bounded complexity measures.

Definition 8.11 A complexity measure ψ is called *1-computable* if there exists an algorithm, which, for a given attribute $f_i \in P$, returns the value $\psi(f_i)$.

The simplest example of a 1-computable bounded complexity measure is the function h.

Definition 8.12 Let ψ be a partially bounded complexity measure. We extend it to the set of all finite subsets of the set P. Let D be a finite subset of the set P. If $D = \emptyset$, then $\psi(D) = 0$. Let $D = \{f_{i_1}, \ldots, f_{i_m}\}$ and $m \geq 1$. Then $\psi(D) = \psi(f_{i_1} \cdots f_{i_m})$.

8.2.4 Parameters of Decision Trees and Tables

Let $\Gamma \in \mathcal{T}_k$ and $\tau = v_1, d_1, \ldots, v_m, d_m, v_{m+1}$ be a complete path of Γ. We correspond to the path τ a word $F(\tau) \in P^*$: if $m = 1$, then $F(\tau) = \lambda$, and if $m > 1$ and, for $j = 2, \ldots, m$, the node v_j is labeled with the attribute f_{i_j}, then $F(\tau) = f_{i_2} \cdots f_{i_m}$.

Definition 8.13 Let ψ be a partially bounded complexity measure. We extend the function ψ to the set $\mathcal{T}_k$. Let $\Gamma \in \mathcal{T}_k$. Then $\psi(\Gamma) = \max\{\psi(F(\tau))\}$, where the maximum is taken over all complete paths τ of the decision tree Γ. For a given partially bounded complexity measure ψ, the value $\psi(\Gamma)$ will be called the *complexity of the decision tree* Γ. The value $h(\Gamma)$ will be called the *depth* of the decision tree Γ.

Let ψ be a partially bounded complexity measure. We now describe the functions ψ^d, ψ^a, Sep, W_ψ, S_ψ, $\hat{S}_\psi$, M_ψ, N, and R defined on the set $\mathcal{M}_k$ and taking values from the set ω. By definition, the value of each of these functions for Λ is equal 0. Let $T \in \mathcal{M}_k \setminus \{\Lambda\}$.

- $\psi^d(T) = \min\{\psi(\Gamma)\}$, where the minimum is taken over all deterministic decision trees Γ for the table T.

- $\psi^a(T) = \min\{\psi(\Gamma)\}$, where the minimum is taken over all nondeterministic decision trees Γ for the table T.
- A set $D \subseteq \mathrm{At}(T)$ is called a *separating set* for the table T if in the set of columns labeled with attributes from D rows of the table T are pairwise different. Then $\mathrm{Sep}(T)$ is the minimum cardinality of a separating set for the table T.
- $W_\psi(T) = \psi(\mathrm{At}(T))$. This is *the complexity of the set of attributes attached to columns of the table* T.
- Let $\bar{\delta}$ be a row of the table T. Denote $S_\psi(T, \bar{\delta}) = \min\{\psi(D)\}$, where the minimum is taken over all subsets D of the set $\mathrm{At}(T)$ such that in the set of columns of T labeled with attributes from D the row $\bar{\delta}$ is different from all other rows of the table T. Then $S_\psi(T) = \max\{S_\psi(T, \bar{\delta})\}$, where the maximum is taken over all rows $\bar{\delta}$ of the table T.
- $\hat{S}_\psi(T) = \max\{S_\psi(T^*) : T^* \in [T]\}$.
- If $T \in \mathscr{M}_k^c$, then $M_\psi(T) = 0$. Let $T \notin \mathscr{M}_k^c$, $|\mathrm{At}(T)| = n$, and columns of the table T be labeled with the attributes $f_{t_1}, \ldots, f_{t_n}$. Let $\bar{\delta} = (\delta_1, \ldots, \delta_n) \in E_k^n$. Denote by $M_\psi(T, \bar{\delta})$ the minimum number $p \in \omega$ for which there exist attributes $f_{t_{i_1}}, \ldots, f_{t_{im}} \in \mathrm{At}(T)$ such that $T(f_{t_{i_1}}, \delta_{i_1}) \cdots (f_{t_{im}}, \delta_{i_m}) \in \mathscr{M}_k^c$ and $\psi(f_{t_{i_1}} \cdots f_{t_{im}}) = p$. Then $M_\psi(T) = \max\{M_\psi(T, \bar{\delta}) : \bar{\delta} \in E_k^n\}$.
- $N(T)$ is the number of rows in the table T.
- $R(T)$ is the number of unordered pairs of rows of the table T labeled with different decisions.

For the bounded complexity measure h, we denote $W(T) = W_h(T)$, $S(T) = S_h(T)$, $\hat{S}(T) = \hat{S}_h(T)$, and $M(T) = M_h(T)$. Note that $W(T)$ is the number of columns in the table T.

Example 8.6 We denote by T_0 the decision table shown in Fig. 8.1. One can show that $h^d(T_0) = 2$, $h^a(T_0) = 2$, $\mathrm{Sep}(T_0) = 3$, $W(T_0) = 3$, $N(T_0) = 7$, $R(T_0) = 14$, $S(T_0) = 3$, $\hat{S}(T_0) = 3$, and $M(T_0) = 2$.

8.3 Main Results

In this section, we consider results obtained for the functions $\mathscr{F}_{\psi,A}$, $\mathscr{G}_{\psi,A}$, $\mathscr{H}_{\psi,A}$, $\mathscr{U}_{\psi,A}$, and $\mathscr{V}_{\psi,A}$ and discuss closed classes of decision tables generated by information systems.

8.3.1 *Function* $\mathscr{F}_{\psi,A}$

Let ψ be a bounded complexity measure and A be a nonempty closed class of decision tables from $\mathscr{M}_k$. We now define a function $\mathscr{F}_{\psi,A} : \omega \to \omega$. Let $n \in \omega$. Then

$$\mathscr{F}_{\psi,A}(n) = \max\{\psi^d(T) : T \in A, W_\psi(T) \le n\} .$$

The function $\mathscr{F}_{\psi,A}$ characterizes the growth in the worst case of the minimum complexity of a deterministic decision tree for a decision table from A with the growth of the complexity of the set of attributes attached to columns of this table.

Let $D = \{n_i : i \in \omega\}$ be an infinite subset of the set ω in which, for any $i \in \omega$, $n_i < n_{i+1}$. Let us define a function $H_D : \omega \to \omega$. Let $n \in \omega$. If $n < n_0$, then $H_D(n) = 0$. If, for some $i \in \omega$, $n_i \le n < n_{i+1}$, then $H_D(n) = n_i$.

Theorem 8.1 *Let ψ be a bounded complexity measure and A be a nonempty closed class of decision tables from $\mathscr{M}_k$. Then $\mathscr{F}_{\psi,A}$ is an everywhere defined nondecreasing function such that $\mathscr{F}_{\psi,A}(n) \le n$ for any $n \in \omega$ and $\mathscr{F}_{\psi,A}(0) = 0$. For this function, one of the following statements holds:*

(a) *If the functions S_ψ and N are bounded from above on the class A, then there exists a positive constant c_0 such that $\mathscr{F}_{\psi,A}(n) \le c_0$ for any $n \in \omega$.*

(b) *If the function S_ψ is bounded from above on the class A and the function N is not bounded from above on the class A, then there exist positive constants c_1, c_2, c_3, c_4 such that $c_1 \log_2 n - c_2 \le \mathscr{F}_{\psi,A}(n) \le c_3 \log_2 n + c_4$ for any $n \in \omega \setminus \{0\}$.*

(c) *If the function S_ψ is not bounded from above on the class A, then there exists an infinite subset D of the set ω such that $H_D(n) \le \mathscr{F}_{\psi,A}(n)$ for any $n \in \omega$.*

For the bounded complexity measure h, the bound from the statement (c) of Theorem 8.1 can be improved.

Theorem 8.2 *Let A be a nonempty closed class of decision tables from $\mathscr{M}_k$ for which the function S is not bounded from above. Then $\mathscr{F}_{h,A}(n) = n$ for any $n \in \omega$.*

8.3.2 Function $\mathscr{G}_{\psi,A}$

Let ψ be a bounded complexity measure and A be a nonempty closed class of decision tables from $\mathscr{M}_k$. We now define a function $\mathscr{G}_{\psi,A}$. Let $n \in \omega$. Then

$$\mathscr{G}_{\psi,A}(n) = \max\{\psi^a(T) : T \in A,\ W_\psi(T) \le n\}\ .$$

The function $\mathscr{G}_{\psi,A}$ characterizes the growth in the worst case of the minimum complexity of a nondeterministic decision tree for a decision table from A with the growth of the complexity of the set of attributes attached to columns of this table.

Theorem 8.3 *Let ψ be a bounded complexity measure and A be a nonempty closed class of decision tables from $\mathscr{M}_k$. Then $\mathscr{G}_{\psi,A}$ is an everywhere defined nondecreasing function such that $\mathscr{G}_{\psi,A}(n) \le n$ for any $n \in \omega$ and $\mathscr{G}_{\psi,A}(0) = 0$. For this function, one of the following statements holds:*

(a) *If the function S_ψ is bounded from above on the class A, then there exists a positive constant c such that $\mathscr{G}_{\psi,A}(n) \le c$ for any $n \in \omega$.*

(b) *If the function S_ψ is not bounded from above on the class A, then there exists an infinite subset D of the set ω such that $H_D(n) \le \mathscr{G}_{\psi,A}(n)$ for any $n \in \omega$.*

For the bounded complexity measure h, the bound from the statement (b) of Theorem 8.3 can be improved.

Theorem 8.4 *Let A be a nonempty closed class of decision tables from $\mathscr{M}_k$ for which the function S is not bounded from above. Then $\mathscr{G}_{h,A}(n) = n$ for any $n \in \omega$.*

8.3.3 Function $\mathscr{H}_{\psi,A}$

Let ψ be a bounded complexity measure and A be a nonempty closed class of decision tables from $\mathscr{M}_k$. We now define possibly partial function $\mathscr{H}_{\psi,A} : \omega \to \omega$. Let $n \in \omega$. If the set $\{\psi^d(T) : T \in A, \psi^a(T) \leq n\}$ is infinite, then the value $\mathscr{H}_{\psi,A}(n)$ is undefined. Otherwise, $\mathscr{H}_{\psi,A}(n) = \max\{\psi^d(T) : T \in A, \psi^a(T) \leq n\}$.

The function $\mathscr{H}_{\psi,A}$ characterizes the growth in the worst case of the minimum complexity of a deterministic decision tree for a decision table from A with the growth of the minimum complexity of a nondeterministic decision tree for this table.

Theorem 8.5 *Let ψ be a bounded complexity measure and A be a nonempty closed class of decision tables from $\mathscr{M}_k$. Then $\mathscr{H}_{\psi,A}(0) = 0$ and $\mathscr{H}_{\psi,A}$ is a nondecreasing function in its domain.*

If $\mathscr{H}_{\psi,A}$ is not an everywhere defined function, then its domain coincides with the set $\{n : n \in \omega, n \leq n_0\}$ for some $n_0 \in \omega$.

If the function $\mathscr{H}_{\psi,A}$ is everywhere defined, then one of the following statements holds:

(a) *If the function ψ^d is bounded from above on the class A, then there is a nonnegative constant c such that $\mathscr{H}_{\psi,A}(n) \leq c$ for any $n \in \omega$.*

(b) *If the function ψ^d is not bounded from above on the class A, then there exists an infinite subset D of the set ω such that $\mathscr{H}_{\psi,A}(n) \geq H_D(n)$ for any $n \in \omega$.*

Remark 8.1 From Theorem 8.1 it follows that the function ψ^d is bounded from above on the class A if and only if the functions S_ψ and N are bounded from above on the class A.

8.3.4 Function $\mathscr{U}_{\psi,A}$

Let ψ be a 1-computable bounded complexity measure. We now describe an algorithm $U_{R,\psi}$ from [1], which, for a given decision table T from $\mathscr{M}_k \setminus \{\Lambda\}$, constructs a deterministic decision tree $U_{R,\psi}(T)$ for the table T.

Algorithm $U_{R,\psi}$

Step 1. Construct a tree G consisting of two nodes v_0 and v_1 and an edge leaving the node v_0 and entering the node v_1. Label the node v_1 with the decision table T.

Step 2. If the tree G does not contain nodes labeled with decision tables, then return G as the decision tree $U_{R,\psi}(T)$.

Otherwise, choose in the tree G a node v labeled with a decision table Q. If all rows of the table Q are labeled with the same decision, then label the node v with this decision instead of the table Q and move on to Step 2.

Otherwise, for each $f_i \in \mathrm{At}(T)$, find the minimum number $\sigma_i \in E_k$ such that $R(Q(f_i, \sigma_i)) = \max\{R(Q(f_i, \sigma)) : \sigma \in E_k\}$. Set $K(Q) = \{f_i : f_i \in \mathrm{At}(T), R(Q) > R(Q(f_i, \sigma_i))\}$. For each $f_i \in K(Q)$, set

$$d(f_i) = \max\{\psi(f_i), R(Q)/(R(Q) - R(Q(f_i, \sigma_i)))\} .$$

Let f_{i_0} be the attribute from $K(Q)$ with the minimum index i_0 for which $d(f_{i_0}) = \min\{d(f_i) : f_i \in K(Q)\}$. Label the node v with the attribute f_{i_0} instead of the table Q. For each $\sigma \in E_k$ such that $Q(f_{i_0}, \sigma) \neq \Lambda$, add to the tree G a node $v(\sigma)$ and an edge that leaves the node v and enters the node $v(\sigma)$. Label this edge with the number σ and label the node $v(\sigma)$ with the table $Q(f_{i_0}, \sigma)$. Move on to Step 2.

We will not consider the time complexity of this algorithm in detail. We only evaluate the number of repetitions of Step 2 during the construction of the decision tree $U_{R,\psi}(T)$.

Theorem 8.6 *Let ψ be a 1-computable bounded complexity measure. Then, for any decision table T from $\mathscr{M}_k \setminus \{\Lambda\}$, the algorithm $U_{R,\psi}$ during the construction of the decision tree $U_{R,\psi}(T)$ repeats Step 2 no more than $2N(T)$ times.*

Let ψ be a 1-computable bounded complexity measure and A be a nonempty closed class of decision tables from $\mathscr{M}_k$. By definition, $\psi(U_{R,\psi}(\Lambda)) = 0$. We now define a function $\mathscr{U}_{\psi,A} : \omega \to \omega$. Let $n \in \omega$. Then

$$\mathscr{U}_{\psi,A}(n) = \max\{\psi(U_{R,\psi}(T)) : T \in A, W_\psi(T) \leq n\} .$$

The function $\mathscr{U}_{\psi,A}$ characterizes the growth in the worst case of the complexity of the deterministic decision tree constructed by the algorithm $U_{R,\psi}$ for a decision table from A with the growth of the complexity of the set of attributes attached to columns of this table.

The next statement shows lower and upper bounds for the function $\mathscr{U}_{\psi,A}(n)$ depending on the function $\mathscr{F}_{\psi,A}(n)$.

Theorem 8.7 *Let ψ be a 1-computable bounded complexity measure and A be a nonempty closed class of decision tables from $\mathscr{M}_k$. Then $\mathscr{U}_{\psi,A}$ is an everywhere defined nondecreasing function such that $\mathscr{F}_{\psi,A}(n) \leq \mathscr{U}_{\psi,A}(n) \leq n$ for any $n \in \omega$. For this function, one of the following statements holds:*

(a) *If the functions S_ψ and N are bounded from above on the class A, then there exists a positive constant c such that $\mathscr{U}_{\psi,A}(n) \leq \mathscr{F}_{\psi,A}(n) + c$ for any $n \in \omega$.*

(b) *If the function S_ψ is bounded from above on the class A and the function N is not bounded from above on the class A, then there exist positive constants c and d such that $\mathscr{U}_{\psi,A}(n) \leq c\mathscr{F}_{\psi,A}(n) + d$ for any $n \in \omega \setminus \{0\}$.*

(c) *If the function S_ψ is not bounded from above on the class A, then there exists a positive constant c such that $\mathscr{U}_{\psi,A}(n) \le c\mathscr{F}_{\psi,A}(n)^3$ for any $n \in \omega$.*

8.3.5 Function $\mathscr{V}_{\psi,A}$

We now describe an algorithm V^a, which, for a given decision table T from $\mathscr{M}_k \setminus \{\Lambda\}$, constructs a nondeterministic decision tree $V^a(T)$ for the table T. Denote $\Delta(T)$ the set of rows of the table T. A subset D of the set $\mathrm{At}(T)$ will be called a *local separating set* for a row $\bar{\delta} \in \Delta(T)$ if, in the set of columns labeled with attributes from D, the row $\bar{\delta}$ is different from all other rows of T. If $N(T) = 1$, then the empty set is a local separating set for the row $\bar{\delta}$. A local separating set D for the row $\bar{\delta}$ will be called *minimal* if any proper subset of the set D is not a local separating set for $\bar{\delta}$.

Algorithm V^a

Step 1. For each row $\bar{\delta}$ of the decision table T, we construct a complete path $\rho(\bar{\delta})$ of the tree $V^a(T)$. We remove attributes from the set $\mathrm{At}(T)$ until we obtain a minimal local separating set $D(\bar{\delta})$ for the row $\bar{\delta}$. Let $D(\bar{\delta}) = \{f_{i_1}, \ldots, f_{i_m}\}$. Then $\rho(\bar{\delta}) = v_0, d_0, v_1, d_1, \ldots, v_m, d_m, v_{m+1}$, where v_0 is the root, v_{m+1} is the terminal node labeled with the decision attached to the row $\bar{\delta}$ and, for $j = 1, \ldots, m$, the node v_j is labeled with the attribute f_{i_j} and the edge d_j is labeled with the value of this attribute in the row $\bar{\delta}$.

Step 2. We merge roots of all complete paths $\rho(\bar{\delta})$, $\bar{\delta} \in \Delta(T)$. As a result, we obtain a nondeterministic decision tree $V^a(T)$ for the table T.

Let ψ be a bounded complexity measure and A be a nonempty closed class of decision tables from $\mathscr{M}_k$. By definition, $\psi(V^a(\Lambda)) = 0$. We now define a function $\mathscr{V}_{\psi,A} : \omega \to \omega$. Let $n \in \omega$. Then

$$\mathscr{V}_{\psi,A}(n) = \max\{\psi(V^a(T)) : T \in A, W_\psi(T) \le n\} .$$

The function $\mathscr{V}_{\psi,A}$ characterizes the growth in the worst case of the complexity of the nondeterministic decision tree constructed by the algorithm V^a for a decision table from A with the growth of the complexity of the set of attributes attached to columns of this table.

Theorem 8.8 *Let ψ be a bounded complexity measure and A be a nonempty closed class of decision tables from $\mathscr{M}_k$. Then $\mathscr{V}_{\psi,A}(n) = \mathscr{G}_{\psi,A}(n)$ for any $n \in \omega$.*

8.3.6 Family of Closed Classes of Decision Tables

Let U be a set and $\Phi = \{f_0, f_1, \ldots\}$ be a finite or countable set of functions (attributes) defined on U and taking values from E_k. The pair (U, Φ) is called a

k*-information system.* A *conventional problem* over (U, Φ) is an arbitrary tuple $z = (U, \nu, f_{i_1}, \ldots, f_{i_n})$, where $n \in \omega$, $\nu : E_k^n \to \omega$ and $f_{i_1}, \ldots, f_{i_n}$ are functions from Φ with pairwise different indices $i_1, \ldots, i_n$.

Let $n > 0$. The problem z is to determine the value $\nu(f_{i_1}(u), \ldots, f_{i_n}(u))$ for a given $u \in U$. Various examples of k-information systems and problems over these systems can be found in [3].

We denote by $T(z)$ a decision table from $\mathscr{M}_k$ with n columns labeled with attributes $f_{i_1}, \ldots, f_{i_n}$. A row $(\delta_1, \ldots, \delta_n) \in E_k^n$ belongs to the table $T(z)$ if and only if the system of equations $\{f_{i_1}(x) = \delta_1, \ldots, f_{i_n}(x) = \delta_n\}$ has a solution from the set U. This row is labeled with the decision $\nu(\delta_1, \ldots, \delta_n)$.

Let the algorithms for the problem z solving be algorithms in which each elementary operation consists in calculating the value of some attribute from the set $\{f_{i_1}, \ldots, f_{i_n}\}$ on a given element $u \in U$. Then, as a model of the problem z, we can use the decision table $T(z)$, and as models of algorithms for the problem z solving–deterministic and nondeterministic decision trees for the table $T(z)$.

If $n = 0$, then we set $T(z) = \Lambda$.

Denote by $\mathrm{Probl}(U, \Phi)$ the set of conventional problems over (U, Φ) and

$$\mathrm{Tab}(U, \Phi) = \{T(z) : z \in \mathrm{Probl}(U, \Phi)\} .$$

One can show that $\mathrm{Tab}(U, \Phi) = [\mathrm{Tab}(U, \Phi)]$, i.e., $\mathrm{Tab}(U, \Phi)$ is a closed class of decision tables from $\mathscr{M}_k$ *generated* by the information system (U, Φ).

Closed classes of decision tables generated by k-information systems are the most natural examples of closed classes. However, the notion of a closed class is essentially wider. In particular, the union $\mathrm{Tab}(U_1, \Phi_1) \cup \mathrm{Tab}(U_2, \Phi_2)$, where (U_1, Φ_1) and (U_2, Φ_2) are k-information systems, is a closed class, but generally, we cannot find an information system (U, Φ) such that $\mathrm{Tab}(U, \Phi) = \mathrm{Tab}(U_1, \Phi_1) \cup \mathrm{Tab}(U_2, \Phi_2)$.

8.3.7 Example of Information System

Let $\mathbb{R}$ be the set of real numbers and $F = \{f_i : i \in \omega\}$ be the set of functions defined on $\mathbb{R}$ and taking values from the set E_2 such that, for any $i \in \omega$ and $a \in \mathbb{R}$,

$$f_i(a) = \begin{cases} 0, & a < i \,, \\ 1, & a \geq i \,. \end{cases}$$

Let ψ be a bounded complexity measure and $A = \mathrm{Tab}(\mathbb{R}, F)$. One can prove the following statements:

- The function N is not bounded from above on the set A.
- The function S_ψ is bounded from above on the set A if and only if there exists a constant $c_0 > 0$ such that $\psi(f_i) \leq c_0$ for any $i \in \omega$.
- The function ψ^d is not bounded from above on the set A.

- The function $\mathscr{H}_{\psi,A}$ is everywhere defined if and only if, for any $n \in \omega$, the set $\{f_i : i \in \omega, \psi(f_i) \leq n\}$ is finite.

8.3.8 Joint Behavior of Functions $\mathscr{F}_{\psi,A}$ and $\mathscr{G}_{\psi,A}$

In this section, we study joint behavior of functions $\mathscr{F}_{\psi,A}$ and $\mathscr{G}_{\psi,A}$. Let ψ be a bounded complexity measure and A be a nonempty closed class of decision tables from $\mathscr{M}_k$. From Theorems 8.1–8.4 it follows that there are three possible types of the join behavior of functions $\mathscr{F}_{\psi,A}$ and $\mathscr{G}_{\psi,A}$:

- If the functions S_ψ and N are bounded from above on the class A, then $\mathscr{F}_{\psi,A}(n) = O(1)$ and $\mathscr{G}_{\psi,A}(n) = O(1)$.
- If the function S_ψ is bounded from above on the class A and the function N is not bounded from above on the class A, then $\mathscr{F}_{\psi,A}(n) = \Theta(\log n)$ and $\mathscr{G}_{\psi,A}(n) = O(1)$.
- If the function S_ψ is not bounded from above on the class A, then there exist infinite subsets D_1 and D_2 of the set ω such that $H_{D_1}(n) \leq \mathscr{F}_{\psi,A}(n) \leq n$ and $H_{D_2}(n) \leq \mathscr{G}_{\psi,A}(n) \leq n$ for any $n \in \omega$. If $\psi = h$, then $\mathscr{F}_{h,A}(n) = \mathscr{G}_{h,A}(n) = n$ for any $n \in \omega$.

All these three types are realizable for the bounded complexity measure h. Let us consider three 2-information systems (ω, F_1), (ω, F_2), and (ω, F_3), where

- $F_1 = \{q\}$ and, for any $j \in \omega$, $q(j) = 0$ if $j = 0$ and $q(j) = 1$ if $j > 0$.
- $F_2 = \{l_i : i \in \omega\}$ and, for any $i, j \in \omega$, $l_i(j) = 0$ if $j \leq i$ and $l_i(j) = 1$ if $j > i$.
- $F_3 = \{p_i : i \in \omega\}$ and, for any $i, j \in \omega$, $p_i(j) = 1$ if $i = j$ and $p_i(j) = 0$ if $i \neq j$.

One can show, that

- The functions S and N are bounded from above on the closed class Tab(ω, F_1).
- The function S is bounded from above on the closed class Tab(ω, F_2) and the function N is not bounded from above on this class.
- The function S is not bounded from above on the closed class Tab(ω, F_3).

8.4 Auxiliary Statements

This section contains auxiliary statements.

It is not difficult to prove the following upper bound on the minimum complexity of deterministic decision trees for a table.

Lemma 8.1 *For any partially bounded complexity measure ψ and any table T from $\mathscr{M}_k$,*

$$\psi^d(T) \leq W_\psi(T) .$$

The next statement follows from Lemma 1.3 and Theorem 2.2 [1].

Lemma 8.2 *For any partially bounded complexity measure ψ and any table T from $\mathscr{M}_k$,*

$$\psi^d(T) \le \begin{cases} 0, & M_\psi(T) = 0\,, \\ M_\psi(T)\log_2 N(T), & M_\psi(T) \ge 1\,. \end{cases}$$

Lemma 8.3 *For any partially bounded complexity measure ψ and any table T from $\mathscr{M}_k$,*

$$M_\psi(T) \le 2\hat{S}_\psi(T)\,.$$

Proof Let $T \in \mathscr{M}_k^c$. Then $M_\psi(T) = 0$. Therefore $M_\psi(T) \le 2\hat{S}_\psi(T)$.

Let $T \notin \mathscr{M}_k^c$, $W(T) = n$ and $f_{t_1}, \ldots, f_{t_n}$ be attributes attached to columns of the table T. Denote $D = \{f_i : f_i \in \mathrm{At}(T), \psi(f_i) \le S_\psi(T)\}$ and $T^* = I(\mathrm{At}(T) \setminus D, T)$. Evidently, $T^* \in [T]$. Taking into account that the function ψ has the nondecreasing property, we obtain that any two rows of T are different in columns labeled with attributes from the set D. Let for the definiteness, $D = \{f_{t_1}, \ldots, f_{t_m}\}$.

Let $\bar{\delta} = (\delta_1, \ldots, \delta_n) \in E_k^n$. Let $(\delta_1, \ldots, \delta_m)$ be a row of T^*. Since $T^* \in [T]$, there exist attributes $f_{t_{j_1}}, \ldots, f_{t_{j_s}}$ of the table T^* such that the row $(\delta_1, \ldots, \delta_m)$ is different from all other rows of T^* in columns labeled with these attributes and $\psi(f_{t_{j_1}} \cdots f_{t_{j_s}}) \le \hat{S}_\psi(T)$. It is clear that $T(f_{t_{j_1}}, \delta_{j_1}) \cdots (f_{t_{j_s}}, \delta_{j_s}) \in \mathscr{M}_k^c$. Therefore $M_\psi(T, \bar{\delta}) \le \hat{S}_\psi(T)$.

Let $(\delta_1, \ldots, \delta_m)$ be not a row of T^*. We consider m tables

$$T^*(f_{t_1}, \delta_1), T^*(f_{t_1}, \delta_1)(f_{t_2}, \delta_2), ..., T^*(f_{t_1}, \delta_1) \cdots (f_{t_m}, \delta_m)\,.$$

If $T^*(f_{t_1}, \delta_1) = \Lambda$, then $T(f_{t_1}, \delta_1) = \Lambda$. Since $f_{t_1} \in D$, $\psi(f_{t_1}) \le S_\psi(T)$. Taking into account that $S_\psi(T) \le \hat{S}_\psi(T)$, we obtain $M_\psi(T, \bar{\delta}) \le \hat{S}_\psi(T)$. Let $T^*(f_{t_1}, \delta_1) \ne \Lambda$. Then there exists $p \in \{1, \ldots, m-1\}$ such that

$$T^*(f_{t_1}, \delta_1) \cdots (f_{t_p}, \delta_p) \ne \Lambda$$

and $T^*(f_{t_1}, \delta_1) \cdots (f_{t_{p+1}}, \delta_{p+1}) = \Lambda$. Denote $C = \{f_{t_{p+1}}, f_{t_{p+2}}, \ldots, f_{t_n}\}$ and $T^0 = I(C, T)$. Evidently, $T^0 \in [T]$. Therefore in T^0 there are attributes $f_{t_{i_1}}, \ldots, f_{t_{i_l}}$ such that the row $(\delta_1, \ldots, \delta_p)$ is different from all other rows of the table T^0 in columns labeled with these attributes and $\psi(f_{t_{i_1}} \cdots f_{t_{i_l}}) \le \hat{S}_\psi(T)$. One can show that

$$T^*(f_{t_{i_1}}, \delta_{i_1}) \cdots (f_{t_{i_l}}, \delta_{i_l}) = T^*(f_{t_1}, \delta_1) \cdots (f_{t_p}, \delta_p)\,.$$

Therefore $T^*(f_{t_{i_1}}, \delta_{i_1}) \cdots (f_{t_{i_l}}, \delta_{i_l})(f_{t_{p+1}}, \delta_{p+1}) = \Lambda$. Hence

$$T(f_{t_{i_1}}, \delta_{i_1}) \cdots (f_{t_{i_l}}, \delta_{i_l})(f_{t_{p+1}}, \delta_{p+1}) = \Lambda\,.$$

Since $f_{t_{p+1}} \in D$, $\psi(f_{t_{p+1}}) \leq S_\psi(T) \leq \hat{S}_\psi(T)$. Using boundedness from above property of the function ψ, we obtain $\psi(f_{t_{i_1}} \cdots f_{t_{i_l}} f_{t_{p+1}}) \leq 2\hat{S}_\psi(T)$. Therefore

$$M_\psi(T, \bar{\delta}) \leq 2\hat{S}_\psi(T) .$$

Thus, for any $\bar{\delta} \in E_k^n$, $M_\psi(T, \bar{\delta}) \leq 2\hat{S}_\psi(T)$. As a result, we obtain $M_\psi(T) \leq 2\hat{S}_\psi(T)$. □

Lemma 8.4 *For any table T from $\mathscr{M}_k \setminus \{\Lambda\}$,*

$$N(T) \leq (kW(T))^{S(T)} .$$

Proof If $N(T) = 1$, then $S(T) = 0$ and the considered inequality holds. Let $N(T) > 1$. Then $S(T) > 0$. Denote $m = S(T)$. Evidently, for any row $\bar{\delta}$ of the table T, there exist attributes $f_{i_1}, \ldots, f_{i_m} \in \mathrm{At}(T)$ and numbers $\sigma_1, \ldots, \sigma_m \in E_k$ such that the table $T(f_{i_1}, \sigma_1) \cdots (f_{i_m}, \sigma_m)$ contains only the row $\bar{\delta}$. Therefore there is a one-to-one mapping of rows of the table T onto some set G of pairs of tuples of the kind $((f_{i_1}, \ldots, f_{i_m}), (\sigma_1, \ldots, \sigma_m))$, where $f_{i_1}, \ldots, f_{i_m} \in \mathrm{At}(T)$ and $\sigma_1, \ldots, \sigma_m \in E_k$. Evidently, $|G| \leq W(T)^m k^m$. Therefore $N(T) \leq (kW(T))^{S(T)}$. □

Lemma 8.5 *For any table T from $\mathscr{M}_k \setminus \{\Lambda\}$, there exists a mapping $\nu : E_k^{W(T)} \to \omega$ such that*

$$h^d(J(\nu, T)) \geq \log_k N(T) .$$

Proof Let $T \in \mathscr{M}_k \setminus \{\Lambda\}$ and $\nu : E_k^{W(T)} \to \omega$ be a mapping for which $\nu(\bar{\delta}) \neq \nu(\bar{\sigma})$ for any $\bar{\delta}, \bar{\sigma} \in E_k^{W(T)}$ such that $\bar{\delta} \neq \bar{\sigma}$. Denote $T^* = J(\nu, T)$. Let Γ be a deterministic decision tree for the table T^* such that $h(\Gamma) = h^d(T^*)$. Denote by $L_t(\Gamma)$ the number of terminal nodes of Γ. Evidently, $N(T) \leq L_t(\Gamma)$. One can show that $L_t(\Gamma) \leq k^{h(\Gamma)}$. Therefore $N(T) \leq k^{h(\Gamma)}$. Since $T \neq \Lambda$, $N(T) > 0$. Hence $h(\Gamma) \geq \log_k N(T)$. Taking into account that $h(\Gamma) = h^d(T^*)$, we obtain $h^d(T^*) \geq \log_k N(T)$. □

It is not difficult to prove the following upper bound on the minimum cardinality of a separating set for a table by the induction on the number of rows in the table.

Lemma 8.6 *For any table T from $\mathcal{M}_k \setminus \{\Lambda\}$,*

$$\mathrm{Sep}(T) \leq N(T) - 1 .$$

Lemma 8.7 *For any partially bounded complexity measure ψ and any table T from $\mathcal{M}_k$,*

$$\psi^a(T) \leq \psi^d(T) .$$

Proof Let $T \in \mathcal{M}_k$. If $T = \Lambda$, then $\psi^a(T) = \psi^d(T) = 0$. Let $T \in \mathcal{M}_k \setminus \{\Lambda\}$. It is clear that each deterministic decision tree for the table T is a nondeterministic decision tree for the table T. Therefore $\psi^a(T) \leq \psi^d(T)$. □

Lemma 8.8 *For any partially bounded complexity measure ψ and any table T from $\mathcal{M}_k$, which contains at least two rows, there exists a table $T^* \in [T]$ such that*

$$\psi^a(T^*) = \psi^d(T^*) = W_\psi(T^*) = S_\psi(T^*) = S_\psi(T) .$$

Proof Let $\bar{\delta}$ be a row of the table T such that $S_\psi(T, \bar{\delta}) = S_\psi(T)$. Let D be a subset of the set $\mathrm{At}(T)$ with the minimum cardinality such that $\psi(D) = S_\psi(T, \bar{\delta})$ and in the set of columns labeled with attributes from D the row $\bar{\delta}$ is different from all other rows of the table T. Let $\bar{\sigma}$ be the tuple obtained from the row $\bar{\delta}$ by the removal of all numbers that are in the intersection with columns labeled with attributes from the set $\mathrm{At}(T) \setminus D$. Let $\nu : E_k^{|D|} \to E_2$ and, for any $\bar{\gamma} \in E_k^{|D|}$, if $\bar{\gamma} = \bar{\sigma}$, then $\nu(\bar{\gamma}) = 1$ and if $\bar{\gamma} \neq \bar{\sigma}$, then $\nu(\bar{\gamma}) = 0$. Denote $T^* = J(\nu, I(\mathrm{At}(T) \setminus D, T))$.

From the fact that D has the minimum cardinality and from the properties of the function ψ it follows that, for any attribute from the set D, there exists a row of T, which is different from the row $\bar{\delta}$ only in the column labeled with the considered attribute among attributes from D. Therefore, for any attribute of the table T^*, there exists a row of T^*, which is different from the row $\bar{\sigma}$ only in the column labeled with this attribute. Thus,

$$S_\psi(T^*, \bar{\sigma}) = W_\psi(T^*) . \tag{8.2}$$

Using properties of the function ψ, we obtain $S_\psi(T^*) \leq W_\psi(T^*)$. From this inequality and from (8.2) it follows that

$$S_\psi(T^*) = W_\psi(T^*) . \tag{8.3}$$

Let Γ be a nondeterministic decision tree for the table T^* such that $\psi(\Gamma) = \psi^a(T^*)$, τ be a complete path of Γ such that the row $\bar{\sigma}$ belongs to the table $T(\tau)$, and $F(\tau) = f_{i_1} \cdots f_{i_t}$. It is clear that $\bar{\sigma}$ is the only row of the table $T(\tau)$. Therefore in columns labeled with attributes $f_{i_1}, \ldots, f_{i_t}$ the row $\bar{\sigma}$ is different from all other rows of the table T^*. Thus, $\psi(F(\tau)) \geq S_\psi(T^*, \bar{\sigma}) = W_\psi(T^*)$. Therefore $\psi(\Gamma) \geq W_\psi(T^*)$ and $\psi^a(T^*) \geq W_\psi(T^*)$. Using Lemmas 8.1 and 8.7, we obtain $\psi^a(T^*) \leq \psi^d(T^*) \leq W_\psi(T^*)$. Therefore

$$\psi^a(T^*) = \psi^d(T^*) = W_\psi(T^*) . \tag{8.4}$$

By the choice of the set D, $W_\psi(T^*) = S_\psi(T)$. From this equality and from (8.3) and (8.4) it follows that $\psi^a(T^*) = \psi^d(T^*) = W_\psi(T^*) = S_\psi(T^*) = S_\psi(T)$. □

It is not difficult to prove the following upper bounds on the minimum complexity of nondeterministic decision trees for a table.

Lemma 8.9 *For any partially bounded complexity measure ψ and any table T from $\mathcal{M}_k$,*

$$\psi^a(T) \le S_\psi(T) \le W_\psi(T) .$$

The next statement follows from Theorem 2.1 [1].

Lemma 8.10 *For any partially bounded complexity measure ψ and any table T from $\mathcal{M}_k$,*

$$\psi^d(T) \ge M_\psi(T) .$$

The next statement follows from Lemma 1.3 and Theorem 3.1 [1].

Lemma 8.11 *Let ψ be a 1-computable bounded complexity measure. Then, for any decision table T from $\mathcal{M}_k \setminus \{\Lambda\}$,*

$$\psi(U_{R,\psi}(T)) \le \begin{cases} M_\psi(T), & M_\psi(T) \le 1 , \\ M_\psi(T)^2 \ln R(T) + M_\psi(T), & M_\psi(T) \ge 2 . \end{cases}$$

For a k-decision tree Γ, we denote by $L(\Gamma)$ the number of nodes in Γ with the exception of the root and $L_0(\Gamma)$ the number of terminal nodes in Γ.

Lemma 8.12 *Let Γ be a k-decision tree such that each node of Γ, which is neither the root nor a terminal node, has at least two leaving edges. Then*

$$L(\Gamma) \le 2L_0(\Gamma) - 1 .$$

Proof We prove this statement by the induction on $L(\Gamma)$. If $L(\Gamma) = 1$, then the statement is obvious. Let $n \ge 2$ and the statement hold for each Γ with $L(\Gamma) < n$. Let $L(\Gamma) = n$. Choose in Γ a node v such that all edges $e_1, \dots, e_t$ leaving v enter the terminal nodes $v_1, \dots, v_t$, respectively. Let us remove from Γ edges $e_1, \dots, e_t$ and nodes $v_1, \dots, v_t$ and label the node v with the number 0. We denote the obtained k-decision tree by Γ^*. Evidently, $L(\Gamma^*) = L(\Gamma) - t$ and $L_0(\Gamma^*) = L_0(\Gamma) - t + 1$. Since $L(\Gamma^*) < n$, by inductive hypothesis, $L(\Gamma^*) \le 2L_0(\Gamma^*) - 1$. Therefore $L(\Gamma) - t \le 2L_0(\Gamma) - 2t + 2 - 1$ and $L(\Gamma) \le 2L_0(\Gamma) - (t - 1)$. Taking into account that $t \ge 2$, we obtain $L(\Gamma) \le 2L_0(\Gamma) - 1$. □

8.5 Proofs of Theorems 8.1–8.8

Proof of Theorem 8.1. Since A is a closed class, $\Lambda \in A$. By definition, $W_\psi(\Lambda) = 0$. Using this fact and Lemma 8.1, we obtain that $\mathcal{F}_{\psi,A}$ is an everywhere defined function and $\mathcal{F}_{\psi,A}(n) \le n$ for any $n \in \omega$. Evidently, $\mathcal{F}_{\psi,A}$ is a nondecreasing function. Let $T \in A$ and $W_\psi(T) \le 0$. Using positivity property of the function ψ, we obtain $T = \Lambda$. Therefore $\mathcal{F}_{\psi,A}(0) = 0$.

(a) Let the functions S_ψ and N be bounded from above on the class A. Then there are constants $a \ge 1$ and $b \ge 2$ such that $S_\psi(T) \le a$ and $N(T) \le b$ for any table $T \in A$. Let $T \in A$. Taking into account that A is a closed class, we obtain $\hat{S}_\psi(T) \le a$. By Lemma 8.3, $M_\psi(T) \le 2a$. From this inequality, inequality $N(T) \le b$ and from Lemma 8.2 it follows that $\psi^d(T) \le 2a \log_2 b$. Denote $c_0 = 2a \log_2 b$. Taking into account that T is an arbitrary table from the class A, we obtain that $\mathcal{F}_{\psi,A}(n) \le c_0$ for any $n \in \omega$.

(b) Let the function S_ψ be bounded from above on the class A and the function N be not bounded from above on A. Then there exists a constant $a \ge 2$ such that, for any table $Q \in A$,

$$S_\psi(Q) \le a \ . \tag{8.5}$$

Let $n \in \omega \setminus \{0\}$ and T be an arbitrary table from A such that $W_\psi(T) \le n$. If $T \in \mathcal{M}_k^c$, then, evidently, $\psi^d(T) = 0$. Let $T \notin \mathcal{M}_k^c$. Using the boundedness from below property of the function ψ, we obtain $W(T) \le n$ and $S(T) \le a$. From these inequalities and Lemma 8.4 it follows that $N(T) \le (kn)^a$. From (8.5) and Lemma 8.3 it follows that $M_\psi(T) \le 2a$. Using the last two inequalities and Lemma 8.2, we obtain

$$\psi^d(T) \le 2a^2 \log_2 n + 2a^2 \log_2 k \ . \tag{8.6}$$

Denote $c_3 = 2a^2$ and $c_4 = 2a^2 \log_2 k$. Then $\psi^d(T) \le c_3 \log_2 n + c_4$. Taking into account that n is an arbitrary number from $\omega \setminus \{0\}$ and T is an arbitrary table from A such that $W_\psi(T) \le n$, we obtain that, for any $n \in \omega \setminus \{0\}$,

$$\mathcal{F}_{\psi,A}(n) \le c_3 \log_2 n + c_4 \ . \tag{8.7}$$

Let $n \in \omega \setminus \{0\}$. Denote $c_1 = 1/\log_2 k$ and $c_2 = \log_k a$. We now show that

$$\mathcal{F}_{\psi,A}(n) \ge c_1 \log_2 n - c_2 \ . \tag{8.8}$$

Denote $m = \lfloor n/a \rfloor$. Taking into account that the function N is not bounded from above on the set A, we obtain that there exists a table $T \in A$ such that $N(T) \ge k^m$. Let C be a separating set for the table T with the minimum cardinality such that $\psi(f_i) \le a$ for any $f_i \in C$. The existence of such set follows from the inequality (8.5) and properties of commutativity and nondecreasing of the function ψ. Evidently, $|C| \ge m$. Let D be a subset of the set C such that $|D| = m$. Denote $T^0 = I(\mathrm{At}(T) \setminus D, T)$. One can show that, for any attribute of T^0, there are two rows of the table T^0 that differ only in the column labeled with this attribute. Therefore D is a separating set for the table T^0 with the minimum cardinality. By Lemma 8.6, $N(T^0) \geq m + 1 \geq n/a$.

Using Lemma 8.5, we obtain that there exists a mapping $\nu : E_k^m \to \omega$ such that

$$h^d(J(\nu, T^0)) \geq \log_k(n/a) .$$

Denote $T^* = J(\nu, T^0)$. Since ψ is a bounded complexity measure, $\psi^d(T^*) \geq \log_k(n/a)$. Using boundedness from above property of the function ψ, we obtain $W_\psi(T^*) \leq \lfloor n/a \rfloor a \leq n$. Hence the inequality (8.8) holds. From (8.7) and (8.8) it follows that $c_1 \log_2 n - c_2 \leq \mathcal{F}_{\psi,A}(n) \leq c_3 \log_2 n + c_4$ for any $n \in \omega \setminus \{0\}$.

(c) Let the function S_ψ be not bounded from above on the class A. Using Lemma 8.8, we obtain that the set $D = \{W_\psi(T) : T \in A, \psi^d(T) = W_\psi(T)\}$ is infinite. Since the class A is closed, $\Lambda \in A$ and, since $\psi^d(\Lambda) = W_\psi(\Lambda) = 0, 0 \in D$. Evidently, for any $n \in D, \mathcal{F}_{\psi,A}(n) \geq n$. Taking into account that $\mathcal{F}_{\psi,A}$ is a nondecreasing function, we obtain that $\mathcal{F}_{\psi,A}(n) \geq H_D(n)$ for any $n \in \omega \setminus \{0\}$. □

Proof of Theorem 8.2. From Theorem 8.1 it follows that $\mathcal{F}_{h,A}(n) \leq n$ for any $n \in \omega$. In particular, $\mathcal{F}_{h,A}(0) = 0$. Let $n \in \omega \setminus \{0\}$. Since the function S is not bounded from above on the class A, there exists a table $T \in A$ such that $S(T) \geq n$. Let $\bar{\delta}$ be a row of the table T for which $S(T, \bar{\delta}) = S(T)$. Let C be a set of attributes of T with the minimum cardinality such that in columns labeled with attributes from C the row $\bar{\delta}$ is different from all other rows of the table T. Evidently, $|C| \geq n$. Let C^* be a subset of the set C with $|C^*| = n$. Denote $T^0 = I(\mathrm{At}(T) \setminus C^*, T)$. One can show that $S(T^0) = n$. From Lemma 8.8 it follows that there exists a table $T^* \in [T^0]$ for which $h^d(T^*) = W(T^*) = n$. Therefore $\mathcal{F}_{h,A}(n) \geq n$ and $\mathcal{F}_{h,A}(n) = n$. □

Proof of Theorem 8.3. Since A is a closed class, $\Lambda \in A$. By definition, $W_\psi(\Lambda) = 0$. Using this fact and Lemma 8.9, we obtain that $\mathcal{G}_{\psi,A}$ is an everywhere defined function and $\mathcal{G}_{\psi,A}(n) \leq n$ for any $n \in \omega$. Evidently, $\mathcal{G}_{\psi,A}$ is a nondecreasing function. Let $T \in A$ and $W_\psi(T) \leq 0$. Using positivity property of the function ψ, we obtain $T = \Lambda$. Therefore $\mathcal{G}_{\psi,A}(0) = 0$.

(a) Let the function S_ψ be bounded from above on the class A. Then there is a constant $c > 0$ such that $S_\psi(T) \leq c$ for any $T \in A$. By Lemma 8.9, $\psi^a(T) \leq S_\psi(T) \leq c$ for any $T \in A$. Therefore, for any $n \in \omega$, $\mathcal{G}_{\psi,A}(n) \leq c$.

(b) Let the function S_ψ be not bounded from above on the class A. Using Lemma 8.8, we obtain that the set $D = \{W_\psi(T) : T \in A, \psi^a(T) = W_\psi(T)\}$ is infinite. Since the class A is closed, $\Lambda \in A$ and, since $\psi^a(\Lambda) = W_\psi(\Lambda) = 0, 0 \in D$. Evidently, for any $n \in D, \mathcal{G}_{\psi,A}(n) \geq n$. Taking into account that $\mathcal{G}_{\psi,A}$ is a nondecreasing function, we obtain that $\mathcal{G}_{\psi,A}(n) \geq H_D(n)$ for any $n \in \omega \setminus \{0\}$. □

Proof of Theorem 8.4. From Theorem 8.3 it follows that $\mathcal{G}_{h,A}(n) \leq n$ for any $n \in \omega$. In particular, $\mathcal{G}_{h,A}(0) = 0$. Let $n \in \omega \setminus \{0\}$. Since the function S is not bounded from above on the class A, there exists a table $T \in A$ such that $S(T) \geq n$. Let $\bar{\delta}$ be a row of the table T such that $S(T, \bar{\delta}) = S(T)$. Let C be a set of attributes of T with the minimum cardinality such that in columns labeled with the attributes from C the row $\bar{\delta}$ is different from all other rows of the table T. Evidently, $|C| \geq n$. Let C^* be a subset of the set C with $|C^*| = n$. Denote $T^0 = I(\mathrm{At}(T) \setminus C^*, T)$. One can show that $S(T^0) = n$. From Lemma 8.8 it follows that there exists a table $T^* \in [T^0]$ for which $h^a(T^*) = W(T^*) = n$. Therefore $\mathscr{G}_{h,A}(n) \geq n$ and $\mathscr{G}_{h,A}(n) = n$. □

Proof of Theorem 8.5. Let $n \in \omega$ and both n and $n+1$ belong to the domain of the function $\mathscr{H}_{\psi,A}$. Immediately from the definition of this function it follows that $\mathscr{H}_{\psi,A}(n) \leq \mathscr{H}_{\psi,A}(n+1)$. Let $T \in A$ and $\psi^a(T) \leq 0$. Using positivity property of the function ψ, one can show that $T \in \mathscr{M}_k^c$. From Lemma 8.2 it follows that $\psi^d(T) = 0$. Therefore $\mathscr{H}_{\psi,A}(0) = 0$.

Let $\mathscr{H}_{\psi,A}$ be not an everywhere defined function and m be the minimum number from ω such that the value $\mathscr{H}_{\psi,A}(m)$ is not defined. It is clear that $m > 0$ and the value $\mathscr{H}_{\psi,A}(n)$ is not defined for each $n \in \omega$ such that $n \geq m$. Denote $n_0 = m - 1$. Then the domain of the function $\mathscr{H}_{\psi,A}$ is equal to $\{n : n \in \omega, n \leq n_0\}$.

We now consider the case when $\mathscr{H}_{\psi,A}$ is an everywhere defined function.

(a) Let there exist a nonnegative constant c such that $\psi^d(T) \leq c$ for any table $T \in A$. Then, evidently, $\mathscr{H}_{\psi,A}(n) \leq c$ for any $n \in \omega$.

(b) Let there be no a nonnegative constant c such that $\psi^d(T) \leq c$ for any table $T \in A$. Let us assume that there exists a nonnegative constant d such that $\psi^a(T) \leq d$ for any table $T \in A$. Then the value $\mathscr{H}_{\psi,A}(d)$ is not defined but this is impossible. Therefore the set $D = \{\psi^a(T) : T \in A\}$ is infinite. Since $\Lambda \in A$, $0 \in D$. By Lemma 8.7, $\psi^a(T) \leq \psi^d(T)$ for any table $T \in A$. Hence $\mathscr{H}_{\psi,A}(n) \geq n$ for any $n \in D$. Taking into account that $\mathscr{H}_{\psi,A}$ is a nondecreasing function, we obtain that $\mathscr{H}_{\psi,A}(n) \geq H_D(n)$ for any $n \in \omega$. □

Proof of Theorem 8.6. One can show that the number of repetitions of Step 2 is equal to $L(U_{R,\psi}(T)) + 1$. From the description of the algorithm $U_{R,\psi}$ it follows that, for any complete path τ of the decision tree $U_{R,\psi}(T)$, $T(\tau) \neq \Lambda$. Evidently, for any row of the table T, there exists exactly one complete path τ of the decision tree $U_{R,\psi}(T)$ such that the considered row belongs to the table $T(\tau)$. Therefore the number of complete paths of $U_{R,\psi}(T)$ is at most $N(T)$ and $L_0(U_{R,\psi}(T)) \leq N(T)$. From the description of the algorithm $U_{R,\psi}$ it follows that each node of $U_{R,\psi}(T)$, which is neither the root nor a terminal node, has at least two leaving edges. By Lemma 8.12, $L(U_{R,\psi}(T)) \leq 2N(T) - 1$. Thus, the algorithm $U_{R,\psi}$ during the construction of the decision tree $U_{R,\psi}(T)$ repeats Step 2 no more than $2N(T)$ times. □

Proof of Theorem 8.7. Let $n \in \omega$. Since A is a closed class, $\Lambda \in A$. By definition, $\psi^d(\Lambda) = \psi(U_{R,\psi}(\Lambda)) = W_\psi(\Lambda) = 0$. Therefore the set $A(n) = \{T : T \in A, W_\psi(T) \leq n\}$ is nonempty. Let $A(n) \setminus \{\Lambda\} \neq \emptyset$ and $T \in A(n) \setminus \{\Lambda\}$. It is clear that $\psi^d(T) \leq \psi(U_{R,\psi}(T))$. It is easy to see that, in any complete path of the decision tree $U_{R,\psi}(T)$, different nodes that are neither the root nor terminal node are labeled with different attributes. Since ψ is a bounded complexity measure, we have $\psi(U_{R,\psi}(T)) \leq W_\psi(T)$. Now it is easy to show that the value $\mathscr{U}_{\psi,A}(n)$ is defined and $\mathscr{F}_{\psi,A}(n) \leq \mathscr{U}_{\psi,A}(n) \leq n$. Evidently, $\mathscr{U}_{\psi,A}$ is a nondecreasing function.

(a) Let the functions S_ψ and N be bounded from above on the class A by constants $a > 1$ and $b > 1$, respectively. Let $T \in A \setminus \{\Lambda\}$. Using Lemma 8.3, we obtain that $M_\psi(T) \leq 2a$. Evidently, $R(T) \leq b^2$. By Lemma 8.11, $\psi(U_{R,\psi}(T)) \leq c$, where $c =$

$(2a)^2 \ln b^2 + 2a$. It is clear that $\psi(U_{R,\psi}(\Lambda)) \le c$. Therefore $\mathscr{U}_{\psi,A}(n) \le \mathscr{F}_{\psi,A}(n) + c$ for any $n \in \omega$.

(b) Let the function S_ψ be bounded from above on the class A by a constant $t \ge 1$ and the function N be not bounded from above on the class A. Using Theorem 8.1, we obtain that there exist positive constants a and b such that $\mathscr{F}_{\psi,A}(n) \ge a \log_2 n - b$ for any $n \in \omega \setminus \{0\}$. Let $T \in A \setminus \{\Lambda\}$. Using Lemma 8.3, we obtain that $M_\psi(T) \le 2t$. Since ψ is a bounded complexity measure, $S(T) \le S_\psi(T)$. By Lemma 8.4, $N(T) \le (kW(T))^t$ and $R(T) \le (kW(T))^{2t}$. From here and from Lemma 8.11 it follows that $\psi(U_{R,\psi}(T)) \le (2t)^2 2t \ln(W(T)k) + 2t$. Since ψ is a bounded complexity measure, $W(T) \le W_\psi(T)$. Using the equality $\psi(U_{R,\psi}(\Lambda)) = 0$, we obtain $\mathscr{U}_{\psi,A}(n) \le \frac{8t^3 \log_2 n}{\log_2 e} + 8t^3 \ln k + 2t$ for any $n \in \omega \setminus \{0\}$. Taking into account that, for any $n \in \omega \setminus \{0\}$, $\log_2 n \le \frac{\mathscr{F}_{\psi,A}(n)+b}{a}$, we obtain that, for any $n \in \omega \setminus \{0\}$, $\mathscr{U}_{\psi,A}(n) \le \frac{8t^3}{a \log_2 e} \mathscr{F}_{\psi,A}(n) + \frac{8t^3 b}{a \log_2 e} + 8t^3 \ln k + 2t$. Thus, for any $n \in \omega \setminus \{0\}$, $\mathscr{U}_{\psi,A}(n) \le c\mathscr{F}_{\psi,A}(n) + d$, where $c = \frac{8t^3}{a \log_2 e}$ and $d = \frac{8t^3 b}{a \log_2 e} + 8t^3 \ln k + 2t$.

(c) Let the function S_ψ be not bounded from above on the class A. Let $n \in \omega$, T be a table from $A \setminus \{\Lambda\}$ and $W_\psi(T) \le n$. Using Lemma 8.10, we obtain $M_\psi(T) \le \psi^d(T) \le \mathscr{F}_{\psi,A}(n)$.

Let $M_\psi(T) \ge 2$. By Lemma 8.5, there exists a mapping $\nu : E_k^{W(T)} \to \omega$ such that $h^d(J(\nu, T)) \ge \log_k N(T)$. Denote $T^* = J(\nu, T)$. It is clear that $T^* \in A$ and $W_\psi(T^*) = W_\psi(T) \le n$. Since ψ is a bounded complexity measure, $h^d(T^*) \le \psi^d(T^*)$. Therefore $\log_k N(T) \le \mathscr{F}_{\psi,A}(n)$ and $\ln N(T) \le \mathscr{F}_{\psi,A}(n)/\log_k e$. Since $R(T) \le N(T)^2$, we have $\ln R(T) \le 2\mathscr{F}_{\psi,A}(n)/\log_k e$. Using Lemma 8.11, we obtain $\psi(U_{R,\psi}(T)) \le M_\psi(T)^2 \ln R(T) + M_\psi(T) \le 2\mathscr{F}_{\psi,A}(n)^3/\log_k e + \mathscr{F}_{\psi,A}(n) \le c\mathscr{F}_{\psi,A}(n)^3$, where $c = 2/\log_k e + 1$.

Let $M_\psi(T) \le 1$. Then, by Lemma 8.11, $\psi(U_{R,\psi}(T)) \le M_\psi(T)$. Therefore

$$\psi(U_{R,\psi}(T)) \le \mathscr{F}_{\psi,A}(n) \le c\mathscr{F}_{\psi,A}(n)^3.$$

It is clear that $\psi(U_{R,\psi}(\Lambda)) \le c\mathscr{F}_{\psi,A}(n)^3$. As a result, we obtain $\mathscr{U}_{\psi,A}(n) \le c\mathscr{F}_{\psi,A}(n)^3$. □

Proof of Theorem 8.8. Let $n \in \omega$. Since A is a closed class, $\Lambda \in A$. By definition, $\psi^a(\Lambda) = \psi(V^a(\Lambda)) = W_\psi(\Lambda) = 0$. Therefore the set $A(n) = \{T : T \in A, W_\psi(T) \le n\}$ is nonempty. Let $A(n) \setminus \{\Lambda\} \ne \emptyset$ and $T \in A(n) \setminus \{\Lambda\}$. It is clear that $\psi^a(T) \le \psi(V^a(T))$. It is easy to see that in any complete path of the decision tree $V^a(T)$, different nodes, which are neither the root nor terminal node, are labeled with different attributes. Since ψ is a bounded complexity measure, we have $\psi(V^a(T)) \le W_\psi(T)$. Therefore the value $\mathscr{V}_{\psi,A}(n)$ is defined and $\mathscr{G}_{\psi,A}(n) \le \mathscr{V}_{\psi,A}(n)$.

Let us prove that $\psi(V^a(T)) \le \mathscr{G}_{\psi,A}(n)$. To this end, it is enough to show that $\psi(D(\bar{\delta})) \le \mathscr{G}_{\psi,A}(n)$ for any row $\bar{\delta}$ of the table T. Denote by $\bar{\delta}'$ the tuple obtained from $\bar{\delta}$ by the removal of digits corresponding to the attributes that do not belong to the set $D(\bar{\delta})$. Denote $T_0 = I(\mathrm{At}(T) \setminus D(\bar{\delta}), T)$ and $T' = J(\nu, T_0)$, where $\nu : E_k^{|\mathrm{At}(T)\setminus D(\bar{\delta})|} \to \{0, 1\}$ and, for any $\bar{\sigma} \in E_k^{|\mathrm{At}(T)\setminus D(\bar{\delta})|}$, $\nu(\bar{\sigma}) = 1$ if and only if $\bar{\sigma} = \bar{\delta}'$. Evidently, $T' \in A$. Since ψ is a bounded complexity measure,

$W_\psi(T') \leq W_\psi(T) \leq n$. Since $D(\bar{\delta})$ is a minimal local separating set for the row $\bar{\delta}$, for any attribute $f_{i_j} \in D(\bar{\delta}) = \mathrm{At}(T')$, there is a row $\bar{\sigma}_{i_j}$ of the table T' that is labeled with the decision 0 and is different from the row $\bar{\delta}'$ of the table T' labeled with the decision 1 only in the column labeled with the attribute f_{i_j}. Using this fact and the fact that ψ is a bounded complexity measure, it is not difficult to show that $\psi^a(T') \geq \psi(D(\bar{\delta}))$. Since $W_\psi(T') \leq n$, $\psi^a(T') \leq \mathcal{G}_{\psi,A}(n)$ and $\psi(D(\bar{\delta})) \leq \mathcal{G}_{\psi,A}(n)$. Therefore $\psi(V^a(T)) \leq \mathcal{G}_{\psi,A}(n)$ and $\mathcal{V}_{\psi,A}(n) \leq \mathcal{G}_{\psi,A}(n)$. Thus, $\mathcal{V}_{\psi,A}(n) = \mathcal{G}_{\psi,A}(n)$. □

References

1. Moshkov, M.: Conditional tests. In: Yablonskii, S.V. (ed.) Problemy Kibernetiki, vol. 40, pp. 131–170. Nauka Publishers, Moscow (1983) (in Russian)
2. Moshkov, M.: On depth of conditional tests for tables from closed classes. In: Markov, A.A. (ed.) Combinatorial-Algebraic and Probabilistic Methods of Discrete Analysis, pp. 78–86. Gorky University Press, Gorky (1989) (in Russian)
3. Moshkov, M.: Time complexity of decision trees. In: Peters, J.F., Skowron, A. (eds.) Transactions on Rough Sets III. Lecture Notes in Computer Science, vol. 3400, pp. 244–459. Springer (2005)
4. Moshkov, M., Zielosko, B.: Combinatorial Machine Learning—A Rough Set Approach. In: Studies in Computational Intelligence, vol. 360. Springer (2011)
5. Ostonov, A., Moshkov, M.: Greedy algorithm for construction of deterministic decision trees for conventional decision tables from closed classes. In: Hu, M., Cornelis, C., Zhang, Y., Lingras, P., Slezak, D., Yao, J.T. (eds.). Rough Sets – International Joint Conference, IJCRS 2024, Halifax, NS, Canada, May 17–20, 2024, Proceedings, Part I. Lecture Notes in Computer Science. vol. 14839, pp 93–104. Springer (2024)
6. Ostonov, A., Moshkov, M.: On complexity of deterministic and nondeterministic decision trees for conventional decision tables from closed classes. Entropy **25**(10), 1411 (2023)
7. Pawlak, Z.: Information systems theoretical foundations. Inf. Syst. **6**(3), 205–218 (1981)

Chapter 9
Complexity of Deterministic and Strongly Nondeterministic Decision Trees for Decision Tables with 0-1-Decisions from Closed Classes

9.1 Introduction

In this chapter, we consider closed classes of decision tables with 0-1-decisions and study relationships among four parameters of these tables: the minimum complexity of a deterministic decision tree, the minimum complexity of a strongly nondeterministic decision tree, the complexity of the set of attributes attached to columns (in particular, for the depth, the complexity of a set of attributes is equal to its cardinality), and the minimum complexity of a test–a set of attributes that separate any two rows with different decisions.

A decision table with 0-1-decisions is a rectangular table in which columns are labeled with attributes, rows are pairwise different and each row is labeled with a decision from the set {0, 1}. Rows are interpreted as tuples of values of the attributes. For a given row, it is required to find the decision attached to the row. To this end, we can use the following queries: we can choose an attribute and ask what is the value of this attribute in the considered row. We study two types of algorithms based on these queries: deterministic and strongly nondeterministic decision trees. One can interpret strongly nondeterministic decision trees for a decision table as a way to represent an arbitrary system of true decision rules for this table that cover all rows labeled with the decision 1. We consider so-called bounded complexity measures, for example, the depth of decision trees.

Conventional decision tables including decision tables with 0-1-decisions appear in data analysis and in such areas as combinatorial optimization, computational geometry, and fault diagnosis, where they are used to represent and explore problems [3, 5].

Deterministic decision trees and decision rule systems are widely used as classifiers, as a means for knowledge representation, and as algorithms for solving various problems of combinatorial optimization, fault diagnosis, etc. Strongly nondeterministic decision trees for decision tables with 0-1-decisions are less known. Note that

K. Durdymyradov et al., *Decision Trees Versus Systems of Decision Rules*, Studies in Big Data 160, https://doi.org/10.1007/978-3-031-71586-0_9

the depth of strongly nondeterministic decision trees for computation Boolean functions (variables of a function are considered as attributes) was studied in [2]. Strongly nondeterministic decision trees for 0-1-problems over information systems (see the next paragraph for the explanations) were studied in the book [4].

We study classes of decision tables with 0-1-decisions closed under two operations: removal of columns and changing of decisions. The most natural examples of such classes are closed classes of decision tables with 0-1-decisions generated by information systems [7]. An information system consists of a set of objects (universe) and a set of attributes (functions) defined on the universe and with values from a finite set. A 0-1-problem over an information system is specified by a finite number of attributes that divide the universe into nonempty domains in which these attributes have fixed values. A decisions from the set $\{0, 1\}$ is attached to each domain. For a given object from the universe, it is required to find the decision attached to the domain containing this object.

A decision table with 0-1-decisions corresponds to this problem in a natural way: columns of this table are labeled with the considered attributes, rows correspond to domains and are labeled with decisions attached to domains. The set of decision tables corresponding to 0-1-problems over an information system forms a closed class generated by this system. Note that the family of all closed classes is essentially wider than the family of closed classes generated by information systems. In particular, the union of two closed classes generated by two information systems is a closed class. However, generally, there is no an information system that generates this class.

Let A be a closed class of decision tables with 0-1-decisions and ψ be a bounded complexity measure. In this chapter, we study four functions: $\mathcal{F}^W_{\psi,A}(n)$, $\mathcal{F}^\Theta_{\psi,A}(n)$, $\mathcal{K}_{\psi,A}(n)$, and $\mathcal{S}_{\psi,A}(n)$.

The function $\mathcal{F}^W_{\psi,A}(n)$ characterizes the growth in the worst case of the minimum complexity of a deterministic decision tree for a decision table from the class A with the growth of the complexity of the set of attributes attached to columns of the table. The function $\mathcal{F}^\Theta_{\psi,A}(n)$ characterizes the growth in the worst case of the minimum complexity of a deterministic decision tree for a decision table from the class A with the growth of the minimum complexity of a test of the table. In particular, we proved that the functions $\mathcal{F}^W_{\psi,A}(n)$ and $\mathcal{F}^\Theta_{\psi,A}(n)$ are either bounded from above by a constant, or grow as a logarithm of n, or grow almost linearly depending on n (they are bounded from above by n and are equal to n for infinitely many n).

The function $\mathcal{K}_{\psi,A}(n)$ characterizes the growth in the worst case of the minimum complexity of a deterministic decision tree for a decision table from the class A with the growth of the minimum complexity of a strongly nondeterministic decision tree for the table. We indicated the condition for the function $\mathcal{K}_{\psi,A}(n)$ to be defined everywhere. Let $\mathcal{K}_{\psi,A}(n)$ be everywhere defined. We proved that this function is either bounded from above by a constant, or is greater than or equal to n for infinitely many n. In particular, the function $\mathcal{K}_{\psi,A}(n)$ can grow as an arbitrary nondecreasing function φ such that $\varphi(n) \geq n$ and $\varphi(0) = 0$. We indicated also conditions for the function $\mathcal{K}_{\psi,A}(n)$ to be bounded from above by a polynomial on n.

The function $\mathcal{S}_{\psi,A}(n)$ characterizes the growth in the worst case of the minimum complexity of a strongly nondeterministic decision tree for a decision table from the class A with the growth of the complexity of the set of attributes attached to columns of the table. In particular, we proved that the function $\mathscr{S}_{\psi,A}(n)$ is either bounded from above by a constant or grows almost linearly depending on n (it is bounded from above by n and is equal to n for infinitely many n).

In this chapter, we also study the joint behavior of the functions $\mathscr{F}^{W}_{\psi,A}$ and $\mathscr{S}_{\psi,A}$, and list all three possible types of this behavior.

There is a similarity between some results obtained in this chapter for closed classes of decision tables with 0-1-decisions and results from the book [4] obtained for 0-1-problems over information systems. However, the results obtained in the present chapter are more general.

This chapter is based mainly on the paper [6]. The chapter consists of seven sections. In Sect. 9.2, main definitions and notation are considered. In Sect. 9.3, we provide the main results. Section 9.4 contains auxiliary statements. In Sects. 9.5–9.7, we prove the main results.

9.2 Main Definitions and Notation

Denote $\omega = \{0, 1, 2, \ldots\}$ and, for any $k \in \omega \setminus \{0, 1\}$, denote $E_k = \{0, 1, \ldots, k-1\}$. Let $P = \{f_i : i \in \omega\}$ be the set of *attributes* (really names of attributes). Two attributes $f_i, f_j \in P$ are considered different if $i \neq j$.

9.2.1 Decision Tables

First, we define the notion of a decision table with 0-1-decisions.

Definition 9.1 Let $k \in \omega \setminus \{0, 1\}$. Denote by $\mathscr{M}_k^{0\text{-}1}$ the set of rectangular tables filled with numbers from E_k in each of which rows are pairwise different, each row is labeled with a number from E_2 (decision), and columns are labeled with pairwise different attributes from P. Rows are interpreted as tuples of values of these attributes. Empty tables without rows belong also to the set $\mathscr{M}_k^{0\text{-}1}$. We will use the same notation Λ for these tables. Tables from $\mathscr{M}_k^{0\text{-}1}$ will be called *decision tables with 0-1-decisions* or simply *decision tables*.

Example 9.1 Figure 9.1 shows a decision table from $\mathscr{M}_k^{0\text{-}1}$.

Denote by $\mathscr{M}_k^{0\text{-}1c}$ the set of tables from $\mathscr{M}_k^{0\text{-}1}$ in each of which all rows are labeled with the same decision. Let $\Lambda \in \mathscr{M}_k^{0\text{-}1c}$.

Let T be a nonempty table from $\mathscr{M}_k^{0\text{-}1}$. Denote by At$(T)$ the set of attributes attached to columns of the table T. Let $f_{i_1}, \ldots, f_{i_m} \in \text{At}(T)$ and $\delta_1, \ldots, \delta_m \in E_k$. We denote by $T(f_{i_1}, \delta_1) \cdots (f_{i_m}, \delta_m)$ the table obtained from T by removal of all rows that do not satisfy the following condition: in columns labeled with attributes $f_{i_1}, \ldots, f_{i_m}$, the row has numbers $\delta_1, \ldots, \delta_m$, respectively.

Fig. 9.1 Decision table from $\mathcal{M}_2^{0\text{-}1}$

f_2	f_4	f_3	
1	1	1	0
0	1	1	0
1	1	0	1
0	0	1	1
1	0	0	1
0	0	0	1

We now define two operations on decision tables: removal of columns and changing of decisions. Let $T \in \mathcal{M}_k^{0\text{-}1}$.

Definition 9.2 (*Removal of columns*) Let $D \subseteq \mathrm{At}(T)$. We remove from T all columns labeled with the attributes from the set D. In each group of rows equal on the remaining columns, we keep one with the minimum decision. Denote the obtained table by $I(D, T)$. In particular, $I(\emptyset, T) = T$ and $I(\mathrm{At}(T), T) = \Lambda$. It is obvious that $I(D, T) \in \mathcal{M}_k^{0\text{-}1}$.

Definition 9.3 (*Changing of decisions*) Let $\nu : E_k^{|\mathrm{At}(T)|} \to E_2$ (by definition, $E_k^0 = \emptyset$). For each row $\bar{\delta}$ of the table T, we replace the decision attached to this row with $\nu(\bar{\delta})$. We denote the obtained table by $J(\nu, T)$. It is obvious that $J(\nu, T) \in \mathcal{M}_k^{0\text{-}1}$.

Definition 9.4 Denote $[T] = \{J(\nu, I(D, T)) : D \subseteq \mathrm{At}(T), \nu : E_k^{|\mathrm{At}(T)\setminus D|} \to E_2\}$. The set $[T]$ is the *closure of the table* T under the operations of removal of columns and changing of decisions.

Example 9.2 Figure 9.2 shows the table $J(\nu, I(D, T))$, where T is the table shown in Fig. 9.1, $D = \{f_4\}$ and $\nu(x_1, x_2) = x_1 \vee x_2$.

Definition 9.5 Let $A \subseteq \mathcal{M}_k^{0\text{-}1}$ and $A \neq \emptyset$. Denote $[A] = \bigcup_{T \in A} [T]$. The set $[A]$ is the *closure of the set* A under the considered two operations. The class (the set) of decision tables A will be called a *closed class* if $[A] = A$.

Let A_1 and A_2 be closed classes of decision tables from $\mathcal{M}_k^{0\text{-}1}$. Then $A_1 \cup A_2$ is a closed class of decision tables from $\mathcal{M}_k^{0\text{-}1}$.

9.2.2 *Tests*

We now define the notion of a test for a decision table.

Fig. 9.2 Decision table obtained from the decision table shown in Fig. 9.1 by removal of a column and changing of decisions

f_2	f_3	
1	1	1
0	1	1
1	0	1
0	0	0

Definition 9.6 Let $T \in \mathscr{M}_k^{0\text{-}1}$. A set of attributes $D \subseteq \text{At}(T)$ will be called a *test* of the decision table T if in columns labeled with attributes from D any two rows of T with different decisions are different. By definition, for any table $T \in \mathscr{M}_k^{0\text{-}1c}$, any subset of the set $\text{At}(T)$ including the empty set is a test of T.

Example 9.3 The set $\{f_4, f_3\}$ is a test of the decision table shown in Fig. 9.1.

9.2.3 *Deterministic and Strongly Nondeterministic Decision Trees*

A *finite tree with root* is a finite directed tree in which exactly one node called the *root* has no entering edges. The nodes without leaving edges are called *terminal* nodes.

Definition 9.7 A *k-0-1-decision tree* is a finite tree with root, which has at least two nodes and in which

- The root and edges leaving the root are not labeled.
- Each terminal node is labeled with a decision from the set E_2.
- Each node, which is neither the root nor a terminal node, is labeled with an attribute from the set P. Each edge leaving such node is labeled with a number from the set E_k.

Example 9.4 Figures 9.3 and 9.4 show 2-0-1-decision trees.

We denote by $\mathscr{T}_k^{0\text{-}1}$ the set of all k-0-1-decision trees. Let $\Gamma \in \mathscr{T}_k^{0\text{-}1}$. We denote by $\text{At}(\Gamma)$ the set of attributes attached to nodes of Γ that are neither the root nor terminal nodes. A *complete path* of Γ is a sequence $\tau = v_1, d_1, \ldots, v_m, d_m, v_{m+1}$ of nodes and edges of Γ in which v_1 is the root of Γ, v_{m+1} is a terminal node of Γ and, for $j = 1, \ldots, m$, the edge d_j leaves the node v_j and enters the node v_{j+1}. Let $T \in \mathscr{M}_k^{0\text{-}1}$. If $\text{At}(\Gamma) \subseteq \text{At}(T)$, then we correspond to the table T and the complete path τ a decision table $T(\tau)$. If $m = 1$, then $T(\tau) = T$. If $m > 1$ and, for $j = 2, \ldots, m$, the node v_j is labeled with the attribute f_{i_j} and the edge d_j is labeled with the number δ_j, then $T(\tau) = T(f_{i_2}, \delta_2) \cdots (f_{i_m}, \delta_m)$.

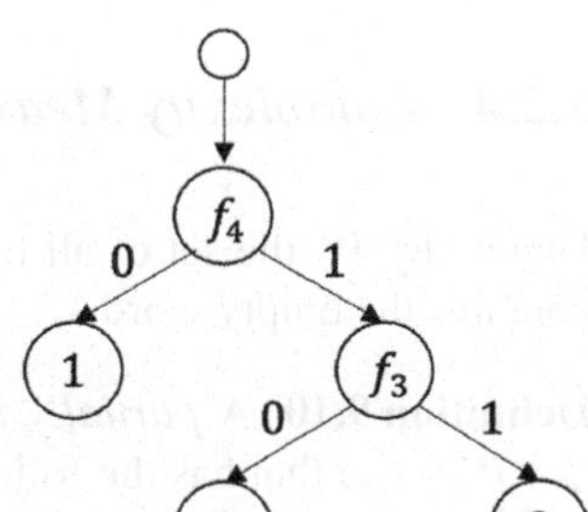

Fig. 9.3 A deterministic decision tree for the decision table shown in Fig. 9.1

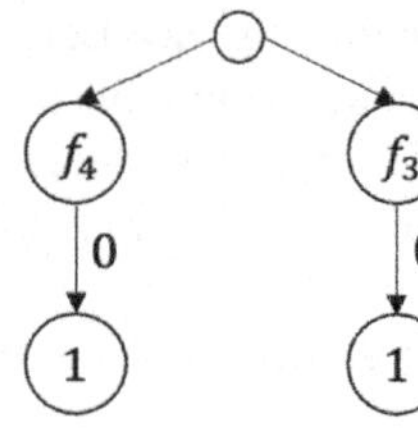

Fig. 9.4 A strongly nondeterministic decision tree for the decision table shown in Fig. 9.1

Definition 9.8 Let $T \in \mathscr{M}_k^{0\text{-}1} \setminus \{\Lambda\}$. A *deterministic decision tree* for the table T is a k-0-1-decision tree Γ satisfying the following conditions:

- Only one edge leaves the root of Γ.
- For any node, which is neither the root nor a terminal node, edges leaving this node are labeled with pairwise different numbers.
- $\text{At}(\Gamma) \subseteq \text{At}(T)$.
- For any row of T, there exists a complete path τ of Γ such that the considered row belongs to the table $T(\tau)$.
- For any complete path τ of Γ, either $T(\tau) = \Lambda$ or all rows of $T(\tau)$ are labeled with the decision attached to the terminal node of τ.

Example 9.5 The 2-0-1-decision tree shown in Fig. 9.3 is a deterministic decision tree for the decision table shown in Fig. 9.1.

Definition 9.9 Let $T \in \mathscr{M}_k^{0\text{-}1} \setminus \mathscr{M}_k^{0\text{-}1c}$. A *strongly nondeterministic decision tree* for the table T is a k-0-1-decision tree Γ satisfying the following conditions:

- Each terminal node is labeled with the decision 1.
- $\text{At}(\Gamma) \subseteq \text{At}(T)$.
- For any row of T labeled with the decision 1, there exists a complete path τ of Γ such that the considered row belongs to the table $T(\tau)$.
- For any complete path τ of Γ, either $T(\tau) = \Lambda$ or all rows of $T(\tau)$ are labeled with the decision 1.

Example 9.6 The 2-0-1-decision tree shown in Fig. 9.4 is a strongly nondeterministic decision tree for the decision table shown in Fig. 9.1.

9.2.4 Complexity Measures

Denote by P^* the set of all finite words over the alphabet $P = \{f_i : i \in \omega\}$, which contains the empty word λ.

Definition 9.10 A *partially bounded complexity measure* is an arbitrary function $\psi : P^* \to \omega$ that has the following properties: for any words $\alpha_1, \alpha_2 \in P^*$,

- $\psi(\alpha_1) = 0$ if and only if $\alpha_1 = \lambda$–*positivity* property.

- $\psi(\alpha_1) = \psi(\alpha_1')$ for any word α_1' obtained from α_1 by permutation of letters–*commutativity* property.
- $\psi(\alpha_1) \leq \psi(\alpha_1\alpha_2)$–*nondecreasing* property.
- $\psi(\alpha_1\alpha_2) \leq \psi(\alpha_1) + \psi(\alpha_2)$–*boundedness from above* property.

The following functions are partially bounded complexity measures:

- Function h for which, for any word $\alpha \in P^*$, $h(\alpha) = |\alpha|$, where $|\alpha|$ is the length of the word α.
- An arbitrary function $\varphi : P^* \to \omega$ such that $\varphi(\lambda) = 0$, for any $f_i \in P$, $\varphi(f_i) > 0$ and, for any nonempty word $f_{i_1} \cdots f_{i_m} \in P^*$,

$$\varphi(f_{i_1} \cdots f_{i_m}) = \sum_{j=1}^{m} \varphi(f_{i_j}) . \tag{9.1}$$

- An arbitrary function $\rho : P^* \to \omega$ such that $\rho(\lambda) = 0$, for any $f_i \in P$, $\rho(f_i) > 0$, and, for any nonempty word $f_{i_1} \cdots f_{i_m} \in P^*$, $\rho(f_{i_1} \cdots f_{i_m}) = \max\{\rho(f_{i_j}) : j = 1, \ldots, m\}$.

Definition 9.11 A *bounded* complexity measure is a partially bounded complexity measure ψ, which has the *boundedness from below* property: for any word $\alpha \in P^*$, $\psi(\alpha) \geq |\alpha|$.

Any partially bounded complexity measure satisfying the equality (9.1), in particular the function h, is a bounded complexity measure. One can show that if functions ψ_1 and ψ_2 are partially bounded complexity measures, then the functions ψ_3 and ψ_4 are partially bounded complexity measures, where for any $\alpha \in P^*$, $\psi_3(\alpha) = \psi_1(\alpha) + \psi_2(\alpha)$ and $\psi_4(\alpha) = \max(\psi_1(\alpha), \psi_2(\alpha))$. If the function ψ_1 is a bounded complexity measure, then the functions ψ_3 and ψ_4 are bounded complexity measures.

Definition 9.12 Let ψ be a partially bounded complexity measure. We extend it to the set of all finite subsets of the set P. Let D be a finite subset of the set P. If $D = \emptyset$, then $\psi(D) = 0$. Let $D = \{f_{i_1}, \ldots, f_{i_m}\}$ and $m \geq 1$. Then $\psi(D) = \psi(f_{i_1} \cdots f_{i_m})$.

9.2.5 Parameters of Decision Trees and Tables

Let $\Gamma \in \mathcal{T}_k^{0\text{-}1}$ and $\tau = v_1, d_1, \ldots, v_m, d_m, v_{m+1}$ be a complete path of Γ. We correspond to the path τ a word $F(\tau) \in P^*$: if $m = 1$, then $F(\tau) = \lambda$, and if $m > 1$ and, for $j = 2, \ldots, m$, the node v_j is labeled with the attribute f_{i_j}, then $F(\tau) = f_{i_2} \cdots f_{i_m}$.

Definition 9.13 Let ψ be a partially bounded complexity measure. We extend the function ψ to the set $\mathcal{T}_k^{0\text{-}1}$. Let $\Gamma \in \mathcal{T}_k^{0\text{-}1}$. Then $\psi(\Gamma) = \max\{\psi(F(\tau))\}$, where the maximum is taken over all complete paths τ of the decision tree Γ. For a given partially bounded complexity measure ψ, the value $\psi(\Gamma)$ will be called the *complexity* *of the decision tree* Γ. The value $h(\Gamma)$ will be called the *depth* of the decision tree Γ.

Let ψ be a partially bounded complexity measure. We now describe the functions ψ^d, ψ^s, Θ_ψ, W_ψ, m_ψ, S_ψ, $\hat{S}_\psi$, M_ψ, and N defined on the set $\mathcal{M}_k^{0\text{-}1}$ and taking values from the set ω. By definition, the value of each of these functions for Λ is equal 0. Let $T \in \mathcal{M}_k^{0\text{-}1} \setminus \{\Lambda\}$.

- $\psi^d(T) = \min\{\psi(\Gamma)\}$, where the minimum is taken over all deterministic decision trees Γ for the table T.
- If $T \in \mathcal{M}_k^{0\text{-}1c}$, then $\psi^s(T) = 0$. If $T \notin \mathcal{M}_k^{0\text{-}1c}$, then $\psi^s(T) = \min\{\psi(\Gamma)\}$, where the minimum is taken over all strongly nondeterministic decision trees Γ for the table T.
- $\Theta_\psi(T) = \min\{\psi(D)\}$, where the minimum is taken over all tests D of the table T.
- $W_\psi(T) = \psi(\mathrm{At}(T))$. This parameter is called *the complexity of the set of attributes attached to columns of the table* T.
- $m_\psi(T) = \max\{\psi(f_i) : f_i \in \mathrm{At}(T)\}$.
- Let $\bar{\delta}$ be a row of the table T. Denote $S_\psi(T, \bar{\delta}) = \min\{\psi(D)\}$, where the minimum is taken over all subsets D of the set $\mathrm{At}(T)$ such that in the set of columns of T labeled with attributes from D the row $\bar{\delta}$ is different from all other rows of the table T. Then $S_\psi(T) = \max\{S_\psi(T, \bar{\delta})\}$, where the maximum is taken over all rows $\bar{\delta}$ of the table T.
- $\hat{S}_\psi(T) = \max\{S_\psi(T^*) : T^* \in [T]\}$.
- If $T \in \mathcal{M}_k^{0\text{-}1c}$, then $M_\psi(T) = 0$. Let $T \notin \mathcal{M}_k^{0\text{-}1c}$, $|\mathrm{At}(T)| = n$, and columns of the table T be labeled with the attributes $f_{t_1}, \ldots, f_{t_n}$. Let $\bar{\delta} = (\delta_1, \ldots, \delta_n) \in E_k^n$. Denote by $M_\psi(T, \bar{\delta})$ the minimum number $p \in \omega$ for which there exist attributes $f_{t_{i_1}}, \ldots, f_{t_{im}} \in \mathrm{At}(T)$ such that $T(f_{t_{i_1}}, \delta_{i_1}) \cdots (f_{t_{im}}, \delta_{i_m}) \in \mathcal{M}_k^{0\text{-}1c}$ and $\psi(f_{t_{i_1}} \cdots f_{t_{im}}) = p$. Then $M_\psi(T) = \max\{M_\psi(T, \bar{\delta}) : \bar{\delta} \in E_k^n\}$.
- $N(T)$ is the number of rows in the table T.

For the bounded complexity measure h, we denote $\Theta(T) = \Theta_h(T)$, $W(T) = W_h(T)$, $S(T, \bar{\delta}) = S_h(T, \bar{\delta})$, $S(T) = S_h(T)$, $\hat{S}(T) = \hat{S}_h(T)$, $M(T, \bar{\delta}) = M_h(T, \bar{\delta})$, and $M(T) = M_h(T)$. Note that $W(T)$ is the number of columns in the table T.

Example 9.7 We denote by T_0 the decision table shown in Fig. 9.1. One can show that $h^d(T_0) = 2$, $h^s(T_0) = 1$, $\Theta(T_0) = 2$, $W(T_0) = 3$, $N(T_0) = 6$, $S(T_0) = 2$, $\hat{S}(T_0) = 2$, and $M(T_0) = 2$.

9.3 Main Results

In this section, we consider results obtained for the functions $\mathcal{F}^W_{\psi,A}$, $\mathcal{F}^\Theta_{\psi,A}$, $\mathcal{K}_{\psi,A}$, and $\mathcal{S}_{\psi,A}$ and discuss closed classes of decision tables generated by information systems.

9.3.1 Functions $\mathscr{F}^{W}_{\psi,A}$ and $\mathscr{F}^{\Theta}_{\psi,A}$

Let ψ be a bounded complexity measure and A be a nonempty closed class of decision tables from $\mathscr{M}^{0\text{-}1}_{k}$. We now define functions $\mathscr{F}^{W}_{\psi,A} : \omega \to \omega$ and $\mathscr{F}^{\Theta}_{\psi,A} : \omega \to \omega$. Let $n \in \omega$. Then

$$\mathscr{F}^{W}_{\psi,A}(n) = \max\{\psi^{d}(T) : T \in A, W_{\psi}(T) \leq n\},$$
$$\mathscr{F}^{\Theta}_{\psi,A}(n) = \max\{\psi^{d}(T) : T \in A, \Theta_{\psi}(T) \leq n\}.$$

The function $\mathscr{F}^{W}_{\psi,A}$ characterizes the growth in the worst case of the minimum complexity of a deterministic decision tree for a decision table from A with the growth of the complexity of the set of attributes attached to columns of the table. The function $\mathscr{F}^{\Theta}_{\psi,A}$ characterizes the growth in the worst case of the minimum complexity of a deterministic decision tree for a decision table from A with the growth of the minimum complexity of a test of the table.

Let $D = \{n_i : i \in \omega\}$ be an infinite subset of the set ω in which, for any $i \in \omega$, $n_i < n_{i+1}$. Let us define a function $H_D : \omega \to \omega$. Let $n \in \omega$. If $n < n_0$, then $H_D(n) = 0$. If, for some $i \in \omega$, $n_i \leq n < n_{i+1}$, then $H_D(n) = n_i$.

Theorem 9.1 *Let ψ be a bounded complexity measure and A be a nonempty closed class of decision tables from $\mathscr{M}^{0\text{-}1}_{k}$. Then $\mathscr{F}^{W}_{\psi,A}$ and $\mathscr{F}^{\Theta}_{\psi,A}$ are everywhere defined nondecreasing functions such that $\mathscr{F}^{W}_{\psi,A}(n) \leq \mathscr{F}^{\Theta}_{\psi,A}(n) \leq n$ for any $n \in \omega$ and $\mathscr{F}^{W}_{\psi,A}(0) = \mathscr{F}^{\Theta}_{\psi,A}(0) = 0$. For these functions, one of the following statements holds:*

(a) *If the functions S_{ψ} and N are bounded from above on the class A, then there exists a nonnegative constant c_0 such that $\mathscr{F}^{W}_{\psi,A}(n) \leq \mathscr{F}^{\Theta}_{\psi,A}(n) \leq c_0$ for any $n \in \omega$.*

(b) *If the function S_{ψ} is bounded from above on the class A and the function N is not bounded from above on the class A, then $\mathscr{F}^{W}_{\psi,A}(0) = \mathscr{F}^{\Theta}_{\psi,A}(0) = 0$ and there exist positive constants c_1, c_2, c_3, c_4 such that $c_1 \log_2 n - c_2 \leq \mathscr{F}^{W}_{\psi,A}(n) \leq \mathscr{F}^{\Theta}_{\psi,A}(n) \leq c_3 \log_2 n + c_4$ for any $n \in \omega \setminus \{0\}$.*

(c) *If the function S_{ψ} is not bounded from above on the class A, then there exists an infinite subset D of the set ω such that $H_D(n) \leq \mathscr{F}^{W}_{\psi,A}(n) \leq \mathscr{F}^{\Theta}_{\psi,A}(n) \leq n$ for any $n \in \omega$.*

For the bounded complexity measure h, the bounds from the statement (c) of Theorem 9.1 can be improved.

Theorem 9.2 *Let A be a nonempty closed class of decision tables from $\mathscr{M}^{0\text{-}1}_{k}$ for which the function S is not bounded from above. Then $\mathscr{F}^{W}_{h,A}(n) = \mathscr{F}^{\Theta}_{h,A}(n) = n$ for any $n \in \omega$.*

In the general case, the upper bound from the statement (c) of Theorem 9.1 is attained not for all $n \in \omega$.

Theorem 9.3 *For any infinite subset D of the set ω such that $0 \notin D$, there exist a bounded complexity measure ψ and a nonempty closed class A of decision tables from the set $\mathcal{M}_k^{0\text{-}1}$ for which $\mathcal{F}^W_{\psi,A}(n) = \mathcal{F}^\Theta_{\psi,A}(n) = H_D(n)$ for any $n \in \omega$.*

9.3.2 Function $\mathcal{K}_{\psi,A}$

Let ψ be a bounded complexity measure and A be a nonempty closed class of decision tables from the set $\mathcal{M}_k^{0\text{-}1}$. We now define possibly partial function $\mathcal{K}_{\psi,A} : \omega \to \omega$. Let $n \in \omega$. If the set $\{\psi^d(T) : T \in A, \psi^s(T) \leq n\}$ is infinite, then the value $\mathcal{K}_{\psi,A}(n)$ is undefined. Otherwise, $\mathcal{K}_{\psi,A}(n) = \max\{\psi^d(T) : T \in A, \psi^s(T) \leq n\}$.

The function $\mathcal{K}_{\psi,A}$ characterizes the growth in the worst case of the minimum complexity of a deterministic decision tree for a decision table from A with the growth of the minimum complexity of a strongly nondeterministic decision tree for the table.

Let us define possibly partial functions $S_{\psi,A} : \omega \to \omega$ and $N_{\psi,A} : \omega \to \omega$. Let $n \in \omega$. Denote $A_\psi(n) = \{T : T \in A, m_\psi(T) \leq n\}$. If the set $\{S(T) : T \in A_\psi(n)\}$ is infinite, then the value $S_{\psi,A}(n)$ is undefined. Otherwise, $S_{\psi,A}(n) = \max\{S(T) : T \in A_\psi(n)\}$. If the set $\{N(T) : T \in A_\psi(n)\}$ is infinite, then the value $N_{\psi,A}(n)$ is undefined. Otherwise, $N_{\psi,A}(n) = \max\{N(T) : T \in A_\psi(n)\}$.

Let us indicate the condition for the function $\mathcal{K}_{\psi,A}$ to be defined everywhere.

Theorem 9.4 *Let ψ be a bounded complexity measure and A be a nonempty closed class of decision tables from $\mathcal{M}_k^{0\text{-}1}$. Then the function $\mathcal{K}_{\psi,A}$ is defined everywhere if and only if the function $N_{\psi,A}$ is defined everywhere.*

The following statement characterizes the possible behavior of the function $\mathcal{K}_{\psi,A}$.

Theorem 9.5 (a) *Let ψ be a bounded complexity measure, A be a nonempty closed class of decision tables from $\mathcal{M}_k^{0\text{-}1}$ and the function $\mathcal{K}_{\psi,A}$ be defined everywhere. Then $\mathcal{K}_{\psi,A}$ is a nondecreasing function and either there exists a nonnegative constant c for which $\mathcal{K}_{\psi,A}(n) \leq c$ for any $n \in \omega$ or there exists an infinite subset D of the set ω for which $\mathcal{K}_{\psi,A}(n) \geq H_D(n)$ for any $n \in \omega$.*

(b) *Let $\varphi : \omega \to \omega$ be a nondecreasing function for which $\varphi(n) \geq n$ for any $n \in \omega$ and $\varphi(0) = 0$. Then there exist a bounded complexity measure ψ and a nonempty closed class A of decision tables from $\mathcal{M}_k^{0\text{-}1}$ for which, for any $n \in \omega$, the value $\mathcal{K}_{\psi,A}(n)$ is defined and coincides with $\varphi(n)$.*

Of particular interest is the case when $\mathcal{K}_{\psi,A}$ is an everywhere defined function bounded from above by some polynomial.

Theorem 9.6 *Let ψ be a bounded complexity measure, A be a nonempty closed class of decision tables from $\mathcal{M}_k^{0\text{-}1}$ and the function $\mathcal{K}_{\psi,A}$ be defined everywhere. Then a polynomial p_0 such that $\mathcal{K}_{\psi,A}(n) \leq p_0(n)$ for any $n \in \omega$ exists if an only if polynomials p_1 and p_2 exist such that $S_{\psi,A}(n) \leq p_1(n)$ and $N_{\psi,A}(n) \leq 2^{p_2(n)}$ for any $n \in \omega$.*

9.3.3 Function $\mathscr{S}_{\psi,A}$

Let ψ be a bounded complexity measure and A be a nonempty closed class of decision tables from $\mathscr{M}_k^{0\text{-}1}$. We now define a function $\mathscr{S}_{\psi,A}$. Let $n \in \omega$. Then

$$\mathscr{S}_{\psi,A}(n) = \max\{\psi^s(T) : T \in A,\ W_\psi(T) \le n\}\ .$$

The function $\mathscr{S}_{\psi,A}$ characterizes the growth in the worst case of the minimum complexity of a strongly nondeterministic decision tree for a decision table from A with the growth of the complexity of the set of attributes attached to columns of the table.

Theorem 9.7 *Let ψ be a bounded complexity measure and A be a nonempty closed class of decision tables from $\mathscr{M}_k^{0\text{-}1}$. Then $\mathscr{S}_{\psi,A}$ is an everywhere defined nondecreasing function such that $\mathscr{S}_{\psi,A}(n) \le n$ for any $n \in \omega$ and $\mathscr{S}_{\psi,A}(0) = 0$. For this function, one of the following statements holds:*

(a) *If the function S_ψ is bounded from above on the class A, then there exists a nonnegative constant c such that $\mathscr{S}_{\psi,A}(n) \le c$ for any $n \in \omega$.*

(b) *If the function S_ψ is not bounded from above on the class A, then there exists an infinite subset D of the set ω such that $H_D(n) \le \mathscr{S}_{\psi,A}(n) \le n$ for any $n \in \omega$.*

For the bounded complexity measure h, the bounds from the statement (b) of Theorem 9.7 can be improved.

Theorem 9.8 *Let A be a nonempty closed class of decision tables from $\mathscr{M}_k^{0\text{-}1}$ for which the function S is not bounded from above. Then $\mathscr{S}_{h,A}(n) = n$ for any $n \in \omega$.*

In the general case, the upper bound from the statement (b) of Theorem 9.7 is attained not for all $n \in \omega$.

Theorem 9.9 *For any infinite subset D of the set ω such that $0 \notin D$, there exist a bounded complexity measure ψ and a nonempty closed class A of decision tables from the set $\mathscr{M}_k^{0\text{-}1}$ for which $\mathscr{S}_{\psi,A}(n) = H_D(n)$ for any $n \in \omega$.*

9.3.4 Family of Closed Classes of Decision Tables

Let U be a set and $\Phi = \{f_0, f_1, \ldots\}$ be a finite or countable set of functions (attributes) defined on U and taking values from E_k. The pair (U, Φ) is called a *k-information system.*

A *0-1-problem* over (U, Φ) is an arbitrary tuple $z = (U, \nu, f_{i_1}, \ldots, f_{i_n})$, where $n \in \omega$, $\nu : E_k^n \to E_2$ and $f_{i_1}, \ldots, f_{i_n}$ are functions from Φ with pairwise different indices $i_1, \ldots, i_n$.

Let $n > 0$. The problem z is to determine the value $\nu(f_{i_1}(u), \ldots, f_{i_n}(u))$ for a given $u \in U$. Various examples of k-information systems and problems over these systems including 0-1-problems can be found in [3].

We denote by $T(z)$ a decision table from $\mathcal{M}_k^{0\text{-}1}$ with n columns labeled with attributes $f_{i_1}, \ldots, f_{i_n}$. A row $(\delta_1, \ldots, \delta_n) \in E_k^n$ belongs to the table $T(z)$ if and only if the system of equations $\{f_{i_1}(x) = \delta_1, \ldots, f_{i_n}(x) = \delta_n\}$ has a solution from the set U. This row is labeled with the decision $\nu(\delta_1, \ldots, \delta_n)$.

Let the algorithms for the problem z solving be algorithms in which each elementary operation consists in calculating the value of some attribute from the set $\{f_{i_1}, \ldots, f_{i_n}\}$ on the element $u \in U$. Then, as a model of the problem z, we can use the decision table $T(z)$, and as models of algorithms for the problem z solving–deterministic and strongly nondeterministic decision trees for the table $T(z)$.

If $n = 0$, then we set $T(z) = \Lambda$.

Denote by $\text{Probl}^{0\text{-}1}(U, \Phi)$ the set of 0-1-problems over (U, Φ) and

$$\text{Tab}^{0\text{-}1}(U, \Phi) = \{T(z) : z \in \text{Probl}^{0\text{-}1}(U, \Phi)\} .$$

One can show that $\text{Tab}^{0\text{-}1}(U, \Phi) = [\text{Tab}^{0\text{-}1}(U, \Phi)]$, i.e., $\text{Tab}^{0\text{-}1}(U, \Phi)$ is a closed class of decision tables from $\mathcal{M}_k^{0\text{-}1}$ *generated* by the information system (U, Φ).

Closed classes of decision tables generated by k-information systems are the most natural examples of closed classes. However, the notion of a closed class is essentially wider. In particular, the union $\text{Tab}^{0\text{-}1}(U_1, \Phi_1) \cup \text{Tab}^{0\text{-}1}(U_2, \Phi_2)$, where (U_1, Φ_1) and (U_2, Φ_2) are k-information systems, is a closed class, but generally, we cannot find an information system (U, Φ) such that $\text{Tab}^{0\text{-}1}(U, \Phi) = \text{Tab}^{0\text{-}1}(U_1, \Phi_1) \cup \text{Tab}^{0\text{-}1}(U_2, \Phi_2)$.

9.3.5 Example of Information System

Let $\mathbb{R}$ be the set of real numbers and $F = \{f_i : i \in \omega\}$ be the set of functions defined on $\mathbb{R}$ and taking values from the set E_2 such that, for any $i \in \omega$ and $a \in \mathbb{R}$,

$$f_i(a) = \begin{cases} 0, & a < i , \\ 1, & a \geq i . \end{cases}$$

Let ψ be a bounded complexity measure and $A = \text{Tab}^{0\text{-}1}(\mathbb{R}, F)$. One can prove the following statements:

- The function N is not bounded from above on the set A.
- The function S_ψ is bounded from above on the set A if and only if there exists a constant $c_0 \geq 0$ such that $\psi(f_i) \leq c_0$ for any $i \in \omega$.
- The function $N_{\psi,A}$ is everywhere defined if and only if the set $\{i : i \in \omega, \psi(f_i) \leq n\}$ is finite for any $n \in \omega$.
- Polynomials p_1 and p_2 such that $S_{\psi,A}(n) \leq p_1(n)$ and $N_{\psi,A}(n) \leq 2^{p_2(n)}$ for any $n \in \omega$ exist if and only if there exists a polynomial p_3 such that

$$|\{i : i \in \omega, \psi(f_i) \leq n\}| \leq 2^{p_3(n)}$$

for any $n \in \omega$.

9.3.6 Joint Behavior of Functions $\mathcal{F}^W_{\psi,A}$ and $\mathcal{S}_{\psi,A}$

In this section, we study joint behavior of functions $\mathcal{F}^W_{\psi,A}$ and $\mathcal{S}_{\psi,A}$. Let ψ be a bounded complexity measure and A be a nonempty closed class of decision tables from $\mathcal{M}_k^{0\text{-}1}$. From Theorems 9.1, 9.2, 9.7, and 9.8 it follows that there are three possible types of the joint behavior of functions $\mathcal{F}^W_{\psi,A}$ and $\mathcal{S}_{\psi,A}$:

- If the functions S_ψ and N are bounded from above on the class A, then $\mathcal{F}^W_{\psi,A}(n) = O(1)$ and $\mathcal{S}_{\psi,A}(n) = O(1)$.
- If the function S_ψ is bounded from above on the class A and the function N is not bounded from above on the class A, then $\mathcal{F}^W_{\psi,A}(n) = \Theta(\log n)$ and $\mathcal{S}_{\psi,A}(n) = O(1)$.
- If the function S_ψ is not bounded from above on the class A, then there exist infinite subsets D_1 and D_2 of the set ω such that $H_{D_1}(n) \le \mathcal{F}^W_{\psi,A}(n) \le n$ and $H_{D_2}(n) \le \mathcal{S}_{\psi,A}(n) \le n$ for any $n \in \omega$. If $\psi = h$, then $\mathcal{F}^W_{h,A}(n) = \mathcal{S}_{h,A}(n) = n$ for any $n \in \omega$.

All these three types are realizable for the bounded complexity measure h. Let us consider three 2-information systems (ω, F_1), (ω, F_2), and (ω, F_3), where

- $F_1 = \{q\}$ and, for any $j \in \omega$, $q(j) = 0$ if $j = 0$ and $q(j) = 1$ if $j > 0$.
- $F_2 = \{l_i : i \in \omega\}$ and, for any $i, j \in \omega$, $l_i(j) = 0$ if $j \le i$ and $l_i(j) = 1$ if $j > i$.
- $F_3 = \{p_i : i \in \omega\}$ and, for any $i, j \in \omega$, $p_i(j) = 1$ if $i = j$ and $p_i(j) = 0$ if $i \ne j$.

One can show, that

- The functions S and N are bounded from above on the closed class $\mathrm{Tab}^{0\text{-}1}(\omega, F_1)$.
- The function S is bounded from above on the closed class $\mathrm{Tab}^{0\text{-}1}(\omega, F_2)$ and the function N is not bounded from above on this class.
- The function S is not bounded from above on the closed class $\mathrm{Tab}^{0\text{-}1}(\omega, F_3)$.

9.4 Auxiliary Statements

This section contains auxiliary statements. Some of them are of independent interest.

Lemma 9.1 *Let $T \in \mathcal{M}_k^{0\text{-}1} \setminus \{\Lambda\}$ and Γ be a deterministic decision tree for the table T or a strongly nondeterministic decision tree for the table T. Then the set* $\mathrm{At}(\Gamma)$ *is a test of the table T.*

Proof Let $T \in \mathcal{M}_k^{0\text{-}1c}$. Then, by definition, $\mathrm{At}(\Gamma)$ is a test of the table T.

Let $T \notin \mathcal{M}_k^{0\text{-}1c}$ and $\bar{\delta}_0, \bar{\delta}_1$ be arbitrary rows of the table T that are labeled with the decisions 0 and 1, respectively. Since Γ is a deterministic decision tree for the table T or a strongly nondeterministic decision tree for the table T, then there exists a complete path τ of Γ for which the row $\bar{\delta}_1$ belongs to the table $T(\tau)$ and all rows of this table are labeled with the decision 1. Evidently, $\bar{\delta}_0$ is not a row of the table $T(\tau)$. Therefore the rows $\bar{\delta}_0$ and $\bar{\delta}_1$ are different in a column labeled with an attribute f_i such that $f_i \in \mathrm{At}(\Gamma)$. Therefore $\mathrm{At}(\Gamma)$ is a test of the table T. □

It is not difficult to prove the following statement.

Lemma 9.2 *Let $T \in \mathcal{M}_k^{0\text{-}1} \setminus \{\Lambda\}$, D be a test of the table T such that $D \neq \emptyset$, $T^* = I(\mathrm{At}(T) \setminus D, T)$ and Γ be a deterministic decision tree for the table T^*. Then Γ is a deterministic decision tree for the table T.*

The notions of a decision table and a deterministic decision tree used in this chapter are somewhat different from the corresponding notions used in [1]. Taking into account these differences, it is easy to prove the following two statements, which follow almost directly from Theorem 2.1, Lemma 1.3 and Theorem 2.2 from [1].

Lemma 9.3 *For any partially bounded complexity measure ψ and any table T from $\mathcal{M}_k^{0\text{-}1}$,*

$$\psi^d(T) \geq M_\psi(T) .$$

Lemma 9.4 *For any partially bounded complexity measure ψ and any table T from $\mathcal{M}_k^{0\text{-}1}$,*

$$\psi^d(T) \leq \begin{cases} 0, & M_\psi(T) = 0 , \\ M_\psi(T)\log_2 N(T), & M_\psi(T) \geq 1 . \end{cases}$$

It is not difficult to prove the following statement.

Lemma 9.5 *For any partially bounded complexity measure ψ and any table T from $\mathcal{M}_k^{0\text{-}1}$,*

$$\psi^d(T) \leq \Theta_\psi(T) .$$

Lemma 9.6 *For any table $T \in \mathcal{M}_k^{0\text{-}1} \setminus \mathcal{M}_k^{0\text{-}1c}$,*

$$h^d(T) > \log_k \Theta(T) .$$

Proof Let Γ be a deterministic decision tree for the table T such that $h(\Gamma) = h^d(T)$. By Lemma 9.1, $\mathrm{At}(\Gamma)$ is a test of the table T. Therefore $\Theta(T) \leq |\mathrm{At}(\Gamma)|$. Denote by $L_w(\Gamma)$ the number of nodes of Γ, which are neither the root nor a terminal node. Evidently, $|\mathrm{At}(\Gamma)| \leq L_w(\Gamma)$. One can show that $L_w(\Gamma) \leq \sum_{i=0}^{h(\Gamma)-1} k^i$. Therefore $\Theta(T) \leq \sum_{i=0}^{h(\Gamma)-1} k^i = \frac{k^{h(\Gamma)}-1}{k-1} < k^{h(\Gamma)}$. Since $T \notin \mathcal{M}_k^{0\text{-}1c}$, $\Theta(T) > 0$. Hence $h(\Gamma) > \log_k \Theta(T)$. Taking into account that $h(\Gamma) = h^d(T)$, we obtain $h^d(T) > \log_k \Theta(T)$. □

Lemma 9.7 *For any partially bounded complexity measure ψ and any table T from $\mathscr{M}_k^{0\text{-}1}$,*

$$M_\psi(T) \le 2\hat{S}_\psi(T) .$$

Proof Let $T \in \mathscr{M}_k^{0\text{-}1c}$. Then $M_\psi(T) = 0$. Therefore $M_\psi(T) \le 2\hat{S}_\psi(T)$.

Let $T \notin \mathscr{M}_k^{0\text{-}1c}$, $W(T) = n$ and $f_{t_1}, \dots, f_{t_n}$ be attributes attached to columns of the table T. Denote $D = \{f_i : f_i \in \mathrm{At}(T), \psi(f_i) \le S_\psi(T)\}$ and $T^* = I(\mathrm{At}(T) \setminus D, T)$. Evidently, $T^* \in [T]$. Taking into account that the function ψ has the nondecreasing property, we obtain that any two rows of T are different in columns labeled with attributes from the set D. Let for the definiteness, $D = \{f_{t_1}, \dots, f_{t_m}\}$.

Let $\bar{\delta} = (\delta_1, \dots, \delta_n) \in E_k^n$. Let $(\delta_1, \dots, \delta_m)$ be a row of T^*. Since $T^* \in [T]$, there exist attributes $f_{t_{i_1}}, \dots, f_{t_{i_s}}$ of the table T^* such that the row $(\delta_1, \dots, \delta_m)$ is different from all other rows of T^* in columns labeled with these attributes and $\psi(f_{t_{i_1}} \cdots f_{t_{i_s}}) \le \hat{S}_\psi(T)$. It is clear that $T(f_{t_{i_1}}, \delta_{i_1}) \cdots (f_{t_{i_s}}, \delta_{i_s}) \in \mathscr{M}_k^{0\text{-}1c}$. Therefore $M_\psi(T, \bar{\delta}) \le \hat{S}_\psi(T)$.

Let $(\delta_1, \dots, \delta_m)$ be not a row of T^*. We consider m tables

$$T^*(f_{t_1}, \delta_1), T^*(f_{t_1}, \delta_1)(f_{t_2}, \delta_2), \dots, T^*(f_{t_1}, \delta_1) \cdots (f_{t_m}, \delta_m) .$$

If $T^*(f_{t_1}, \delta_1) = \Lambda$, then $T(f_{t_1}, \delta_1) = \Lambda$. Since $f_{t_1} \in D$, $\psi(f_{t_1}) \le S_\psi(T)$. Taking into account that $S_\psi(T) \le \hat{S}_\psi(T)$, we obtain $M_\psi(T, \bar{\delta}) \le \hat{S}_\psi(T)$. Let $T^*(f_{t_1}, \delta_1) \neq \Lambda$. Then there exists $p \in \{1, \dots, m-1\}$ such that $T^*(f_{t_1}, \delta_1) \cdots (f_{t_p}, \delta_p) \neq \Lambda$ and $T^*(f_{t_1}, \delta_1) \cdots (f_{t_{p+1}}, \delta_{p+1}) = \Lambda$. Denote $C = \{f_{t_{p+1}}, f_{t_{p+2}}, \dots, f_{t_n}\}$ and $T^0 = I(C, T)$. Evidently, $T^0 \in [T]$. Therefore in T^0 there are attributes $f_{t_{i_1}}, \dots, f_{t_{i_l}}$ such that the row $(\delta_1, \dots, \delta_p)$ is different from all other rows of the table T^0 in columns labeled with these attributes and $\psi(f_{t_{i_1}} \cdots f_{t_{i_l}}) \le \hat{S}_\psi(T)$. One can show that

$$T^*(f_{t_{i_1}}, \delta_{i_1}) \cdots (f_{t_{i_l}}, \delta_{i_l}) = T^*(f_{t_1}, \delta_1) \cdots (f_{t_p}, \delta_p) .$$

Therefore $T^*(f_{t_{i_1}}, \delta_{i_1}) \cdots (f_{t_{i_l}}, \delta_{i_l})(f_{t_{p+1}}, \delta_{p+1}) = \Lambda$. Hence

$$T(f_{t_{i_1}}, \delta_{i_1}) \cdots (f_{t_{i_l}}, \delta_{i_l})(f_{t_{p+1}}, \delta_{p+1}) = \Lambda .$$

Since $f_{t_{p+1}} \in D$, $\psi(f_{t_{p+1}}) \le S_\psi(T) \le \hat{S}_\psi(T)$. Using boundedness from above property of the function ψ, we obtain $\psi(f_{t_{i_1}} \cdots f_{t_{i_l}} f_{t_{p+1}}) \le 2\hat{S}_\psi(T)$. Therefore

$$M_\psi(T, \bar{\delta}) \le 2\hat{S}_\psi(T) .$$

Thus, for any $\bar{\delta} \in E_k^n$, $M_\psi(T, \bar{\delta}) \le 2\hat{S}_\psi(T)$. As a result, we obtain $M_\psi(T) \le 2\hat{S}_\psi(T)$. □

It is not difficult to prove the following two upper bounds on the minimum complexity of strongly nondeterministic decision trees for a table.

Lemma 9.8 *For any partially bounded complexity measure ψ and any table T from $\mathcal{M}_k^{0\text{-}1}$,*

$$\psi^s(T) \leq \psi^d(T) .$$

Lemma 9.9 *For any partially bounded complexity measure ψ and any table T from $\mathcal{M}_k^{0\text{-}1}$,*

$$\psi^s(T) \leq S_\psi(T) .$$

Lemma 9.10 *For any table T from $\mathcal{M}_k^{0\text{-}1} \setminus \{\Lambda\}$,*

$$N(T) \leq (kW(T))^{S(T)} .$$

Proof If $N(T) = 1$, then $S(T) = 0$ and the considered inequality holds. Let $N(T) > 1$. Then $S(T) > 0$. Denote $m = S(T)$. Evidently, for any row $\bar{\delta}$ of the table T, there exist attributes $f_{i_1}, \ldots, f_{i_m} \in \mathrm{At}(T)$ and numbers $\sigma_1, \ldots, \sigma_m \in E_k$ such that the table $T(f_{i_1}, \sigma_1) \cdots (f_{i_m}, \sigma_m)$ contains only the row $\bar{\delta}$. Therefore there is a one-to-one mapping of rows of the table T onto some set B of pairs of tuples of the kind $((f_{i_1}, \ldots, f_{i_m}), (\sigma_1, \ldots, \sigma_m))$, where $f_{i_1}, \ldots, f_{i_m} \in \mathrm{At}(T)$ and $\sigma_1, \ldots, \sigma_m \in E_k$. Evidently, $|B| \leq W(T)^m k^m$. Therefore $N(T) \leq (kW(T))^{S(T)}$. □

It is not difficult to prove the following statement by the induction on the number of rows in the table.

Lemma 9.11 *For any table T from $\mathcal{M}_k^{0\text{-}1} \setminus \{\Lambda\}$,*

$$\Theta(T) \leq N(T) - 1 .$$

Lemma 9.12 *For any partially bounded complexity measure ψ and any table T from $\mathcal{M}_k^{0\text{-}1}$, which contains at least two rows, there exists a table $T^* \in [T]$ such that $\psi^d(T^*) = W_\psi(T^*) = S_\psi(T^*) = S_\psi(T)$ and $\psi^s(T^*) \leq m_\psi(T)$.*

Proof Let $\bar{\delta}$ be a row of the table T such that $S_\psi(T, \bar{\delta}) = S_\psi(T)$. Let D be a subset of the set $\mathrm{At}(T)$ with the minimum cardinality such that $\psi(D) = S_\psi(T, \bar{\delta})$ and in the set of columns labeled with attributes from D the row $\bar{\delta}$ is different from all other rows of the table T. Let $\bar{\sigma}$ be a tuple obtained from the row $\bar{\delta}$ by the removal of all numbers that are in the intersections with columns labeled with attributes from the set $\mathrm{At}(T) \setminus D$. Let $\nu : E_k^{|D|} \to E_2$ and, for any $\bar{\gamma} \in E_k^{|D|}$, if $\bar{\gamma} = \bar{\sigma}$, then $\nu(\bar{\gamma}) = 0$ and if $\bar{\gamma} \neq \bar{\sigma}$, then $\nu(\bar{\gamma}) = 1$. Denote $T^* = J(\nu, I(\mathrm{At}(T) \setminus D, T))$. From the fact that D has the minimum cardinality and from the properties of the function ψ it follows that, for any attribute from the set D, there exists a row of T, which is different from the row $\bar{\delta}$ only in the column labeled with the considered attribute among attributes from D. Therefore, for any attribute of the table T^*, there exists a row of T^*, which is different from the row $\bar{\sigma}$ only in the column labeled with the considered attribute. Thus,

$$S_\psi(T^*, \bar{\sigma}) = W_\psi(T^*) . \quad (9.2)$$

Using properties of the function ψ, we obtain $S_\psi(T^*) \leq W_\psi(T^*)$. From this inequality and from (9.2) it follows that

$$S_\psi(T^*) = W_\psi(T^*) . \tag{9.3}$$

One can show that $M_\psi(T^*, \bar{\sigma}) = S_\psi(T^*, \bar{\sigma})$. Therefore $M_\psi(T^*) \geq S_\psi(T^*, \bar{\sigma})$. From this inequality and from Lemma 9.3 it follows that $\psi^d(T^*) \geq S_\psi(T^*, \bar{\sigma})$. Using (9.2), we obtain $\psi^d(T^*) \geq W_\psi(T^*)$. Using Lemma 9.5 and nondecreasing property of the function ψ, we obtain $\psi^d(T^*) \leq W_\psi(T^*)$. Hence

$$\psi^d(T^*) = W_\psi(T^*) . \tag{9.4}$$

By the choice of the set D, $W_\psi(T^*) = S_\psi(T)$. From this equality and from (9.3) and (9.4) it follows that $\psi^d(T^*) = W_\psi(T^*) = S_\psi(T^*) = S_\psi(T)$. One can show that $\psi^s(T^*) \leq m_\psi(T)$. □

Let G be an undirected graph whose nodes are colored in two colors. An edge of the graph G will be called a *multi-colored* edge, if it is incident to nodes having different colors. By induction on the number of nodes of the graph, it is easy to prove the following statement.

Lemma 9.13 *Let G be an arbitrary undirected graph without loops and multiple edges. Then the nodes of the graph G can be colored in two colors such that at least half of the edges of G will be multi-colored.*

A table T from $\mathscr{M}_k^{0\text{-}1}$ will be called *critical* if $T \neq \Lambda$ and, for any attribute of T, there are two rows of the table T that differ only in the column labeled with this attribute.

Lemma 9.14 *For any critical table T from $\mathscr{M}_k^{0\text{-}1}$, there exists a mapping $\nu : E_k^{W(T)} \to E_2$ such that*

$$h^d(J(\nu, T)) > \log_k(W(T)/2) .$$

Proof We correspond to any attribute $f_i \in \mathrm{At}(T)$ two rows $\bar{\delta}_0(f_i)$ and $\bar{\delta}_1(f_i)$ of the table T, which differ only in the column labeled with this attribute. Denote by G the undirected graph with the set of nodes $\{\bar{\delta}_0(f_i), \bar{\delta}_1(f_i) : f_i \in \mathrm{At}(T)\}$ and the set of edges $\{(\bar{\delta}_0(f_i), \bar{\delta}_1(f_i)) : f_i \in \mathrm{At}(T)\}$. Using Lemma 9.13, we obtain that the nodes of G can be colored in two colors (for example, blue and green) such that at least $W(T)/2$ edges of G will be multi-colored. Let us define a mapping $\nu : E_k^{W(T)} \to E_2$. Let $\bar{\delta} \in E_k^{W(T)}$. If $\bar{\delta}$ is a blue node of G, then $\nu(\bar{\delta}) = 0$. Otherwise, $\nu(\bar{\delta}) = 1$. Denote $T^* = J(\nu, T)$. Let D be an arbitrary test of the table T^*. Evidently, if $f_i \in \mathrm{At}(T)$ and the edge $(\bar{\delta}_0(f_i), \bar{\delta}_1(f_i))$ is multi-colored, then $f_i \in D$. Therefore $\Theta(T^*) \geq W(T)/2$. Using Lemma 9.6, we obtain $h^d(T^*) > \log_k \Theta(T^*) \geq \log_k(W(T)/2)$. □

Lemma 9.15 *For any table $T \in \mathscr{M}_k^{0\text{-}1} \setminus \{\Lambda\}$, there exists a table $T^* \in [T]$ such that*

$$h^d(T^*) \geq \frac{\log_k N(T)}{\hat{S}(T)} - 2 .$$

Proof If $N(T) = 1$, then, evidently, the considered statement holds. Let $N(T) \geq 2$. Let D be a subset of $\mathrm{At}(T)$ with the minimum cardinality such that in columns labeled with attributes from D any two rows of T are different. Denote $T^0 = I(\mathrm{At}(T) \setminus D, T)$. One can show that T^0 is a critical table. By Lemma 9.14, there exists a mapping $\nu : E_k^{W(T^0)} \to E_2$ such that $h^d(J(\nu, T^0)) > \log_k(W(T^0)/2) \geq \log_k W(T^0) - 1$. Denote $T^* = J(\nu, T^0)$. Evidently, $T^* \in [T]$. Taking into account that $W(T^0) = W(T^*)$, we obtain

$$h^d(T^*) > \log_k W(T^*) - 1 . \tag{9.5}$$

Using Lemma 9.10, we obtain $N(T){=}N(T^*) \leq (kW(T^*))^{S(T^*)} \leq (kW(T^*))^{\hat{S}(T)}$. Therefore $\hat{S}(T)(\log_k W(T^*) + 1) \geq \log_k N(T)$ and $\log_k W(T^*) \geq \frac{\log_k N(T)}{\hat{S}(T)} - 1$. Using (9.5), we obtain $h^d(T^*) \geq \frac{\log_k N(T)}{\hat{S}(T)} - 2$. □

9.5 Proofs of Theorems 9.1, 9.2, and 9.3

Proof of Theorem 9.1. Evidently, $\mathscr{F}^W_{\psi,A}$ and $\mathscr{F}^\Theta_{\psi,A}$ are nondecreasing functions. Using properties of commutativity and nondecreasing of the function ψ, we obtain that, for any table T from $\mathscr{M}_k^{0\text{-}1}$, $\Theta_\psi(T) \leq W_\psi(T)$. From this inequality and from Lemma 9.5 it follows that, for any $n \in \omega$,

$$0 \leq \mathscr{F}^W_{\psi,A}(n) \leq \mathscr{F}^\Theta_{\psi,A}(n) \leq n . \tag{9.6}$$

Since A is a closed class, $\Lambda \in A$. By definition, $\Theta_\psi(\Lambda) = W_\psi(\Lambda) = 0$. Using this fact, we obtain that $\mathscr{F}^W_{\psi,A}$ and $\mathscr{F}^\Theta_{\psi,A}$ are everywhere defined functions. Let $T \in A$ and $\Theta_\psi(T) \leq 0$. One can show that in this case $T \in \mathscr{M}_k^{0\text{-}1c}$ and therefore $\psi^d(T) = 0$. Thus, $\mathscr{F}^\Theta_{\psi,A}(0) \leq 0$. From this inequality and (9.6) it follows that

$$\mathscr{F}^W_{\psi,A}(0) = \mathscr{F}^\Theta_{\psi,A}(0) . \tag{9.7}$$

(a) Let the functions S_ψ and N be bounded from above on the class A. Then there are constants $a > 0$ and $b > 0$ such that $S_\psi(T) \leq a$ and $N(T) \leq b$ for any table $T \in A$. Let $T \in A$. Taking into account that A is a closed class, we obtain $\hat{S}_\psi(T) \leq a$. By Lemma 9.7, $M_\psi(T) \leq 2a$. From this inequality, inequality $N(T) \leq b$ and from Lemma 9.4 it follows that $\psi^d(T) \leq 2a \log_2 b$. Denote $c_0 = 2a \log_2 b$. Taking into account that T is an arbitrary table from the class A and using (9.6), we obtain that, for any $n \in \omega$,

$$\mathscr{F}^W_{\psi,A}(n) \leq \mathscr{F}^\Theta_{\psi,A}(n) \leq c_0 .$$

(b) Let the function S_ψ be bounded from above on the closed class A and the function N be not bounded from above on A. Then there exists a constant $a \geq 1$ such that, for any table $T \in A$,

$$S_\psi(T) \leq a\ . \tag{9.8}$$

Let $n \in \omega \setminus \{0\}$ and T be an arbitrary table from A such that $\Theta_\psi(T) \leq n$. If $\Theta_\psi(T) = 0$, then using Lemma 9.5, we obtain

$$\psi^d(T) = 0\ . \tag{9.9}$$

Let $\Theta_\psi(T) \neq 0$ and D be a test of the table T such that $\psi(D) = \Theta_\psi(T)$. Denote $T^* = I(\mathrm{At}(T) \setminus D, T)$. Evidently, $T^* \in A$ and $T^* \neq \Lambda$. Using (9.8) and the boundedness from below property of the function ψ, we obtain

$$S(T^*) \leq a\ . \tag{9.10}$$

It is clear that $W(T^*) = |D|$. Using the boundedness from below property of the function ψ, we obtain $|D| \leq \psi(D) = \Theta_\psi(T)$. Therefore $W(T^*) \leq n$. From this inequality, inequality (9.10), relation $T^* \neq \Lambda$, and Lemma 9.10 it follows that $N(T^*) \leq (kn)^a$. From (9.8) and Lemma 9.7 it follows that $M(T^*) \leq 2a$. Using the last two inequalities and Lemma 9.4, we obtain

$$\psi^d(T^*) \leq 2a^2 \log_2 n + 2a^2 \log_2 k\ . \tag{9.11}$$

Denote $c_3 = 2a^2$ and $c_4 = 2a^2 \log_2 k$. From Lemma 9.2 it follows that $\psi^d(T) \leq \psi^d(T^*)$. From this inequality and (9.11) it follows that $\psi^d(T) \leq c_3 \log_2 n + c_4$. Taking into account that n is an arbitrary number from $\omega \setminus \{0\}$ and T is an arbitrary table from A such that $0 < \Theta_\psi(T) \leq n$ and using (9.9), we obtain that, for any $n \in \omega \setminus \{0\}$,

$$\mathscr{F}^{\Theta}_{\psi,A}(n) \leq c_3 \log_2 n + c_4\ . \tag{9.12}$$

Let $n \in \omega \setminus \{0\}$. Denote $c_1 = 1/\log_2 k$ and $c_2 = \log_k(4a)$. We now show that

$$\mathscr{F}^{W}_{\psi,A}(n) \geq c_1 \log_2 n - c_2\ . \tag{9.13}$$

If $n \leq 4a$, then, evidently, the considered inequality holds. Let $n > 4a$. Denote $m = \lfloor n/a \rfloor$. Taking into account that the function N is not bounded from above on the set A, we obtain that there exists a table $T \in A$ such that $N(T) \geq k^m$. Let B be a subset of the set $\mathrm{At}(T)$ with the minimum cardinality such that in the columns labeled with attributes from B any two rows of the table T are different and, for any $f_i \in B$, $\psi(f_i) \leq a$. The existence of such set follows from the inequality (9.8) and properties of commutativity and nondecreasing of the function ψ. Evidently, $|B| \geq m$. Let D be a subset of the set B such that $|D| = m$. Denote $T^0 = I(\mathrm{At}(T) \setminus D, T)$. One can show that T^0 is a critical table. Using Lemma 9.14, we obtain that there exists a mapping $\nu : E_k^m \to E_2$ such that $h^d(J(\nu, T^0)) > \log_k(m/2) = \log_k(\lfloor n/a \rfloor /2)$. Denote

$T^* = J(\nu, T^0)$. Taking into account that $n > 4a$, we obtain $\lfloor n/a \rfloor /2 \geq n/(4a)$. Therefore $h^d(T^*) > \log_k(n/(4a))$. Since ψ is a bounded complexity measure, $\psi^d(T^*) > \log_k(n/(4a))$ and $W_\psi(T^*) \leq \lfloor n/a \rfloor a \leq n$. Hence the inequality (9.13) holds. From (9.6), (9.7), (9.12), and (9.13) it follows that $\mathscr{F}^W_{\psi,A}(0) = \mathscr{F}^\Theta_{\psi,A}(0) = 0$ and, for any $n \in \omega \setminus \{0\}$,

$$c_1 \log_2 n - c_2 \leq \mathscr{F}^W_{\psi,A}(n) \leq \mathscr{F}^\Theta_{\psi,A}(n) \leq c_3 \log_2 n + c_4 .$$

(c) Let the function S_ψ be not bounded from above on the class A. Let $n \in \omega \setminus \{0\}$ and T be a table from A such that $\Theta_\psi(T) \leq n$. Using Lemma 9.5, we obtain that $\psi^d(T) \leq \Theta_\psi(T)$. Therefore

$$\mathscr{F}^\Theta_{\psi,A}(n) \leq n . \tag{9.14}$$

Using Lemma 9.12, we obtain that the set $D = \{W_\psi(T) : T \in A, \psi^d(T) = W_\psi(T)\}$ is infinite. Evidently, for any $n \in D$, $\mathscr{F}^W_{\psi,A}(n) \geq n$. Taking into account that $\mathscr{F}^W_{\psi,A}$ is a nondecreasing function, we obtain that, for any $n \in \omega \setminus \{0\}$,

$$\mathscr{F}^W_{\psi,A}(n) \geq H_D(n) .$$

From this inequality and from (9.6), (9.7), and (9.14) it follows that, for any $n \in \omega$, $H_D(n) \leq \mathscr{F}^W_{\psi,A}(n) \leq \mathscr{F}^\Theta_{\psi,A}(n) \leq n$. □

Proof of Theorem 9.2. From Theorem 9.1 it follows that, for any $n \in \omega$, $\mathscr{F}^W_{h,A}(n) \leq \mathscr{F}^\Theta_{h,A}(n) \leq n$. In particular, $\mathscr{F}^W_{h,A}(0) \leq \mathscr{F}^\Theta_{h,A}(0) = 0$. Let $n \in \omega \setminus \{0\}$. Since the function S is not bounded from above on the class A, there exists a table $T \in A$ such that $S(T) \geq n$. Let $\bar{\delta}$ be a row of the table T for which $S(T, \bar{\delta}) = S(T)$. Let C be a set of attributes of T with the minimum cardinality such that in columns labeled with attributes from C the row $\bar{\delta}$ is different from all other rows of the table T. Evidently, $|C| \geq n$. Let C^* be a subset of the set C with $|C^*| = n$. Denote $T^0 = I(\mathrm{At}(T) \setminus C^*, T)$. One can show that $S(T^0) = n$. From Lemma 9.12 it follows that there exists a table $T^* \in [T^0]$ for which $h^d(T^*) = W(T^*) = n$. Therefore $\mathscr{F}^W_{h,A}(n) \geq n$. □

Proof of Theorem 9.3. Let $D = \{n_i : i \in \omega\}$, $0 \notin D$ and, for any $i \in \omega$, $n_i < n_{i+1}$. For $i \in \omega$, we denote by T_i a decision table with one column labeled with the attribute f_i and two rows (0) and (1) labeled with decisions 0 and 1, respectively. Denote $A = [\{T_i : i \in D\}]$. We now define a complexity measure ψ. Let $\psi(\lambda) = 0$, $\psi(f_0) = 1$, $\psi(f_i) = i$ for any $i \in \omega \setminus \{0\}$, and $\psi(f_{i_1} \cdots f_{i_m}) = \sum_{j=1}^m \psi(f_{i_j})$ for any nonempty word $f_{i_1} \cdots f_{i_m} \in P^*$. Evidently, ψ is a bounded complexity measure. Let $n < n_0$. Then $\mathscr{F}^W_{\psi,A}(n) = \mathscr{F}^\Theta_{\psi,A}(n) = 0$. Let $n_i \leq n < n_{i+1}$. One can show that in this case $\mathscr{F}^W_{\psi,A}(n) = \mathscr{F}^\Theta_{\psi,A}(n) = n_i$. □

9.6 Proofs of Theorems 9.4, 9.5, and 9.6

Proof of Theorem 9.4. (a) Let the function $N_{\psi,A}$ be everywhere defined. Let $n \in \omega$, $T \in A$ and $\psi^s(T) \le n$. Let $T \notin \mathcal{M}_k^{0\text{-}1c}$ and Γ be a strongly nondeterministic decision tree for the table T such that $\psi(\Gamma) = \psi^s(T)$. Using Lemma 9.1, we obtain that $\mathrm{At}(\Gamma)$ is a test of the table T. Denote $T^* = I(\mathrm{At}(T) \setminus \mathrm{At}(\Gamma), T)$. Using the properties of commutativity and nondecreasing of the function ψ, we obtain that, for any $f_i \in \mathrm{At}(\Gamma)$, $\psi(f_i) \le \psi(\Gamma)$ and hence

$$\psi(f_i) \le n \,. \tag{9.15}$$

Therefore $T^* \in A_\psi(n)$. Using this relation and Lemma 9.11, we obtain that $\Theta(T^*) < N(T^*) - 1 < N_{\psi,A}(n)$. From these inequalities and from the boundedness from above property of the function ψ it follows that

$$\Theta_\psi(T^*) < nN_{\psi,A}(n) \,. \tag{9.16}$$

Using Lemma 9.5, we obtain that $\psi^d(T^*) \le \Theta_\psi(T^*)$. Taking into account that $\mathrm{At}(\Gamma)$ is a test of the table T and using Lemma 9.2, we obtain that $\psi^d(T) \le \psi^d(T^*)$. From the last two inequalities and from (9.16) it follows that $\psi^d(T) < nN_{\psi,A}(n)$. If $T \in \mathcal{M}_k^{0\text{-}1c}$, then $\psi^d(T) = 0$. Therefore the last inequality holds. Hence the set $\{\psi^d(T) : T \in A, \psi^s(T) \le n\}$ is finite and the value $\mathcal{K}_{\psi,A}(n)$ is defined.

(b) Let the function $N_{\psi,A}$ be not everywhere defined and the function $S_{\psi,A}$ be everywhere defined. Then there exists $n \in \omega$ such that the set $\Omega = \{N(T) : T \in A_\psi(n)\}$ is infinite. Let $T \in A_\psi(n) \setminus \{\Lambda\}$. Taking into account that $[T] \subseteq A_\psi(n)$ and using Lemma 9.15, we obtain that there exists a table T^* from $A_\psi(n)$ for which $h^d(T^*) \ge \frac{\log_k N(T)}{\hat{S}(T)} - 2$. Using the boundedness from below property of the function ψ, we obtain $\psi^d(T^*) \ge h^d(T^*)$. Evidently, $\hat{S}(T) \le S_{\psi,A}(n)$. Hence $\psi^d(T^*) \ge \frac{\log_k N(T)}{S_{\psi,A}(n)} - 2$. Taking into account that the set Ω is infinite, we obtain that the set $\Delta = \{\psi^d(T) : T \in A_\psi(n)\}$ is infinite. Using Lemma 9.9, we obtain that $\psi^s(T) \le nS_{\psi,A}(n)$ for any table $T \in A_\psi(n)$. From this inequality and from the fact that the set Δ is infinite it follows that the set $\{\psi^d(T) : T \in A, \psi^s(T) \le nS_{\psi,A}(n)\}$ is infinite. Therefore the value $\mathcal{K}_{\psi,A}(nS_{\psi,A}(n))$ is not defined.

(c) Let the functions $N_{\psi,A}$ and $S_{\psi,A}$ be not everywhere defined. Then there exists $n \in \omega$ for which the set $\{S(T) : T \in A_\psi(n)\}$ is infinite. Therefore the set $\{S_\psi(T) : T \in A_\psi(n)\}$ is infinite. Using Lemma 9.12, we obtain that the set $\{\psi^d(T) : T \in A, \psi^s(T) \le n\}$ is infinite. Hence, the value $\mathcal{K}_{\psi,A}(n)$ is not defined. □

Proof of Theorem 9.5. (a) From the definition of the function $\mathcal{K}_{\psi,A}$ it follows that it is nondecreasing. Let the function ψ^d be bounded from above on the set A by a constant c. Then, evidently, $\mathcal{K}_{\psi,A}(n) \le c$ for any $n \in \omega$. Let the function ψ^d be not bounded from above on the set A. Let us assume that the function ψ^s is bounded from above on the set A by a constant $d \in \omega$. Then the value $\mathcal{K}_{\psi,A}(d)$ is not defined, which is impossible. Thus, the set $D = \{\psi^s(T) : T \in A\}$ is infinite. Using Lemma

9.8, we obtain that $\psi^d(T) \geq \psi^s(T)$ for any $T \in A$. Therefore $\mathscr{K}_{\psi,A}(n) \geq n$ for any $n \in D$. Taking into account that $\mathscr{K}_{\psi,A}$ is a nondecreasing function, we obtain that $\mathscr{K}_{\psi,A}(n) \geq H_D(n)$ for any $n \in \omega$.

(b) Let $n \in \omega \setminus \{0\}$. Denote by T_n the decision table depicted in Fig. 9.5, where $t(n) = \sum_{i=1}^{n-1} \left\lceil \frac{\varphi(i)}{i} \right\rceil$ if $n > 1$ and $t(1) = 0$. Denote $A = [\{T_n : n \in \omega \setminus \{0\}\}]$. We now define a bounded complexity measure ψ. Let $i \in \omega$. If $i = 0$, then $\psi(f_i) = 1$. Let for some $n \in \omega \setminus \{0\}$, $t(n) < i < t(n+1)$. Then $\psi(f_i) = n$. Let, for some $n \in \omega \setminus \{0\}$, $i = t(n+1)$ and $\varphi(n) = ln + j$, where $l \in \omega \setminus \{0\}$ and $0 \leq j < n$. If $j = 0$, then $\psi(f_i) = n$. If $j > 0$, then $\psi(f_i) = j$. For any nonempty word $f_{i_1} \cdots f_{i_m} \in P^*$, $\psi(f_{i_1} \cdots f_{i_m}) = \sum_{j=1}^{m} \psi(f_{i_j})$. It is easy to show that $W_\psi(T_n) = \varphi(n)$ for any $n \in \omega \setminus \{0\}$.

One can show that the function $N_{\psi,A}$ is everywhere defined. Using Theorem 9.4, we obtain that the function $\mathscr{K}_{\psi,A}$ is everywhere defined. Evidently, $\mathscr{K}_{\psi,A}(0) = 0 = \varphi(0)$. Let $n \in \omega \setminus \{0\}$. One can show that $\psi^s(T_n) = n$ and $M_\psi(T_n, \bar{\delta}) = \varphi(n)$, where $\bar{\delta}$ is the first row of the table T_n. Using Lemma 9.3, we obtain $\psi^d(T_n) \geq \varphi(n)$. Hence

$$\mathscr{K}_{\psi,A}(n) \geq \varphi(n) . \tag{9.17}$$

We now show that, for any $n \in \omega \setminus \{0\}$,

$$\mathscr{K}_{\psi,A}(n) \leq \varphi(n) . \tag{9.18}$$

Let $n \in \omega \setminus \{0\}$, $T \in A$ and $\psi^s(T) \leq n$. Let $T \in [\{T_1, \ldots, T_n\}]$. Using Lemma 9.5, we obtain that $\psi^d(T) \leq \Theta_\psi(T)$. From here and from the properties of commutativity and nondecreasing of the function ψ it follows that $\psi^d(T) \leq W_\psi(T)$. One can show that $W_\psi(T) \leq \max\{W_\psi(T_1), \ldots, W_\psi(T_n)\} \leq \varphi(n)$. Hence $\psi^d(T) \leq \varphi(n)$.

Let $T \in [\{T_{n+1}, T_{n+2}, \ldots\}]$. Let Γ be a strongly nondeterministic decision tree for the table T such that $\psi(\Gamma) \leq n$. Using properties of commutativity and nondecreasing of the function ψ, we obtain that $\psi(f_i) \leq n$ for any $f_i \in \mathrm{At}(\Gamma)$. Evidently, $|\{f_i : f_i \in \mathrm{At}(T), \psi(f_i) \leq n\}| \leq 1$. Using the property of boundedness from above of the function ψ, we obtain that $\psi(\mathrm{At}(\Gamma)) \leq n$. From Lemma 9.1 it follows that $\mathrm{At}(\Gamma)$ is a test of the table T. Therefore $\Theta_\psi(T) \leq n$. Using Lemma 9.5, we obtain that $\psi^d(T) \leq n$. Since $\varphi(n) \geq n$, $\psi^d(T) \leq \varphi(n)$. Thus, the inequality (9.18) holds. From (9.17) and (9.18) it follows that $\mathscr{K}_{\psi,A}(n) = \varphi(n)$. □

Fig. 9.5 Decision table T_n

$f_{t(n)+1}$	$f_{t(n)+2}$	$f_{t(n)+3}$	$\cdots$	$f_{t(n)+\lceil \frac{\varphi(n)}{n} \rceil}$	
0	0	0	$\cdots$	0	0
1	0	0	$\cdots$	0	1
0	1	0	$\cdots$	0	1
0	0	1	$\cdots$	0	1
$\vdots$	$\vdots$	$\vdots$	$\vdots$	$\vdots$	$\vdots$
0	0	0	$\cdots$	1	1

Proof of Theorem 9.6. (a) Let there exist polynomials p_1 and p_2 such that, for any $n \in \omega$, $S_{\psi,A}(n) \le p_1(n)$ and $N_{\psi,A}(n) \le 2^{p_2(n)}$. Without loss of generality, we can assume that, for any $n \in \omega$, $p_1(n) \ge 0$ and $p_2(n) \ge 0$. Let $n \in \omega$, $T \in A$ and $\psi^s(T) \le n$. Let us show that

$$\psi^d(T) \le 2np_1(n)p_2(n) . \tag{9.19}$$

Let $T \in \mathscr{M}_k^{0\text{-}1c}$. Then, as it is easy to check, $\psi^d(T) = 0$. Therefore the inequality (9.19) holds. Let $T \notin \mathscr{M}_k^{0\text{-}1c}$ and Γ be a strongly nondeterministic decision tree for the table T such that $\psi(\Gamma) \le n$. Using Lemma 9.1, we obtain that $\mathrm{At}(\Gamma)$ is a test of the table T. Taking into account the properties of commutativity and nondecreasing of the function ψ, we obtain that, for any $f_i \in \mathrm{At}(\Gamma)$, $\psi(f_i) \le n$. Denote $T^* = I(\mathrm{At}(T) \setminus \mathrm{At}(\Gamma), T)$. Evidently, $T^* \in A_\psi(n)$. Using Lemma 9.7, we obtain that $M(T^*) \le 2\hat{S}(T^*) \le 2S_{\psi,A}(n) \le 2p_1(n)$. Using the boundedness from above property of the function ψ, we obtain that $M_\psi(T^*) \le nM(T^*) \le 2np_1(n)$. It is clear that $N(T^*) \le N_{\psi,A}(n) \le 2^{p_2(n)}$. From these inequalities and from Lemma 9.4 it follows that $\psi^d(T^*) \le 2np_1(n)p_2(n)$. Using Lemma 9.2, we obtain that $\psi^d(T) \le \psi^d(T^*)$. Hence the inequality (9.19) holds. Taking into account that T is an arbitrary table from A for which $\psi^s(T) \le n$, we obtain $\mathscr{K}_{\psi,A}(n) \le 2np_1(n)p_2(n)$.

(b) Let there be no a polynomial p_1 such that $S_{\psi,A}(n) \le p_1(n)$ for any $n \in \omega$. Since the function $\mathscr{K}_{\psi,A}$ is everywhere defined, from Theorem 9.4 it follows that $N_{\psi,A}$ is an everywhere defined function. Let $n \in \omega$ and $T \in A_\psi(n)$. Using Lemma 9.11, one can show that $S(T) \le N(T) - 1 \le N_{\psi,A}(n) - 1$. Therefore the value $S_{\psi,A}(n)$ is defined. We now show that $\mathscr{K}_{\psi,A}(n) \ge S_{\psi,A}(n)$. If $S_{\psi,A}(n) = 0$, then the considered inequality holds. Let $S_{\psi,A}(n) > 0$ and T^0 be a table from $A_\psi(n)$ for which $S(T^0) = S_{\psi,A}(n)$. From Lemma 9.12 it follows that there exists a table $T^* \in [T^0]$ for which $\psi^d(T^*) = S_\psi(T^0)$ and $\psi^s(T^*) \le m_\psi(T^0)$. Taking into account that $T \in A_\psi(n)$, we obtain $\psi^s(T^*) \le n$. Using the boundedness from below property of the function ψ, we obtain that $\psi^d(T^*) = S_\psi(T^0) \ge S(T^0) = S_{\psi,A}(n)$. Therefore $\mathscr{K}_{\psi,A}(n) \ge S_{\psi,A}(n)$. Thus, there is no a polynomial p_0 such that $\mathscr{K}_{\psi,A}(n) \le p_0(n)$ for any $n \in \omega$.

(c) Let there exist a polynomial p_1 such that $S_{\psi,A}(n) \le p_1(n)$ for any $n \in \omega$ and there be no a polynomial p_2 such that $N_{\psi,A}(n) \le 2^{p_2(n)}$ for any $n \in \omega$. Taking into account that $\mathscr{K}_{\psi,A}$ is an everywhere defined function and using Theorem 9.4 we obtain that $N_{\psi,A}$ is everywhere defined. Let $n \in \omega$ and T be a table from $A_\psi(n)$ for which $N(T) = N_{\psi,A}(n)$. Let $N_{\psi,A}(n) > 0$. Using Lemma 9.15, we obtain that there exists a table $T^* \in A_\psi(n)$ for which $h^d(T^*) \ge \frac{\log_k N(T) - 2}{\hat{S}(T)}$. Then $\log_k N(T) \le h^d(T^*)\hat{S}(T) + 2$. Taking into account that ψ is a bounded complexity measure, we obtain that $\hat{S}(T) \le S_{\psi,A}(n) \le p_1(n)$ and $h^d(T^*) \le \psi^d(T^*)$. Using Lemma 9.9, we obtain that $\psi^s(T^*) \le S_\psi(T^*) \le nS_{\psi,A}(n) \le np_1(n)$. Therefore $\psi^d(T^*) \le \mathscr{K}_{\psi,A}(np_1(n))$. Hence $\log_k N_{\psi,A}(n) \le \mathscr{K}_{\psi,A}(np_1(n))p_1(n) + 2$. Let us assume that there exists a polynomial p_0 such that $\mathscr{K}_{\psi,A}(n) \le p_0(n)$ for any $n \in \omega$. Then, for any $n \in \omega$, $N_{\psi,A}(n) \le 2^{(p_0(np_1(n))p_1(n)+2)\log_2 k}$, which is impossible. Thus, there is no a polynomial p_0 such that $\mathscr{K}_{\psi,A}(n) \le p_0(n)$ for any $n \in \omega$. □

9.7 Proofs of Theorems 9.7, 9.8, and 9.9

Proof of Theorem 9.7. Since A is a closed class, $\Lambda \in A$. By definition, $W_\psi(\Lambda) = 0$. Using this fact, we obtain that $\mathscr{S}_{\psi,A}$ is an everywhere defined function. It is easy to show that $\psi^a(T) \leq W_\psi(T)$ for any $T \in A$. Therefore $\mathscr{S}_{\psi,A}(n) \leq n$ for any $n \in \omega$. Evidently, $\mathscr{S}_{\psi,A}$ is a nondecreasing function. Let $T \in A$ and $W_\psi(T) \leq 0$. Using positivity property of the function ψ, we obtain $T = \Lambda$. Therefore $\mathscr{S}_{\psi,A}(0) = 0$.

(a) Let the function S_ψ be bounded from above on the class A. Then there is a constant $c \geq 0$ such that $S_\psi(T) \leq c$ for any $T \in A$. By Lemma 9.9, $\psi^s(T) \leq S_\psi(T) \leq c$ for any $T \in A$. Therefore, for any $n \in \omega$, $\mathscr{S}_{\psi,A}(n) \leq c$.

(b) Let the function S_ψ be not bounded from above on the class A. Then the set $D = \{S_\psi(T) : T \in A\}$ is infinite. We now show that $\mathscr{S}_{\psi,A}(n) \geq H_D(n)$ for any $n \in \omega$. Let $m \in D$, $T \in A$, $S_\psi(T) = m$ and $\bar{\delta}$ be a row of T such that $S_\psi(T, \bar{\delta}) = m$. Let Q be a subset of the set $\text{At}(T)$ with the minimum cardinality such that $\psi(Q) = m$ and in columns labeled with the attributes from the set Q the row $\bar{\delta}$ is different from all other rows of the table T. Denote $T^0 = I(\text{At}(T) \setminus Q, T)$ and $\bar{\sigma}$ the tuple obtained from $\bar{\delta}$ by the removal of numbers in the intersection with columns labeled with the attributes from $\text{At}(T) \setminus Q$. One can show that $\bar{\sigma}$ is a row of T^0 and $S_\psi(T^0, \bar{\sigma}) = m$. Let a mapping $\nu : E_k^{|Q|} \to E_2$ be defined in the following way: for any $\bar{\gamma} \in E_k^{|Q|}$, $\nu(\bar{\gamma}) = 1$ if and only if $\bar{\gamma} = \bar{\sigma}$. Denote $T^* = J(\nu, T^0)$. One can show that $\psi^s(T^*) = W_\psi(T^*) = m$. Therefore $\mathscr{S}_{\psi,A}(m) \geq m$. Taking into account that m is an arbitrary number from D and $\mathscr{S}_{\psi,A}$ is a nondecreasing function, we obtain that $\mathscr{S}_{\psi,A}(n) \geq H_D(n)$ for any $n \in \omega$. Let $T \in A$. By Lemma 9.9, $\psi^s(T) \leq S_\psi(T)$. It is clear that $S_\psi(T) \leq W_\psi(T)$. Therefore $\mathscr{S}_{\psi,A}(n) \leq n$ for any $n \in \omega$. □

Proof of Theorem 9.8. From Theorem 9.7 it follows that, for any $n \in \omega$, $\mathscr{S}_{h,A}(n) \leq n$. In particular, $\mathscr{S}_{h,A}(0) = 0$. Let $n \in \omega \setminus \{0\}$. Since the function S is not bounded from above on the class A, there exists a table $T \in A$ such that $S(T) \geq n$. Let $\bar{\delta}$ be a row of the table T such that $S(T, \bar{\delta}) = S(T)$. Let C be a set of attributes of T with the minimum cardinality such that in columns labeled with the attributes from C the row $\bar{\delta}$ is different from all other rows of the table T. Evidently, $|C| \geq n$. Let C^* be a subset of the set C with $|C^*| = n$. Denote $T^0 = I(\text{At}(T) \setminus C^*, T)$. One can show that $S(T^0) = n$. From Lemma 9.12 it follows that there exists a table $T^1 \in [T^0]$ for which $S(T^1) = W(T^1) = n$. Let $\bar{\sigma}$ be a row of the table T^1 such that $S(T^1, \bar{\sigma}) = S(T^1)$. Let a mapping $\nu : E_k^{W(T^1)} \to E_2$ be defined in the following way: for any $\bar{\gamma} \in E_k^{W(T^1)}$, $\nu(\bar{\gamma}) = 1$ if and only if $\bar{\gamma} = \bar{\sigma}$. Denote $T^* = J(\nu, T^1)$. One can show that $\psi^s(T^*) = W(T^*) = n$. Therefore $\mathscr{S}_{h,A}(n) \geq n$. □

Proof of Theorem 9.9. Let $D = \{n_i : i \in \omega\}$, $0 \notin D$ and, for any $i \in \omega$, $n_i < n_{i+1}$. For $i \in \omega$, we denote by T_i a decision table with one column labeled with the attribute f_i and two rows (0) and (1) labeled with decisions 0 and 1, respectively. Denote $A = [\{T_i : i \in D\}]$. We now define a complexity measure ψ. Let $\psi(f_0) = 1$, $\psi(f_i) = i$ for any $i \in \omega \setminus \{0\}$, and $\psi(f_{i_1} \cdots f_{i_m}) = \sum_{j=1}^{m} \psi(f_{i_j})$ for any nonempty word $f_{i_1} \cdots f_{i_m} \in P^*$. Evidently, ψ is a bounded complexity measure. Let $n < n_0$. Then $\mathscr{S}_{\psi,A}(n) = 0$. Let $n_i \leq n < n_{i+1}$. One can show that in this case $\mathscr{S}_{\psi,A}(n) = n_i$. □

References

1. Moshkov, M.: Conditional tests. In: Yablonskii, S.V. (ed.) Problemy Kibernetiki, vol. 40, pp. 131–170. Nauka Publishers, Moscow (1983) (in Russian)
2. Moshkov, M.: About the depth of decision trees computing Boolean functions. Fundam. Inform. **22**(3), 203–215 (1995)
3. Moshkov, M.: Time complexity of decision trees. In: Peters, J.F., Skowron, A. (eds.) Transactions on Rough Sets III. Lecture Notes in Computer Science, vol. 3400, pp. 244–459. Springer (2005)
4. Moshkov, M.: Comparative Analysis of Deterministic and Nondeterministic Decision Trees. In: Intelligent Systems Reference Library, vol. 179. Springer (2020)
5. Moshkov, M., Zielosko, B.: Combinatorial Machine Learning—A Rough Set Approach. In: Studies in Computational Intelligence, vol. 360. Springer (2011)
6. Ostonov, A., Moshkov, M.: Complexity of deterministic and strongly nondeterministic decision trees for decision tables from closed classes. IEEE Access, **12**, 164979–164988 (2024). https://doi.org/10.1109/ACCESS.2024.3487514
7. Pawlak, Z.: Information systems theoretical foundations. Inf. Syst. **6**(3), 205–218 (1981)

Part III
Recognition and Membership Problems for Formal Languages

In this part, we study formal languages over a finite alphabet Σ. Let L be a language over the alphabet Σ and $L(n)$ be the set of words of this language of the length n. For this set, we investigate the minimum depth of decision trees solving the recognition and the membership problems deterministically and nondeterministically. In the case of recognition problem, for a given word from $L(n)$, we should recognize it using queries (attributes) each of which, for some $i \in \{1, \ldots, n\}$, returns the ith letter of the word. In the case of membership problem, for a given word over the alphabet Σ of the length n, we should recognize if it belongs to $L(n)$ using the same queries. This part consists of two chapters.

In Chap. 10, for arbitrary binary subword-closed languages, we investigate the depth of decision trees solving the recognition and the membership problems for words of the length n deterministically and nondeterministically. For each problem and each type of decision trees, we list all types of behavior of the minimum depth of decision trees. We also study the joint behavior of the minimum depths of decision trees for the considered four cases and describe five complexity classes of binary subword-closed languages.

In Chap. 11, for arbitrary regular factorial languages, we investigate the depth of decision trees solving the recognition and the membership problems for words of the length n deterministically and nondeterministically. For each problem and each type of decision trees, we list all types of behavior of the smoothed minimum depth of decision trees. As corollaries of the obtained results, we study joint behavior of smoothed minimum depths of decision trees for the considered four cases and describe five complexity classes of regular factorial languages. We also investigate the class of regular factorial languages over the alphabet $\{0, 1\}$ each of which is given by one forbidden word.

Chapter 10
Decision Trees for Binary Subword-Closed Languages

10.1 Introduction

In this chapter, we study arbitrary binary languages (languages over the alphabet $E = \{0, 1\}$) that are subword-closed: if a word $w_1u_1w_2 \cdots w_mu_mw_{m+1}$ belongs to a language, then the word $u_1 \cdots u_m$ belongs to this language. Subword-closed languages have attracted the attention of researchers in the field of formal languages for many years [1–4, 6].

For the set of words $L(n)$ of the length n belonging to a binary subword-closed language L, we investigate the depth of decision trees solving the recognition and the membership problems deterministically and nondeterministically. In the case of the recognition problem, for a given word from $L(n)$, we should recognize it using queries each of which, for some $i \in \{1, \ldots, n\}$, returns the ith letter of the word. In the case of the membership problem, for a given word over the alphabet E of the length n, we should recognize if it belongs to $L(n)$ using the same queries.

We proved that, for an arbitrary binary subword-closed language, with the growth of n, the minimum depth of decision trees solving the problem of recognition deterministically is either bounded from above by a constant or grows as a logarithm, or linearly. For other types of trees and problems (decision trees solving the problem of recognition nondeterministically, and decision trees solving the membership problem deterministically and nondeterministically), with the growth of n, the minimum depth of decision trees is either bounded from above by a constant or grows linearly.

We also study joint behavior of the minimum depths of the considered four types of decision trees and describe five complexity classes of binary subword-closed languages.

This chapter is based on the paper [5]. The rest of the chapter is organized as follows. In Sect. 10.2, we consider main notions, in Sect. 10.3–main results, and in Sect. 10.4–proofs.

K. Durdymyradov et al., *Decision Trees Versus Systems of Decision Rules*, Studies in Big Data 160, https://doi.org/10.1007/978-3-031-71586-0_10

10.2 Main Notions

Let $\omega = \{0, 1, 2, \ldots\}$ be the set of nonnegative integers and $E = \{0, 1\}$. By E^*, we denote the set of all finite words over the alphabet E, including the empty word λ.

Definition 10.1 Any subset L of the set E^* is called a *binary language*. This language is called *subword-closed* if, for any word $w_1 u_1 w_2 \cdots w_m u_m w_{m+1}$ belonging to L, the word $u_1 \cdots u_m$ belongs to L, where $w_i, u_j \in E^*$, $i = 1, \ldots, m+1$, $j = 1, \ldots, m$.

For any natural n, we denote by $L(n)$ the set of words from L, which length is equal to n. We consider two problems related to the set $L(n)$.

Definition 10.2 The *problem of recognition*: for a given word from $L(n)$, we should recognize it using attributes (queries) $l_1^n, \ldots, l_n^n$, where $l_i^n, i \in \{1, \ldots, n\}$, is a function from $E^*(n)$ to E such that $l_i^n(a_1 \cdots a_n) = a_i$ for any word $a_1 \cdots a_n \in E^*(n)$.

Definition 10.3 The *problem of membership*: for a given word from $E^*(n)$, we should recognize if this word belongs to the set $L(n)$ using the same attributes.

To solve these problems, we use decision trees over $L(n)$.

Definition 10.4 A *decision tree over* $L(n)$ is a marked finite directed tree with the root, which has the following properties:

- The root and the edges leaving the root are not labeled.
- Each node, which is not the root or terminal node, is labeled with an attribute from the set $\{l_1^n, \ldots, l_n^n\}$.
- Each edge leaving a node, which is not a root, is labeled with a number from E.

Definition 10.5 A decision tree over $L(n)$ is called *deterministic* if it satisfies the following conditions:

- Exactly one edge leaves the root.
- For any node, which is not the root nor terminal node, the edges leaving this node are labeled with pairwise different numbers.

Let Γ be a decision tree over $L(n)$. *A complete path* in Γ is any sequence $\xi = v_0, e_0, \ldots, v_m, e_m, v_{m+1}$ of nodes and edges of Γ such that v_0 is the root, v_{m+1} is a terminal node, and v_i is the initial and v_{i+1} is the terminal node of the edge e_i for $i = 0, \ldots, m$. We define a subset $E(n, \xi)$ of the set $E^*(n)$ in the following way: if $m = 0$, then $E(n, \xi) = E^*(n)$. Let $m > 0$, the attribute $l_{i_j}^n$ be assigned to the node v_j and b_j be the number assigned to the edge e_j, $j = 1, \ldots, m$. Then

$$E(n, \xi) = \{a_1 \cdots a_n \in E^*(n) : a_{i_1} = b_1, \ldots, a_{i_m} = b_m\} .$$

Definition 10.6 Let $L(n) \neq \emptyset$. We say that a decision tree Γ over $L(n)$ *solves the problem of recognition for* $L(n)$ *nondeterministically* if Γ satisfies the following conditions:

- Each terminal node of Γ is labeled with a word from $L(n)$.
- For any word $w \in L(n)$, there exists a complete path ξ in the tree Γ such that $w \in E(n, \xi)$.
- For any word $w \in L(n)$ and for any complete path ξ in the tree Γ such that $w \in E(n, \xi)$, the terminal node of the path ξ is labeled with the word w.

Definition 10.7 We say that a decision tree Γ over $L(n)$ *solves the problem of recognition for* $L(n)$ *deterministically* if Γ is a deterministic decision tree, which solves the problem of recognition for $L(n)$ nondeterministically.

Definition 10.8 We say that a decision tree Γ over $L(n)$ *solves the problem of membership for* $L(n)$ *nondeterministically* if Γ satisfies the following conditions:

- Each terminal node of Γ is labeled with a number from E.
- For any word $w \in E^*(n)$, there exists a complete path ξ in the tree Γ such that $w \in E(n, \xi)$.
- For any word $w \in E^*(n)$ and for any complete path ξ in the tree Γ such that $w \in E(n, \xi)$, the terminal node of the path ξ is labeled with the number 1 if $w \in L(n)$ and with the number 0, otherwise.

Definition 10.9 We say that a decision tree Γ over $L(n)$ *solves the problem of membership for* $L(n)$ *deterministically* if Γ is a deterministic decision tree, which solves the problem of membership for $L(n)$ nondeterministically.

Let Γ be a decision tree over $L(n)$. We denote by $h(\Gamma)$ the maximum number of nodes in a complete path in Γ that are not the root nor terminal node. The value $h(\Gamma)$ is called the *depth* of the decision tree Γ.

We denote by $h_L^{ra}(n)$ ($h_L^{rd}(n)$) the minimum depth of a decision tree, which solves the problem of recognition for $L(n)$ nondeterministically (deterministically). If $L(n) = \emptyset$, then $h_L^{ra}(n) = h_L^{rd}(n) = 0$.

We denote by $h_L^{ma}(n)$ ($h_L^{md}(n)$) the minimum depth of a decision tree, which solves the problem of membership for $L(n)$ nondeterministically (deterministically). If $L(n) = \emptyset$, then $h_L^{ma}(n) = h_L^{md}(n) = 0$.

10.3 Main Results

Let L be a binary subword-closed language. For any $a \in E$ and $i \in \omega$, we denote by a^i the word $a \cdots a$ of the length i (if $i = 0$, then $a^i = \lambda$). For any $a \in E$, let $\bar{a} = 1$ if $a = 0$ and $\bar{a} = 0$ if $a = 1$.

Definition 10.10 We define the parameter $\mathrm{Hom}(L)$ of the language L, which is called the *homogeneity dimension* of the language L. If for each natural number m, there exists $a \in E$ such that the word $a^m \bar{a} a^m$ belongs to L, then $\mathrm{Hom}(L) = \infty$. Otherwise, $\mathrm{Hom}(L)$ is the maximum number $m \in \omega$ such that there exists $a \in E$ for which the word $a^m \bar{a} a^m$ belongs to L. If $L = \emptyset$, then $\mathrm{Hom}(L) = 0$.

Table 10.1 Joint behavior of functions h_L^{rd}, h_L^{ra}, h_L^{md}, and h_L^{ma} for binary subword-closed languages

	Hom(L)	Het(L)	$\|L\|$	L^C	h_L^{rd}	h_L^{ra}	h_L^{md}	h_L^{ma}
$\mathscr{L}_1$	$=\infty$			$\neq\emptyset$	$\Theta(n)$	$\Theta(n)$	$\Theta(n)$	$\Theta(n)$
$\mathscr{L}_2$	$=\infty$			$=\emptyset$	$\Theta(n)$	$\Theta(n)$	$O(1)$	$O(1)$
$\mathscr{L}_3$	$<\infty$	$=\infty$			$\Theta(\log n)$	$O(1)$	$\Theta(n)$	$\Theta(n)$
$\mathscr{L}_4$	$<\infty$	$<\infty$	$=\infty$		$O(1)$	$O(1)$	$\Theta(n)$	$\Theta(n)$
$\mathscr{L}_5$	$<\infty$	$<\infty$	$<\infty$		$O(1)$	$O(1)$	$O(1)$	$O(1)$

Definition 10.11 We now define the parameter Het(L) of the language L, which is called the *heterogeneity dimension* of the language L. If for each natural number m, there exists $a \in E$ such that the word $a^m\bar{a}^m$ belongs to L, then Het(L) $= \infty$. Otherwise, Het(L) is the maximum number $m \in \omega$ such that there exists $a \in E$ for which the word $a^m\bar{a}^m$ belongs to L. If $L = \emptyset$, then Het(L) $= 0$.

Theorem 10.1 *Let L be a binary subword-closed language.*
(a) *If* Hom(L) $= \infty$, *then* $h_L^{rd}(n) = \Theta(n)$ *and* $h_L^{ra}(n) = \Theta(n)$.
(b) *If* Hom(L) $< \infty$ *and* Het(L) $= \infty$, *then* $h_L^{rd}(n) = \Theta(\log n)$ *and* $h_L^{ra}(n) = O(1)$.
(c) *If* Hom(L) $< \infty$ *and* Het(L) $< \infty$, *then* $h_L^{rd}(n) = O(1)$ *and* $h_L^{ra}(n) = O(1)$.

Example 10.1 Let us consider the binary subword-closed language $L_0 = \{1^i0^j : i, j \in \omega\}$. One can show that Hom($L_0$) $= 0$ and Het(L_0) $= \infty$. By Theorem 10.1, $h_{L_0}^{rd}(n) = \Theta(\log n)$ and $h_{L_0}^{ra}(n) = O(1)$.

For a binary subword-closed language L, we denote by L^C its *complementary language* $E^* \setminus L$. The notation $|L| = \infty$ means that L is an infinite language, and the notation $|L| < \infty$ means that L is a finite language.

Theorem 10.2 *Let L be a binary subword-closed language.*
(a) *If* $|L| = \infty$ *and* $L^C \neq \emptyset$, *then* $h_L^{md}(n) = \Theta(n)$ *and* $h_L^{ma}(n) = \Theta(n)$.
(b) *If* $|L| < \infty$ *or* $L^C = \emptyset$, *then* $h_L^{md}(n) = O(1)$ *and* $h_L^{ma}(n) = O(1)$.

Example 10.2 One can show that, for the binary subword-closed language $L_0 = \{1^i0^j : i, j \in \omega\}$ considered in Example 10.1, $|L_0| = \infty$ and $L_0^C \neq \emptyset$. By Theorem 10.2, $h_{L_0}^{md}(n) = \Theta(n)$ and $h_{L_0}^{ma}(n) = \Theta(n)$.

To study all possible types of joint behavior of functions $h_L^{rd}(n)$, $h_L^{ra}(n)$, $h_L^{md}(n)$, and $h_L^{ma}(n)$ for binary subword-closed languages L, we consider five classes of languages $\mathscr{L}_1, \ldots, \mathscr{L}_5$ described in the columns 2–5 of Table 10.1. In particular, $\mathscr{L}_1$ consists of all binary subword-closed languages L with Hom(L) $= \infty$ and $L^C \neq \emptyset$. It is easy to show that the complexity classes $\mathscr{L}_1, \ldots, \mathscr{L}_5$ are pairwise disjoint, and each binary subword-closed language belongs to one of these classes. The behavior of functions $h_L^{rd}(n)$, $h_L^{ra}(n)$, $h_L^{md}(n)$, and $h_L^{ma}(n)$ for languages from these classes is described in the last four columns of Table 10.1. For each class, the results considered in Table 10.1 follow from Theorems 10.1 and 10.2, and the next three remarks: (i) from the condition $\mathrm{Hom}(L) = \infty$ it follows $|L| = \infty$, (ii) from the condition $\mathrm{Het}(L) = \infty$ it follows $|L| = \infty$, and (iii) from the condition $\mathrm{Hom}(L) < \infty$ it follows $L^C \neq \emptyset$.

We now show that the classes $\mathscr{L}_1, \ldots, \mathscr{L}_5$ are nonempty. To this end, we consider the following five binary subword-closed languages:

$$
\begin{aligned}
L_1 &= \{0^i 1 0^j, 0^i : i, j \in \omega\} ,\\
L_2 &= E^*,\\
L_3 &= \{0^i 1^j : i, j \in \omega\} ,\\
L_4 &= \{0^i : i \in \omega\} ,\\
L_5 &= \{0\} .
\end{aligned}
$$

It is easy to see that $L_i \in \mathscr{L}_i$ for $i = 1, \ldots, 5$.

10.4 Proofs of Theorems 10.1 and 10.2

In this section, we prove Theorems 10.1 and 10.2. First, we consider two auxiliary statements. For a word w, we denote by $|w|$ its length.

Lemma 10.1 *Let L be a binary subword-closed language for which* $\mathrm{Hom}(L) < \infty$. *Then any word w from L can be represented in the form*

$$w_1 a^i w_2 \bar{a}^j w_3 , \tag{10.1}$$

where $a \in E$, $i, j \in \omega$, and w_1, w_2, w_3 are words from E^* with length at most $2\,\mathrm{Hom}(L)$ *each.*

Proof Denote $m = \mathrm{Hom}(L)$. Then the words $0^{m+1}10^{m+1}$ and $1^{m+1}01^{m+1}$ do not belong to L. Let w be a word from L. Then, for any $a \in E$, any entry of the letter a in w has at most m $\bar{a}$s to the left of this entry (we call it l-entry of a) or at most m $\bar{a}$s to the right of this entry (we call it r-entry of a). Let $a \in E$. We say that w is (i) a-l-word if any entry of a in w is l-entry; (ii) a-r-word if any entry of a in w is r-entry; and (iii) a-b-word if w is not a-l-word and is not a-r-word. Let $c, d \in \{l, r, b\}$. We say that w is cd-word if w is 0-c-word and 1-d-word. There are nine possible pairs cd. We divide them into four groups: (a) ll and rr, (b) lr and rl, (c) lb, rb, bl, and br, and (d) bb, and consider them separately. Let

$$w = a_1 \cdots a_n .$$

We assume that w contains both 0s and 1s. Otherwise, w can be represented in the form (10.1).

(a) Let w be ll-word. Let $a_n = 0$ and a_i be the rightmost entry of 1 in w. Since w is ll-word, there are at most m 1 s to the left of a_n and at most m 0 s to the left of a_i. Denote $w_1 = a_1 \cdots a_i$. Then w_1 contains at most m 0 s and at most m 1 s, i.e., the length of w_1 is at most $2m$. Moreover, to the right of a_i there are only 0 s. Thus, $w = w_1 0^{n-i}$, where $|w_1| = i \leq 2m$, i.e., w can be represented in the form (10.1).

Let $a_n = 1$ and a_i be the rightmost entry of 0 in w. Denote $w_1 = a_1 \cdots a_i$. Then w_1 contains at most m 0 s and at most m 1s, i.e., $|w_1| \leq 2m$. Moreover, to the right of a_i there are only 1 s. Thus, $w = w_1 1^{n-i}$, i.e., w can be represented in the form (10.1).

One can prove in a similar way that any rr-word can be represented in the form (10.1).

(b) Let w be lr-word, a_i be the rightmost entry of 0 and a_j be the leftmost entry of 1. Then either $j = i + 1$ or $j < i$. Let $j = i + 1$. Then $w = 0^i 1^{n-i}$, i.e., w can be represented in the form (10.1). Let now $j < i$. Denote $w_2 = a_j \cdots a_i$. The word w has at most m 0 s to the right of a_j and at most m 1 s to the left of a_i. Therefore $|w_2| \leq 2m$ and $w = 0^{j-1} w_2 1^{n-i}$, i.e., w can be represented in the form (10.1).

One can prove in a similar way that any rl-word can be represented in the form (10.1).

(c) Let w be lb-word, a_i be the rightmost entry of 1 such that to the left of this entry, we have at most m 0 s and a_j be the next after a_i entry of 1. It is clear that to the right of a_j there are at most m 0 s, $j \geq i + 2$, and all letters $a_{i+1}, \ldots, a_{j-1}$ are equal to 0. Let a_k be the rightmost entry of 0. Then to the left of a_k there are at most m 1 s. It is clear that either $k = j - 1$ or $k > j$. Denote $w_1 = a_1 \cdots a_i$. Then $|w_1| \leq 2m$. Let $k = j - 1$. In this case, $w = w_1 0^{j-i-1} 1^{n-j+1}$, i.e., w can be represented in the form (10.1). Let $k > j$. Denote $w_2 = a_j \cdots a_k$. Then $|w_2| \leq 2m$. We have $w = w_1 0^{j-i-1} w_2 1^{n-k}$, i.e., w can be represented in the form (10.1).

One can prove in a similar way that any rb- or bl-, or br-word can be represented in the form (10.1).

(d) Let w be bb-word, a_i be the rightmost entry of 0 such that there are at most m 1 s to the left of this entry and a_j be the next after a_i entry of 0. Then there are at most m 1 s to the right of a_j, $j \geq i + 2$, and $w = a_1 \cdots a_i 1 \cdots 1 a_j \cdots a_n$. Denote $A = \{1, \ldots, i\}$, $B = \{i + 1, \ldots, j - 1\}$, and $C = \{j, \ldots, n\}$. Let a_k be the rightmost entry of 1 such that there are at most m 0 s to the left of this entry and a_l be the next after a_k entry of 1. Then there are at most m 0 s to the right of a_l, $l \geq k + 2$, and $w = a_1 \cdots a_k 0 \cdots 0 a_l \cdots a_n$.

There are four possible types of location of a_k and a_l: (i) $k \in A$ and $l \in A$, (ii) $k \in A$ and $l \in B$ (the combination $k \in A$ and $l \in C$ is impossible since all letters with indices from B are 1 s but all letters between a_k and a_l are 0s), (iii) $k \in B$ and $l \in C$ (the combination $k \in B$ and $l \in B$ is impossible since all letters with indices from B are 1 s but all letters between a_k and a_l are 0 s), and (iv) $k \in C$ and $l \in C$. We now consider cases (i)–(iv) in detail.

(i) Let $k \in A$ and $l \in A$. Then $w = a_1 \cdots a_k 0 \cdots 0 a_l \cdots a_i 1 \cdots 1 a_j \cdots a_n$. Denote $w_1 = a_1 \cdots a_k$, $w_2 = a_l \cdots a_i$, and $w_3 = a_j \cdots a_n$. The length of w_1 is at most $2m$ since from the left of a_k there are at most m 0 s and from the left of a_i there are at most m 1 s. We can prove in a similar way that $|w_2| \leq 2m$ and $|w_3| \leq 2m$. Therefore w can be represented in the form (10.1).

(ii) Let $k \in A$ and $l \in B$. Then $l = i + 1$ and

$$w = a_1 \cdots a_k 0 \cdots 0 a_i a_{i+1} 1 \cdots 1 a_j \cdots a_n ,$$

where $a_i = 0$ and $a_{i+1} = 1$. Denote $w_1 = a_1 \cdots a_k$ and $w_3 = a_j \cdots a_n$. It is easy to show that $|w_1| \leq 2m$ and $|w_3| \leq 2m$. Therefore w can be represented in the form (10.1).

(iii) Let $k \in B$ and $l \in C$. Then $k = j - 1$ and

$$w = a_1 \cdots a_i 1 \cdots 1 a_{j-1} a_j 0 \cdots 0 a_l \cdots a_n ,$$

where $a_{j-1} = 1$ and $a_j = 0$. Denote $w_1 = a_1 \cdots a_i$ and $w_3 = a_l \cdots a_n$. It is easy to show that $|w_1| \leq 2m$ and $|w_3| \leq 2m$. Therefore w can be represented in the form (10.1).

(iv) Let $k \in C$ and $l \in C$. Then $w = a_1 \cdots a_i 1 \cdots 1 a_j \cdots a_k 0 \cdots 0 a_l \cdots a_n$. Denote $w_1 = a_1 \cdots a_i$, $w_2 = a_j \cdots a_k$, and $w_3 = a_l \cdots a_n$. It is easy to show that $|w_1| \leq 2m$, $|w_2| \leq 2m$, and $|w_3| \leq 2m$. Therefore w can be represented in the form (10.1). □

Lemma 10.2 *Let L be a binary subword-closed language for which* Hom$(L) < \infty$ *and* Het$(L) < \infty$. *Then there exists natural p such that $|L(n)| \leq p$ for any natural n.*

Proof Denote $m = \max(\text{Hom}(L), \text{Het}(L))$. Then words $0^{m+1}1^{m+1}$ and $1^{m+1}0^{m+1}$ do not belong to L. Using Lemma 10.1, we obtain that each word w from L can be represented in the form $w_1 a^i w_2 \bar{a}^j w_3$, where $a \in E$, the length of w_k is at most $t = 2m$ for $k = 1, 2, 3, i, j \in \omega$, and $i \leq m$ or $j \leq m$. We now evaluate the number of such words, which length is equal to n. Let $k \in \{1, 2, 3\}$. Then the number of different words w_k is at most $2^0 + 2^1 + \cdots + 2^t < 2^{t+1}$. Let us assume that the words w_1, w_2, and w_3 are fixed and $|w_1| + |w_2| + |w_3| \leq n$. Then the number of different words $a^i \bar{a}^j$ of the length $n - |w_1| - |w_2| - |w_3|$ is at most $4(m + 1)$ since $i \leq m$ or $j \leq m$. Thus, the number of words in $L(n)$ is at most $p = 2^{3t+3}(2t + 4)$.□

Proof of Theorem 10.1. It is clear that $h_L^{ra}(n) \leq h_L^{rd}(n)$ for any natural n.

(a) Let Hom$(L) = \infty$ and n be a natural number. Then there exists $a \in E$ such that $a^n \bar{a} a^n \in L$. Therefore $a^n, a^i \bar{a} a^{n-i-1} \in L(n)$ for $i = 0, \ldots, n - 1$. Let Γ be a decision tree over $L(n)$, which solves the problem of recognition for $L(n)$ nondeterministically and has the minimum depth $h_L^{ra}(n)$, and ξ be a complete path in Γ such that $a^n \in E(n, \xi)$. Let us assume that there is $i \in \{0, \ldots, n - 1\}$ such that the attribute l_{i+1}^n is not attached to any node of ξ, which is not the root nor the terminal node. Then $a^i \bar{a} a^{n-i-1} \in E(n, \xi)$, which is impossible. Therefore $h(\Gamma) \geq n$ and $h_L^{ra}(n) \geq n$. It is easy to show that $h_L^{rd}(n) \leq n$. Thus, $h_L^{ra}(n) = h_L^{rd}(n) = n$ for any natural n.

(b) Let $\text{Hom}(L) < \infty$ and $\text{Het}(L) = \infty$. By Lemma 10.1, each word from L can be represented in the form $w_1 a^i w_2 \bar{a}^j w_3$, where $a \in E$, the length of w_k is at most $t = 2\,\text{Hom}(L)$ for $k = 1, 2, 3$, and $i, j \in \omega$. Note that either $w_2 = \lambda$ or w_2 is a word of the kind $\bar{a} \cdots a$.

Let n be a natural number such that $n \geq 10t$. We now describe the work of a decision tree over $L(n)$, which solves the problem of recognition for $L(n)$ deterministically. Let $w \in L(n)$. We represent this word as follows: $w = L_1 L_2 L_3 A R_3 R_2 R_1$, where the length of each word $L_1, L_2, L_3, R_3, R_2, R_1$ is equal to t. First, we recognize all letters in the words L_1, L_2, R_2, R_1 using $4t$ queries (attributes). We now consider four cases.

(i) Let $L_2 = R_2 = a^t$ for some $a \in E$. Then $L_3 A R_3 = a^{n-4t}$ and the word w is recognized.

(ii) Let $L_2 = a^t$ for some $a \in E$ and R_2 contain both 0 and 1. Then R_2 has an intersection with the word w_2. It is clear that w_2 has no intersection with the word A and $L_3 A = a^{n-5t}$. We recognize all letters of the word R_3. As a result, the word w will be recognized.

(iii) Let $R_2 = a^t$ for some $a \in E$ and L_2 contain both 0 and 1. Then L_2 has an intersection with the word w_2. It is clear that w_2 has no intersection with the word A and $A R_3 = a^{n-5t}$. We recognize all letters of the word L_3. As a result, the word w will be recognized.

(iv) Let $L_2 = a^t$ and $R_2 = \bar{a}^t$ for some $a \in E$. Then we need to recognize the position of the word w_2 and the word w_2 itself. Beginning with the left, we divide $L_3 A R_3$ and, probably, a prefix of R_2 into blocks of the length t. As a result, we have $k \leq n/t$ blocks. We recognize all letters in the block with the number $r = \lceil k/2 \rceil$. If all letters in this block are equal to $\bar{a}$, then we apply the same procedure to the blocks with numbers $1, \ldots, r-1$. If all letters in this block are equal to a, then we apply the same procedure to the blocks with numbers $r+1, \ldots, k$. If the considered block contains both 0 and 1, then we recognize t letters before this block and t letters after this block and, as a result, recognize both the word w_2 and its position. After each iteration, the number of blocks is at most one-half of the previous number of blocks. Let q be the whole number of iterations. Then after the iteration $q-1$ we have at least one unchecked block. Therefore $k/2^{q-1} \geq 1$ and $q \leq \log_2 k + 1$.

In case (i), to recognize the word w we make $4t$ queries. In cases (ii) and (iii), we make $5t$ queries. In case (iv), we make at most $t \log_2(n/t) + 7t$ queries. As a result, we have $h_L^{rd}(n) = O(\log n)$.

Since $\text{Het}(L) = \infty$, for any natural n, the set $L(n)$ contains for some $a \in E$ words $a^i \bar{a}^{n-i}$ for $i = 0, \ldots, n$. Then $|L(n)| \geq n+1$, and each decision tree Γ over $L(n)$ solving the problem of recognition for $L(n)$ deterministically has at least $n+1$ terminal nodes. One can show that the number of terminal nodes in Γ is at most $2^{h(\Gamma)}$. Therefore $h(\Gamma) \geq \log_2(n+1)$. Thus $h_L^{rd}(n) = \Omega(\log n)$ and $h_L^{rd}(n) = \Theta(\log n)$.

We now prove that $h_L^{ra}(n) = O(1)$. To this end, it is enough to show that there is a natural number c such that, for each natural n and for each word $w \in L(n)$, there exists a subset B_w of the set of attributes $\{l_1^n, \ldots, l_n^n\}$ such that $|B_w| \leq c$ and, for any word $u \in L(n)$ different from w, there exists an attribute $l_i^n \in B_w$ for which $l_i^n(w) \neq l_i^n(u)$. We now show that as c, we can use the number $7t$. In case (i), in the capacity of the set B_w, we can choose all attributes corresponding to $4t$ letters from the subwords L_1, L_2, R_2, and R_1. In case (ii), we can choose all attributes corresponding to $5t$ letters from the subwords L_1, L_2, R_3, R_2, and R_1. In case (iii), we can choose all attributes corresponding to $5t$ letters from the subwords L_1, L_2, L_3, R_2, and R_1. In case (iv), in the capacity of the set B_w we can choose all attributes corresponding to $4t$ letters from the subwords L_1, L_2, R_2, and R_1, and $3t$ letters from the block containing both 0 and 1 and from the blocks that are its left and right neighbors.

(c) Let $\mathrm{Hom}(L) < \infty$ and $\mathrm{Het}(L) < \infty$. By Lemma 10.2, there exists natural p such that $|L(n)| \le p$ for any natural n. Let n be a natural number. Then the set $L(n)$ contains at most p words, and there exists a subset B of the set of attributes $\{l_1^n, \dots, l_n^n\}$ such that $|B| \le p^2$ and, for any two different words $u, w \in L(n)$, there exists an attribute $l_i^n \in B$ for which $l_i^n(w) \ne l_i^n(u)$. It is easy to construct a decision tree over $L(n)$, which solves the problem of recognition for $L(n)$ deterministically by sequential computing attributes from B. The depth of this tree is at most p^2. Therefore $h_L^{rd}(n) = O(1)$ and $h_L^{ra}(n) = O(1)$. □

Proof of Theorem 10.2. It is clear that $h_L^{ma}(n) \le h_L^{md}(n)$ for any natural n.

(a) Let $|L| = \infty$, $L^C \ne \emptyset$, and w_0 be a word with the minimum length from L^C. Since $|L| = \infty$, $L(n) \ne \emptyset$ for any natural n. Let n be a natural number such that $n > |w_0|$ and Γ be a decision tree over $L(n)$ that solves the problem of membership for $L(n)$ nondeterministically and has the minimum depth. Let $w \in L(n)$ and ξ be a complete path in Γ such that $w \in E(n, \xi)$. Then the terminal node of ξ is labeled with the number 1. Let us assume that the number of nodes labeled with attributes in ξ is at most $n - |w_0|$. Then we can change at most $|w_0|$ letters in the word w such that the obtained word w' will satisfy the following conditions: w_0 is a subword of w' and $w' \in E(n, \xi)$. However, it is impossible since in this case $w' \notin L(n)$ and $w' \in E(n, \xi)$ but the terminal node of ξ is labeled with the number 1. Therefore the depth of Γ is greater than $n - |w_0|$. Thus $h_L^{ma}(n) = \Omega(n)$. It is easy to construct a decision tree over $L(n)$ that solves the problem of membership for $L(n)$ deterministically and has a depth equals to n. Therefore $h_L^{md}(n) = O(n)$. Thus, $h_L^{md}(n) = \Theta(n)$ and $h_L^{ma}(n) = \Theta(n)$.

(b) Let $|L| < \infty$. Then there exists natural m such that $L(n) = \emptyset$ for any natural $n \ge m$. Therefore, for each natural $n \ge m$, $h_L^{md}(n) = 0$ and $h_L^{ma}(n) = 0$.

Let $L^C = \emptyset$, n be a natural number, and Γ be a decision tree over $L(n)$, which consists of the root, a terminal node labeled with 1, and an edge that leaves the root and enters the terminal node. One can show that Γ solves the problem of membership for $L(n)$ deterministically and has a depth equals to 0. Therefore $h_L^{md}(n) = 0$ and $h_L^{ma}(n) = 0$. □

References

1. Atminas, A., Lozin, V.V.: Deciding atomicity of subword-closed languages. In: Diekert, V., Volkov, M.V. (eds.) Developments in Language Theory—26th International Conference, DLT 2022, Tampa, FL, USA, May 9–13, 2022, Proceedings. Lecture Notes in Computer Science, vol. 13257, pp. 69–77. Springer (2022)
2. Brzozowski, J.A., Jirásková, G., Zou, C.: Quotient complexity of closed languages. Theory Comput. Syst. **54**(2), 277–292 (2014)
3. Haines, L.H.: On free monoids partially ordered by embedding. J. Comb. Theory **6**, 94–98 (1969)
4. Hospodár, M.: Power, positive closure, and quotients on convex languages. Theor. Comput. Sci. **870**, 53–74 (2021)
5. Moshkov, M.: Decision trees for binary subword-closed languages. Entropy **25**(2), 349 (2023)
6. Okhotin, A.: On the state complexity of scattered substrings and superstrings. Fundam. Inform. **99**(3), 325–338 (2010)

Chapter 11
Decision Trees for Regular Factorial Languages

11.1 Introduction

In this chapter, we study arbitrary regular factorial languages over a finite alphabet Σ, which are well known in the field of formal languages [1–3]. A factorial language satisfies the following condition: if a word $w_1 u w_2$ belongs to the language, then the word u also belongs to it. For the set of words $L(n)$ of the length n belonging to a regular factorial language L, we investigate the depth of decision trees solving the recognition and the membership problems deterministically and nondeterministically. In the case of recognition problem, for a given word from $L(n)$, we should recognize it using queries each of which, for some $i \in \{1, \ldots, n\}$, returns the ith letter of the word. In the case of membership problem, for a given word over the alphabet Σ of the length n, we should recognize if it belongs to $L(n)$ using the same queries.

For a given problem (problem of recognition or membership problem) and type of trees (solving the problem deterministically or nondeterministically), instead of the minimum depth $h(n)$ of a decision tree of the considered type solving the problem for $L(n)$, we study the smoothed minimum depth $H(n) = \max\{h(m) : m \leq n\}$.

For an arbitrary regular factorial language, with the growth of n, the smoothed minimum depth of decision trees solving the problem of recognition deterministically is either bounded from above by a constant, or grows as a logarithm, or linearly. These results follow immediately from more general, obtained in [7] for arbitrary regular languages.

For other cases (decision trees solving the problem of recognition nondeterministically, and decision trees solving the membership problem deterministically and nondeterministically), with the growth of n, the smoothed minimum depth of decision trees is either bounded from above by a constant, or grows linearly. In the conference paper [6], a classification of arbitrary regular languages depending on the smoothed

K. Durdymyradov et al., *Decision Trees Versus Systems of Decision Rules*, Studies in Big Data 160, https://doi.org/10.1007/978-3-031-71586-0_11

minimum depth of decision trees solving the problem of recognition nondeterministically was announced without proofs. In the present paper, we consider a simpler classification for regular factorial languages with full proofs.

As corollaries of the obtained results, we study joint behavior of the smoothed minimum depths of decision trees for the considered four cases and describe five complexity classes of regular factorial languages. We also investigate the class of regular factorial languages over the alphabet $E = \{0, 1\}$ each of which is given by one forbidden word.

As we already mentioned in Chap. 1, a well-known approach to evaluate complexity of an infinite language L over a finite alphabet Σ is to study its so-called combinatorial complexity (known also as counting function) $f_L(n)$ that is the number of words of the length n in L [4, 9]. In this and in the previous chapters, we consider additional ways to evaluate the complexity of the language L based on the study how the depth of decision trees solving the recognition and the membership problems deterministically and nondeterministically depends on the length of words. This way is more complicated, but can give more detailed classification of languages. To show this, we compare languages generated by diagrams I_3 and I_4 depicted in Figs. 11.5 and 11.6. For both languages, the counting function grows linearly. For the first language, the minimum depth of decision trees solving the problem of recognition deterministically grows as a logarithm, but for the second language, the minimum depth of decision trees solving the problem of recognition deterministically grows linearly.

This chapter is based on the paper [8]. The rest of the chapter is organized as follows. In Sect. 11.2, we consider main notions, in Sect. 11.3—main results, and in Sect. 11.4—two corollaries of these results.

11.2 Main Notions

In this section, we discuss the notions related to regular factorial languages and decision trees solving problems of recognition and membership for these languages.

11.2.1 Regular Factorial Languages

Let $\omega = \{0, 1, 2, \ldots\}$ be the set of nonnegative integers and Σ be a finite alphabet with at least two letters. By Σ^*, we denote the set of all finite words over the alphabet Σ, including the empty word λ.

Definition 11.1 A word $w \in \Sigma^*$ is called a *factor* of a word $u \in \Sigma^*$ if $u = v_1 w v_2$ and $v_1, v_2 \in \Sigma^*$.

Definition 11.2 A language $L \subseteq \Sigma^*$ is called *factorial* if it contains all factors of its words.

Definition 11.3 A word $w \in \Sigma^*$ is called a *minimal forbidden word* for L if $w \notin L$ and all proper factors of w belong to L.

We denote by $MF(L)$ the language of minimal forbidden words for L. It is known [3] that a factorial language L is regular if and only if the language $MF(L)$ is regular. In particular, a factorial language L with a finite set of minimal forbidden words $MF(L)$ is regular. In this chapter, we study arbitrary nonempty regular factorial languages.

It is well known that each regular language can be represented by a deterministic finite automaton (DFA) [10]. As in [10], we will consider not only complete DFA with total transition function but also partial DFA with partial transition function. Such DFA can be represented by its transition diagram (diagram for short) [5].

Definition 11.4 A *diagram over the alphabet* Σ is a triple $I = (G, q_0, Q)$, where G is a finite directed graph, possibly with multiple edges and loops, in which each edge is labeled with a letter from Σ and edges leaving each node are labeled with pairwise different letters, q_0 is a node of G called *starting*, and Q is a nonempty set of the graph G nodes called *final*.

A *path of the diagram* I is an arbitrary sequence $\xi = v_1, d_1, \ldots, v_m, d_m, v_{m+1}$ of nodes and edges of G such that the edge d_i leaves the node v_i and enters the node v_{i+1} for $i = 1, \ldots, m$. We now define a word $w(\xi)$ from Σ^* in the following way: if $m = 0$, then $w(\xi) = \lambda$. Let $m > 0$ and let δ_j be the letter attached to the edge d_j, $j = 1, \ldots, m$. Then $w(\xi) = \delta_1 \cdots \delta_m$. We say that the path ξ *generates* the word $w(\xi)$. Note that different paths which start in the same node generate different words.

Definition 11.5 We denote by $\Xi(I)$ the set of all paths of the diagram I each of which starts in the node q_0 and finishes in a node from Q. Let

$$L_I = \{w(\xi) : \xi \in \Xi(I)\} .$$

We say that the diagram I *generates the language* L_I. It is well known that L_I is a regular language.

The diagram I is called *complete over the alphabet* Σ if exactly $|\Sigma|$ edges leave each node of G. Note that these edges are labeled with pairwise different letters from Σ. Such diagram corresponds to a complete DFA [10]. The diagram I is called *reduced* if, for each node of G, there exists a path from $\Xi(I)$, which contains this node. Such diagram corresponds to a reduced DFA [10]. It is known [10] that, for each regular language over the alphabet Σ, there exists a complete over the alphabet Σ diagram, which generates this language. Therefore, for each nonempty regular language, there exists a reduced diagram, which generates this language.

Let L be a regular factorial language and $I = (G, q_0, Q)$ be a reduced diagram that generates the language L. Since the language L is factorial, we can assume additionally that each node of the graph G is final – it will not change the language generated by I since with each word the language L contains each prefix of this word.

Definition 11.6 The diagram I will be called *f-reduced* if it is reduced and each node of the graph G is final.

Further we will assume that a considered regular factorial language L is nonempty and it is given by an f-reduced diagram, which generates this language.

We will not consider nondeterministic finite automata (NFA) to represent regular factorial languages since the study of NFA is essentially more complicated task.

11.2.2 Decision Trees for Recognition and Membership Problems

Let L be a regular factorial language over the alphabet Σ. For any natural n, denote $L(n) = L \cap \Sigma^n$, where Σ^n is the set of words over the alphabet Σ, which length is equal to n. We consider two problems related to the set $L(n)$.

Definition 11.7 The *problem of recognition*: for a given word from $L(n)$, we should recognize it using attributes (queries) $l_1^n, \dots, l_n^n$, where l_i^n, $i \in \{1, \dots, n\}$, is a function from Σ^n to Σ such that $l_i^n(a_1 \cdots a_n) = a_i$ for any word $a_1 \cdots a_n \in \Sigma^n$.

Definition 11.8 The *problem of membership*: for a given word from Σ^n, we should recognize if this word belongs to the set $L(n)$ using the same attributes.

To solve these problems, we use decision trees over $L(n)$.

Definition 11.9 A *decision tree over* $L(n)$ is a marked finite directed tree with root, which has the following properties:

- The root and the edges leaving the root are not labeled.
- Each node, which is not the root nor terminal node, is labeled with an attribute from the set $\{l_1^n, \dots, l_n^n\}$.
- Each edge leaving a node, which is not a root, is labeled with a letter from the alphabet Σ.

Definition 11.10 A decision tree over $L(n)$ is called *deterministic* if it satisfies the following conditions:

- Exactly one edge leaves the root.
- For any node, which is not the root nor terminal node, the edges leaving this node are labeled with pairwise different letters.

Let Γ be a decision tree over $L(n)$. A *complete path* in Γ is any sequence $\xi = v_0, e_0, \dots, v_m, e_m, v_{m+1}$ of nodes and edges of Γ such that v_0 is the root, v_{m+1} is a terminal node, and v_i is the initial and v_{i+1} is the terminal node of the edge e_i

for $i = 0, \ldots, m$. We define a subset $\Sigma(n, \xi)$ of the set Σ^n in the following way: if $m = 0$, then $\Sigma(n, \xi) = \Sigma^n$. Let $m > 0$, the attribute $l^n_{i_j}$ be attached to the node v_j, and b_j be the letter attached to the edge e_j, $j = 1, \ldots, m$. Then

$$\Sigma(n, \xi) = \{a_1 \cdots a_n \in \Sigma^n : a_{i_1} = b_1, \ldots, a_{i_m} = b_m\} .$$

Definition 11.11 Let $L(n) \neq \emptyset$. We say that a decision tree Γ over $L(n)$ *solves the problem of recognition for* $L(n)$ *nondeterministically* if Γ satisfies the following conditions:

- Each terminal node of Γ is labeled with a word from $L(n)$.
- For any word $w \in L(n)$, there exists a complete path ξ in the tree Γ such that $w \in \Sigma(n, \xi)$.
- For any word $w \in L(n)$ and for any complete path ξ in the tree Γ such that $w \in \Sigma(n, \xi)$, the terminal node of the path ξ is labeled with the word w.

Definition 11.12 We say that a decision tree Γ over $L(n)$ *solves the problem of recognition for* $L(n)$ *deterministically* if Γ is a deterministic decision tree, which solves the problem of recognition for $L(n)$ nondeterministically.

Examples of decision trees illustrating the considered notions are presented in Fig. 11.1.

Definition 11.13 We say that a decision tree Γ over $L(n)$ *solves the problem of membership for* $L(n)$ *nondeterministically* if Γ satisfies the following conditions:

- Each terminal node of Γ is labeled with a number from the set $\{0, 1\}$.
- For any word $w \in \Sigma^n$, there exists a complete path ξ in the tree Γ such that $w \in \Sigma(n, \xi)$.
- For any word $w \in \Sigma^n$ and for any complete path ξ in the tree Γ such that $w \in \Sigma(n, \xi)$, the terminal node of the path ξ is labeled with the number 1 if $w \in L(n)$ and with the number 0, otherwise.

Definition 11.14 We say that a decision tree Γ over $L(n)$ *solves the problem of membership for* $L(n)$ *deterministically* if Γ is a deterministic decision tree which solves the problem of membership for $L(n)$ nondeterministically.

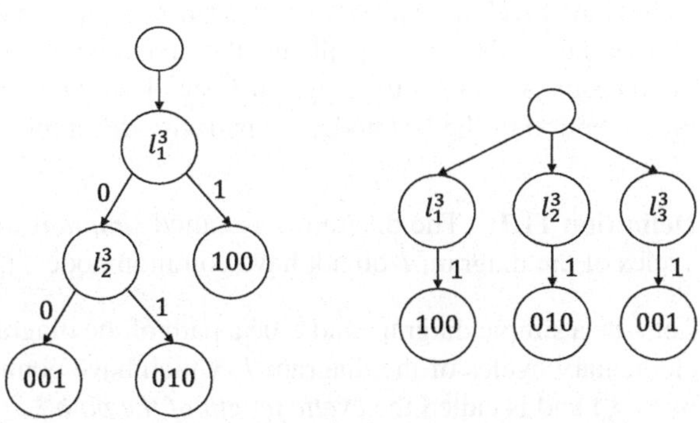

Fig. 11.1 Decision trees that solve the problem of recognition for the set of words {100, 010, 001} deterministically and nondeterministically

Let Γ be a decision tree over $L(n)$. We denote by $h(\Gamma)$ the maximum number of nodes in a complete path in Γ that are not the root nor terminal node. The value $h(\Gamma)$ is called the *depth* of the decision tree Γ.

We denote by $h_L^{ra}(n)$ ($h_L^{rd}(n)$) the minimum depth of a decision tree over $L(n)$, which solves the problem of recognition for $L(n)$ nondeterministically (deterministically). If $L(n) = \emptyset$, then $h_L^{ra}(n) = h_L^{rd}(n) = 0$.

We denote by $h_L^{ma}(n)$ ($h_L^{md}(n)$) the minimum depth of a decision tree over $L(n)$, which solves the problem of membership for $L(n)$ nondeterministically (deterministically). If $L(n) = \emptyset$, then $h_L^{ma}(n) = h_L^{md}(n) = 0$.

11.3 Bounds on Decision Tree Depth

Let L be a nonempty regular factorial language. In this section, we consider the behavior of four functions H_L^{ra}, H_L^{rd}, H_L^{ma}, and H_L^{md} defined on the set $\omega \setminus \{0\}$ and with values from ω. For any natural n,

$$H_L^{ra}(n) = \max\{h_L^{ra}(m) : 1 \leq m \leq n\} ,$$
$$H_L^{rd}(n) = \max\{h_L^{rd}(m) : 1 \leq m \leq n\} ,$$
$$H_L^{ma}(n) = \max\{h_L^{ma}(m) : 1 \leq m \leq n\} ,$$
$$H_L^{md}(n) = \max\{h_L^{md}(m) : 1 \leq m \leq n\} .$$

For any pair $bc \in \{ra, rd, ma, md\}$, the function $H_L^{bc}(n)$ is a smoothed analog of the function $h_L^{bc}(n)$.

11.3.1 Decision Trees Solving Recognition Problem Deterministically

Let $I = (G, q_0, Q)$ be an f-reduced diagram over the alphabet Σ.

Definition 11.15 A path of the diagram I is called a *cycle* of the diagram I if there is at least one edge in this path, and the first node of this path is equal to the last node of this path. A cycle of the diagram I is called *elementary* if nodes of this cycle, with the exception of the last node, are pairwise different.

Definition 11.16 The diagram I is called *simple* if every two different elementary cycles of the diagram I do not have common nodes.

Let I be a simple diagram and ξ be a path of the diagram I. The number of different elementary cycles of the diagram I, which have common nodes with ξ, is denoted by $cl(\xi)$ and is called the *cyclic length of the path* ξ.

Definition 11.17 The value

$$cl(I) = \max\{cl(\xi) : \xi \in \Xi(I)\}$$

is called the *cyclic length of the diagram I*.

Let I be a simple diagram, C be an elementary cycle of the diagram I, and v be a node of the cycle C. Beginning with the node v, the cycle C generates an infinite periodic word over the alphabet Σ. This word will be denoted by $W(I, C, v)$. We denote by $r(I, C, v)$ the minimum period of the word $W(I, C, v)$.

Definition 11.18 The diagram I is called *dependent* if there exist two different elementary cycles C_1 and C_2 of the diagram I, nodes v_1 and v_2 of the cycles C_1 and C_2, respectively, and a path π of the diagram I from v_1 to v_2, which satisfy the following conditions: $W(I, C_1, v_1) = W(I, C_2, v_2)$ and the length of the path π is a number divisible by $r(I, C_1, v_1)$. If the diagram I is not dependent, then it is called *independent*.

Next theorem follows immediately from Theorem 2.1 [7], which is a similar statement that holds for all regular languages.

Theorem 11.1 *Let L be a nonempty regular factorial language over the alphabet* Σ *and I be an f-reduced diagram, which generates the language L. Then the following statements hold:*
(a) *If I is an independent simple diagram and* $cl(I) \leq 1$, *then* $H_L^{rd}(n) = O(1)$.
(b) *If I is an independent simple diagram and* $cl(I) \geq 2$, *then* $H_L^{rd}(n) = \Theta(\log n)$.
(c) *If I is not an independent simple diagram, then* $H_L^{rd}(n) = \Theta(n)$.

11.3.2 Decision Trees Solving Recognition Problem Nondeterministically

Let L be a nonempty regular factorial language over the alphabet Σ.

Definition 11.19 For any natural n, we define a parameter $T_L(n)$ of the language L. If $L(n) = \emptyset$, then $T_L(n) = 0$. Let $L(n) \neq \emptyset$, $w = a_1 \cdots a_n \in L(n)$, and $J \subseteq \{1, \ldots, n\}$. Denote $L(w, J) = \{b_1 \cdots b_n \in L(n) : b_j = a_j, j \in J\}$ (if $J = \emptyset$, then $L(w, J) = L(n)$) and $M_L(n, w) = \min\{|J| : J \subseteq \{1, \ldots, n\}, |L(w, J)| = 1\}$. Then

$$T_L(n) = \max\{M_L(n, w) : w \in L(n)\}.$$

Note that, for any word $w \in L(n)$, $M_L(n, w)$ is the minimum number of letters of the word w, which allow us to distinguish it from all other words belonging to $L(n)$.

Lemma 11.1 *Let L be a nonempty regular factorial language over the alphabet* Σ. *Then* $h_L^{ra}(n) = T_L(n)$ *for any natural n.*

Proof First, we prove that $h_L^{ra}(n) \geq T_L(n)$. Let Γ be a decision tree over $L(n)$, which solves the problem of recognition for $L(n)$ nondeterministically and for which $h(\Gamma) = h_L^{ra}(n)$. Let w be a word from $L(n)$ for which $T_L(n) = M_L(n, w)$. Then the decision tree Γ contains a complete path ξ such that $w \in \Sigma(n, \xi)$ and the terminal node of the path ξ is labeled with the word w. It is clear that $\Sigma(n, \xi) \cap L(n) = \{w\}$. Let ξ contain m nodes that are not the root nor terminal node and $l_{i_1}^n, \ldots, l_{i_m}^n$ be attributes attached to these nodes. Denote $J = \{i_1, \ldots, i_m\}$. Then $L(w, J) = \{w\}$. Therefore $m \geq M_L(n, w) = T_L(n)$. It is clear that $h(\Gamma) \geq m$. Thus, $h_L^{ra}(n) = h(\Gamma) \geq m \geq M_L(n, w) = T_L(n)$.

We now prove that $h_L^{ra}(n) \leq T_L(n)$. One can show that, for each $w \in L(n)$, we can construct a complete path ξ_w, which satisfies the following conditions: the number of nodes in ξ_w that are not the root nor terminal node is equal to $M_L(n, w)$, $\Sigma(n, \xi_w) \cap L(n) = \{w\}$, and the terminal node of ξ_w is labeled with the word w. If we merge roots of all paths ξ_w, $w \in L(n)$, we obtain a decision tree, which solves the problem of recognition for $L(n)$ nondeterministically and which depth is equal to $T_L(n)$. Thus, $h_L^{ra}(n) \leq T_L(n)$ and $h_L^{ra}(n) = T_L(n)$. □

Theorem 11.2 *Let L be a nonempty regular factorial language over the alphabet Σ and $I = (G, q_0, Q)$ be an f-reduced diagram, which generates the language L. Then the following statements hold:*

(a) *If I is an independent simple diagram, then $H_L^{ra}(n) = O(1)$.*

(b) *If I is not an independent simple diagram, then $H_L^{ra}(n) = \Theta(n)$.*

Proof (a) Let I be an independent simple diagram and $cl(I) \leq 1$. By Theorem 11.1, $H_L^{rd}(n) = O(1)$. It is clear that $H_L^{ra}(n) \leq H_L^{rd}(n)$. Therefore $H_L^{ra}(n) = O(1)$.

Let I be an independent simple diagram and $cl(I) \geq 2$. Let n be a natural number. If $L(n) = \emptyset$, then $T_L(n) = 0$. Let $L(n) \neq \emptyset$. Denote by d the number of nodes in the graph G. In the proof of Lemma 4.5 [7], it was proved that $M_L(n, w) \leq d(4d + 1)$ for any word $w \in L(n)$. Therefore $T_L(n) \leq d(4d + 1)$. Thus, by Lemma 11.1, $h_L^{ra}(n) \leq d(4d + 1)$ for any natural n and $H_L^{ra}(n) = O(1)$.

(b) Let I be not simple diagram and C_1, C_2 be different elementary cycles of the diagram I, which have a common node v. Since I is an f-reduced diagram, it contains a path ξ from the node q_0 to the node v, and v is a final node. Let the length of the path ξ be equal to a, the length of the cycle C_1 be equal to b, and the length of the cycle C_2 be equal to c. Let α be the word generated by the path ξ, β be the word generated by a path from v to v obtained by the passage c times along the cycle C_1, and γ be the word generated by a path from v to v obtained by the passage b times along the cycle C_2. The words β and γ are different and they have the same length bc.

Consider the sequence of numbers $n_i = a + ibc$, $i = 1, 2, \ldots$. Let $i \in \omega \setminus \{0\}$. The set $L(n_i)$ contains the word $\alpha\gamma^i$ and the words $\alpha\gamma^j\beta\gamma^{i-j-1}$ for $j = 0, \ldots, i-1$. It is easy to show that $M_L(n_i, \alpha\gamma^i) \geq i$: to distinguish the word $\alpha\gamma^i$ from the words $\alpha\gamma^j\beta\gamma^{i-j-1}$, $j = 0, \ldots, i-1$, we need to use at least one letter from each of i words γ appearing in $\alpha\gamma^i$. Therefore $T_L(n_i) \geq i$ and, by Lemma 11.1, $h_L^{ra}(n_i) \geq i = (n_i - a)/(bc)$. Let $n \geq n_1$ and let i be the maximum natural number such that $n \geq n_i$.

Evidently, $n - n_i \le bc$. Hence $H_L^{ra}(n) \ge h_L^{ra}(n_i) \ge (n - bc - a)/(bc)$. Therefore $H_L^{ra}(n) \ge n/(2bc)$ for large enough n. The inequality $H_L^{ra}(n) \le n$ is obvious. Thus, $H_L^{ra}(n) = \Theta(n)$.

Let I be a dependent simple diagram. Then there exist two different elementary cycles C_1 and C_2 of the diagram I, nodes v_1 and v_2 of the cycles C_1 and C_2, respectively, and a path π of the diagram I from v_1 to v_2, which satisfy the following conditions: $W(I, C_1, v_1) = W(I, C_2, v_2)$ and the length of the path π is a number divisible by $r(I, C_1, v_1)$. Let us remind that, for $i = 1, 2$, $W(I, C_i, v_i)$ is the infinite periodic word over the alphabet Σ generated by the cycle C_i beginning with the node v_i, and $r(I, C_1, v_1)$ is the minimum period of the word $W(I, C_1, v_1)$. Since I is an f-reduced diagram, it contains a path ξ from the node q_0 to the node v_1, and all nodes of the graph G are final. Let the path ξ generate the word α of the length a. Denote $r = r(I, C_1, v_1)$. Let the length of the cycle C_1 be equal to br, the length of the path π be equal to cr, and the path π generate the word β. Denote by γ the prefix of the length r of the word $W(I, C_1, v_1)$. We now define two words of the length rbc: $u = \gamma^{bc}$ and $w = \beta\gamma^{c(b-1)}$. It is clear that $u \ne w$.

Consider the sequence of numbers $n_i = a + irbc$, $i = 1, 2, \ldots$. Let $i \in \omega \setminus \{0\}$. The set $L(n_i)$ contains the word αu^i and the words $\alpha u^j w u^{i-j-1}$ for $j = 0, \ldots, i-1$. It is easy to show that $M_L(n, \alpha u^i) \ge i$: to distinguish the word αu^i from the words $\alpha u^j w u^{i-j-1}$, $j = 0, \ldots, i-1$, we need to use at least one letter from each of i words u appearing in αu^i. Therefore $T_L(n_i) \ge i$ and, by Lemma 11.1, $h_L^{ra}(n_i) \ge i = (n_i - a)/(rbc)$. Let $n \ge n_1$ and let i be the maximum natural number such that $n \ge n_i$. Evidently, $n - n_i \le rbc$. Hence $H_L^{ra}(n) \ge h_L^{ra}(n_i) \ge (n - rbc - a)/(rbc)$. Therefore $H_L^{ra}(n) \ge n/(2rbc)$ for large enough n. The inequality $H_L^{ra}(n) \le n$ is obvious. Thus, $H_L^{ra}(n) = \Theta(n)$. □

Note that in general case (when we consider not only factorial languages) the classification of reduced diagrams depending on the minimum depth of decision trees solving the problem of recognition nondeterministically is more complicated [6]. In particular, there exists a dependent simple reduced diagram I_0 (see Fig. 11.2) with the starting node labeled with the symbol $+$ and the unique final node labeled with the symbol $*$ that generates the regular language $L_0 = \{0^i 1 0^j : i, j \in \omega\}$ over the alphabet $\{0, 1\}$, which is not factorial and for which $H_{L_0}^{ra}(n) = O(1)$.

Fig. 11.2 Diagram I_0

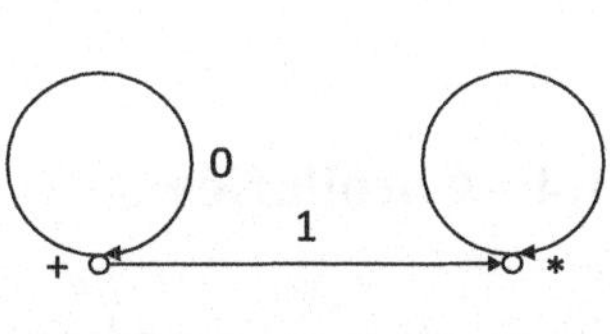

11.3.3 Decision Trees Solving Membership Problem

For a regular factorial language L, the notation $|L| = \infty$ means that L is an infinite language, and the notation $|L| < \infty$ means that L is a finite language.

Theorem 11.3 *Let L be a regular factorial language over the alphabet Σ.*
(a) *If $|L| = \infty$ and $L \neq \Sigma^*$, then $H_L^{md}(n) = \Theta(n)$ and $H_L^{ma}(n) = \Theta(n)$.*
(b) *If $|L| < \infty$ or $L = \Sigma^*$, then $H_L^{md}(n) = O(1)$ and $H_L^{ma}(n) = O(1)$.*

Proof It is clear that $h_L^{ma}(n) \leq h_L^{md}(n)$ for any natural n.

(a) Let $|L| = \infty$, $L \neq \Sigma^*$, and w_0 be a word with the minimum length from $\Sigma^* \setminus L$. Denote by t the length of w_0. Since $|L| = \infty$, $L(n) \neq \emptyset$ for any natural n. Let n be a natural number such that $n > t$ and Γ be a decision tree over $L(n)$ that solves the problem of membership for $L(n)$ nondeterministically and has the minimum depth. Let $w \in L(n)$ and ξ be a complete path in Γ such that $w \in \Sigma(n, \xi)$. Then the terminal node of ξ is labeled with the number 1. Beginning with the first letter, we divide the word w into $\lfloor n/t \rfloor$ blocks with t letters in each and the suffix of the length $n - t\lfloor n/t \rfloor$. Let us assume that the number of nodes labeled with attributes in ξ is less than $\lfloor n/t \rfloor$. Then there is a block such that queries (attributes) attached to nodes of ξ does not ask about letters from the block. We replace this block in the word w with the word w_0 and denote by w' the obtained word. It is clear that $w' \notin L$ and $w' \in \Sigma(n, \xi)$, but this is impossible since the terminal node of the path ξ is labeled with the number 1. Therefore the depth of Γ is greater than or equal to $\lfloor n/t \rfloor$. Thus, $h_L^{ma}(n) \geq \lfloor n/t \rfloor$. It is easy to construct a decision tree over $L(n)$ that solves the problem of membership for $L(n)$ deterministically and has the depth equals to n. Therefore $h_L^{md}(n) \leq n$. Thus, $H_L^{md}(n) = \Theta(n)$ and $H_L^{ma}(n) = \Theta(n)$.

(b) Let $|L| < \infty$. Then there exists natural m such that $L(n) = \emptyset$ for any natural $n \geq m$. Therefore, for each natural $n \geq m$, $h_L^{md}(n) = 0$ and $h_L^{ma}(n) = 0$. Thus, $H_L^{md}(n) = O(1)$ and $H_L^{ma}(n) = O(1)$.

Let $L = \Sigma^*$, n be a natural number, and Γ be a decision tree over $L(n)$, which consists of the root, a terminal node labeled with 1, and an edge that leaves the root and enters the terminal node. One can show that Γ solves the problem of membership for $L(n)$ deterministically and has the depth equals to 0. Therefore $h_L^{md}(n) = 0$ and $h_L^{ma}(n) = 0$. Thus, $H_L^{md}(n) = O(1)$ and $H_L^{ma}(n) = O(1)$. □

11.4 Corollaries

In this section, we consider two corollaries of Theorems 11.1, 11.2, and 11.3.

11.4.1 Joint Behavior of Functions H_L^{ra}, H_L^{rd}, H_L^{ma}, and H_L^{md}

In this section, we assume that each regular factorial language over the alphabet Σ is given by an f-reduced diagram I, which generates the considered language denoted by L_I. To study all possible types of joint behavior of functions $H_{L_I}^{rd}$, $H_{L_I}^{ra}$, $H_{L_I}^{md}$, and $H_{L_I}^{ma}$, we consider five classes of regular factorial languages $\mathscr{F}_1, \ldots, \mathscr{F}_5$ described in the columns 2-4 of Table 11.1. In particular, $\mathscr{F}_1$ consists of all regular factorial languages L_I for which the diagram I is an independent simple diagram and $cl(I) = 0$. It is easy to show that the complexity classes $\mathscr{F}_1, \ldots, \mathscr{F}_5$ are pairwise disjoint, and each regular factorial language L_I belongs to one of these classes. The behavior of functions $H_{L_I}^{rd}$, $H_{L_I}^{ra}$, $H_{L_I}^{md}$, and $H_{L_I}^{ma}$ for languages from these classes is described in the last four columns of Table 11.1. For each class, the results considered in Table 11.1 for the functions $H_{L_I}^{rd}$ and $H_{L_I}^{ra}$ follow directly from Theorems 11.1 and 11.2.

We now consider the behavior of the functions $H_{L_I}^{md}$ and $H_{L_I}^{ma}$ for each of the classes $\mathscr{F}_1, \ldots, \mathscr{F}_5$. Let $I = (G, q_0, Q)$ be an f-reduced diagram over the alphabet Σ, which generates a regular factorial language.

Let $L_I \in \mathscr{F}_1$. Since $cl(I) = 0$, G is a directed acyclic graph, and the language L_I is finite. Using Theorem 11.3, we obtain $H_{L_I}^{md}(n) = O(1)$ and $H_{L_I}^{ma}(n) = O(1)$.

Let $L_I \in \mathscr{F}_2$. Since $cl(I) = 1$, G is a graph containing a cycle, and the language L_I is infinite. By Lemma 4.2 [7], $|L_I(n)| = O(1)$. Therefore $L_I \neq \Sigma^*$. Using Theorem 11.3, we obtain $H_{L_I}^{md}(n) = \Theta(n)$ and $H_{L_I}^{ma}(n) = \Theta(n)$.

Let $L_I \in \mathscr{F}_3$. Since $cl(I) \geq 2$, G is a graph containing a cycle, and the language L_I is infinite. By Lemma 4.2 [7], $|L_I(n)| = O(n^{cl(I)})$. Therefore $L_I \neq \Sigma^*$. Using Theorem 11.3, we obtain $H_{L_I}^{md}(n) = \Theta(n)$ and $H_{L_I}^{ma}(n) = \Theta(n)$.

Let $L_I \in \mathscr{F}_4$. Since I is not an independent simple diagram, G is a graph containing a cycle, and the language L_I is infinite. We know that $L_I \neq \Sigma^*$. Using Theorem 11.3 we obtain $H_{L_I}^{md}(n) = \Theta(n)$ and $H_{L_I}^{ma}(n) = \Theta(n)$.

Table 11.1 Complexity classes $\mathscr{F}_1, \ldots, \mathscr{F}_5$

	I is independent simple diagram	$cl(I)$	L_I	$H_{L_I}^{rd}$	$H_{L_I}^{ra}$	$H_{L_I}^{md}$	$H_{L_I}^{ma}$
$\mathscr{F}_1$	Yes	$=0$		$O(1)$	$O(1)$	$O(1)$	$O(1)$
$\mathscr{F}_2$	Yes	$=1$		$O(1)$	$O(1)$	$\Theta(n)$	$\Theta(n)$
$\mathscr{F}_3$	Yes	≥ 2		$\Theta(\log n)$	$O(1)$	$\Theta(n)$	$\Theta(n)$
$\mathscr{F}_4$	No		$\neq \Sigma^*$	$\Theta(n)$	$\Theta(n)$	$\Theta(n)$	$\Theta(n)$
$\mathscr{F}_5$	No		$= \Sigma^*$	$\Theta(n)$	$\Theta(n)$	$O(1)$	$O(1)$

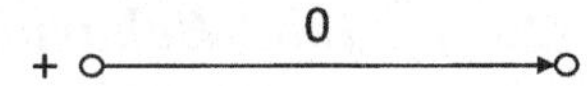

Fig. 11.3 Diagram I_1

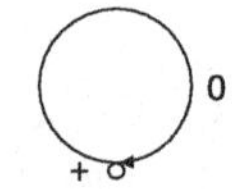

Fig. 11.4 Diagram I_2

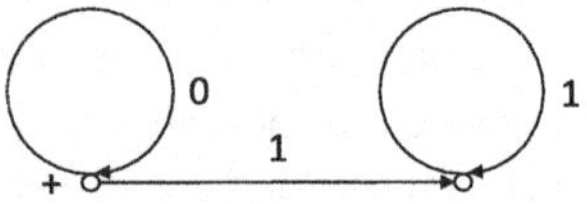

Fig. 11.5 Diagram I_3

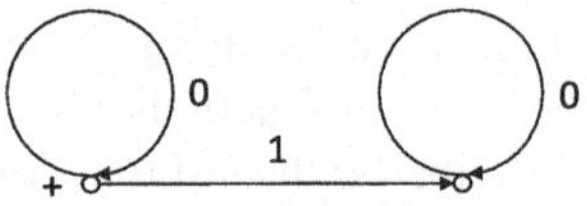

Fig. 11.6 Diagram I_4

Let $L_I \in \mathscr{F}_5$. Then $L_I = \Sigma^*$. Using Theorem 11.3, we obtain $H^{md}_{L_I}(n) = O(1)$ and $H^{ma}_{L_I}(n) = O(1)$.

We now show that the classes $\mathscr{F}_1, \ldots, \mathscr{F}_5$ are nonempty. For simplicity, we assume that $\Sigma = E$, where $E = \{0, 1\}$. It is easy to generalize the considered examples to the case of an arbitrary finite alphabet Σ with at least two letters. In the examples of diagrams, the starting node is labeled with the symbol $+$, and all nodes are final.

Denote by I_1 the diagram over the alphabet E depicted in Fig. 11.3. One can show that I_1 is an independent simple f-reduced diagram and $cl(I_1) = 0$. This diagram generates the language $L_{I_1} = \{\lambda, 0\}$, which is factorial. Therefore $L_{I_1} \in \mathscr{F}_1$.

Denote by I_2 the diagram over the alphabet E depicted in Fig. 11.4. One can show that I_2 is an independent simple f-reduced diagram and $cl(I_2) = 1$. This diagram generates the language $L_{I_2} = \{0^i : i \in \omega\}$, which is factorial. Therefore $L_{I_2} \in \mathscr{F}_2$.

Denote by I_3 the diagram over the alphabet E depicted in Fig. 11.5. One can show that I_3 is an independent simple f-reduced diagram and $cl(I_1) = 2$. This diagram generates the language $L_{I_3} = \{0^i 1^j : i, j \in \omega\}$, which is factorial. Therefore $L_{I_3} \in \mathscr{F}_3$.

Denote by I_4 the diagram over the alphabet E depicted in Fig. 11.6. One can show that I_4 is a dependent simple f-reduced diagram generating the language $L_{I_4} = \{0^i 1^j 0^k : i, k \in \omega, j \in \{0, 1\}\}$, which is factorial. It is clear that $L_{I_4} \neq E^*$. Therefore $L_{I_4} \in \mathscr{F}_4$.

Denote by I_5 the diagram over the alphabet E depicted in Fig. 11.7. One can show that I_5 is an f-reduced diagram that is not simple. This diagram generates the language $L_{I_5} = E^*$, which is factorial. It is clear that $L_{I_5} = E^*$. Therefore $L_{I_5} \in \mathscr{F}_5$.

A regular factorial language L can have different f-reduced diagrams, which generate it. However, for each of such diagrams I, the language $L_I = L$ will belong to the same complexity class. Let us assume the contrary: there exist a regular factorial language L and two f-reduced diagrams I_1 and I_2, which generate it and for which languages L_{I_1} and L_{I_2} belong to different complexity classes. Then, for some pair

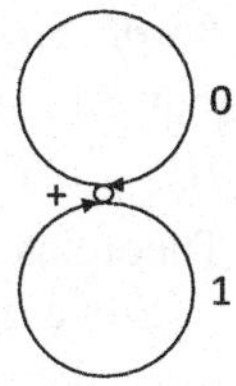

Fig. 11.7 Diagram I_5

$bc \in \{rd, ra, md, ma\}$, the functions $H^{bc}_{L_{I_1}}$ and $H^{bc}_{L_{I_2}}$ have different behavior, but this is impossible since $H^{bc}_{L_{I_1}}(n) = H^{bc}_{L_{I_2}}(n)$ for any natural n.

11.4.2 Languages Over Alphabet $\{0, 1\}$ *Given by One Forbidden Word*

Let $E = \{0, 1\}$, $\alpha \in E^*$, and $\alpha \neq \lambda$. We denote by $L(\alpha)$ the language over the alphabet E, which consists of all words from E^* that do not contain α as a factor. This is a regular factorial language with $MF(L(\alpha)) = \{\alpha\}$. The following theorem indicates for each nonempty word $\alpha \in E^*$ the complexity class $\mathscr{F}_i$ to which the language $L(\alpha)$ belongs.

Theorem 11.4 *Let* $\alpha \in E^*$ *and* $\alpha \neq \lambda$.
(a) *If* $\alpha \in \{0, 1\}$, *then* $L(\alpha) \in \mathscr{F}_2$.
(b) *If* $\alpha \in \{01, 10\}$, *then* $L(\alpha) \in \mathscr{F}_3$.
(c) *If* $\alpha \notin \{0, 1, 01, 10\}$, *then* $L(\alpha) \in \mathscr{F}_4$.

We now describe an f-reduced diagram $I(\alpha)$ that generates the language $L(\alpha)$ for a nonempty word $\alpha \in E^*$. Let $\alpha = a_1 \cdots a_n$, $\alpha_0 = \lambda$, and $\alpha_i = a_1 \cdots a_i$ for $i = 1, \ldots, n-1$. The set $P(\alpha) = \{\alpha_0, \alpha_1, \ldots, \alpha_{n-1}\}$ is the set of all proper prefixes of the word α. Then $I(\alpha) = (G, q_0, Q)$, where the set of nodes of the graph G is equal to $P(\alpha)$, $q_0 = \alpha_0$, and $Q = P(\alpha)$. For $i = 0, \ldots, n-2$, an edge leaves the node α_i and enters the node α_{i+1}. This edge is labeled with the letter a_{i+1}. For $i = 0, \ldots, n-1$, an edge leaves the node α_i and enters the node $\alpha_j \in P(\alpha)$ such that α_j is the longest suffix of the word $\alpha_i \bar{a}_{i+1}$, where $\bar{a}_{i+1} = 0$ if $a_{i+1} = 1$ and $\bar{a}_{i+1} = 1$ if $a_{i+1} = 0$. This edge is labeled with the letter $\bar{a}_{i+1}$. It is easy to show that $I(\alpha)$ is an f-reduced diagram over the alphabet E. From Theorem 10 [3] it follows that the diagram $I(\alpha)$ generates the language $L(\alpha)$.

Let $\alpha \in E^* \setminus \{\lambda\}$ and $\alpha = a_1 \cdots a_n$. We denote by $\bar{\alpha}$ the word $\bar{a}_1 \cdots \bar{a}_n$. It is easy to prove the following statement.

Lemma 11.2 *Let* $\alpha \in E^*$ *and* $\alpha \neq \lambda$. *Then* $H^{bc}_{L(\bar{\alpha})}(n) = H^{bc}_{L(\alpha)}(n)$ *for any pair* $bc \in \{rd, ra, md, ma\}$ *and any natural* n.

Lemma 11.3 *Let* $\alpha \in E^* \setminus \{\lambda\}$, $\beta \in E^*$, *and* $L(\alpha) \in \mathscr{F}_4$. *Then* $L(\alpha\beta) \in \mathscr{F}_4$.

Proof Since $L(\alpha) \in \mathscr{F}_4$, $H^{rd}_{L(\alpha)}(n) = \Theta(n)$ and $H^{ra}_{L(\alpha)}(n) = \Theta(n)$. One can show that $L(\alpha) \subseteq L(\alpha\beta)$. Using this fact it is not difficult to prove that $H^{rd}_{L(\alpha)}(n) \leq H^{rd}_{L(\alpha\beta)}(n)$ and $H^{ra}_{L(\alpha)}(n) \leq H^{ra}_{L(\alpha\beta)}(n)$ for any natural n. From here and from Theorems 11.1 and 11.2 it follows that $H^{rd}_{L(\alpha\beta)}(n) = \Theta(n)$ and $H^{ra}_{L(\alpha\beta)}(n) = \Theta(n)$.

Since $\alpha\beta \notin L(\alpha\beta)$, $L(\alpha\beta) \neq E^*$. The diagram $I(\alpha\beta)$ contains at least one circle formed by the edge that leaves and enters the node λ and is labeled with the letter $\bar{a}_1$, where a_1 is the first letter of the word α. Therefore the language $L(\alpha\beta)$ is infinite. By Theorem 11.3, $H^{md}_{L(\alpha\beta)}(n) = \Theta(n)$ and $H^{ma}_{L(\alpha\beta)}(n) = \Theta(n)$. Thus, $L(\alpha\beta) \in \mathscr{F}_4$. □

Proof (Proof of Theorem 11.4.) In each figure depicting a diagram $I(\alpha)$, $\alpha \in E^* \setminus \{\lambda\}$, we label each node with a corresponding prefix of the word α.

(a) The diagram $I(0)$ is depicted in Fig. 11.8. This is an independent simple f-reduced diagram with $cl(I(0)) = 1$. Therefore $L(0) \in \mathscr{F}_2$. By Lemma 11.2, $L(1) \in \mathscr{F}_2$.

(b) The diagram $I(01)$ is depicted in Fig. 11.9. This is an independent simple f-reduced diagram with $cl(I(01)) = 2$. Therefore $L(01) \in \mathscr{F}_3$. By Lemma 11.2, $L(10) \in \mathscr{F}_3$.

(c) The diagram $I(00)$ is depicted in Fig. 11.10. This is not a simple diagram. It is clear that $L(00) \neq E^*$. Therefore $L(00) \in \mathscr{F}_4$. By Lemma 11.2, $L(11) \in \mathscr{F}_4$. Using Lemma 11.3, we obtain $L(000)$, $L(001)$, $L(110)$, $L(111) \in \mathscr{F}_4$.

The diagram $I(010)$ is depicted in Fig. 11.11. This is not a simple diagram. It is clear that $L(010) \neq E^*$. Therefore $L(010) \in \mathscr{F}_4$. By Lemma 11.2, $L(101) \in \mathscr{F}_4$.

The diagram $I(011)$ is depicted in Fig. 11.12. This is not a simple diagram. It is clear that $L(011) \neq E^*$. Therefore $L(011) \in \mathscr{F}_4$. By Lemma 11.2, $L(100) \in \mathscr{F}_4$.

We proved that, for any word $\alpha \in E^*$ of the length three, $L(\alpha) \in \mathscr{F}_4$. Using Lemma 11.3, we obtain that, for any word $\alpha \in E^*$ of the length greater than or equal to four, $L(\alpha) \in \mathscr{F}_4$. □

Fig. 11.8 Diagram $I(0)$

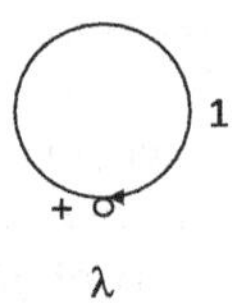

Fig. 11.9 Diagram $I(01)$

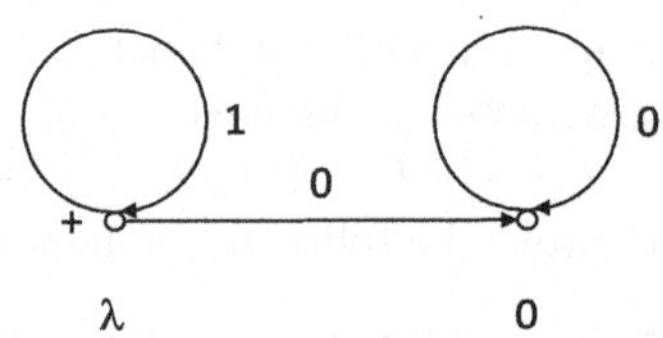

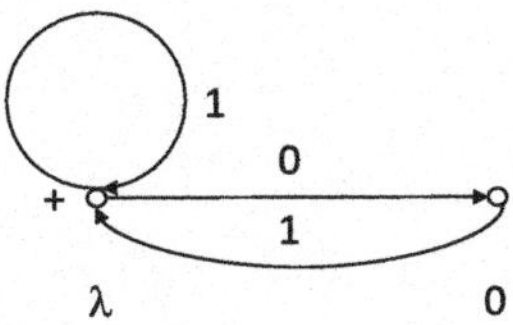

Fig. 11.10 Diagram $I(00)$

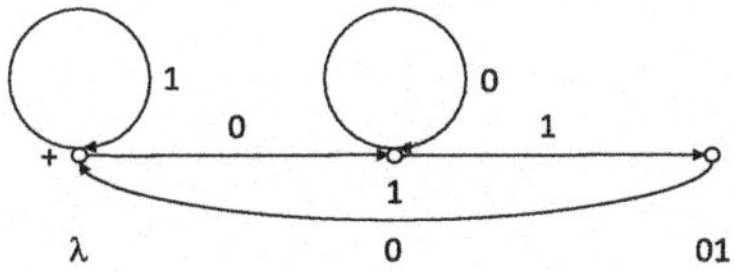

Fig. 11.11 Diagram $I(010)$

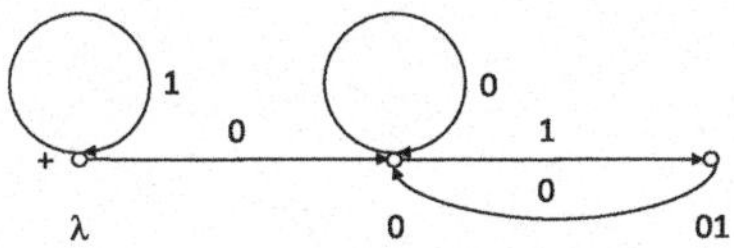

Fig. 11.12 Diagram $I(011)$

References

1. Atminas, A., Lozin, V.V., Moshkov, M.: WQO is decidable for factorial languages. Inf. Comput. **256**, 321–333 (2017)
2. Avgustinovich, S.V., Frid, A.E.: Canonical decomposition of a regular factorial language. In: Grigoriev, D., Harrison, J., Hirsch, E.A. (eds.) Computer Science - Theory and Applications, First International Symposium on Computer Science in Russia, CSR 2006, St. Petersburg, Russia, June 8–12, 2006, Proceedings, Lecture Notes in Computer Science, vol. 3967, pp. 18–22. Springer (2006)
3. Crochemore, M., Mignosi, F., Restivo, A.: Automata and forbidden words. Inf. Process. Lett. **67**(3), 111–117 (1998)
4. D'Alessandro, F., Intrigila, B., Varricchio, S.: On the structure of the counting function of sparse context-free languages. Theor. Comput. Sci. **356**(1–2), 104–117 (2006)
5. Hopcroft, J.E., Motwani, R., Ullman, J.D.: Introduction to Automata Theory, Languages, and Computation, 3rd edn. Addison-Wesley, Pearson international edition (2007)
6. Moshkov, M.: Complexity of deterministic and nondeterministic decision trees for regular language word recognition. In: Bozapalidis, S. (ed.) Proceedings of the 3rd International Conference Developments in Language Theory, DLT 1997, Thessaloniki, Greece, July 20–23, 1997, pp. 343–349. Aristotle University of Thessaloniki (1997)
7. Moshkov, M.: Decision trees for regular language word recognition. Fundam. Inform. **41**(4), 449–461 (2000)
8. Moshkov, M.: Decision trees for regular factorial languages. Array **15**, 100203 (2022)
9. Shur, A.M.: Combinatorial complexity of regular languages. In: Hirsch, E.A., Razborov, A.A., Semenov, A.L., Slissenko, A. (eds.) Computer Science - Theory and Applications, Third International Computer Science Symposium in Russia, CSR 2008, Moscow, Russia, June 7–12, 2008, Proceedings, Lecture Notes in Computer Science, vol. 5010, pp. 289–301. Springer (2008)
10. Yu, S.: Regular languages. In: Rozenberg, G., Salomaa, A. (eds.) Handbook of Formal Languages, Volume 1: Word, Language, Grammar, pp. 41–110. Springer (1997)

Part IV
Transforming Decision Rule Systems into Deterministic Decision Trees

In this part, we study the problem of transforming a decision rule system into a deterministic decision tree. We show that the minimum depth of a decision tree derived from a system of decision rules can be much smaller than the number of different attributes in the system's rules. For such systems of decision rules, it is reasonable to use decision trees. We prove that in many cases the minimum number of nodes in decision trees can grow as a superpolynomial function depending on the size of the decision rule systems. We show that this issue can be resolved using two kinds of acyclic decision graphs representing decision trees. However, in this case, it is necessary to simultaneously minimize the depth and the number of nodes in acyclic decision graphs, which is a difficult bi-criteria optimization problem. We leave this problem for future research and study a different approach to deal with decision trees. Instead of building the entire decision tree, we model the work of this tree for a given tuple of attribute values using a polynomial time algorithm. This part consists of two chapters.

In Chap. 12, we study unimprovable upper and lower bounds on the minimum depth of decision trees derived from decision rule systems depending on the various parameters of these systems. To illustrate the process of transformation of decision rule systems into decision trees, we generalize a well known result for Boolean functions to the case of functions of k-valued logic.

In Chap. 13, we study the complexity of constructing decision trees and acyclic decision graphs from decision rule systems, and we discuss the possibility of not building the entire decision tree, but describing the computation path in this tree for a given input. We also describe dynamic programming algorithms for the computation of the minimum depth of decision trees.

Chapter 12
Bounds on Depth of Decision Trees Derived from Decision Rule Systems

12.1 Introduction

The study of relationships between deterministic decision trees and systems of decision rules is an important task of computer science. Regular methods for extracting decision rules from decision trees have been known for a long time: first, each path in the decision tree from the root to a terminal node is assigned a decision rule, and then the resulting rules are simplified. In this chapter, we study issues related to the more complex problem of transforming a decision rule system into a deterministic decision tree.

One of the authors of this book begun in [5, 6] the development of the so-called syntactic approach to the study of the problem of transforming a decision rule system into a decision tree, which assumes that we do not know input data but only have a system of decision rules that must be transformed into a decision tree. The present chapter continues work in this direction. The results obtained in [5, 6] are described in remarks after Theorems 12.1 and 12.2.

Let there be a system S of decision rules of the form

$$(a_{i_1} = \delta_1) \wedge \cdots \wedge (a_{i_m} = \delta_m) \rightarrow \sigma \ ,$$

where $a_{i_1}, \ldots, a_{i_m}$ are attributes, $\delta_1, \ldots, \delta_m$ are values of these attributes, and σ is a decision. We study three algorithmic problems associated with this system:

- For a given input (a tuple of values of all attributes included in S), it is necessary to find at least one rule that is realizable for this input (having a true left-hand side) or show that there are no such rules.
- For a given input, it is necessary to find all the rules that are realizable for this input, or show that there are no such rules.
- For a given input, it is necessary to find all the right-hand sides of rules that are realizable for this input (we should return realizable rules with these right-hand sides), or show that there are no such rules.

K. Durdymyradov et al., *Decision Trees Versus Systems of Decision Rules*,
Studies in Big Data 160, https://doi.org/10.1007/978-3-031-71586-0_12

For each algorithmic problem, we consider two its variants. The first assumes that in the input each attribute can have only those values that occur for this attribute in the system S. In the second case, we assume that in the input any attribute can have any value.

Our goal is to minimize the number of queries for attribute values in the given input. For this purpose, deterministic decision trees are studied as algorithms for solving the considered six problems.

In this chapter, we study unimprovable upper and lower bounds on the minimum depth of decision trees depending on three parameters of the decision rule system S: the total number $n(S)$ of different attributes in the rules belonging to the system S, the maximum length $d(S)$ of a decision rule, and the maximum number $k(S)$ of attribute values.

To illustrate the results obtained, consider the problem $AR(S)$ for a system of decision rules S: for a given input containing only values of attributes that occur in the system S, it is necessary to find all the rules from S that are realizable for this input.

For this problem, we study lower $h_{AR}(n, d, k)$ and upper $H_{AR}(n, d, k)$ unimprovable bounds on the minimum depth of decision trees solving the problem $AR(S)$ for systems of decision rules S such that $n(S) = n$, $d(S) = d$, and $k(S) = k$. Let $n, d, k \in \omega \setminus \{0\}$ and $d \leq n$, where ω is the set of nonnegative integers. We prove that $H_{AR}(n, d, k) = n$, and $h_{AR}(n, d, k) = n$ if $k = 1$ or $d = 1$ and $\max\left\{d, \frac{n(k-1)}{k^d}\right\} \leq h_{AR}(n, d, k) \leq d + \frac{n}{k^{d-1}}$ if $k \geq 2$ and $d \geq 2$.

We show that not only for the problem $AR(S)$, but for each of the considered problems, there are systems of decision rules for which the minimum depth of the decision trees that solve the problem is much less than the total number of attributes in the rule system. For such systems of decision rules, it is advisable to use decision trees.

To illustrate the process of transformation of decision rule systems into decision trees, we generalize well known inequality $h(f) \leq C(f)^2$ [1, 4, 7], where $h(f)$ is the minimum depth of a deterministic decision tree computing Boolean function f and $C(f)$ is the certificate complexity of f (see details in [2]), to the case of functions of k-valued logic. Note that the certificate complexity $C(f)$ of the Boolean function f is equal to the minimum depth of a nondeterministic decision tree computing this function.

This chapter is based on the paper [3]. It consists of six sections. Section 12.2 discusses the main definitions and notation. Sections 12.3–12.5 are devoted to the study of unimprovable lower and upper bounds on the depth of decision trees depending on the parameters of the system of decision rules: Sect. 12.3 contains auxiliary statements, and Sects. 12.4 and 12.5 contain upper and lower bounds, respectively. In Sect. 12.6, we consider functions of k-valued logic.

12.2 Main Definitions and Notation

In this section, we discuss the main definitions and notation related to decision rule systems, decision trees, and functions that characterize the depth of decision trees derived from decision rule systems.

12.2.1 Decision Rule Systems

Let $\omega = \{0, 1, 2, \ldots\}$ and $A = \{a_i : i \in \omega\}$. Elements of the set A will be called *attributes*.

Definition 12.1 A *decision rule* is an expression of the form

$$(a_{i_1} = \delta_1) \wedge \cdots \wedge (a_{i_m} = \delta_m) \to \sigma ,$$

where $m \in \omega$, $a_{i_1}, \ldots, a_{i_m}$ are pairwise different attributes from A and $\delta_1, \ldots, \delta_m, \sigma \in \omega$.

We denote this decision rule by r. The expression $(a_{i_1} = \delta_1) \wedge \cdots \wedge (a_{i_m} = \delta_m)$ will be called the *left-hand side*, and the number σ will be called the *right-hand side* of the rule r. The number m will be called the *length* of the decision rule r. Denote $A(r) = \{a_{i_1}, \ldots, a_{i_m}\}$ and $K(r) = \{a_{i_1} = \delta_1, \ldots, a_{i_m} = \delta_m\}$. If $m = 0$, then $A(r) = K(r) = \emptyset$.

Definition 12.2 Two decision rules r_1 and r_2 are *equal* if $K(r_1) = K(r_2)$ and the right-hand sides of the rules r_1 and r_2 are equal.

Definition 12.3 A *system of decision rules* S is a finite nonempty set of decision rules.

Denote $A(S) = \bigcup_{r \in S} A(r)$, $n(S) = |A(S)|$, $D(S)$ the set of the right-hand sides of decision rules from S, and $d(S)$ the maximum length of a decision rule from S. Let $n(S) > 0$. For $a_i \in A(S)$, let $V_S(a_i) = \{\delta : a_i = \delta \in \bigcup_{r \in S} K(r)\}$ and $EV_S(a_i) = V_S(a_i) \cup \{*\}$, where the symbol $*$ is interpreted as a number that does not belong to the set $V_S(a_i)$. Denote $k(S) = \max\{|V_S(a_i)| : a_i \in A(S)\}$. If $n(S) = 0$, then, by definition, $k(S) = 0$. We denote by Σ the set of systems of decision rules.

Example 12.1 Let us consider a decision rule system $S = \{(a_1 = 0) \wedge (a_2 = 0) \wedge (a_3 = 0) \to 3, (a_1 = 1) \wedge (a_4 = 0) \to 4, (a_1 = 2) \to 5\}$. Then $A(r_1) = \{a_1, a_2, a_3\}$, $K(r_1) = \{a_1 = 0, a_2 = 0, a_3 = 0\}$, where r_1 denotes the first rule from S. $A(S) = \bigcup_{r \in S} A(r) = \{a_1, a_2, a_3, a_4\}$, $n(S) = |A(S)| = 4$, $D(S) = \{3, 4, 5\}$, $d(S) = 3$, $V_S(a_1) = \{\delta : a_1 = \delta \in \bigcup_{r \in S} K(r)\} = \{0, 1, 2\}$, $EV_S(a_1) = V_S(a_1) \cup \{*\} = \{0, 1, 2, *\}$ and $k(S) = \max\{|V_S(a_i)| : a_i \in A(S)\} = |V_S(a_1)| = 3$.

Let $S \in \Sigma$, $n(S) > 0$, and $A(S) = \{a_{j_1}, \ldots, a_{j_n}\}$, where $j_1 < \cdots < j_n$. Denote $V(S) = V_S(a_{j_1}) \times \cdots \times V_S(a_{j_n})$ and $EV(S) = EV_S(a_{j_1}) \times \cdots \times EV_S(a_{j_n})$. For $\bar{\delta} = (\delta_1, \ldots, \delta_n) \in EV(S)$, denote $K(S, \bar{\delta}) = \{a_{j_1} = \delta_1, \ldots, a_{j_n} = \delta_n\}$.

We now can define rules for which the left-hand side is true for a given tuple of attribute values.

Definition 12.4 We will say that a decision rule r from S is *realizable* for a tuple $\bar{\delta} \in EV(S)$ if $K(r) \subseteq K(S, \bar{\delta})$.

It is clear that any rule with an empty left-hand side is realizable for the tuple $\bar{\delta}$.

Example 12.2 Let us consider a decision rule system $S = \{r_1 : (a_1 = 0) \wedge (a_2 = 0) \rightarrow 0, r_2 : (a_1 = 0) \wedge (a_3 = 1) \rightarrow 1\}$ and a tuple $\bar{\delta} = (0, 0, 0) \in EV(S)$. Then the decision rule r_1 from S is realizable for the tuple $\bar{\delta}$, but r_2 is not.

Let $V \in \{V(S), EV(S)\}$. We now define three problems related to the rule system S.

Definition 12.5 Problem *All Rules* for the pair (S, V): for a given tuple $\bar{\delta} \in V$, it is required to find the set of rules from S that are realizable for the tuple $\bar{\delta}$.

Definition 12.6 Problem *All Decisions* for the pair (S, V): for a given tuple $\bar{\delta} \in V$, it is required to find a set Z of decision rules from S satisfying the following conditions:

- All decision rules from Z are realizable for the tuple $\bar{\delta}$.
- For any $\sigma \in D(S) \setminus D(Z)$, any decision rule from S with the right-hand side equal to σ is not realizable for the tuple $\bar{\delta}$.

Definition 12.7 Problem *Some Rules* for the pair (S, V): for a given tuple $\bar{\delta} \in V$, it is required to find a set Z of decision rules from S satisfying the following conditions:

- All decision rules from Z are realizable for the tuple $\bar{\delta}$.
- If $Z = \emptyset$, then any decision rule from S is not realizable for the tuple $\bar{\delta}$.

Denote $AR(S)$ and $EAR(S)$ the problems All Rules for pairs $(S, V(S))$ and $(S, EV(S))$, respectively. Denote $AD(S)$ and $EAD(S)$ the problems All Decisions for pairs $(S, V(S))$ and $(S, EV(S))$, respectively. Denote $SR(S)$ and $ESR(S)$ the problems Some Rules for pairs $(S, V(S))$ and $(S, EV(S))$, respectively.

Example 12.3 Let a decision rule system $S = \{(a_1 = 0) \wedge (a_2 = 0) \wedge (a_3 = 0) \rightarrow 0, (a_1 = 0) \wedge (a_2 = 0) \rightarrow 1, (a_1 = 0) \rightarrow 0\}$ and a tuple $\bar{\delta} = (0, 0, 0) \in EV(S)$ are given. Then $\{(a_1 = 0) \wedge (a_2 = 0) \wedge (a_3 = 0) \rightarrow 0, (a_1 = 0) \wedge (a_2 = 0) \rightarrow 1, (a_1 = 0) \rightarrow 0\}$ is the solution for the problem $AR(S)$ and the tuple $\bar{\delta}$, $\{(a_1 = 0) \wedge (a_2 = 0) \rightarrow 1, (a_1 = 0) \rightarrow 0\}$ is a solution for the problem $AD(S)$ and the tuple $\bar{\delta}$, and $\{(a_1 = 0) \rightarrow 0\}$ is a solution for the problem $SR(S)$ and the tuple $\bar{\delta}$.

In the special case, when $n(S) = 0$, all rules from S have an empty left-hand side. In this case, it is natural to consider (i) the set S as the solution to the problems $AR(S)$ and $EAR(S)$, (ii) any subset Z of the set S with $D(Z) = D(S)$ as a solution to the problems $AD(S)$ and $EAD(S)$, and (iii) any nonempty subset Z of the set S as a solution to the problems $SR(S)$ and $ESR(S)$.

12.2.2 Decision Trees

A *finite directed tree with root* is a finite directed tree in which only one node has no entering edges. This node is called the *root*. The nodes without leaving edges are called *terminal* nodes. The nodes that are not terminal will be called *working* nodes. A *complete path* in a finite directed tree with root is a sequence $\xi = v_1, d_1, \ldots, v_m, d_m, v_{m+1}$ of nodes and edges of this tree in which v_1 is the root, v_{m+1} is a terminal node and, for $i = 1, \ldots, m$, the edge d_i leaves the node v_i and enters the node v_{i+1}.

We will consider two types of decision trees: o-decision trees (ordinary decision trees, o-trees in short) and e-decision trees (extended decision trees, e-trees in short).

Definition 12.8 A *decision tree over a decision rule system* S is a labeled finite directed tree with root Γ satisfying the following conditions:

- Each working node of the tree Γ is labeled with an attribute from the set $A(S)$.
- Let a working node v of the tree Γ be labeled with an attribute a_i. If Γ is an o-tree, then exactly $|V_S(a_i)|$ edges leave the node v and these edges are labeled with pairwise different elements from the set $V_S(a_i)$. If Γ is an e-tree, then exactly $|EV_S(a_i)|$ edges leave the node v and these edges are labeled with pairwise different elements from the set $EV_S(a_i)$.
- Each terminal node of the tree Γ is labeled with a subset of the set S.

Let Γ be a decision tree over the decision rule system S. We denote by $CP(\Gamma)$ the set of complete paths in the tree Γ. Let $\xi = v_1, d_1, \ldots, v_m, d_m, v_{m+1}$ be a complete path in Γ. We correspond to this path a set of attributes $A(\xi)$ and an equation system $K(\xi)$. If $m = 0$ and $\xi = v_1$, then $A(\xi) = \emptyset$ and $K(\xi) = \emptyset$. Let $m > 0$ and, for $j = 1, \ldots, m$, the node v_j be labeled with the attribute a_{i_j} and the edge d_j be labeled with the element $\delta_j \in \omega \cup \{*\}$. Then $A(\xi) = \{a_{i_1}, \ldots, a_{i_m}\}$ and $K(\xi) = \{a_{i_1} = \delta_1, \ldots, a_{i_m} = \delta_m\}$. We denote by $\tau(\xi)$ the set of decision rules attached to the node v_{m+1}.

Example 12.4 Let us consider a decision rule system $S = \{(a_1 = 0) \wedge (a_2 = 0) \rightarrow 0, (a_1 = 1) \wedge (a_3 = 0) \rightarrow 0, (a_1 = 1) \rightarrow 0\}$. Then examples of o-tree Γ_1 and e-tree Γ_2 over the decision rule system S are given in Fig. 12.1, where r_1, r_2 and r_3 are the first, second and third decision rules in S, respectively.

Let ξ be a complete path in the o-tree, which is finished in the terminal node labeled with the set of rules $\{r_2, r_3\}$. Then $A(\xi) = \{a_1, a_3\}$, $K(\xi) = \{a_1 = 1, a_3 = 0\}$ and $\tau(\xi) = \{r_2, r_3\}$.

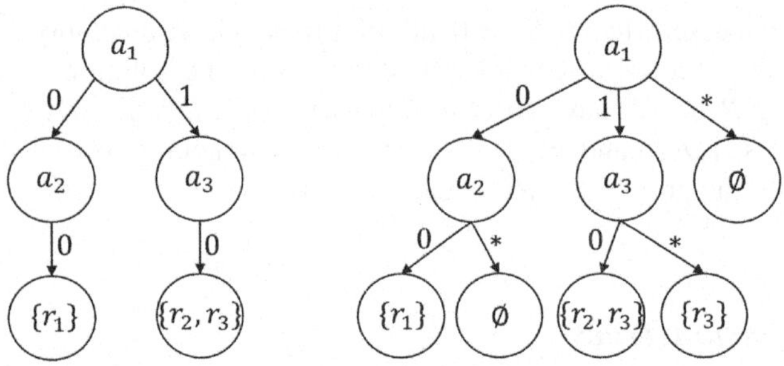

Fig. 12.1 O-decision tree Γ_1 and e-decision tree Γ_2 from Example 12.4

We now define the notions of inconsistent and consistent systems of equations that are based on syntactic properties of these systems.

Definition 12.9 A system of equations $\{a_{i_1} = \delta_1, \ldots, a_{i_m} = \delta_m\}$, $\delta_1, \ldots, \delta_m \in \omega \cup \{*\}$, will be called *inconsistent* if there exist $l, k \in \{1, \ldots, m\}$ such that $l \neq k$, $i_l = i_k$, and $\delta_l \neq \delta_k$. If the system of equations is not inconsistent then it will be called *consistent.*

Let S be a decision rule system and Γ be a decision tree over S.

Definition 12.10 We will say that Γ *solves* the problem $AR(S)$ (problem $EAR(S)$, respectively) if Γ is an o-tree (an e-tree, respectively) and any path $\xi \in CP(\Gamma)$ with consistent system of equations $K(\xi)$ satisfies the following conditions:

- For any decision rule $r \in \tau(\xi)$, the relation $K(r) \subseteq K(\xi)$ holds.
- For any decision rule $r \in S \backslash \tau(\xi)$, the system of equations $K(r) \cup K(\xi)$ is inconsistent.

Example 12.5 Let S be a decision rule system from Example 12.4. Then the decision trees Γ_1 and Γ_2 depicted in Fig. 12.1 solve the problems $AR(S)$ and $EAR(S)$, respectively.

Definition 12.11 We will say that Γ *solves* the problem $AD(S)$ (problem $EAD(S)$, respectively) if Γ is an o-tree (an e-tree, respectively) and any path $\xi \in CP(\Gamma)$ with consistent system of equations $K(\xi)$ satisfies the following conditions:

- For any decision rule $r \in \tau(\xi)$, the relation $K(r) \subseteq K(\xi)$ holds.
- If $r \in S \backslash \tau(\xi)$ and the right-hand side of r does not belong to the set $D(\tau(\xi))$, then the system of equations $K(r) \cup K(\xi)$ is inconsistent.

Definition 12.12 We will say that Γ *solves* the problem $SR(S)$ (problem $ESR(S)$, respectively) if Γ is an o-tree (an e-tree, respectively) and any path $\xi \in CP(\Gamma)$ with consistent system of equations $K(\xi)$ satisfies the following conditions:

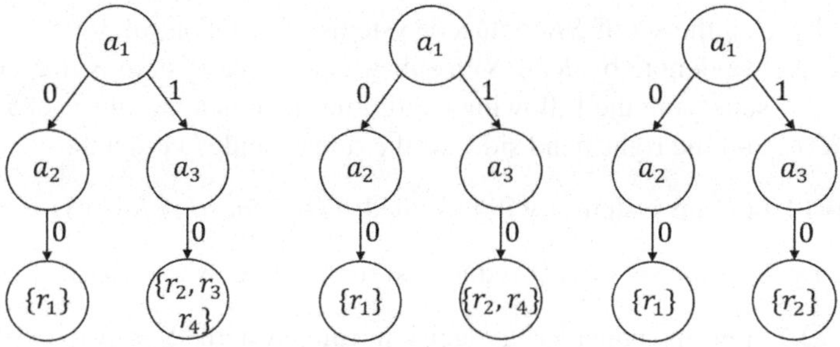

Fig. 12.2 Decision trees Γ_1, Γ_2, and Γ_3 from Example 12.6

- For any decision rule $r \in \tau(\xi)$, the relation $K(r) \subseteq K(\xi)$ holds.
- If $\tau(\xi) = \emptyset$, then, for any decision rule $r \in S$, the system of equations $K(r) \cup K(\xi)$ is inconsistent.

Example 12.6 Let us consider a decision rule system $S = \{(a_1 = 0) \wedge (a_2 = 0) \to 0, (a_1 = 1) \wedge (a_3 = 0) \to 1, (a_1 = 1) \to 1, (a_1 = 1) \to 2\}$ and decision trees Γ_1, Γ_2, and Γ_3 depicted in Fig. 12.2. Then the decision tree Γ_1 solves the problem $AR(S)$, Γ_2 solves $AD(S)$, and Γ_3 solves $SR(S)$, where r_1, r_2, r_3 and r_4 are the first, second, third and fourth decision rules in S, respectively.

For any complete path $\xi \in CP(\Gamma)$, we denote by $h(\xi)$ the number of working nodes in ξ. The value $h(\Gamma) = \max\{h(\xi) : \xi \in CP(\Gamma)\}$ is called the *depth* of the decision tree Γ.

Let S be a decision rule system and $C \in \{AR, EAR, AD, EAD, SR, ESR\}$. We denote by $h_C(S)$ the minimum depth of a decision tree over S, which solves the problem $C(S)$.

Let $n(S) = 0$. If $C \in \{AR, EAR\}$, then there is only one decision tree solving the problem $C(S)$. This tree consists of one node labeled with the set of rules S. If $C \in \{AD, EAD\}$, then the set of decision trees solving the problem $C(S)$ coincides with the set of trees each of which consists of one node labeled with a subset Z of the set S with $D(Z) = D(S)$. If $C \in \{SR, ESR\}$, then the set of decision trees solving the problem $C(S)$ coincides with the set of trees each of which consists of one node labeled with a nonempty subset Z of the set S. Therefore if $n(S) = 0$, then $h_C(S) = 0$ for any $C \in \{AR, EAR, AD, EAD, SR, ESR\}$.

12.2.3 Functions Characterizing Depth of Decision Trees

Let $S \in \Sigma$, where Σ is the set of decision rule systems. We denote by $R_{SR}(S)$ a subsystem of the system S that consists of all rules $r \in S$ satisfying the following condition: there is no a rule $r' \in S$ such that $K(r') \subset K(r)$.

Definition 12.13 The system S will be called *SR-reduced* if $R_{SR}(S) = S$.

Denote by Σ_{SR} the set of SR-reduced systems of decision rules.

For $S \in \Sigma$, we denote by $R_{AD}(S)$ a subsystem of the system S that consists of all rules $r \in S$ satisfying the following condition: there is no a rule $r' \in S$ such that $K(r') \subset K(r)$ and the right-hand sides of the rules r and r' coincide.

Definition 12.14 The system S will be called *AD-reduced* if $R_{AD}(S) = S$.

Denote by Σ_{AD} the set of AD-reduced systems of decision rules.

Example 12.7 Let us consider a decision rule system $S = \{(a_1 = 0) \wedge (a_2 = 0) \wedge (a_3 = 0) \to 0, (a_1 = 0) \wedge (a_2 = 0) \to 0, (a_1 = 0) \to 1\}$. For this system, $R_{AD}(S) = \{(a_1 = 0) \wedge (a_2 = 0) \to 0, (a_1 = 0) \to 1\}$ and $R_{SR}(S) = \{(a_1 = 0) \to 1\}$.

It is easy to show that

$$\begin{aligned}\{(n(S), d(S), k(S)) : S \in \Sigma\} &= \{(n(S), d(S), k(S)) : S \in \Sigma_{SR}\}\\ &= \{(n(S), d(S), k(S)) : S \in \Sigma_{AD}\}\\ &= \{(0, 0, 0)\} \cup \{(n, d, k) : n, d, k \in \omega \setminus \{0\}, d \le n\}\,.\end{aligned}$$

Really, if $n(S) = 0$, i.e., each rule from S has empty left-hand side, then $d(S) = k(S) = 0$. If $n(S) > 0$, then $d(S) > 0$, $k(S) > 0$, and $d(S) \le n(S)$ since, in any decision rule, attributes from different equations in the left-hand side are different. Let $n, d, k \in \omega \setminus \{0\}$, $d \le n$, and $S = \{(a_1 = 0) \wedge \cdots \wedge (a_d = 0) \to 0, (a_{d+1} = 0) \to 0, \ldots, (a_n = 0) \to 0, (a_1 = 1) \to 0, \ldots, (a_1 = k - 1) \to 0\}$. One can show, that S is AD-reduced and SR-reduced, $n(S) = n$, $d(S) = d$, and $k(S) = k$.

We will not study the case, when $(n(S), d(S), k(S)) = (0, 0, 0)$, since if $n(S) = 0$, then $h_C(S) = 0$ for any $C \in \{SR, ESR, AD, EAD, AR, EAR\}$.

Definition 12.15 Let $C \in \{SR, ESR, AD, EAD, AR, EAR\}$, $n, d, k \in \omega \setminus \{0\}$, and $d \le n$. Denote

$$\begin{aligned}h_C(n, d, k) &= \min\{h_C(S) : S \in \Sigma, n(S) = n, d(S) = d, k(S) = k\}\,,\\ H_C(n, d, k) &= \max\{h_C(S) : S \in \Sigma, n(S) = n, d(S) = d, k(S) = k\}\,.\end{aligned}$$

The considered parameters are lower ($h_C(n, d, k)$) and upper ($H_C(n, d, k)$) unimprovable bounds on the minimum depth of decision trees solving the problem $C(S)$ for systems of decision rules S such that $n(S) = n$, $d(S) = d$, and $k(S) = k$.

We also consider similar parameters for SR-reduced and for AD-reduced systems of decision rules. Interest in the study of such rule systems is due to the fact that the transformation of a system S into systems $R_{SR}(S)$ and $R_{AD}(S)$ is not difficult and the systems $R_{SR}(S)$ and $R_{AD}(S)$ can be much simpler than the system S. In addition, as Lemma 12.9 shows, equalities $h_{ESR}(S) = h_{ESR}(R_{SR}(S))$ and $h_{EAD}(S) = h_{EAD}(R_{AD}(S))$ hold, and the transformation of decision trees solving problems $ESR(R_{SR}(S))$ and $EAD(R_{AD}(S))$ into decision trees solving problems $ESR(S)$ and $EAD(S)$, respectively, can be carried out relatively simply.

Definition 12.16 Let $C \in \{SR, ESR, AD, EAD\}$, $n, d, k \in \omega \setminus \{0\}$, and $d \le n$. Let $C' = SR$ if $C \in \{SR, ESR\}$ and $C' = AD$ if $C \in \{AD, EAD\}$. Denote

$$h_C^R(n, d, k) = \min\{h_C(S) : S \in \Sigma_{C'}, n(S) = n, d(S) = d, k(S) = k\} ,$$
$$H_C^R(n, d, k) = \max\{h_C(S) : S \in \Sigma_{C'}, n(S) = n, d(S) = d, k(S) = k\} .$$

The considered parameters are lower ($h_C^R(n, d, k)$) and upper ($H_C^R(n, d, k)$) unimprovable bounds on the minimum depth of decision trees solving the problem $C(S)$ for C'-reduced systems of decision rules S such that $n(S) = n$, $d(S) = d$, and $k(S) = k$.

12.3 Auxiliary Statements

In this section, we prove several auxiliary statements that will be used later.

Let S be a decision rule system and Γ be an e-decision tree over S. We denote by $o(\Gamma)$ an o-tree over S, which is obtained from the tree Γ by the removal of all nodes v such that the path from the root to the node v in Γ contains an edge labeled with $*$. Together with a node v, we remove all edges entering or leaving v. One can prove the following two lemmas.

Lemma 12.1 *Let S be a decision rule system, $C \in \{AR, AD, SR\}$, and Γ be an e-decision tree over S solving the problem $EC(S)$. Then the decision tree $o(\Gamma)$ solves the problem $C(S)$.*

Lemma 12.2 *Let S be a decision rule system and Γ be a decision tree over S. Then*

(a) *If the tree Γ solves the problem $AR(S)$ ($EAR(S)$, respectively), then the tree Γ solves the problem $AD(S)$ ($EAD(S)$, respectively).*

(b) *If the tree Γ solves the problem $AD(S)$ ($EAD(S)$, respectively), then the tree Γ solves the problem $SR(S)$ ($ESR(S)$, respectively).*

Lemma 12.3 *Let S be a decision rule system. Then the following inequalities hold:*

$$\begin{array}{ccccccc} h_{ESR}(S) & \le & h_{EAD}(S) & \le & h_{EAR}(S) & \le & n(S) , \\ \text{\rotatebox{90}{$\le$}} & & \text{\rotatebox{90}{$\le$}} & & \text{\rotatebox{90}{$\le$}} & & \\ h_{SR}(S) & \le & h_{AD}(S) & \le & h_{AR}(S) . & & \end{array}$$

Proof It is clear that the considered inequalities hold if $n(S) = 0$. Let $n(S) > 0$. Let Γ be an e-decision tree over S. It is clear that $h(o(\Gamma)) \le h(\Gamma)$. Using this inequality and Lemma 12.1 we obtain $h_{SR}(S) \le h_{ESR}(S)$, $h_{AD}(S) \le h_{EAD}(S)$, and $h_{AR}(S) \le h_{EAR}(S)$. By Lemma 12.2, $h_{ESR}(S) \le h_{EAD}(S) \le h_{EAR}(S)$ and $h_{SR}(S) \le h_{AD}(S) \le h_{AR}(S)$. One can construct an o-decision tree Γ over S, which solves the problem $EAR(S)$ by sequential computation of values of all attributes from $A(S)$. Therefore $h_{EAR}(S) \le n(S)$. □

Let S be a decision rule system, Γ be a decision tree over S, and $\alpha = \{a_{i_1} = \delta_1, \ldots, a_{i_m} = \delta_m\}$ be a consistent equation system such that $a_{i_1}, \ldots, a_{i_m} \in A$ and $\delta_1, \ldots, \delta_m \in \omega \cup \{*\}$. We now define a decision rule system S_α. Let r be a decision rule for which the equation system $K(r) \cup \alpha$ is consistent. We denote by r_α the decision rule obtained from r by the removal from the left-hand side of r all equations that belong to α. Then S_α is the set of decision rules r_α such that $r \in S$ and the equation system $K(r) \cup \alpha$ is consistent.

Let us assume that $n(S) > 0$, $a_{i_1}, \ldots, a_{i_m} \in A(S)$, $\delta_j \in EV_S(a_{i_j})$ for $j = 1, \ldots, m$ and Γ is an e-tree. We now define an e-tree Γ_α. First, we define a tree Γ'_α. Let an edge d in the tree Γ enter a node v. We will say that a subtree of Γ with the root v *corresponds to the edge* d. Denote $A(\alpha) = \{a_{i_1}, \ldots, a_{i_m}\}$. Beginning with the root of Γ, we will process all working nodes of the tree Γ. Let v be a working node of Γ labeled with an attribute a_t. Let $a_t \in A(S_\alpha)$. We keep all edges that leave v and are labeled with elements from the set $EV_{S_\alpha}(a_t)$. We remove all other edges leaving v together with the subtrees corresponding to these edges. Let $a_t \notin A(S_\alpha)$. We choose an element $\delta \in EV_S(a_t)$ in the following way. If $a_t \in A(\alpha)$ and $t = i_j$, then $\delta = \delta_j$. If $a_t \notin A(\alpha)$, then δ is the minimum number from $EV_S(a_t)$. We remove all edges that leave v and are labeled with elements different from δ together with the subtrees corresponding to these edges. Denote by Γ'_α the obtained tree.

We now process all nodes in this tree. Let v be a working node that is labeled with an attribute a_t. If $a_t \in A(S_\alpha)$, then we keep this node untouched. Let $a_t \notin A(S_\alpha)$. If v is the root, then remove the node v and the edge leaving v. Let v be not the root, d_0 be the edge entering v, and d_1 be the edge leaving v and entering a node v_1. We remove from Γ'_α the node v and the edge d_1, and join the edge d_0 to the node v_1. Let v be a terminal node that is labeled with a set Z of decision rules. We replace the set Z with the set Z_α. We denote the obtained tree Γ_α.

Let us assume that $n(S) > 0$, $a_{i_1}, \ldots, a_{i_m} \in A(S)$, $\delta_j \in V_S(a_{i_j})$ for $j = 1, \ldots, m$ and Γ be an o-tree. The tree Γ'_α and the o-tree Γ_α are defined in almost the same way as in the previous case. The only difference is that instead of the sets $EV_S(a_t)$ and $EV_{S_\alpha}(a_t)$ we consider the sets $V_S(a_t)$ and $V_{S_\alpha}(a_t)$.

Example 12.8 Let us consider a decision rule system $S = \{r^1 : (a_1 = 0) \wedge (a_2 = 0) \rightarrow 0, r^2 : (a_1 = 1) \wedge (a_3 = 0) \rightarrow 0, r^3 : (a_2 = 1) \wedge (a_3 = 1) \rightarrow 0\}$ and $\alpha = \{a_3 = 0\}$. Then $S_\alpha = \{r^1_\alpha : (a_1 = 0) \wedge (a_2 = 0) \rightarrow 0, r^2_\alpha : (a_1 = 1) \rightarrow 0\}$. In Fig. 12.3, decision trees Γ, Γ'_α, and Γ_α are depicted, where Γ is an o-decision tree over S.

Lemma 12.4 *Let S be a decision rule system with $n(S) > 0$, Γ be a decision tree over S, $C \in \{EAR, EAD, ESR\}$, and $\alpha = \{a_{i_1} = \delta_1, \ldots, a_{i_m} = \delta_m\}$ be a consistent equation system such that $a_{i_j} \in A(S)$ and $\delta_j \in EV_S(a_{i_j})$ for $j = 1, \ldots, m$. If the decision tree Γ solves the problem $C(S)$ and $S_\alpha \neq \emptyset$, then the decision tree Γ_α solves the problem $C(S_\alpha)$.*

Proof Let Γ solve the problem $C(S)$ and $S_\alpha \neq \emptyset$. It is clear that Γ_α is an e-decision tree over the decision rule system S_α. It is also clear that each complete path in Γ'_α coincides with some complete path in Γ and there exists one-to-one correspondence

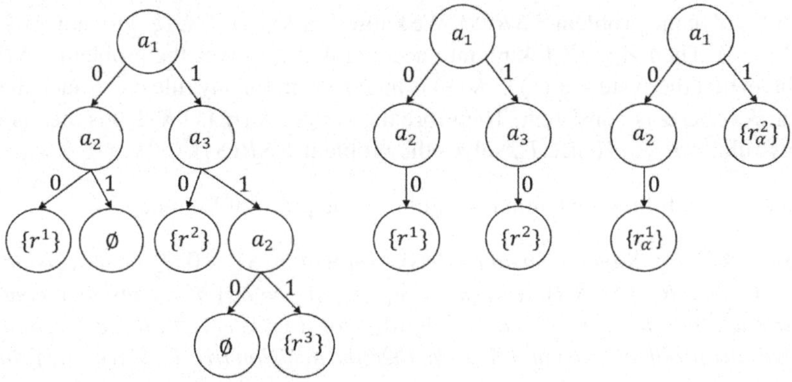

Fig. 12.3 Decision trees Γ, Γ'_α, and Γ_α from Example 12.8

between sets of complete paths in Γ'_α and Γ_α such that each path ξ in Γ'_α corresponds to a path ξ_α in Γ_α obtained from ξ by the removal of some nodes and edges leaving these nodes. One can show that the equation system $K(\xi)$ is inconsistent if and only if the system $K(\xi_\alpha)$ is inconsistent.

Let ξ be a complete path in Γ'_α for which the equation system $K(\xi)$ is consistent. Let the terminal node of the path ξ be labeled with a set of decision rules Z. Then the terminal node of the path ξ_α is labeled with the set of decision rules Z_α. Let $r \in Z$. Since Γ solves the problem $C(S)$, $K(r) \subseteq K(\xi)$. Therefore the system $K(r) \cup \alpha$ is consistent. It is clear that $K(\xi_\alpha) \subseteq K(\xi)$ and the system $K(\xi) \backslash K(\xi_\alpha)$ contains only equations of the form $a_t = \delta$ such that $a_t \notin A(S_\alpha)$. Since $K(r_\alpha) \subseteq K(r)$, $K(r_\alpha) \subseteq K(\xi_\alpha)$. Using these relations, we obtain $Z_\alpha = \{r_\alpha : r \in Z\}$ and $K(r_\alpha) \subseteq K(\xi_\alpha)$ for any $r_\alpha \in Z_\alpha$.

Let $r \in S$, the system $K(r) \cup \alpha$ be consistent, and the system $K(r) \cup K(\xi)$ be inconsistent. Then there exist a_t, δ, σ such that $\delta \neq \sigma$, $a_t = \delta \in K(r)$ and $a_t = \sigma \in K(\xi)$. It is clear that $a_t \notin \{a_{i_1}, \ldots, a_{i_m}\}$. Therefore $a_t = \delta \in K(r_\alpha)$ and $a_t \in A(S_\alpha)$. Thus, $a_t = \sigma \in K(\xi_\alpha)$ and $K(r_\alpha) \cup K(\xi_\alpha)$ is inconsistent.

Let Γ solve the problem $EAR(S)$. We know that $K(r_\alpha) \subseteq K(\xi_\alpha)$ for any $r_\alpha \in Z_\alpha$. Let $r' \in S_\alpha \backslash Z_\alpha$ and r be a rule from S such that the system $K(r) \cup \alpha$ is consistent and $r' = r_\alpha$. It is clear that $r \notin Z$. Taking into account that Γ solves the problem $EAR(S)$, we obtain that the system $K(r) \cup K(\xi)$ is inconsistent. Therefore the system $K(r_\alpha) \cup K(\xi_\alpha)$ is inconsistent. Hence Γ_α solves the problem $EAR(S_\alpha)$.

Let Γ solve the problem $EAD(S)$. We know that $K(r_\alpha) \subseteq K(\xi_\alpha)$ for any $r_\alpha \in Z_\alpha$. Let $r' \in S_\alpha \backslash Z_\alpha$ and the right-hand side of the rule r' do not belong to $D(Z_\alpha)$. We now consider a decision rule $r \in S$ such that the system $K(r) \cup \alpha$ is consistent and $r' = r_\alpha$. It is clear that $r \notin Z$ and $D(Z) = D(Z_\alpha)$. Taking into account that Γ solves the problem $EAD(S)$, we obtain that the system $K(r) \cup K(\xi)$ is inconsistent. Therefore the system $K(r_\alpha) \cup K(\xi_\alpha)$ is inconsistent. Hence Γ_α solves the problem $EAD(S_\alpha)$.

Let Γ solve the problem $ESR(S)$. We know that $K(r_\alpha) \subseteq K(\xi_\alpha)$ for any $r_\alpha \in Z_\alpha$. Let $Z_\alpha = \emptyset$. Then $Z = \emptyset$. Taking into account that Γ solves the problem $ESR(S)$, we obtain that the system $K(r) \cup K(\xi)$ is inconsistent for any rule $r \in S$ such that the system $K(r) \cup \alpha$ is consistent. Therefore the system $K(r_\alpha) \cup K(\xi_\alpha)$ is inconsistent for any rule $r_\alpha \in S_\alpha$. Hence Γ_α solves the problem $ESR(S_\alpha)$. □

Proof of the following lemma is similar to the proof of Lemma 12.4.

Lemma 12.5 *Let S be a decision rule system with $n(S) > 0$, Γ be a decision tree over S, $C \in \{AR, AD, SR\}$, $\alpha = \{a_{i_1} = \delta_1, \ldots, a_{i_m} = \delta_m\}$ be a consistent equation system such that $a_{i_j} \in A(S)$ and $\delta_j \in V_S(a_{i_j})$ for $j = 1, \ldots, m$. If the decision tree Γ solves the problem $C(S)$ and $S_\alpha \neq \emptyset$, then the decision tree Γ_α solves the problem $C(S_\alpha)$.*

Lemma 12.6 *Let S be a decision rule system with $n(S) > 0$, $C \in \{EAR, AR, EAD, AD, ESR, SR\}$, $\alpha = \{a_{i_1} = \delta_1, \ldots, a_{i_m} = \delta_m\}$ be a consistent equation system such that $a_{i_1}, \ldots, a_{i_m} \in A(S)$ and, for $j = 1, \ldots, m$, $\delta_j \in EV_S(a_{i_j})$ if $C \in \{EAR, EAD, ESR\}$ and $\delta_j \in V_S(a_{i_j})$ if $C \in \{AR, AD, SR\}$. Then $h_C(S) \geq h_C(S_\alpha)$.*

Proof Let Γ be a decision tree over S, which solves the problem $C(S)$ and for which $h(\Gamma) = h_C(S)$. Using Lemmas 12.4 and 12.5, we obtain that the decision tree Γ_α solves the problem $C(S_\alpha)$. It is clear that $h(\Gamma_\alpha) \leq h(\Gamma)$. Therefore $h_C(S_\alpha) \leq h_C(S)$. □

We correspond to a decision rule system S a hypergraph $G(S)$ with the set of nodes $A(S)$ and the set of edges $\{A(r) : r \in S\}$. A *node cover* of the hypergraph $G(S)$ is a subset B of the set of nodes $A(S)$ such that $A(r) \cap B \neq \emptyset$ for any rule $r \in S$ such that $A(r) \neq \emptyset$. If $A(S) = \emptyset$, then the empty set is the only node cover of the hypergraph $G(S)$. Denote by $\beta(S)$ the minimum cardinality of a node cover of the hypergraph $G(S)$.

Example 12.9 Let us consider a decision rule system $S = \{(a_1 = 0) \wedge (a_2 = 0) \rightarrow 0, (a_1 = 1) \wedge (a_3 = 0) \rightarrow 0, (a_2 = 1) \wedge (a_3 = 1) \rightarrow 0\}$. One can show, that the set $\{a_1, a_2\}$ is a node cover of the hypergraph $G(S)$ and $\beta(S) = 2$.

We define a subsystem $I_{SR}(S)$ of the system S in the following way. If S does not contain rules of the length 0, then $I_{SR}(S) = S$. Otherwise, $I_{SR}(S)$ consists of all rules from S of the length 0.

Remark 12.1 Note that $I_{SR}(S) \neq S$ if and only if S contains both a rule of the length 0 and a rule of the length greater than 0.

We now define a subsystem $I_{AD}(S)$ of the system S. Denote by $D_0(S)$ the set of the right-hand sides of decision rules from S, which length is equal to 0. Then the subsystem $I_{AD}(S)$ consists of all rules from S of the length 0 and all rules from S for which the right-hand sides do not belong to $D_0(S)$.

Remark 12.2 Note that $I_{AD}(S) \neq S$ if and only if S contains both a rule of the length 0 and a rule of the length greater than 0 with the same right-hand sides.

Example 12.10 Let us consider a decision rule system $S = \{(a_1 = 0) \to 0, (a_1 = 1) \to 1, \to 1, \to 2\}$. For this system, $I_{SR}(S) = \{\to 1, \to 2\}$ and $I_{AD}(S) = \{(a_1 = 0) \to 0, \to 1, \to 2\}$.

Let S be a decision rule system. This system will be called *incomplete* if there exists a tuple $\bar{\delta} \in V(S)$ such that the equation system $K(r) \cup K(S, \bar{\delta})$ is inconsistent for any decision rule $r \in S$, i.e., if there exists a tuple $\bar{\delta} \in V(S)$ for which the system S does not contain realizable rules. Otherwise, the system S will be called *complete*. If $n(S) = 0$, then the system S will be considered as complete.

Lemma 12.7 *Let S be a decision rule system. Then*
(a) *If $C \in \{EAR, AR\}$, then $h_C(S) \geq \beta(S)$.*
(b) *If $C \in \{AD, SR\}$, then $h_{EC}(S) \geq \beta(I_C(S))$.*
(c) *If $C \in \{AD, SR\}$ and the system S is incomplete, then $h_C(S) \geq \beta(S)$.*

Proof It is clear that the statements of the lemma hold if $n(S) = 0$. Let $n(S) > 0$.

(a) Let Γ be a decision tree over S, which solves the problem $AR(S)$ and for which $h(\Gamma) = h_{AR}(S)$. Let ξ be a complete path in Γ for which the equation system $K(\xi)$ is consistent. It is clear that, for any decision rule $r \in S$, either $K(r) \subseteq K(\xi)$ or the equation system $K(r) \cup K(\xi)$ is inconsistent. Hence $A(\xi) \cap A(r) \neq \emptyset$ if $A(r) \neq \emptyset$ and $A(\xi)$ is a node cover of the hypergraph $G(S)$. Thus, $h(\Gamma) \geq h(\xi) \geq |A(\xi)| \geq \beta(S)$ and $h_{AR}(S) \geq \beta(S)$. Using Lemma 12.3, we obtain $h_{EAR}(S) \geq \beta(S)$.

(b) If S contains rules of the length 0, then $d(I_{SR}(S)) = 0$, $\beta(I_{SR}(S)) = 0$ and the inequality $h_{ESR}(S) \geq \beta(I_{SR}(S))$ holds. Let S do not contain rules of the length 0. Then $I_{SR}(S) = S$. Let Γ be a decision tree over S, which solves the problem $ESR(S)$ and for which $h(\Gamma) = h_{ESR}(S)$. Let $\bar{\delta} = (*, \ldots, *) \in EV(S)$ and ξ be a complete path in Γ such that $K(\xi) \subseteq K(S, \bar{\delta})$. It is clear that the terminal node of this path is labeled with the empty set of decision rules. Therefore, for any $r \in S$, the equation system $K(\xi) \cup K(r)$ is inconsistent. Hence $A(\xi) \cap A(r) \neq \emptyset$. Thus, $A(\xi)$ is a node cover of the hypergraph $G(S)$. It is clear that $h(\Gamma) \geq |A(\xi)|$ and $|A(\xi)| \geq \beta(S)$. Therefore $h(\Gamma) \geq \beta(S)$ and $h_{ESR}(S) \geq \beta(S) = \beta(I_{SR}(S))$.

Let Γ be a decision tree over S, which solves the problem $EAD(S)$ and for which $h(\Gamma) = h_{EAD}(S)$. Let $\bar{\delta} = (*, \ldots, *) \in EV(S)$ and ξ be a complete path in Γ such that $K(\xi) \subseteq K(S, \bar{\delta})$. It is clear that $\tau(\xi)$ contains only rules of the length 0 and $D(\tau(\xi)) = D_0(S)$. It is clear also that, for any rule $r \in S$ such that the right-hand side of r does not belong to the set $D_0(\tau(\xi))$, the system of equations $K(r) \cup K(\xi)$ is inconsistent. Therefore $A(\xi)$ is a node cover of the hypergraph $G(I_{AD}(S))$. It is clear that $h(\Gamma) \geq |A(\xi)|$ and $|A(\xi)| \geq \beta(I_{AD}(S))$. Therefore $h(\Gamma) \geq \beta(I_{AD}(S))$ and $h_{EAD}(S) \geq \beta(I_{AD}(S))$.

(c) Let Γ be a decision tree over S, which solves the problem $SR(S)$ and for which $h(\Gamma) = h_{SR}(S)$. Let the decision rule system S be incomplete and $\bar{\delta}$ be a tuple from $V(S)$ for which the equation system $K(r) \cup K(S, \bar{\delta})$ is inconsistent for any decision rule $r \in S$. Let us consider a complete path ξ in Γ such that $K(\xi) \subseteq$

$K(S, \bar{\delta})$. It is clear that the terminal node of this path is labeled with the empty set of decision rules. Therefore, for any rule $r \in S$, the equation system $K(\xi) \cup K(r)$ is inconsistent. Hence $A(\xi)$ is a node cover of the hypergraph $G(S)$. Therefore $h(\Gamma) \geq h(\xi) \geq |A(\xi)| \geq \beta(S)$ and $h_{SR}(S) \geq \beta(S)$. Using Lemma 12.3, we obtain $h_{AD}(S) \geq \beta(S)$. □

Note that the condition of incompleteness in the statement (c) of Lemma 12.7 is essential. Let us consider a decision rule system $S = \{(a_1 = 0) \to 0, \ldots, (a_n = 0) \to 0\}$. It is easy to see that this system is complete and $\beta(S) = n$. We now consider a decision tree Γ that contains two nodes v_1, v_2 and an edge d leaving the node v_1 and entering the node v_2. The node v_1 is labeled with the attribute a_1, the node v_2 is labeled with the decision rule system $\{(a_1 = 0) \to 0\}$, and the edge d is labeled with the number 0. One can show that Γ solves the problems $AD(S)$ and $SR(S)$, and $h(\Gamma) = 1$.

Note also that the incompleteness is not a hereditary property: there exists an incomplete decision rule system S and a consistent system of equations $\alpha = \{a_{i_1} = \delta_1, \ldots, a_{i_m} = \delta_m\}$ with $a_{i_1}, \ldots, a_{i_m} \in A(S)$ and $\delta_1 \in V_S(a_{i_1}), \ldots, \delta_m \in V_S(a_{i_m})$ such that the decision rule system S_α is complete. Let us consider a decision rule system $S = \{(a_1 = 0) \wedge (a_2 = 0) \to 0, (a_1 = 1) \wedge (a_2 = 0) \to 0, (a_1 = 0) \wedge (a_2 = 1) \to 0\}$. This system is incomplete: for the tuple $(1, 1) \in V(S)$, there is no decision rule from S that is realizable for this tuple. Let $\alpha = \{a_2 = 0\}$. Then $S_\alpha = \{(a_1 = 0) \to 0, (a_1 = 1) \to 0\}$. It is clear that the decision rule system S_α is complete.

Lemma 12.8 *Let S be a decision rule system. Then*
(a) $h_{EAR}(S) \geq h_{AR}(S) \geq d(S)$.
(b) *If S is an SR-reduced system, then $h_{ESR}(S) \geq d(S)$.*
(c) *If S is an AD-reduced system, then $h_{EAD}(S) \geq d(S)$.*

Proof It is clear that the statements of the lemma hold if $n(S) = 0$. Let $n(S) > 0$.

(a) Let r be a decision rule from S for which the length is equal to $d(S)$. Let Γ be a decision tree over S, which solves the problem $AR(S)$ and for which $h(\Gamma) = h_{AR}(S)$. It is clear that there exists a complete path ξ in Γ such that the equation system $K(r) \cup K(\xi)$ is consistent. Taking into account that Γ solves the problem $AR(S)$, we obtain $K(r) \subseteq K(\xi)$. Therefore $h(\xi) \geq d(S)$, $h(\Gamma) \geq d(S)$, and $h_{AR}(S) \geq d(S)$. Using Lemma 12.3, we obtain $h_{EAR}(S) \geq h_{AR}(S) \geq d(S)$.

(b) Let S be an SR-reduced system of decision rules, r be a decision rule of the length $d(S)$ from S, and Γ be a decision tree over S, which solves the problem $ESR(S)$ and for which $h(\Gamma) = h_{ESR}(\Gamma)$. We denote by $S(r)$ the set of rules ρ from S such that $K(\rho) = K(r)$. Let us consider a tuple $\bar{\delta} \in EV(S)$ such that $K(r) \subseteq K(S, \bar{\delta})$ and each equation from the system $K(S, \bar{\delta}) \backslash K(r)$ has the form $a_t = *$. Taking into account that S is an SR-reduced system, we obtain that, for any rule $r' \in S \backslash S(r)$, the equation system $K(r') \cup K(S, \bar{\delta})$ is inconsistent. Let ξ be a complete path in Γ such that $K(\xi) \subseteq K(S, \bar{\delta})$. Since Γ solves the problem $ESR(S)$, the terminal node of the path ξ is labeled with a nonempty subset of the set $S(r)$ and $K(\rho) \subseteq K(\xi)$ for any $\rho \in S(r)$. Therefore $h(\xi) \geq d(S)$, $h(\Gamma) \geq d(S)$, and $h_{ESR}(S) \geq d(S)$.

(c) Let S be an AD-reduced system of decision rules, r be a decision rule of the length $d(S)$ from S, and Γ be a decision tree over S, which solves the problem $EAD(S)$ and for which $h(\Gamma) = h_{EAD}(\Gamma)$. We denote by $S'(r)$ the set of rules ρ from S such that ρ and r are equal. Let us consider a tuple $\bar{\delta} \in EV(S)$ such that $K(r) \subseteq K(S, \bar{\delta})$ and each equation from the system $K(S, \bar{\delta}) \backslash K(r)$ has the form $a_t = *$. Taking into account that S is an AD-reduced system, we obtain that, for any rule $r' \in S \backslash S'(r)$ such that $K(r') \subseteq K(S, \bar{\delta})$, the right-hand side of the rule r' is different from the right-hand side of the rule r. Let ξ be a complete path in Γ such that $K(\xi) \subseteq K(S, \bar{\delta})$. Taking into account that Γ solves the problem $EAD(S)$, one can show that at least one rule $r \in S'(r)$ belongs to the set $\tau(\xi)$ and $K(r) \subseteq K(\xi)$. Therefore $h(\xi) \geq d(S)$, $h(\Gamma) \geq d(S)$, and $h_{EAD}(S) \geq d(S)$. □

Note that we cannot obtain for the problems $SR(S)$ and $AD(S)$ bounds similar to those mentioned in statements (b) and (c) of Lemma 12.8. Let us consider a decision rule system $S = \{(a_1 = 0) \rightarrow 0, (a_2 = 0) \wedge \cdots \wedge (a_n = 0) \rightarrow 0\}$. It is easy to see that the system S is SR-reduced and AD-reduced, and $d(S) = n - 1$. We now consider a decision tree Γ that contains two nodes v_1, v_2 and an edge d leaving the node v_1 and entering the node v_2. The node v_1 is labeled with the attribute a_1, the node v_2 is labeled with the decision rule system $\{(a_1 = 0) \rightarrow 0\}$, and the edge d is labeled with the number 0. One can show that Γ solves the problems $AD(S)$ and $SR(S)$, and $h(\Gamma) = 1$.

Lemma 12.9 *Let S be a decision rule system. Then $h_{ESR}(S) = h_{ESR}(R_{SR}(S))$ and $h_{EAD}(S) = h_{EAD}(R_{AD}(S))$.*

Proof It is clear that the considered equalities hold if $n(S) = 0$. Let $n(S) > 0$.

Let us first show that $h_{ESR}(S) \leq h_{ESR}(R_{SR}(S))$. Let Γ be a decision tree over $R_{SR}(S)$, which solves the problem $ESR(R_{SR}(S))$ and for which $h(\Gamma) = h_{ESR}(R_{SR}(S))$. We will process all nodes of the tree Γ beginning with the terminal ones. If v is a terminal node of the tree Γ, then we will keep it untouched. Let v be a working node of the tree Γ and let, for each edge d leaving v, all nodes in the subtree of Γ corresponding to d be already processed. Let v be labeled with an attribute a_i. If $EV_S(a_i) = EV_{R_{SR}(S)}(a_i)$, then we keep the node v untouched. Otherwise, for each $\delta \in EV_S(a_i) \backslash EV_{R_{SR}(S)}(a_i)$, we add to the tree Γ a subtree G corresponding to the edge that leaves v and is labeled with the symbol $*$. We also add an edge that leaves v, enters the root of G, and is labeled with the number δ. We denote by Γ' the tree obtained after processing all nodes of Γ. One can show that Γ' is a decision tree over S that solves the problem $ESR(S)$. It is clear that $h(\Gamma') = h(\Gamma)$. Therefore $h_{ESR}(S) \leq h_{ESR}(R_{SR}(S))$.

We now show that $h_{ESR}(S) \geq h_{ESR}(R_{SR}(S))$. Let Γ be a decision tree over S, which solves the problem $ESR(S)$ and for which $h(\Gamma) = h_{ESR}(S)$. We will process all nodes of the tree Γ beginning with the terminal ones. Let v be a terminal node of Γ and v be labeled with a set of decision rules B. If $B = \emptyset$, then we will keep the node v untouched. Otherwise, replace each rule r from B with a rule $r' \in R_{SR}(S)$ such that $K(r') \subseteq K(r)$. Let v be a working node labeled with an attribute a_i and, for any edge d leaving v, all nodes in the subtree of Γ corresponding to d be already processed.

Let $a_i \notin A(R_{SR}(S))$ and G be a subtree of Γ corresponding to the edge d that leaves v and is labeled with the symbol $*$. If v is the root of Γ, then remove from Γ all nodes and edges with the exception of the subtree G. Let v be not the root and d' be an edge that enters v. We remove from the subtree corresponding to d' all nodes and edges with the exception of the subtree G and join d' to the root of G. Let $a_i \in A(R_{SR}(S))$. We process all edges leaving v. Let an edge d leave v and be labeled with a number δ. If $\delta \in EV_{R_{SR}(S)}(a_i)$, then we will keep d untouched. If $\delta \notin EV_{R_{SR}(S)}(a_i)$, then we remove d and corresponding to it subtree. We denote by Γ' the tree obtained after processing all nodes of Γ. One can show that Γ' is a decision tree over $R_{SR}(S)$, which solves the problem $ESR(R_{SR}(S))$. It is clear that $h(\Gamma') \leq h(\Gamma)$. Therefore $h_{ESR}(R_{SR}(S)) \leq h_{ESR}(S)$. Thus, $h_{ESR}(R_{SR}(S)) = h_{ESR}(S)$.

The second part of the lemma statement, the equality $h_{EAD}(S) = h_{EAD}(R_{AD}(S))$, can be proved similarly. □

12.4 Unimprovable Upper Bounds

In this section, we study unimprovable upper bounds on the minimum depth of decision trees depending on three parameters of the decision rule systems.

First, we prove several lemmas.

Lemma 12.10 *Let $n, d, k \in \omega \setminus \{0\}$ and $d \leq n$. Then the following inequalities hold:*

$$\begin{array}{ccccccc} H_{ESR}(n,d,k) & \leq & H_{EAD}(n,d,k) & \leq & H_{EAR}(n,d,k) & \leq & n\ , \\ \text{\rotatebox{90}{$\geq$}} & & \text{\rotatebox{90}{$\geq$}} & & \text{\rotatebox{90}{$\geq$}} & & \\ H_{SR}(n,d,k) & \leq & H_{AD}(n,d,k) & \leq & H_{AR}(n,d,k)\ , & & \end{array}$$

$$H^R_{SR}(n,d,k) \leq H^R_{ESR}(n,d,k) \leq n,\ H^R_{AD}(n,d,k) \leq H^R_{EAD}(n,d,k) \leq n\ .$$

Proof The considered inequalities follow immediately from Lemma 12.3. □

Lemma 12.11 *Let $n, d, k \in \omega \setminus \{0\}$ and $d \leq n$. Then the following inequalities hold:*

$$H_{SR}(n,d,k) \geq H^R_{SR}(n,d,k),\ \ H_{ESR}(n,d,k) \geq H^R_{ESR}(n,d,k)\ ,$$
$$H_{AD}(n,d,k) \geq H^R_{AD}(n,d,k),\ \ H_{EAD}(n,d,k) \geq H^R_{EAD}(n,d,k)\ .$$

Proof The considered inequalities follow from the obvious inclusions $\Sigma_{SR} \subset \Sigma$ and $\Sigma_{AD} \subset \Sigma$. □

Lemma 12.12 *Let $n, d, k \in \omega \setminus \{0\}$ and $d \leq n$. Then*

$$H^R_{SR}(n,d,k) = H_{SR}(n,d,k) = \begin{cases} 1, & \text{if } d = 1\ , \\ d, & \text{if } k = 1\ , \\ n, & \text{if } k > 1 \text{ and } d > 1\ . \end{cases}$$

Proof Let $S \in \Sigma$ and $d(S) = 1$. Let $a_i \in A(S)$. It is clear that, for any $\delta \in V_S(a_i)$, the system S contains a rule r_δ of the form $(a_i = \delta) \to \sigma$. Let $V_S(a_i) = \{\delta_1, \ldots, \delta_m\}$. We denote by Γ a decision tree over S that consists of a node v_0 labeled with the attribute a_i, a node v_j labeled with the set $\{r_{\delta_j}\}$, and an edge d_j, which leaves the node v_0, enters the node v_j, and is labeled with the number δ_j, $j = 1, \ldots, m$. One can show that Γ solves the problem $SR(S)$ and $h(\Gamma) = 1$. Therefore $h_{SR}(S) \le 1$. Using Lemma 12.11, we obtain $H^R_{SR}(n, 1, k) \le H_{SR}(n, 1, k) \le 1$. Let us consider a decision rule system $S = \{(a_1 = 0) \to 0, \ldots, (a_1 = k - 1) \to 0, (a_2 = 0) \to 0, (a_3 = 0) \to 0, \ldots, (a_n = 0) \to 0\}$. It is clear that $S \in \Sigma_{SR}, n(S) = n, d(S) = 1, k(S) = k$, and $h_{SR}(S) \ge 1$. Therefore $H^R_{SR}(n, 1, k) \ge 1$. Thus, $H^R_{SR}(n, 1, k) = H_{SR}(n, 1, k) = 1$.

Let $S \in \Sigma$ and $k(S) = 1$. One can show that the value $h_{SR}(S)$ is equal to the minimum length of a rule from S. Using this fact and Lemma 12.11, we obtain that $H^R_{SR}(n, d, 1) \le H_{SR}(n, d, 1) \le d$. Let us consider a system of decision rules $S = \{(a_i = 0) \wedge \cdots \wedge (a_{i+d-1} = 0) \to i : i = 1, \ldots, n - d + 1\}$. It is clear that $S \in \Sigma_{SR}$, $n(S) = n$, $d(S) = d$, $k(S) = 1$, and $h_{SR}(S) = d$. Therefore $H^R_{SR}(n, d, 1) \ge d$ and $H^R_{SR}(n, d, 1) = H_{SR}(n, d, 1) = d$.

Let $k > 1$ and $d > 1$. We now show that $H^R_{SR}(n, d, k) \ge n$. Let us consider a decision rule system

$$
\begin{aligned}
S = \{&(a_1 = 1) \wedge \cdots \wedge (a_d = 1) \to 0, (a_1 = 1) \wedge (a_2 = 0) \to 1\,,\\
&(a_2 = 1) \wedge (a_3 = 0) \to 2, \ldots, (a_{n-1} = 1) \wedge (a_n = 0) \to n - 1\,,\\
&(a_n = 1) \wedge (a_1 = 0) \to n, (a_1 = 2) \to 0, \ldots, (a_1 = k - 1) \to 0\}\,.
\end{aligned}
$$

It is clear that $S \in \Sigma_{SR}$, $n(S) = n$, $d(S) = d$, and $k(S) = k$. Let Γ be a decision tree, which solves the problem $SR(S)$ and for which $h(\Gamma) = h_{SR}(S)$. Let $\bar{\delta} = (0, \ldots, 0) \in V(S)$ and ξ be a complete path in Γ such that $K(\xi) \subseteq K(S, \bar{\delta})$. It is clear that there are no rules from S that are realizable for $\bar{\delta}$. Therefore the terminal node of ξ is labeled with the empty set of decision rules and, for any rule $r \in S$, the system of equations $K(r) \cup K(\xi)$ is inconsistent. Let us assume that, for some $i \in \{1, \ldots, n\}$, the equation $a_i = 0$ does not belong to $K(\xi)$. Then the system of equations $K(r_i) \cup K(\xi)$ is consistent, where r_i is the rule from S with the right-hand side equal to i, but this is impossible. Therefore $h(\xi) \ge n$, $h(\Gamma) \ge n$, and $h_{SR}(S) \ge n$. Hence $H^R_{SR}(n, d, k) \ge n$. Using Lemmas 12.10 and 12.11, we obtain $H^R_{SR}(n, d, k) = H_{SR}(n, d, k) = n$. □

Lemma 12.13 *Let $n, d, k \in \omega \setminus \{0\}$ and $d \le n$. Then*

$$
H^R_{ESR}(n, d, k) = n \quad and \quad H^R_{AD}(n, d, k) = n\,.
$$

Proof Let us consider a system of decision rules $S = S_1 \cup S_2$, where $S_1 = \{(a_1 = 0) \wedge \cdots \wedge (a_d = 0) \to 0, (a_{d+1} = 0) \to d + 1, \ldots, (a_n = 0) \to n\}$ and $S_2 = \{(a_1 = 1) \to n + 1, \ldots, (a_1 = k - 1) \to n + k - 1\}$. If $k = 1$, then $S_2 = \emptyset$. It is clear that $S \in \Sigma_{SR} \cap \Sigma_{AD}$, $n(S) = n$, $d(S) = d$, and $k(S) = k$.

Let Γ_1 be a decision tree, which solves the problem $AD(S)$ and for which $h(\Gamma_1) = h_{AD}(S)$. Let $\bar{\delta} = (0, \ldots, 0) \in V(S)$ and ξ be a complete path in Γ_1 such that $K(\xi) \subseteq K(S, \bar{\delta})$. Taking into account that Γ_1 solves the problem $AD(S)$, one can show that the terminal node of this path is labeled with the set S_1 and $K(r) \subseteq K(\xi)$ for any $r \in S_1$. Therefore $h(\xi) \geq n$. Hence $h(\Gamma_1) \geq n$ and $h_{AD}(S) \geq n$. Thus, $H^R_{AD}(n, d, k) \geq n$. Using Lemma 12.10, we obtain that $H^R_{AD}(n, d, k) = n$.

Let Γ_2 be a decision tree, which solves the problem $ESR(S)$ and for which $h(\Gamma_2) = h_{ESR}(S)$. Let $\bar{\delta} = (\delta_1, \ldots, \delta_n)$ be a tuple from $EV(S)$ such that $\delta_1 = \cdots = \delta_d = 0$ and $\delta_{d+1} = \cdots = \delta_n = *$. Let ξ be a complete path in Γ_2 for which $K(\xi) \subseteq K(S, \bar{\delta})$. We denote by r_0 the decision rule $(a_1 = 0) \wedge \cdots \wedge (a_d = 0) \rightarrow 0$. It is clear that the only decision rule r_0 from S is realizable for the tuple $\bar{\delta}$. Taking into account that Γ_2 solves the problem $ESR(S)$, we obtain that $K(r_0) \subseteq K(\xi)$. Let $\xi = v_1, d_1, \ldots, v_m, d_m, v_{m+1}$. For $i = 1, \ldots, d$, let j_i be the minimum number from the set $\{1, \ldots, m\}$ such that the node v_{j_i} is labeled with the attribute a_i. Let i_0 be a number from $\{1, \ldots, d\}$ such that $j_{i_0} = \max\{j_1, \ldots, j_d\}$. Let us consider the tuple $\bar{\sigma} = (\sigma_1, \ldots, \sigma_n)$ from $EV(S)$ such that $\sigma_1 = \cdots = \sigma_{i_0 - 1} = \sigma_{i_0 + 1} = \cdots = \sigma_d = 0$ and $\sigma_{i_0} = \sigma_{d+1} = \cdots = \sigma_n = *$. Let ξ' be a complete path in Γ_2 such that $K(\xi') \subseteq K(S, \bar{\sigma})$. It is clear that there are no decision rules from S that are realizable for the tuple $\bar{\sigma}$. Taking into account that Γ_2 solves the problem $ESR(S)$, we obtain that the system of equations $K(r) \cup K(\xi')$ is inconsistent for any $r \in S$. It is clear that the paths ξ and ξ' have j_{i_0} common nodes $v_1, \ldots, v_{j_{i_0}}$. Therefore $a_1, \ldots, a_d \in A(\xi')$. Let us assume that $a_j \notin A(\xi')$ for some $j \in \{d + 1, \ldots, n\}$. Then the system of equations $K((a_j = 0) \rightarrow j) \cup K(\xi')$ is consistent, which is impossible. Therefore $|A(\xi')| = n$ and $h(\xi') \geq n$. Hence $h(\Gamma_2) \geq n$, $h_{ESR}(S) \geq n$, and $H^R_{ESR}(n, d, k) \geq n$. Using Lemma 12.10, we obtain $H^R_{ESR}(n, d, k) = n$. □

Theorem 12.1 *Let $n, d, k \in \omega \setminus \{0\}$ and $d \leq n$. Then*

$$H^R_{SR}(n, d, k) = H_{SR}(n, d, k) = \begin{cases} 1, & \text{if } d = 1\,, \\ d, & \text{if } k = 1\,, \\ n, & \text{if } k > 1 \text{ and } d > 1 \end{cases}$$

and

$$\begin{aligned} H^R_{ESR}(n, d, k) = H^R_{AD}(n, d, k) = H^R_{EAD}(n, d, k) = H_{AD}(n, d, k) = H_{AR}(n, d, k) \\ = H_{ESR}(n, d, k) = H_{EAD}(n, d, k) = H_{EAR}(n, d, k) = n\,. \end{aligned}$$

Proof The first statement of the theorem follows from Lemma 12.12. The second statement follows from Lemmas 12.10, 12.11, and 12.13. □

Remark 12.3 Note that the equality $H_{AR}(n, d, k) = n$ mentioned in Theorem 12.1 was obtained in [5]. The equality $H_{EAR}(n, d, k) = n$ mentioned in Theorem 12.1 was published in [6] without proof.

12.5 Unimprovable Lower Bounds

In this section, we study unimprovable lower bounds on the minimum depth of decision trees depending on three parameters of the decision rule systems.

First, we consider two simple statements.

Lemma 12.14 *Let $n, d, k \in \omega \setminus \{0\}$ and $d \leq n$. Then the following inequalities hold:*

$$\begin{array}{ccccccc} h_{ESR}(n,d,k) & \leq & h_{EAD}(n,d,k) & \leq & h_{EAR}(n,d,k) & \leq n\ , \\ \text{\rotatebox{90}{$\leq$}} & & \text{\rotatebox{90}{$\leq$}} & & \text{\rotatebox{90}{$\leq$}} & \\ h_{SR}(n,d,k) & \leq & h_{AD}(n,d,k) & \leq & h_{AR}(n,d,k)\ , & \end{array}$$

$$h^R_{SR}(n,d,k) \leq h^R_{ESR}(n,d,k) \leq n,\ h^R_{AD}(n,d,k) \leq h^R_{EAD}(n,d,k) \leq n\ .$$

Proof The considered inequalities follow immediately from Lemma 12.3. □

Lemma 12.15 *Let $n, d, k \in \omega \setminus \{0\}$ and $d \leq n$. Then the following inequalities hold:*

$$h_{SR}(n,d,k) \leq h^R_{SR}(n,d,k),\ \ h_{ESR}(n,d,k) \leq h^R_{ESR}(n,d,k)\ ,$$
$$h_{AD}(n,d,k) \leq h^R_{AD}(n,d,k),\ \ h_{EAD}(n,d,k) \leq h^R_{EAD}(n,d,k)\ .$$

Proof The considered inequalities follow from the obvious inclusions $\Sigma_{SR} \subset \Sigma$ and $\Sigma_{AD} \subset \Sigma$. □

12.5.1 Bounds on $h_{SR}(n,d,k)$, $h_{AD}(n,d,k)$, $h_{ESR}(n,d,k)$, $h_{EAD}(n,d,k)$, $h^R_{SR}(n,d,k)$, and $h^R_{AD}(n,d,k)$

Lemma 12.16 *Let $n, d, k \in \omega \setminus \{0\}$ and $d \leq n$. Then*

$$h_{SR}(n,d,k) = h_{AD}(n,d,k) = h_{ESR}(n,d,k) = h_{EAD}(n,d,k) = 0\ .$$

Proof Let $C \in \{SR, AD, ESR, EAD\}$. We now consider a decision rule system

$$S = \{\rightarrow 0, (a_1 = 0) \wedge \cdots \wedge (a_d = 0) \rightarrow 0, (a_{d+1} = 0) \rightarrow 0, \ldots, (a_n = 0) \rightarrow 0\ ,$$
$$(a_1 = 1) \rightarrow 0, \ldots, (a_1 = k - 1) \rightarrow 0\}\ .$$

It is clear that $n(S) = n$, $d(S) = d$, and $k(S) = k$. Let Γ be a decision tree that consists of one node, which is labeled with the decision rule set $\{\rightarrow 0\}$. It is easy to show that Γ solves the problem $C(S)$. Therefore $h_C(n,d,k) \leq 0$. Evidently, $h_C(n,d,k) \geq 0$. Thus, $h_C(n,d,k) = 0$. □

Lemma 12.17 *Let $n, d, k \in \omega \setminus \{0\}$ and $d \leq n$. Then*

$$h^R_{SR}(n, d, k) = h^R_{AD}(n, d, k) = \begin{cases} n, \text{ if } d = n \ , \\ 1, \text{ if } d \neq n \ . \end{cases}$$

Proof Let $S \in \Sigma_{SR}$, $n(S) = n$, $d(S) = n$, and $k(S) = k$. Let $A(S) = \{a_{i_1}, \ldots, a_{i_n}\}$. Then there exists a decision rule r_0 from S of the form $(a_{i_1} = \delta_1) \wedge \cdots \wedge (a_{i_n} = \delta_n) \rightarrow \sigma$. Let Γ be a decision tree, which solves the problem $SR(S)$ and for which $h(\Gamma) = h_{SR}(S)$. Let $\bar{\delta} = (\delta_1, \ldots, \delta_n)$ and ξ be a complete path in Γ such that $K(\xi) \subseteq K(S, \bar{\delta})$. Since S is an SR-reduced system of decision rules, only decision rules r from S with $K(r) = K(r_0)$ are realizable for the tuple $\bar{\delta}$. Since Γ solves the problem $SR(S)$, $K(r_0) \subseteq K(\xi)$. Hence $h(\xi) \geq n$, $h(\Gamma) \geq n$, $h_{SR}(S) \geq n$, and $h^R_{SR}(n, n, k) \geq n$. Using Lemma 12.14, we obtain $h^R_{SR}(n, n, k) = n$.

Let $S \in \Sigma_{AD}$, $n(S) = n$, $d(S) = n$, and $k(S) = k$. Let $A(S) = \{a_{i_1}, \ldots, a_{i_n}\}$. Then there exists a decision rule r_0 from S of the form $(a_{i_1} = \delta_1) \wedge \cdots \wedge (a_{i_n} = \delta_n) \rightarrow \sigma$. Let Γ be a decision tree, which solves the problem $AD(S)$ and for which $h(\Gamma) = h_{AD}(S)$. Let $\bar{\delta} = (\delta_1, \ldots, \delta_n)$ and ξ be a complete path in Γ such that $K(\xi) \subseteq K(S, \bar{\delta})$. It is clear that r_0 is realizable for the tuple $\bar{\delta}$. Taking into account that S is an AD-reduced system of decision rules, we obtain that any rule $r \in S$, which is different from r_0 and is realizable for the tuple $\bar{\delta}$, has the right-hand side different from σ. Taking into account that Γ solves the problem $AD(S)$, we obtain $K(r_0) \subseteq K(\xi)$. Therefore $h(\Gamma) \geq n$, $h_{AD}(S) \geq n$, and $h^R_{AD}(n, n, k) \geq n$. Using Lemma 12.14, we obtain $h^R_{SR}(n, n, k) = n$.

Let $d < n$. Let us consider a decision rule system $S = S_1 \cup S_2$, where $S_1 = \{(a_1 = 0) \wedge \cdots \wedge (a_d = 0) \rightarrow 0, (a_{d+1} = 0) \rightarrow 0, \ldots, (a_n = 0) \rightarrow 0\}$ and $S_2 = \{(a_1 = 1) \rightarrow 0, \ldots, (a_1 = k - 1) \rightarrow 0\}$. If $k = 1$, then $S_2 = \emptyset$. It is clear that $S \in \Sigma_{SR} \cap \Sigma_{AD}$, $n(S) = n$, $d(S) = d$, and $k(S) = k$. Denote by Γ a decision tree that consists of nodes v_1, v_2 and edge d leaving v_1 and entering v_2. The node v_1 is labeled with the attribute a_n, the node v_2 is labeled with the set of decision rules $\{(a_n = 0) \rightarrow 0\}$, and the edge d is labeled with the number 0. One can show that $h(\Gamma) = 1$ and the decision tree Γ solves the problems $SR(S)$ and $AD(S)$. Therefore $h^R_{SR}(n, d, k) \leq 1$ and $h^R_{AD}(n, d, k) \leq 1$.

Let $S \in \Sigma_{SR}$ and $n(S) > 0$. Then S does not contain decision rules of the length 0. Using this fact, one can show that $h_{SR}(S) \geq 1$. Therefore $h^R_{SR}(n, d, k) = 1$ if $d < n$.

Let $S \in \Sigma_{AD}$ and $n(S) > 0$. Then S contains a decision rule of the form $(a_{i_1} = \delta_1) \wedge \cdots \wedge (a_{i_m} = \delta_m) \rightarrow \sigma$, where $m > 0$. Taking into account that $S \in \Sigma_{AD}$, we obtain S does not contain decision rules of the form $\rightarrow \sigma$. Using these considerations, one can show that $h_{AD}(S) \geq 1$. Therefore $h^R_{AD}(n, d, k) = 1$ if $d < n$. □

12.5.2 Bounds on $h_{AR}(n, d, k)$ and $h_{EAR}(n, d, k)$

Lemma 12.18 *Let S be a decision rule system and S' be a nonempty subsystem of the system S. Then $h_{AR}(S) \geq h_{AR}(S')$.*

Proof It is clear that the considered inequality holds if $n(S) = 0$. Let $n(S) > 0$.

Let Γ be a decision tree, which solves the problem $AR(S)$ and for which $h(\Gamma) = h_{AR}(S)$. Starting from the root we will process all working nodes of the tree Γ. Let v be a working node labeled with an attribute a_t. Let $a_t \in A(S')$. We keep all edges leaving v that are labeled with numbers from the set $V_{S'}(a_t)$. We remove all other edges leaving v together with the subtrees corresponding to these edges. Let $a_t \notin A(S')$ and δ be the minimum number from $V_S(a_t)$. We remove all edges leaving v together with the subtrees corresponding to these edges with the exception of the edge labeled with δ. Denote by Γ_0 the tree obtained from Γ after processing of all working nodes.

We now process all nodes in the tree Γ_0. Let v be a working node labeled with an attribute a_t. If $a_t \in A(S')$, then leave the node v untouched. Let $a_t \notin A(S')$. If v is the root of the tree Γ_0, then remove the node v and the edge leaving it. Let v be not the root. Let d_0 be the edge entering v and d_1 be the edge leaving the node v. Let d_1 enter a node v_1. We remove from Γ_0 the node v and the edge d_1, and join the edge d_0 to the node v_1. Let v be a terminal node labeled with a set of decision rules B. Replace the set B with the set $B \cap S'$. Denote the obtained tree Γ'.

One can show that Γ' is an o-tree over the decision rule system S'. We now prove that Γ' solves the problem $AR(S')$. Let ξ' be a complete path in Γ' with consistent system of equations $K(\xi')$. Then there exists a complete path ξ in Γ satisfying the following conditions: $\tau(\xi') = \tau(\xi) \cap S'$, $K(\xi') \subseteq K(\xi)$, and all equations from the set $K(\xi) \backslash K(\xi')$ have the form $a_t = \delta$, where $a_t \notin A(S')$ and δ is the minimum number from the set $V_S(a_t)$. It is easy to show that the system of equations $K(\xi)$ is consistent. Let $r \in \tau(\xi')$. Then $r \in \tau(\xi)$. Therefore $K(r) \subseteq K(\xi)$ and hence $K(r) \subseteq K(\xi')$. Let $r \in S' \backslash \tau(\xi')$. Then $r \in S \backslash \tau(\xi)$. Thus, the system $K(r) \cup K(\xi)$ is inconsistent. Therefore the system $K(r) \cup K(\xi')$ is also inconsistent. As a result, we obtain that Γ' solves the problem $AR(S')$. It is clear that $h(\Gamma') \le h(\Gamma)$. Thus, $h_{AR}(S') \le h(\Gamma) = h_{AR}(S)$. □

Lemma 12.19 *Let $n, d, k \in \omega \backslash \{0\}$ and $d \le n$. Then*
(a) *If $k = 1$ or $d = 1$, then $h_{AR}(n, d, k) = n$.*
(b) *If $k \ge 2$ and $d \ge 2$, then $h_{AR}(n, d, k) \ge \max \left\{d, \frac{n(k-1)}{k^d}\right\}$.*

Proof (a) Let S be a decision rule system with $n(S) = n$, $d(S) = d$, and $k(S) = k$.

Let $k = 1$ and Γ be a decision tree, which solves the problem $AR(S)$ and for which $h(\Gamma) = h_{AR}(S)$. It is clear that the decision tree Γ has exactly one complete path ξ. The terminal node of this path is labeled with the set S and $K(r) \subseteq K(\xi)$ for any rule $r \in S$. It is clear that $h(\xi) \ge n$, $h(\Gamma) \ge n$, and $h_{AR}(S) \ge n$. Taking into account that S is an arbitrary decision rule system with $k(S) = 1$, we obtain $h_{AR}(n, d, 1) \ge n$. By Lemma 12.14, $h_{AR}(n, d, 1) = n$.

Let $d = 1$. One can show that in this case $\beta(S) = n$. By Lemma 12.7, $h_{AR}(S) \ge n$. Taking into account that S is an arbitrary decision rule system with $d(S) = 1$, we obtain $h_{AR}(n, 1, k) \ge n$. From this inequality and from Lemma 12.14 it follows that $h_{AR}(n, 1, k) = n$.

(b) Let S be a decision rule system. From Lemma 12.8 it follows that $h_{AR}(S) \geq d(S)$. Since S is an arbitrary decision rule system, $h_{AR}(n, d, k) \geq d$.

We will prove by induction on d that, for any $n, d, k \in \omega \setminus \{0\}$ such that $d \leq n$, the following inequality holds:

$$h_{AR}(n, d, k) \geq \frac{n(k-1)}{k^d} . \tag{12.1}$$

From the equality $h_{AR}(n, 1, k) = n$ proved above it follows that the inequality (12.1) holds if $d = 1$. Let us assume that, for some $d \geq 1$, the inequality (12.1) holds for any d', $1 \leq d' \leq d$. We now show that (12.1) holds for $d + 1$. Let S be a decision rule system with $n(S) = n$, $d(S) = d + 1$, and $k(S) = k$. We choose a subsystem S' of the system S such that $A(S') = A(S)$ and $A(S_1) \neq A(S)$ for any system $S_1 \subset S'$. It is clear that $n(S') = n$, $d(S') \leq d + 1$, and $k(S') \leq k$. Using Lemma 12.18, we obtain

$$h_{AR}(S) \geq h_{AR}(S') . \tag{12.2}$$

Let $S' = \{r_1, \ldots, r_p\}$. It is clear that, for $j = 1, \ldots, p$, the equation system $K(r_j)$ contains an equation $a_{i_j} = \sigma_j$ such that $a_{i_j} \notin A(S') \setminus A(r_j)$. Therefore

$$p = |S'| \leq n . \tag{12.3}$$

Let us assume that $|S'| \leq \frac{n(k-1)}{k}$. Denote $\alpha = \{a_{i_1} = \sigma_1, \ldots, a_{i_p} = \sigma_p\}$. It is clear that α is a consistent equation system. We now consider the system of decision rules S'_α (corresponding definition is before Lemma 12.4). Denote $n_0 = n(S'_\alpha)$, $d_0 = d(S'_\alpha)$, and $k_0 = k(S'_\alpha)$. One can show that $n_0 = n - p \geq n - \frac{n(k-1)}{k} = \frac{n}{k}$, $1 \leq k_0 \leq k$, and $1 \leq d_0 \leq d$. Let $k_0 = 1$ or $d_0 = 1$. Using part (a) of the lemma statement, we obtain $h_{AR}(S'_\alpha) \geq n_0 \geq \frac{n}{k} \geq \frac{n}{k} \frac{(k-1)}{k^d} = \frac{n(k-1)}{k^{d+1}}$.

Let $k_0 \geq 2$ and $d_0 \geq 2$. In this case, $d \geq 2$. By inductive hypothesis, $h_{AR}(S'_\alpha) \geq \frac{n_0(k_0-1)}{k_0^{d_0}} \geq \frac{n}{k} \frac{(k_0-1)}{k_0^d}$. We now show that $\frac{k_0-1}{k_0^d} \geq \frac{k-1}{k^d}$. To this end, for $x \geq 2$, we consider the function $f(x) = \frac{x-1}{x^d}$ and its derivative $f'(x) = \frac{1-d}{x^d} + \frac{d}{x^{d+1}}$. It is easy to show that $f'(x) < 0$ if $x > 2$. Using mean value theorem and the inequalities $2 \leq k_0 \leq k$, we obtain $f(k_0) \geq f(k)$. Therefore $h_{AR}(S'_\alpha) \geq \frac{n(k-1)}{k^{d+1}}$. By Lemma 12.6, $h_{AR}(S') \geq h_{AR}(S'_\alpha)$. From the considered relations and (12.2) it follows that $h_{AR}(S) \geq \frac{n(k-1)}{k^{d+1}}$.

Let us assume now that $|S'| > \frac{n(k-1)}{k}$. Denote $m = n - |S'|$. From (12.3) it follows that $m \geq 0$. Let $m = 0$. Then, as it is not difficult to note, $d(S') = 1$. Using part (a) of the lemma statement, we obtain $h_{AR}(S') \geq n(S') = |S'| \geq \frac{n(k-1)}{k} \geq \frac{n(k-1)}{k^{d+1}}$. Using (12.2), we obtain $h_{AR}(S) \geq \frac{n(k-1)}{k^{d+1}}$.

We now assume that $m > 0$. Denote $B = A(S) \setminus \{a_{i_1}, \ldots, a_{i_p}\}$. It is clear that $|B| = m$. Let $B = \{a_{l_1}, \ldots, a_{l_m}\}$ and $l_1 < \cdots < l_m$. Let $j \in \{1, \ldots, m\}$. We now define a set V_j. If $|V_{S'}(a_{l_j})| = k$, then $V_j = V_{S'}(a_{l_j})$. If $|V_{S'}(a_{l_j})| < k$, then V_j is a subset of ω such that $|V_j| = k$ and $V_{S'}(a_{l_j}) \subset V_j$. Denote $V = V_1 \times \cdots \times V_m$. Let $q = d(S') - 1$. It is clear that $q \leq m$ and $q \leq d$. It is also clear that, for any decision rule $r \in S'$, there exist at least k^{m-q} tuples $\bar{\delta} = (\delta_1, \ldots, \delta_m) \in V$ such that the system of equations $K(r) \cup \{a_{l_1} = \delta_1, \ldots, a_{l_m} = \delta_m\}$ is consistent. For each $\bar{\delta} = (\delta_1, \ldots, \delta_m) \in V$, let $N(\bar{\delta})$ be the number of decision rules $r \in S'$ such that the system $K(r) \cup \{a_{l_1} = \delta_1, \ldots, a_{l_m} = \delta_m\}$ is consistent. Denote $N = \sum_{\bar{\delta} \in V} N(\bar{\delta})$. It is clear that $N \geq |S'| \, k^{m-q} \geq \frac{k^m(k-1)n}{k^{q+1}}$. It is also clear that there exists a tuple $\bar{\delta}' \in V$ such that $N(\bar{\delta}') \geq \frac{N}{|V|} = \frac{N}{k^m} \geq \frac{n(k-1)}{k^{q+1}} \geq \frac{n(k-1)}{k^{d+1}}$. Let $\bar{\delta}' = (\delta'_1, \ldots, \delta'_m)$. We now define a tuple $\bar{\delta} = (\delta_1, \ldots, \delta_m)$. Let $j \in \{1, \ldots, m\}$. If $\delta'_j \in V_{S'}(a_{l_j})$, then $\delta_j = \delta'_j$. If $\delta'_j \notin V_{S'}(a_{l_j})$, then δ_j is the minimum number from the set $V_{S'}(a_{l_1})$. It is not difficult to show that $N(\bar{\delta}) \geq N(\bar{\delta}') \geq \frac{n(k-1)}{k^{d+1}}$. Denote $\alpha = \{a_{l_1} = \delta_1, \ldots, a_{l_m} = \delta_m\}$ and consider the system of decision rules S'_α. One can show that $k(S'_\alpha) = 1$, $d(S'_\alpha) = 1$, and $n(S'_\alpha) = N(\bar{\delta}) \geq \frac{n(k-1)}{k^{d+1}}$. Using part (a) of the lemma statement, we obtain $h_{AR}(S'_\alpha) \geq \frac{n(k-1)}{k^{d+1}}$. By Lemma 12.6, $h_{AR}(S') \geq \frac{n(k-1)}{k^{d+1}}$. From this inequality and (12.2) it follows that $h_{AR}(S) \geq \frac{n(k-1)}{k^{d+1}}$. Therefore the inequality (12.1) holds for $d+1$. Thus, the inequality (12.1) holds. □

Lemma 12.20 *Let $n, d, k \in \omega \setminus \{0\}$, $d \leq n$, $d \geq 2$, and $k \geq 2$. Then $h_{EAR}(n, d, k) \leq d + \frac{n}{k^{d-1}}$.*

Proof We now describe a system of decision rules S with $A(S) = \{a_1, \ldots, a_n\}$. Denote $E_k = \{0, 1, \ldots, k-1\}$. Let us consider a partition $\{a_d, \ldots, a_n\} = \bigcup_{\bar{\delta} \in E_k^{d-1}} B(\bar{\delta})$ such that $|B(\bar{\delta})| \leq \lceil \frac{n-d+1}{k^{d-1}} \rceil$ for any $\bar{\delta} \in E_k^{d-1}$ (we assume that $B(\bar{\delta}_1) \cap B(\bar{\delta}_2) = \emptyset$ for any $\bar{\delta}_1, \bar{\delta}_2 \in E_k^{d-1}$, $\bar{\delta}_1 \neq \bar{\delta}_2$). It is clear that some sets in this partition can be empty, but at least one is nonempty. Let $\bar{\delta} = (\delta_1, \ldots, \delta_{d-1}) \in E_k^{d-1}$. Describe a system of decision rules $S(\bar{\delta})$. If $B(\bar{\delta}) = \emptyset$, then $S(\bar{\delta}) = \{(a_1 = \delta_1) \wedge \cdots \wedge (a_{d-1} = \delta_{d-1}) \to 0\}$. If $B(\bar{\delta}) \neq \emptyset$, then $S(\bar{\delta}) = \{(a_1 = \delta_1) \wedge \cdots \wedge (a_{d-1} = \delta_{d-1}) \wedge (a_i = 0) \to 0 : a_i \in B(\bar{\delta})\}$. Denote $S = \bigcup_{\bar{\delta} \in E_k^{d-1}} S(\bar{\delta})$. It is clear that $n(S) = n$, $d(S) = d$, and $k(S) = k$.

Let us describe a decision tree Γ. First, we compute the values of attributes $a_1, \ldots, a_{d-1}$. Let $a_1 = \delta_1, \ldots, a_{d-1} = \delta_{d-1}$. If there exists $i \in \{1, \ldots, d-1\}$ such that $\delta_i = *$, then the output of Γ is the empty set of decision rules. Let $\bar{\delta} = (\delta_1, \ldots, \delta_{d-1}) \in E_k^{d-1}$. If $B(\bar{\delta}) = \emptyset$, then the output of Γ is the set $S(\bar{\delta})$. Let $B(\bar{\delta}) \neq \emptyset$ and $B(\bar{\delta}) = \{a_{i_1}, \ldots, a_{i_m}\}$. We compute the values of attributes $a_{i_1}, \ldots, a_{i_m}$. Let $a_{i_1} = \sigma_1, \ldots, a_{i_m} = \sigma_m$. Then the output of Γ is the set of decision rules $\{(a_1 = \delta_1) \wedge \cdots \wedge (a_{d-1} = \delta_{d-1}) \wedge (a_{i_j} = 0) \to 0 : j \in \{1, \ldots, m\}, \sigma_j = 0\}$. One can show that Γ solves the problem $EAR(S)$ and $h(\Gamma) \leq d - 1 + \lceil \frac{n-d+1}{k^{d-1}} \rceil \leq d + \frac{n}{k^{d-1}}$. Therefore $h_{EAR}(n, d, k) \leq d + \frac{n}{k^{d-1}}$. □

Lemma 12.21 *Let $n, d, k \in \omega \setminus \{0\}$ and $d \leq n$. Then*

(a) *If $k = 1$ or $d = 1$, then $h_{AR}(n, d, k) = h_{EAR}(n, d, k) = n$.*

(b) *If $k \geq 2$ and $d \geq 2$, then*

$$\max\left\{d, \frac{n(k-1)}{k^d}\right\} \leq h_{AR}(n, d, k) \leq h_{EAR}(n, d, k) \leq d + \frac{n}{k^{d-1}} .$$

Proof From Lemma 12.14 it follows that $h_{AR}(n,d,k) \le h_{EAR}(n,d,k) \le n$. Using these inequalities and Lemmas 12.19 and 12.20, we obtain the considered statements. □

12.5.3 Bounds on $h^R_{ESR}(n,d,k)$ and $h^R_{EAD}(n,d,k)$

Lemma 12.22 *Let $C \in \{SR, AD\}$, $n, d, k \in \omega \setminus \{0\}$ and $d \le n$. Then $h^R_{EC}(n,d,k) \ge \max\left\{d, \frac{(nk)^{1/d}}{k} - d\right\}$.*

Proof Let $S \in \Sigma_C$. Using Lemma 12.8, we obtain that $h_{EC}(S) \ge d(S)$. Since S is an arbitrary C-reduced system, the inequality $h^R_{EC}(n,d,k) \ge d$ holds.

We prove by induction on d that, for any $n, d, k \in \omega \setminus \{0\}$ such that $d \le n$, the following statement holds. Let $S' \in \Sigma_C$, $S \subseteq S'$, $n(S) = n$, $k(S) = k$, and $d(S) = d$. Then

$$h_{EC}(S') \ge \frac{(nk)^{1/d}}{k} - d\ .$$

Let $S' \in \Sigma_C$, $S \subseteq S'$, $n(S) = n$, $k(S) = k$, and $d(S) = 1$. It is clear that $\beta(S') \ge \beta(S) = n$. Let us show that $I_C(S') = S'$ (corresponding definitions are before Lemma 12.7). If $C = SR$, then S' does not contain both a rule of the length 0 and a rule of the length greater than 0. If $C = AD$, then S' does not contain both a rule of the length 0 and a rule of the length greater than 0 with the same right-hand sides. Using Remarks 12.1 and 12.2, we obtain $I_C(S') = S'$. By Lemma 12.7, $h_{EC}(S') \ge n$. Thus, for $d = 1$, the considered statement holds.

Let us assume that, for some $d \ge 1$, this statement holds for each d', $1 \le d' \le d$. We prove that it holds for $d+1$. Let $S' \in \Sigma_C$, $S \subseteq S'$, $n(S) = n$, $k(S) = k$, and $d(S) = d+1$. We now show that

$$h_{EC}(S') \ge \frac{(nk)^{\frac{1}{d+1}}}{k} - d - 1\ . \tag{12.4}$$

By proved above, $h_{EC}(S') \ge d(S') \ge d(S) \ge 2$. Therefore if $2 \ge \frac{(nk)^{\frac{1}{d+1}}}{k} - d - 1$, then the inequality (12.4) holds. Let us assume that $\frac{(nk)^{\frac{1}{d+1}}}{k} - d - 1 > 2$. Then

$$\frac{(nk)^{\frac{1}{d+1}}}{k} > 2\ . \tag{12.5}$$

Denote $m = \beta(S)$. It is clear that $\beta(S') \ge \beta(S)$. Using Remarks 12.1 and 12.2 and the fact that $S' \in \Sigma_C$, we obtain $I_C(S') = S'$. From Lemma 12.7 it follows that $h_{EC}(S') \ge m$. Let $m \ge \frac{(nk)^{\frac{1}{d+1}}}{k} - d - 1$. Then the inequality (12.4) holds. Let us assume now that $\frac{(nk)^{\frac{1}{d+1}}}{k} - d - 1 > m$. Then

$$\frac{(nk)^{\frac{1}{d+1}}}{k} > m . \qquad (12.6)$$

Let $B = \{a_{i_1}, \ldots, a_{i_m}\}$ be a node cover of the hypergraph $G(S)$ with the minimum cardinality. For arbitrary $j \in \{1, \ldots, m\}$ and $\delta \in V_S(a_{i_j})$, we denote by $S(a_{i_j} = \delta)$ the set of decision rules from S, which have the equation $a_{i_j} = \delta$ in the left-hand side. Since B is a node cover of the hypergraph $G(S)$, $A(S) = \bigcup_{j\in\{1,\ldots,m\}, \delta\in V_S(a_{i_j})} A(S(a_{i_j} = \delta))$. Therefore there exist $j \in \{1, \ldots, m\}$ and $\delta \in V_S(a_{i_j})$ such that $|A(S(a_{i_j} = \delta))| \geq \frac{n}{mk}$. Using (12.6), we obtain

$$|A(S(a_{i_j} = \delta))| > \frac{n}{(nk)^{\frac{1}{d+1}}} . \qquad (12.7)$$

Denote $\alpha = \{a_{i_j} = \delta\}$. Let us consider the system of decision rules $S'' = R_C(S'_\alpha)$. Denote by S_0 the system of decision rules obtained from the system $S(a_{i_j} = \delta)$ by removing the equation $a_{i_j} = \delta$ from the left-hand sides of the rules included in $S(a_{i_j} = \delta)$. Denote $k_0 = k(S_0)$, $d_0 = d(S_0)$, and $n_0 = n(S_0)$. One can show that $k_0 \leq k$, $d_0 \leq d$, $n_0 = |A(S(a_{i_j} = \delta))| - 1$, and $S_0 \subseteq S''$. Using (12.7), we obtain that $n_0 > \frac{n}{(nk)^{\frac{1}{d+1}}} - 1$. Denote $q = \frac{n}{(nk)^{\frac{1}{d+1}}}$. From (12.5) it follows that $n > 2^{d+1}k^d$. As a result, we have $q = \frac{n^{\frac{d}{d+1}}}{k^{\frac{1}{d+1}}} > \frac{(2^{d+1}k^d)^{\frac{d}{d+1}}}{k^{\frac{1}{d+1}}} = 2^d k^{\frac{d^2-1}{d+1}} = 2^d k^{d-1} \geq 2^d \geq 2$. Using the inductive hypothesis, we obtain

$$h_{EC}(S'') \geq \frac{(n_0k_0)^{1/d_0}}{k_0} - d_0 \geq \frac{((q-1)k)^{1/d}}{k} - d = \frac{((q-1)k)^{1/d}}{k} + 1 - d - 1 .$$

By Lemma 12.9, $h_{EC}(S'') = h_{EC}(S'_\alpha)$. From Lemma 12.6 it follows that $h_{EC}(S') \geq h_{EC}(S'_\alpha) = h_{EC}(S'')$. We now prove that $\frac{((q-1)k)^{\frac{1}{d}}}{k} + 1 \geq \frac{(qk)^{\frac{1}{d}}}{k}$. To this end, we should show that $(q-1)^{\frac{1}{d}} + k^{1-\frac{1}{d}} - q^{\frac{1}{d}} \geq 0$. It is clear that $k^{1-\frac{1}{d}} \geq 1$. We now prove that $(q-1)^{\frac{1}{d}} + 1 - q^{\frac{1}{d}} \geq 0$. For this, for $x \geq 2$, we consider the function $f(x) = (x-1)^{\frac{1}{d}} + 1 - x^{\frac{1}{d}}$ and its derivative $f'(x) = \frac{1}{d(x-1)^{\frac{d-1}{d}}} - \frac{1}{dx^{\frac{d-1}{d}}}$. It is easy to show that $f'(x) > 0$ for any $x \geq 2$. Using mean value theorem, we obtain that $f(x) \geq f(2)$ for any $x \geq 2$. It is clear that $f(2) = 2 - 2^{\frac{1}{d}} \geq 0$. Therefore $h_{EC}(S') \geq \frac{(qk)^{\frac{1}{d}}}{k} - d - 1$. It is easy to show that $\frac{(qk)^{1/d}}{k} = \frac{(nk)^{\frac{1}{d+1}}}{k}$. Hence $h_{EC}(S') \geq \frac{(nk)^{\frac{1}{d+1}}}{k} - d - 1$. Thus, the inequality (12.4) is proved. Therefore the considered statement holds. From this statement it follows immediately that, for any $n, d, k \in \omega\setminus\{0\}$ such that $d \leq n$, the inequality $h^R_{EC}(n, d, k) \geq \frac{(nk)^{1/d}}{k} - d$ holds. □

Lemma 12.23 *Let $n, d, k \in \omega\setminus\{0\}$ and $d \leq n$. Then $h^R_{ESR}(n, d, k) \leq 2d\left\lceil \frac{(nk)^{1/d}}{k} \right\rceil$ and $h^R_{EAD}(n, d, k) \leq 2d\left\lceil \frac{(nk)^{1/d}}{k} \right\rceil$.*

Proof We prove by induction on d that, for any $n, d, k \in \omega \setminus \{0\}$ such that $d \leq n$, there exists an SR-reduced system of decision rules S satisfying the following conditions: the right-hand sides of all rules from S are equal to 0, $n(S) = n, d(S) = d, k(S) = k$, and $h_{ESR}(S) \leq 2d \left\lceil \frac{(nk)^{1/d}}{k} \right\rceil$.

Let $d = 1$. Consider the system $S = \{(a_1 = 0) \to 0, (a_1 = 1) \to 0, \ldots, (a_1 = k-1) \to 0, (a_2 = 0) \to 0, \ldots, (a_n = 0) \to 0\}$. It is clear that $S \in \Sigma_{SR}, n(S) = n$, $d(S) = 1, k(S) = k$, and the right-hand sides of all rules from S are equal to 0. Using Lemma 12.14, we obtain that $h_{ESR}(S) \leq n$. Therefore, for $d = 1$, the considered statement holds. Let us assume that this statement holds for some $d \geq 1$. We will show that it also holds for $d + 1$.

Denote $t = \left\lceil \frac{(nk)^{\frac{1}{d+1}}}{k} \right\rceil$. It is clear that $t \geq 1$. Let $2(d+1)t \geq n$. Let us consider a system of decision rules $S = \{(a_1 = 0) \wedge \cdots \wedge (a_{d+1} = 0) \to 0, (a_1 = 1) \to 0, \ldots, (a_1 = k-1) \to 0, (a_{d+2} = 0) \to 0, \ldots, (a_n = 0) \to 0\}$. It is clear that $n(S) = n, d(S) = d+1, k(S) = k, S \in \Sigma_{SR}$, and the right-hand sides of all rules from S are equal to 0. Using Lemma 12.14, we obtain that $h_{ESR}(S) \leq n \leq 2(d+1)t$. Therefore in this case the considered statement holds.

Let us assume now that $2(d+1)t < n$. Then $n - 2t > 2dt \geq 2d$. Using these inequalities, we obtain that there exist numbers $n_0, n_1, \ldots, n_{2tk} \in \omega$ satisfying the following conditions: $\sum_{i=0}^{2tk} n_i = n$, $n_0 = 2t$, $n_1 = d$, if $\left\lceil \frac{n-2t}{2tk} \right\rceil < d$, and $d \leq n_1 \leq \left\lceil \frac{n-2t}{2tk} \right\rceil$ otherwise, and $n_i \leq \left\lceil \frac{n-2t}{2tk} \right\rceil$ for $i = 2, \ldots, 2tk$. It is clear that there exist sets of attributes $A'_0, A'_1, \ldots, A'_{2tk}$ such that $A'_i \cap A'_j = \emptyset$ for any $i, j \in \{0, 1, \ldots, 2tk\}, i \neq j$, $|A'_i| = n_i$ for $i = 0, 1, \ldots, 2tk$, $\bigcup_{i=0}^{2tk} A'_i = \{a_1, \ldots, a_n\}$, and $A'_0 = \{a_1, \ldots, a_{2t}\}$. We now define sets of attributes $A_0, A_1, \ldots, A_{2tk}$. Let $i \in \{0, 1, \ldots, 2tk\}$. If $A'_i \neq \emptyset$, then $A_i = A'_i$. If $A'_i = \emptyset$, then $A_i = \{a_n\}$. It is clear that $A_0 = \{a_1, \ldots, a_{2t}\}$, $|A_1| \geq d$, $A_0 \cap A_i = \emptyset$ and $A_i \neq \emptyset$ for $i = 1, \ldots, 2tk$, $\bigcup_{i=0}^{2tk} A_i = \{a_1, \ldots, a_n\}$, and, for $i = 2, \ldots, 2tk$,

$$|A_i| \leq \left\lceil \frac{n - 2t}{2tk} \right\rceil . \tag{12.8}$$

For each $i \in \{1, \ldots, 2tk\}$, define a decision rule system S_i such that $A(S_i) = A_i$. Let $|A_i| \leq d$ and $A_i = \{a_{j_1}, \ldots, a_{j_m}\}$. Then $S_i = \{(a_{j_1} = 0) \wedge \cdots \wedge (a_{j_m} = 0) \to 0, (a_{j_1} = 1) \to 0, \ldots, (a_{j_1} = k-1) \to 0\}$. It is clear that $n(S_i) = |A_i|$, $d(S_i) = |A_i|$, $k(S_i) = k$, the right-hand sides of decision rules from S_i are equal to 0, and $S_i \in \Sigma_{SR}$. Using Lemma 12.14, we obtain that $h_{ESR}(S_i) \leq m \leq d$. Therefore

$$h_{ESR}(S_i) \leq 2d \left\lceil \frac{(nk)^{\frac{1}{d+1}}}{k} \right\rceil . \tag{12.9}$$

Let $|A_i| > d$. Then $|A_i| \geq 2$. Using inductive hypothesis, we obtain that there exists a decision rule system S_i satisfying the following conditions: $n(S_i) = |A_i|$, $d(S_i) = d$, $k(S_i) = k$, $S_i \in \Sigma_{SR}$, the right-hand sides of decision rules from S_i are equal to 0, and

$$h_{ESR}(S_i) \le 2d \left\lceil \frac{(|A_i|\,k)^{1/d}}{k} \right\rceil . \tag{12.10}$$

We now show that the system S_i satisfies the inequality (12.9). Using the inequality (12.8), we obtain that

$$2 \le |A_i| \le \left\lceil \frac{n-2t}{2tk} \right\rceil \le \frac{n}{2tk} + 1 \le \frac{n}{tk} \le \frac{nk}{k(nk)^{\frac{1}{d+1}}} = \frac{n}{(nk)^{\frac{1}{d+1}}} .$$

From these relations and the inequality (12.10) it follows that

$$h_{ESR}(S_i) \le 2d \left\lceil \frac{\left(\frac{nk}{(nk)^{\frac{1}{d+1}}} \right)^{1/d}}{k} \right\rceil = 2d \left\lceil \frac{(nk)^{\frac{1}{d+1}}}{k} \right\rceil .$$

Thus, the inequality (12.9) holds.

Let $j \in \{1, \ldots, 2t\}$ and $\sigma \in \{0, \ldots, k-1\}$. Describe a decision rule system $S_{j\sigma}$. Let $i = k(j-1) + \sigma + 1$. We add to the left-hand side of each rule from S_i the equation $a_j = \sigma$. We denote $S_{j\sigma}$ the obtained decision rule system. Denote $S_0 = \{(a_i = \delta) \wedge (a_j = \sigma) \to 0 : 1 \le i < j \le 2t; \delta, \sigma \in \{0, \ldots, k-1\}\}$. We now consider the decision rule system

$$S = S_0 \cup \bigcup_{\substack{j \in \{1,\ldots,2t\} \\ \sigma \in \{0,\ldots,k-1\}}} S_{j\sigma} .$$

One can show that $n(S) = n$, $d(S) = d + 1$, $k(S) = k$, $S \in \Sigma_{SR}$, and the right-hand sides of all rules from S are equal to 0.

Let us describe the operation of a decision tree Γ solving the problem $ESR(S)$. First, we compute the values of the attributes $a_1, \ldots, a_{2t}$. If $a_1 = \cdots = a_{2t} = *$, then no one rule from S is realizable for the considered tuple of attribute values. Let there exist $i, j \in \{1, \ldots, 2t\}$ such that $a_i = \delta$ and $a_j = \sigma$, where $i \ne j$ and $\sigma, \delta \in \{0, \ldots, k-1\}$. Then, for the considered tuple of attribute values, the rule $(a_i = \delta) \wedge (a_j = \sigma) \to 0$ from S is realizable. Let $a_1 = \cdots = a_{j-1} = a_{j+1} = \cdots = a_{2t} = *$ and $a_j = \sigma$, where $\sigma \in \{0, \ldots, k-1\}$. In this case, we obtain that, for the considered tuple of attribute values, only rules from $S_{j\sigma}$ can be realizable. Taking into account that $a_j = \sigma$, we obtain that the problem $ESR(S)$ solving is reduced to the problem $ESR(S_i)$ solving, where $i = k(j-1) + \sigma + 1$. We solve the problem $ESR(S_i)$ using a decision tree with the minimum depth. By (12.9),

$$h(\Gamma) \le 2t + 2dt = 2(d+1) \left\lceil \frac{(nk)^{\frac{1}{d+1}}}{k} \right\rceil .$$

Hence the considered statement is fully proven.

Let $n, d, k \in \omega \setminus \{0\}$ and $d \le n$. Then there exists a decision rule system S satisfying the following conditions: $n(S) = n$, $d(S) = d$, $k(S) = k$, $S \in \Sigma_{SR}$, the right-hand sides of all decision rules from S are equal to 0, and $h_{ESR}(S) \le 2d \left\lceil \frac{(nk)^{1/d}}{k} \right\rceil$. It is clear that $S \in \Sigma_{AD}$ and $h_{EAD}(S) = h_{ESR}(S)$. Therefore the following two inequalities hold:

$$h^R_{ESR}(n, d, k) \le 2d \left\lceil \frac{(nk)^{1/d}}{k} \right\rceil, \; h^R_{EAD}(n, d, k) \le 2d \left\lceil \frac{(nk)^{1/d}}{k} \right\rceil .$$

□

12.5.4 Theorem on Lower Bounds

Theorem 12.2 *Let $n, d, k \in \omega \setminus \{0\}$ and $d \le n$. Then*

(a) $h_{SR}(n, d, k) = h_{AD}(n, d, k) = h_{ESR}(n, d, k) = h_{EAD}(n, d, k) = 0$.

(b) *If $k = 1$ or $d = 1$, then $h_{AR}(n, d, k) = h_{EAR}(n, d, k) = n$. If $k \ge 2$ and $d \ge 2$, then*

$$\max\left\{d, \frac{n(k-1)}{k^d}\right\} \le h_{AR}(n, d, k) \le h_{EAR}(n, d, k) \le d + \frac{n}{k^{d-1}} .$$

(c) $h^R_{SR}(n, d, k) = h^R_{AD}(n, d, k) = \begin{cases} n, & \text{if } d = n, \\ 1, & \text{if } d \ne n. \end{cases}$

(d) $\max\left\{d, \frac{(nk)^{1/d}}{k} - d\right\} \le h^R_{ESR}(n, d, k) \le 2d \left\lceil \frac{(nk)^{1/d}}{k} \right\rceil$.

(e) $\max\left\{d, \frac{(nk)^{1/d}}{k} - d\right\} \le h^R_{EAD}(n, d, k) \le 2d \left\lceil \frac{(nk)^{1/d}}{k} \right\rceil$.

Proof The statements of the theorem follow from Lemmas 12.16, 12.17, 12.21, 12.22, and 12.23. □

Remark 12.4 Note that the results for $h_{AR}(n, d, k)$ mentioned in the theorem were obtained in [5]. The results for $h_{EAR}(n, d, k)$ were published in [6] without proof.

12.6 Decision Rules and Trees for Functions of *k*-Valued Logic

For $k \in \omega \setminus \{0, 1\}$, denote $E_k = \{0, 1, \ldots, k-1\}$. A function $f : E_k^n \to E_k$ is called a *function of k-valued logic*. Let $f = f(x_1, \ldots, x_n)$. Let S be a system of decision rules of the kind $(x_{i_1} = \sigma_1) \wedge \cdots \wedge (x_{i_m} = \sigma_m) \to \sigma$, where $x_{i_1}, \ldots, x_{i_m} \in \{x_1, \ldots, x_n\}$ and $\sigma_1, \ldots, \sigma_m, \sigma \in E_k$. Denote this rule r.

Let $\bar{\delta} = (\delta_1, \ldots, \delta_n) \in E_k^n$. The rule r is called *realizable* for the tuple $\bar{\delta}$ if $\sigma_1 = \delta_{i_1}, \ldots, \sigma_m = \delta_{i_m}$. The rule r is called *true* for the function f if, for any tuple $\bar{\delta} \in E_k^n$

such that r is realizable for $\bar{\delta}$, $f(\bar{\delta}) = \sigma$. The system S is called *complete* for the function f if each rule from S is true for f and, for any tuple $\bar{\delta} \in E_k^n$, there exists a rule from S that is realizable for $\bar{\delta}$. Denote $D(S)$ the set of the right-hand sides of rules from S. For any $i \in D(S)$, denote by $S^{(i)}$ the set of rules from S with the right-hand side equal to i.

Let S be a complete decision rule system for the function f. It is easy to see that, for any two rules r_1 and r_2 from S with different right-hand sides, $K(r_1) \cap K(r_2)$ is an inconsistent system of equations. We will call such rule systems *strongly separable*. It is clear that, for any $\bar{\delta} \in E_k^n$, all realizable for $\bar{\delta}$ rules from S have the same right-hand side.

We consider the following problem: for a given tuple $\bar{\delta} \in E_k^n$, it is required to find the right-hand side of rules from S realizable for $\bar{\delta}$. We now define a decision tree $\Gamma(S, E_k^n)$, which solves this problem. To this end, we describe the work of $\Gamma(S, E_k^n)$ on an arbitrary tuple $\bar{\delta} = (\delta_1, \dots, \delta_n) \in E_k^n$.

Step 0. Set $Q = S$ and i_0 the minimum number from $D(S)$.
Step 1. If Q contains a rule with empty left-hand side, then return the right-hand side of this rule as the value $f(\bar{\delta})$ and stop the work of the decision tree.
Step 2. If, for only one $i \in D(S)$, the set $Q^{(i)}$ is nonempty, then return i as the value $f(\bar{\delta})$ and stop the work of the decision tree.
Step 3. If $Q^{(i_0)} = \emptyset$, then set i_0 the minimum number from $D(Q)$.
Step 4. Choose a rule $r \in Q^{(i_0)}$. Let r be equal to $(x_{i_1} = \sigma_1) \wedge \cdots \wedge (x_{i_m} = \sigma_m) \to \sigma$. Compute values of attributes $x_{i_1}, \dots, x_{i_m}$. We obtain $x_{i_1} = \delta_{i_1}, \dots, x_{i_m} = \delta_{i_m}$. Set $Q = Q_\alpha$, where $\alpha = \{x_{i_1} = \delta_{i_1}, \dots, x_{i_m} = \delta_{i_m}\}$ (see corresponding definition after proof of Lemma 12.3), and go to Step 1.

It is clear that, for any $\bar{\delta} \in E_k^n$, the decision tree $\Gamma(S, E_k^n)$ computes the value $f(\bar{\delta})$. One can show that, during the work of the decision tree $\Gamma(S, E_k^n)$, (i) the set Q continues to be strongly separable, (ii) after each repetition of Step 4, the length of each rule from $Q \setminus Q^{(i_0)}$ remaining in Q_α decreases by at least one, and (iii) during each repetition of Step 4, we compute values of at most $d(S)$ variables. Therefore the number of repetitions of Step 4 is at most $d(S)$ and the depth of the decision tree $\Gamma(S, E_k^n)$ is at most $d(S)^2$.

We do not require to return a rule from S that is realizable for the input tuple $\bar{\delta}$ because the system S is complete for f. We did not consider this possibility in the previous sections of the chapter since the case of a complete system of rules is not general enough. However, this possibility deserves special consideration, which will be done in the future.

The obtained result generalizes well known inequality $h(f) \leq C(f)^2$ proved in [1, 4, 7], where $h(f)$ is the minimum depth of a decision tree computing Boolean function f and $C(f)$ is the certificate complexity of f (see details in [2]). One can show that $C(f)$ is equal to the minimum value of $d(S)$ among all complete systems S of decision rules for f.

References

1. Blum, M., Impagliazzo, R.: Generic oracles and oracle classes (extended abstract). In: 28th Annual Symposium on Foundations of Computer Science, Los Angeles, California, USA, 27–29 1987, pp. 118–126. IEEE Computer Society (1987)
2. Buhrman, H., de Wolf, R.: Complexity measures and decision tree complexity: a survey. Theor. Comput. Sci. **288**(1), 21–43 (2002)
3. Durdymyradov, K., Moshkov, M.: Bounds on depth of decision trees derived from decision rule systems with discrete attributes. Ann. Math. Artif. Intell. **92**(3), 703–732 (2024)
4. Hartmanis, J., Hemachandra, L.A.: One-way functions, robustness, and the non-isomorphism of NP-complete sets. In: Proceedings of the Second Annual Conference on Structure in Complexity Theory, Cornell University, Ithaca, New York, USA, 16–19, 1987. IEEE Computer Society (1987)
5. Moshkov, M.: Some relationships between decision trees and decision rule systems. In: Polkowski, L., Skowron, A. (eds.) Rough Sets and Current Trends in Computing, First International Conference, RSCTC'98, Warsaw, Poland, June 22–26, 1998, Proceedings, Lecture Notes in Computer Science, vol. 1424, pp. 499–505. Springer (1998)
6. Moshkov, M.: On transformation of decision rule systems into decision trees. In: Proceedings of the Seventh International Workshop Discrete Mathematics and its Applications, Moscow, Russia, January 29 – February 2, 2001, Part 1, pp. 21–26. Center for Applied Investigations of Faculty of Mathematics and Mechanics, Moscow State University (2001). (in Russian)
7. Tardos, G.: Query complexity, or why is it difficult to separate $NP^A \cap coNP^A$ from P^A by random oracles A? Comb. **9**(4), 385–392 (1989)

Chapter 13
Construction of Decision Trees and Acyclic Decision Graphs from Decision Rule Systems

13.1 Introduction

The study of the relationships between deterministic decision trees and decision rule systems can be seen as an important task of computer science. Methods for transforming decision trees into systems of decision rules are simple and well known. In this chapter, we consider issues related to the inverse transformation problem, which is not trivial.

The chapter continues the development of the so-called syntactic approach to the study of the considered problem proposed in works [6, 7]. This approach assumes that we do not know input data but only have a system of decision rules that must be transformed into a decision tree. Note that some results considered in this chapter were published in [7] without proofs.

Let there be a system S of decision rules of the form

$$(a_{i_1} = \delta_1) \wedge \cdots \wedge (a_{i_m} = \delta_m) \to \sigma,$$

where $a_{i_1}, \ldots, a_{i_m}$ are attributes, $\delta_1, \ldots, \delta_m$ are values of these attributes, and σ is a decision. We study three algorithmic problems associated with this system:

- For a given input (a tuple of values of all attributes included in S), it is necessary to find at least one rule that is realizable for this input (having a true left-hand side) or show that there are no such rules.
- For a given input, it is necessary to find all the rules that are realizable for this input, or show that there are no such rules.
- For a given input, it is necessary to find all the right-hand sides of rules that are realizable for this input (we should return realizable rules with these right-hand sides), or show that there are no such rules.

K. Durdymyradov et al., *Decision Trees Versus Systems of Decision Rules*, Studies in Big Data 160, https://doi.org/10.1007/978-3-031-71586-0_13

For each problem, we consider two its variants. The first assumes that in the input each attribute can have only those values that occur for this attribute in the system S. In the second case, we assume that in the input any attribute can have any value.

Our goal is to minimize the number of queries for attribute values. For this purpose, deterministic decision trees are studied as algorithms for solving the considered six problems.

We proved in Chap. 12 that, for each problem, there are systems of decision rules for which the minimum depth of the decision trees that solve the problem is much less than the total number of attributes in the rule system. For such systems of decision rules, it is reasonable to use decision trees.

In the present chapter, we study the problems of constructing deterministic decision trees and decision graphs representing these trees from decision rule systems, and the opportunities not to build the entire deterministic decision tree, but to describe the computation path in this tree for a given input.

To illustrate the obtained results, we consider the problem $AR(S)$ for a system of decision rules S: for a given input containing only values of attributes that occur in the system S, it is necessary to find all the rules from S that are realizable for this input. We use the following two parameters of the decision rule system S: the maximum length $d(S)$ of a decision rule from S and the maximum number $k(S)$ of values of an attribute from S.

First, we study possibilities to derive decision trees from decision rule systems using polynomial time algorithms. We denote by $P_{DT}(AR)$ the set of pairs $(k, d) \in (\omega\setminus\{0\})^2$, where ω is the set of nonnegative integers, satisfying the following condition: there exists a polynomial time algorithm that, for an arbitrary decision rule system S with $k(S) = k$ and $d(S) = d$, constructs a decision tree solving the problem $AR(S)$. We prove that $P_{DT}(AR) = \{(1, d) : d \in \omega\setminus\{0\}\}$.

Besides decision trees, we study acyclic decision graphs that are defined similarly to the decision trees, but instead of finite directed trees with roots, finite directed graphs with roots that do not have directed cycles are considered. We denote by $P_{DG}(AR)$ the set of pairs $(k, d) \in (\omega\setminus\{0\})^2$ satisfying the following condition: there exists a polynomial time algorithm that, for an arbitrary decision rule system S with $k(S) = k$ and $d(S) = d$, constructs an acyclic decision graph solving the problem $AR(S)$. We prove that $P_{DG}(AR) = \{(1, d) : d \in \omega\setminus\{0\}\}$, i.e., for the problem $AR(S)$ (unlike some other problems), the use of acyclic decision graphs cannot improve the situation with the existence of polynomial time algorithms.

These difficulties can be circumvented by considering acyclic decision graphs with writing, which, in addition to nodes labeled with attributes and one terminal node, have writing nodes labeled each with a decision rule. We begin the work with the empty set of rules and each time, when we come to a writing node that is labeled with a decision rule, we add this rule to the set under construction. When we reach the terminal node, the set of rules formed by this moment is the result of the work of the considered acyclic decision graph with writing.

We denote by $P_{DGW}(AR)$ the set of pairs $(k, d) \in (\omega\setminus\{0\})^2$ satisfying the following condition: there exists a polynomial time algorithm that, for an arbitrary decision rule system S with $k(S) = k$ and $d(S) = d$, constructs an acyclic decision graph with writing solving the problem $AR(S)$. We prove that $P_{DGW}(AR) = (\omega \backslash \{0\})^2$.

It would be possible to move on to the study of acyclic decision graphs and acyclic decision graphs with writing. However, when constructing them, one should simultaneously try to make the depth as small as possible without allowing an excessive increase in the number of nodes. The possibilities of such bi-criteria optimization are the subject of a special study in the future.

In this chapter, we consider mainly another approach: instead of constructing the entire decision tree, we restrict ourselves to the consideration of polynomial time algorithms that, for a given tuple of attribute values, describe the work of the decision tree on this tuple. To this end, we study bounds on the depth of decision trees and algorithms for the description of the decision tree work based on node covers for the hypergraph corresponding to the considered decision rule system (nodes of this hypergraph are attributes from the rule system and edges correspond to rules).

We mention here only the results for the problem $AR(S)$. We developed a polynomial time algorithm, which models the operation of a decision tree that solves the problem $AR(S)$ for a given tuple of attribute values. The depth of this tree does not exceed the cube of the minimum depth of a decision tree solving the problem $AR(S)$. For the remaining problems considered, we are conducting similar studies.

We consider also algorithms for the description of the decision tree work that are not based on node covers for the hypergraph and dynamic programming algorithms that, for given decision rule systems, return the minimum depth of decision trees for these systems.

To make this chapter autonomous, we almost repeat the main definitions from Chap. 12 and give some lemmas from Chap. 12 without proofs.

The present chapter is based mainly on the papers [1–5]. It consists of 10 sections. Section 13.2 discusses the main definitions and notation. Section 13.3 contains auxiliary statements. Sections 13.4–13.6 consider the problems of constructing decision trees and acyclic decision graphs representing decision trees. Sections 13.7–13.9 discuss the possibility to construct not the entire decision tree, but the computation path in this tree for a given input. Section 13.10 considers dynamic programming algorithms for decision tree construction.

13.2 Main Definitions and Notation

In this section, we discuss the main definitions and notation related to decision rule systems and decision trees. In fact, we almost repeat the definitions and notation from Chap. 12, but consider new examples.

13.2.1 Decision Rule Systems

Let $\omega = \{0, 1, 2, \ldots\}$ and $A = \{a_i : i \in \omega\}$. Elements of the set A will be called *attributes*.

Definition 13.1 A *decision rule* is an expression of the form

$$(a_{i_1} = \delta_1) \wedge \cdots \wedge (a_{i_m} = \delta_m) \rightarrow \sigma,$$

where $m \in \omega$, $a_{i_1}, \ldots, a_{i_m}$ are pairwise different attributes from A and $\delta_1, \ldots, \delta_m, \sigma \in \omega$.

We denote this decision rule by r. The expression $(a_{i_1} = \delta_1) \wedge \cdots \wedge (a_{i_m} = \delta_m)$ will be called the *left-hand side*, and the number σ will be called the *right-hand side* of the rule r. The number m will be called the *length* of the decision rule r. Denote $A(r) = \{a_{i_1}, \ldots, a_{i_m}\}$ and $K(r) = \{a_{i_1} = \delta_1, \ldots, a_{i_m} = \delta_m\}$. If $m = 0$, then $A(r) = K(r) = \emptyset$. We assume that each decision rule has an *identifier* and identifiers are elements of a linearly ordered set.

Definition 13.2 Two decision rules r_1 and r_2 are *equal* if $K(r_1) = K(r_2)$ and the right-hand sides of the rules r_1 and r_2 are equal. Two decision rules are *identical* if they are equal and have equal identifiers.

Definition 13.3 A *system of decision rules* S is a finite nonempty set of decision rules with pairwise different identifiers. The system S can have equal rules with different identifiers. Two decision rule systems S_1 and S_2 are *equal* if there is a one-to-one correspondence $\mu : S_1 \rightarrow S_2$ such that, for any rule $r \in S_1$, the rules r and $\mu(r)$ are identical.

Denote $A(S) = \bigcup_{r \in S} A(r)$, $n(S) = |A(S)|$, $D(S)$ the set of the right-hand sides of decision rules from S, and $d(S)$ the maximum length of a decision rule from S. Let $n(S) > 0$. For $a_i \in A(S)$, let $V_S(a_i) = \{\delta : a_i = \delta \in \bigcup_{r \in S} K(r)\}$ and $EV_S(a_i) = V_S(a_i) \cup \{*\}$, where the symbol $*$ is interpreted as a number that does not belong to the set $V_S(a_i)$. Denote $k(S) = \max\{|V_S(a_i)| : a_i \in A(S)\}$. If $n(S) = 0$, then $k(S) = 0$. We denote by Σ the set of systems of decision rules.

Example 13.1 Let us consider a decision rule system $S = \{(a_1 = 0) \rightarrow 1, (a_1 = 1) \wedge (a_2 = 0) \rightarrow 2, (a_1 = 2) \wedge (a_3 = 0) \wedge (a_4 = 0) \rightarrow 3\}$. Then $A(r_3) = \{a_1, a_3, a_4\}$, $K(r_3) = \{a_1 = 2, a_3 = 0, a_4 = 0\}$, where r_3 denotes the third rule from S. $A(S) = \bigcup_{r \in S} A(r) = \{a_1, a_2, a_3, a_4\}$, $n(S) = |A(S)| = 4$, $D(S) = \{1, 2, 3\}$, $d(S) = 3$, $V_S(a_1) = \{\delta : a_1 = \delta \in \bigcup_{r \in S} K(r)\} = \{0, 1, 2\}$, $EV_S(a_1) = V_S(a_1) \cup \{*\} = \{0, 1, 2, *\}$ and $k(S) = \max\{|V_S(a_i)| : a_i \in A(S)\} = |V_S(a_1)| = 3$.

Let $S \in \Sigma$, $n(S) > 0$, and $A(S) = \{a_{j_1}, \ldots, a_{j_n}\}$, where $j_1 < \cdots < j_n$. Denote $V(S) = V_S(a_{j_1}) \times \cdots \times V_S(a_{j_n})$ and $EV(S) = EV_S(a_{j_1}) \times \cdots \times EV_S(a_{j_n})$. For $\bar{\delta} = (\delta_1, \ldots, \delta_n) \in EV(S)$, denote $K(S, \bar{\delta}) = \{a_{j_1} = \delta_1, \ldots, a_{j_n} = \delta_n\}$.

Definition 13.4 We will say that a decision rule r from S is *realizable* for a tuple $\bar{\delta} \in EV(S)$ if $K(r) \subseteq K(S, \bar{\delta})$.

It is clear that any rule with an empty left-hand side is realizable for the tuple $\bar{\delta}$.

Example 13.2 Let us consider a decision rule system $S = \{r_1 : (a_1 = 0) \wedge (a_2 = 1) \rightarrow 1, r_2 : (a_1 = 1) \wedge (a_3 = 1) \rightarrow 2\}$ and a tuple $\bar{\delta} = (1, 1, 1) \in EV(S)$. Then the decision rule r_2 from S is realizable for the tuple $\bar{\delta}$, but r_1 is not.

Let $V \in \{V(S), EV(S)\}$. We now define three problems related to the rule system S.

Definition 13.5 Problem *All Rules* for the pair (S, V): for a given tuple $\bar{\delta} \in V$, it is required to find the set of rules from S that are realizable for the tuple $\bar{\delta}$.

Definition 13.6 Problem *All Decisions* for the pair (S, V): for a given tuple $\bar{\delta} \in V$, it is required to find a set Z of decision rules from S satisfying the following conditions:

- All decision rules from Z are realizable for the tuple $\bar{\delta}$.
- For any $\sigma \in D(S) \setminus D(Z)$, any decision rule from S with the right-hand side equal to σ is not realizable for the tuple $\bar{\delta}$.

Definition 13.7 Problem *Some Rules* for the pair (S, V): for a given tuple $\bar{\delta} \in V$, it is required to find a set Z of decision rules from S satisfying the following conditions:

- All decision rules from Z are realizable for the tuple $\bar{\delta}$.
- If $Z = \emptyset$, then any decision rule from S is not realizable for the tuple $\bar{\delta}$.

Denote $AR(S)$ and $EAR(S)$ the problems All Rules for pairs $(S, V(S))$ and $(S, EV(S))$, respectively. Denote $AD(S)$ and $EAD(S)$ the problems All Decisions for pairs $(S, V(S))$ and $(S, EV(S))$, respectively. Denote $SR(S)$ and $ESR(S)$ the problems Some Rules for pairs $(S, V(S))$ and $(S, EV(S))$, respectively.

Example 13.3 Let a decision rule system $S = \{(a_1 = 0) \rightarrow 1, (a_1 = 0) \wedge (a_2 = 1) \rightarrow 1, (a_1 = 0) \wedge (a_2 = 1) \wedge (a_3 = 2) \rightarrow 2\}$ and a tuple $\bar{\delta} = (0, 1, 2) \in V(S)$ are given. Then $\{(a_1 = 0) \rightarrow 1, (a_1 = 0) \wedge (a_2 = 1) \rightarrow 1, (a_1 = 0) \wedge (a_2 = 1) \wedge (a_3 = 2) \rightarrow 2\}$ is the solution for the problem $AR(S)$ and the tuple $\bar{\delta}$, $\{(a_1 = 0) \rightarrow 1, (a_1 = 0) \wedge (a_2 = 1) \wedge (a_3 = 2) \rightarrow 2\}$ is a solution for the problem $AD(S)$ and the tuple $\bar{\delta}$, and $\{(a_1 = 0) \rightarrow 1\}$ is a solution for the problem $SR(S)$ and the tuple $\bar{\delta}$.

In the special case, when $n(S) = 0$, all rules from S have an empty left-hand side. In this case, it is natural to consider (i) the set S as the solution to the problems $AR(S)$ and $EAR(S)$, (ii) any subset Z of the set S with $D(Z) = D(S)$ as a solution to the problems $AD(S)$ and $EAD(S)$, and (iii) any nonempty subset Z of the set S as a solution to the problems $SR(S)$ and $ESR(S)$.

Let $S \in \Sigma$, where Σ is the set of decision rule systems. We denote by $R_{SR}(S)$ a subsystem of the system S that consists of all rules $r \in S$ satisfying the following condition: there is no rule $r' \in S$ such that $K(r') \subset K(r)$.

Definition 13.8 The system S will be called *SR-reduced* if $R_{SR}(S) = S$.

Denote by Σ_{SR} the set of SR-reduced systems of decision rules.

For $S \in \Sigma$, we denote by $R_{AD}(S)$ a subsystem of the system S that consists of all rules $r \in S$ satisfying the following condition: there is no a rule $r' \in S$ such that $K(r') \subset K(r)$ and the right-hand sides of the rules r and r' coincide.

Definition 13.9 The system S will be called *AD-reduced* if $R_{AD}(S) = S$.

Denote by Σ_{AD} the set of AD-reduced systems of decision rules.

Example 13.4 Let us consider a decision rule system $S = \{(a_1 = 0) \wedge (a_2 = 1) \wedge (a_3 = 2) \to 1, (a_1 = 0) \wedge (a_2 = 1) \to 2, (a_1 = 0) \to 2\}$. For this system, $R_{AD}(S) = \{(a_1 = 0) \wedge (a_2 = 1) \wedge (a_3 = 2) \to 1, (a_1 = 0) \to 2\}$ and $R_{SR}(S) = \{(a_1 = 0) \to 2\}$.

13.2.2 Decision Trees

A *finite directed tree with root* is a finite directed tree in which only one node has no entering edges. This node is called the *root*. The nodes without leaving edges are called *terminal* nodes. The nodes that are not terminal will be called *working* nodes. A *complete path* in a finite directed tree with root is a sequence $\xi = v_1, d_1, \ldots, v_m, d_m, v_{m+1}$ of nodes and edges of this tree in which v_1 is the root, v_{m+1} is a terminal node and, for $i = 1, \ldots, m$, the edge d_i leaves the node v_i and enters the node v_{i+1}.

We will consider two types of decision trees: o-decision trees (ordinary decision trees, o-trees in short) and e-decision trees (extended decision trees, e-trees in short).

Definition 13.10 A *decision tree over a decision rule system* S is a labeled finite directed tree with root Γ satisfying the following conditions:

- Each working node of the tree Γ is labeled with an attribute from the set $A(S)$.
- Let a working node v of the tree Γ be labeled with an attribute a_i. If Γ is an o-tree, then exactly $|V_S(a_i)|$ edges leave the node v and these edges are labeled with pairwise different elements from the set $V_S(a_i)$. If Γ is an e-tree, then exactly $|EV_S(a_i)|$ edges leave the node v and these edges are labeled with pairwise different elements from the set $EV_S(a_i)$.
- Each terminal node of the tree Γ is labeled with a subset of the set S.

Let Γ be a decision tree over the decision rule system S. We denote by $CP(\Gamma)$ the set of complete paths in the tree Γ. Let $\xi = v_1, d_1, \ldots, v_m, d_m, v_{m+1}$ be a complete path in Γ. We correspond to this path a set of attributes $A(\xi)$ and an equation system $K(\xi)$. If $m = 0$ and $\xi = v_1$, then $A(\xi) = \emptyset$ and $K(\xi) = \emptyset$. Let $m > 0$ and, for $j = 1, \ldots, m$, the node v_j be labeled with the attribute a_{i_j} and the edge d_j be labeled with the element $\delta_j \in \omega \cup \{*\}$. Then $A(\xi) = \{a_{i_1}, \ldots, a_{i_m}\}$ and $K(\xi) = \{a_{i_1} = \delta_1, \ldots, a_{i_m} = \delta_m\}$. We denote by $\tau(\xi)$ the set of decision rules attached to the node v_{m+1}.

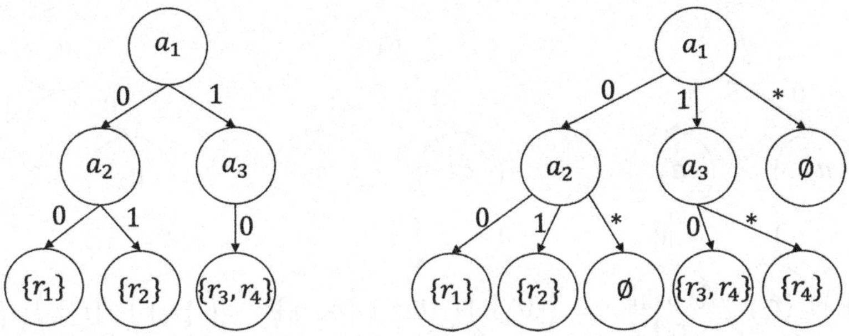

Fig. 13.1 O-decision tree Γ_1 and e-decision tree Γ_2 from Example 13.5

Example 13.5 Let us consider a decision rule system $S = \{(a_1 = 0) \wedge (a_2 = 0) \rightarrow 1, (a_1 = 0) \wedge (a_2 = 1) \rightarrow 2, (a_1 = 1) \wedge (a_3 = 0) \rightarrow 3, (a_1 = 1) \rightarrow 4\}$. Then o-tree Γ_1 and e-tree Γ_2 over the decision rule system S are given in Fig. 13.1, where r_1, r_2, r_3 and r_4 are the first, second, third and last decision rules in S, respectively.

Let ξ be a complete path in the o-tree Γ_1, which is finished in the terminal node labeled with the set of rules $\{r_3, r_4\}$. Then $A(\xi) = \{a_1, a_3\}$, $K(\xi) = \{a_1 = 1, a_3 = 0\}$ and $\tau(\xi) = \{r_3, r_4\}$.

Definition 13.11 A system of equations $\{a_{i_1} = \delta_1, \dots, a_{i_m} = \delta_m\}$, where $a_{i_1}, \dots, a_{i_m} \in A$ and $\delta_1, \dots, \delta_m \in \omega \cup \{*\}$, will be called *inconsistent* if there exist $l, k \in \{1, \dots, m\}$ such that $l \neq k$, $i_l = i_k$, and $\delta_l \neq \delta_k$. If the system of equations is not inconsistent, then it will be called *consistent*.

Let S be a decision rule system and Γ be a decision tree over S.

Definition 13.12 We will say that Γ *solves* the problem $AR(S)$ (problem $EAR(S)$, respectively) if Γ is an o-tree (an e-tree, respectively) and any path $\xi \in CP(\Gamma)$ with consistent system of equations $K(\xi)$ satisfies the following conditions:

- For any decision rule $r \in \tau(\xi)$, the relation $K(r) \subseteq K(\xi)$ holds.
- For any decision rule $r \in S \backslash \tau(\xi)$, the system of equations $K(r) \cup K(\xi)$ is inconsistent.

Example 13.6 Let S be a decision rule system from Example 13.5. Then the decision trees Γ_1 and Γ_2 depicted in Fig. 13.1 solve the problems $AR(S)$ and $EAR(S)$, respectively.

Definition 13.13 We will say that Γ *solves* the problem $AD(S)$ (problem $EAD(S)$, respectively) if Γ is an o-tree (an e-tree, respectively) and any path $\xi \in CP(\Gamma)$ with consistent system of equations $K(\xi)$ satisfies the following conditions:

- For any decision rule $r \in \tau(\xi)$, the relation $K(r) \subseteq K(\xi)$ holds.
- If $r \in S \backslash \tau(\xi)$ and the right-hand side of r does not belong to the set $D(\tau(\xi))$, then the system of equations $K(r) \cup K(\xi)$ is inconsistent.

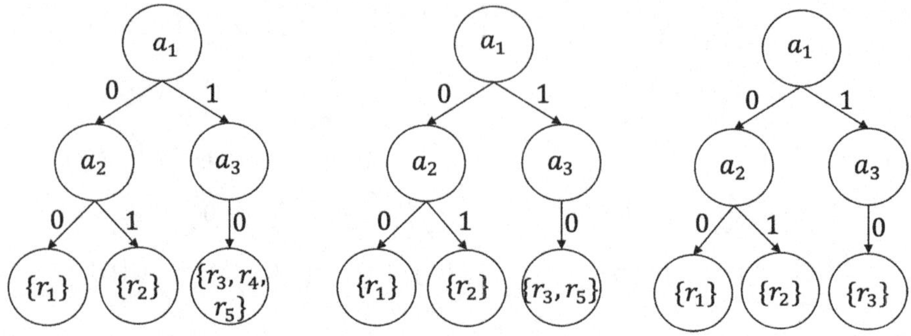

Fig. 13.2 Decision trees Γ_1, Γ_2, and Γ_3 from Example 13.7

Definition 13.14 We will say that Γ *solves* the problem $SR(S)$ (problem $ESR(S)$, respectively) if Γ is an o-tree (an e-tree, respectively) and any path $\xi \in CP(\Gamma)$ with consistent system of equations $K(\xi)$ satisfies the following conditions:

- For any decision rule $r \in \tau(\xi)$, the relation $K(r) \subseteq K(\xi)$ holds.
- If $\tau(\xi) = \emptyset$, then, for any decision rule $r \in S$, the system of equations $K(r) \cup K(\xi)$ is inconsistent.

Example 13.7 Let us consider a decision rule system $S = \{(a_1 = 0) \wedge (a_2 = 0) \rightarrow 1, (a_1 = 0) \wedge (a_2 = 1) \rightarrow 2, (a_1 = 1) \wedge (a_3 = 0) \rightarrow 3, (a_1 = 1) \rightarrow 3, (a_1 = 1) \rightarrow 4\}$ and decision trees Γ_1, Γ_2, and Γ_3 depicted in Fig. 13.2. Then the decision tree Γ_1 solves the problem $AR(S)$, Γ_2 solves $AD(S)$, and Γ_3 solves $SR(S)$, where r_1, r_2, r_3, r_4 and r_5 are the first, second, third, fourth and fifth decision rules in S, respectively.

For any complete path $\xi \in CP(\Gamma)$, we denote by $h(\xi)$ the number of working nodes in ξ. The value $h(\Gamma) = \max\{h(\xi) : \xi \in CP(\Gamma)\}$ is called the *depth* of the decision tree Γ.

Let S be a decision rule system and $C \in \{AR, EAR, AD, EAD, SR, ESR\}$. We denote by $h_C(S)$ the minimum depth of a decision tree over S, which solves the problem $C(S)$.

Let $n(S) = 0$. If $C \in \{AR, EAR\}$, then there is only one decision tree solving the problem $C(S)$. This tree consists of one node labeled with the set of rules S. If $C \in \{AD, EAD\}$, then the set of decision trees solving the problem $C(S)$ coincides with the set of trees each of which consists of one node labeled with a subset Z of the set S with $D(Z) = D(S)$. If $C \in \{SR, ESR\}$, then the set of decision trees solving the problem $C(S)$ coincides with the set of trees each of which consists of one node labeled with a nonempty subset Z of the set S. Therefore if $n(S) = 0$, then $h_C(S) = 0$ for any $C \in \{AR, EAR, AD, EAD, SR, ESR\}$.

13.3 Auxiliary Statements

In this section, we will first give some statements from Chap. 12 and then we will prove some new ones.

Let S be a decision rule system and Γ be an e-decision tree over S. We denote by $o(\Gamma)$ an o-tree over S, which is obtained from the tree Γ by the removal of all nodes v such that the path from the root to the node v in Γ contains an edge labeled with $*$. Together with a node v, we remove all edges entering or leaving v.

Lemma 13.1 (Lemma 12.1) *Let S be a decision rule system, $C \in \{AR, AD, SR\}$, and Γ be an e-decision tree over S solving the problem $EC(S)$. Then the decision tree $o(\Gamma)$ solves the problem $C(S)$.*

Lemma 13.2 (Lemma 12.2) *Let S be a decision rule system and Γ be a decision tree over S. Then*

(a) *If the tree Γ solves the problem $AR(S)$ ($EAR(S)$, respectively), then the tree Γ solves the problem $AD(S)$ ($EAD(S)$, respectively).*

(b) *If the tree Γ solves the problem $AD(S)$ ($EAD(S)$, respectively), then the tree Γ solves the problem $SR(S)$ ($ESR(S)$, respectively).*

Lemma 13.3 (Lemma 12.3) *Let S be a decision rule system. Then the following inequalities hold:*

$$\begin{array}{ccccccc} h_{ESR}(S) & \leq & h_{EAD}(S) & \leq & h_{EAR}(S) & \leq & n(S), \\ \text{\rotatebox{90}{$\geq$}} & & \text{\rotatebox{90}{$\geq$}} & & \text{\rotatebox{90}{$\geq$}} & & \\ h_{SR}(S) & \leq & h_{AD}(S) & \leq & h_{AR}(S). & & \end{array}$$

Let S be a decision rule system and $\alpha = \{a_{i_1} = \delta_1, \ldots, a_{i_m} = \delta_m\}$ be a consistent equation system such that $a_{i_1}, \ldots, a_{i_m} \in A$ and $\delta_1, \ldots, \delta_m \in \omega \cup \{*\}$. We now define a decision rule system S_α. Let r be a decision rule for which the equation system $K(r) \cup \alpha$ is consistent. We denote by r_α the decision rule obtained from r by the removal from the left-hand side of r all equations that belong to α. The rules r and r_α have the same identifiers. We will say that the rule r_α *corresponds* to the rule r. Then S_α is the set of decision rules r_α such that $r \in S$ and the equation system $K(r) \cup \alpha$ is consistent.

Lemma 13.4 (Lemma 12.6) *Let S be a decision rule system with $n(S) > 0$, $C \in \{EAR, AR, EAD, AD, ESR, SR\}$, $\alpha = \{a_{i_1} = \delta_1, \ldots, a_{i_m} = \delta_m\}$ be a consistent equation system such that $a_{i_1}, \ldots, a_{i_m} \in A(S)$ and, for $j = 1, \ldots, m$, $\delta_j \in EV_S(a_{i_j})$ if $C \in \{EAR, EAD, ESR\}$ and $\delta_j \in V_S(a_{i_j})$ if $C \in \{AR, AD, SR\}$. Then $h_C(S) \geq h_C(S_\alpha)$.*

We correspond to a decision rule system S a hypergraph $G(S)$ with the set of nodes $A(S)$ and the set of edges $\{A(r) : r \in S\}$. A *node cover* of the hypergraph $G(S)$ is a subset B of the set of nodes $A(S)$ such that $A(r) \cap B \neq \emptyset$ for any rule $r \in S$ such that $A(r) \neq \emptyset$. If $A(S) = \emptyset$, then the empty set is the only node cover of the hypergraph $G(S)$. Denote by $\beta(S)$ the minimum cardinality of a node cover of the hypergraph $G(S)$.

Example 13.8 Let us consider a decision rule system $S = \{(a_1 = 0) \wedge (a_2 = 1) \rightarrow 1, (a_1 = 0) \wedge (a_3 = 1) \rightarrow 2, (a_4 = 0) \rightarrow 3\}$. One can show, that the set $\{a_1, a_4\}$ is a node cover of the hypergraph $G(S)$ and $\beta(S) = 2$.

We define a subsystem $I_{SR}(S)$ of the system S in the following way. If S does not contain rules of the length 0, then $I_{SR}(S) = S$. Otherwise, $I_{SR}(S)$ consists of all rules from S of the length 0.

Remark 13.1 Note that $I_{SR}(S) \neq S$ if and only if S contains both a rule of the length 0 and a rule of the length greater than 0.

We now define a subsystem $I_{AD}(S)$ of the system S. Denote by $D_0(S)$ the set of the right-hand sides of decision rules from S, which length is equal to 0. Then the subsystem $I_{AD}(S)$ consists of all rules from S of the length 0 and all rules from S for which the right-hand sides do not belong to $D_0(S)$.

Remark 13.2 Note that $I_{AD}(S) \neq S$ if and only if S contains both a rule of the length 0 and a rule of the length greater than 0 with the same right-hand sides.

Example 13.9 Let us consider a decision rule system $S = \{(a_1 = 0) \rightarrow 1, (a_2 = 0) \rightarrow 2, \rightarrow 2\}$. For this system, $I_{SR}(S) = \{\rightarrow 2\}$ and $I_{AD}(S) = \{(a_1 = 0) \rightarrow 1, \rightarrow 2\}$.

Let S be a decision rule system. This system will be called *incomplete* if there exists a tuple $\bar{\delta} \in V(S)$ such that the equation system $K(r) \cup K(S, \bar{\delta})$ is inconsistent for any decision rule $r \in S$. Otherwise, the system S will be called *complete*. If $n(S) = 0$, then the system S will be considered as complete.

Lemma 13.5 (Lemma 12.7) *Let S be a decision rule system. Then*
(a) *If $C \in \{EAR, AR\}$, then $h_C(S) \geq \beta(S)$.*
(b) *If $C \in \{AD, SR\}$, then $h_{EC}(S) \geq \beta(I_C(S))$.*
(c) *If $C \in \{AD, SR\}$ and the system S is incomplete, then $h_C(S) \geq \beta(S)$.*

Lemma 13.6 (Lemma 12.8) *Let S be a decision rule system. Then*
(a) $h_{EAR}(S) \geq h_{AR}(S) \geq d(S)$.
(b) *If S is an SR-reduced system, then $h_{ESR}(S) \geq d(S)$.*
(c) *If S is an AD-reduced system, then $h_{EAD}(S) \geq d(S)$.*

Lemma 13.7 (Lemma 12.9) *Let S be a decision rule system. Then*

$$h_{ESR}(S) = h_{ESR}(R_{SR}(S)) \text{ and } h_{EAD}(S) = h_{EAD}(R_{AD}(S)).$$

Let $C \in \{AR, EAR, AD, EAD, SR, ESR\}$, S be a decision rule system, and Γ be a decision tree over S. We denote by $L(\Gamma)$ the number of nodes in the tree Γ and by $T(\Gamma)$ we denote the number of terminal nodes in Γ that are labeled with pairwise different sets of decision rules. Let $L_C(S) = \min L(\Gamma)$ and $T_C(S) = \min T(\Gamma)$, where the minimum is taken over all decision trees Γ over S that solve the problem $C(S)$. It is clear that $T_C(S) \leq L_C(S)$.

Lemma 13.8 *Let S be a decision rule system. Then the following inequalities hold:*

$$\begin{array}{ccccc} L_{ESR}(S) & \leq & L_{EAD}(S) & \leq & L_{EAR}(S), \\ \mathrel{\rotatebox{90}{$\geq$}} & & \mathrel{\rotatebox{90}{$\geq$}} & & \mathrel{\rotatebox{90}{$\geq$}} \\ L_{SR}(S) & \leq & L_{AD}(S) & \leq & L_{AR}(S), \end{array}$$

$$\begin{array}{ccccc} T_{ESR}(S) & \leq & T_{EAD}(S) & \leq & T_{EAR}(S), \\ \mathrel{\rotatebox{90}{$\geq$}} & & \mathrel{\rotatebox{90}{$\geq$}} & & \mathrel{\rotatebox{90}{$\geq$}} \\ T_{SR}(S) & \leq & T_{AD}(S) & \leq & T_{AR}(S). \end{array}$$

Proof It is clear that the considered inequalities hold if $n(S) = 0$. Let $n(S) > 0$.

Let Γ be an e-decision tree over S. It is clear that $L(o(\Gamma)) \leq L(\Gamma)$ and $T(o(\Gamma)) \leq T(\Gamma)$. Using these inequalities and Lemma 13.1, we obtain that $L_{SR}(S) \leq L_{ESR}(S)$, $L_{AD}(S) \leq L_{EAD}(S)$, $L_{AR}(S) \leq L_{EAR}(S)$, $T_{SR}(S) \leq T_{ESR}(S)$, $T_{AD}(S) \leq T_{EAD}(S)$, and $T_{AR}(S) \leq T_{EAR}(S)$.

Using Lemma 13.2, we obtain that $L_{ESR}(S) \leq L_{EAD}(S) \leq L_{EAR}(S)$, $L_{SR}(S) \leq L_{AD}(S) \leq L_{AR}(S)$, $T_{ESR}(S) \leq T_{EAD}(S) \leq T_{EAR}(S)$, and $T_{SR}(S) \leq T_{AD}(S) \leq T_{AR}(S)$. □

A system of decision rules S can be represented by a word over the alphabet

$$\{(,), a, =, \wedge, \rightarrow, 0, 1, ;\}$$

in which numbers from ω (attribute indexes, attribute values, and right-hand sides of decision rules) are in binary representation (are represented by words over the alphabet $\{0, 1\}$) and the symbol ";" is used to separate two rules. The length of this word will be called the *size* of the decision rule system S and will be denoted $size(S)$.

Definition 13.15 A decision rule system S will be called *reduced* if it satisfies the following conditions:

- If $n(s) = n$, then $A(S) \subseteq \{a_0, \ldots, a_n\}$.
- If $|D(S)| = t$, then $D(S) \subseteq \{0, \ldots, t\}$.
- If $k(S) = k$, then $V_S(a_i) \subseteq \{0, \ldots, k\}$ for any $a_i \in A(S)$.
- $d(S) \geq 1$.

Let S be a reduced system of decision rules. It is clear that $n(S) \leq d(S)\,|S|$, $k(S) \leq |S|$, and $|D(S)| \leq |S|$. Therefore the maximum number from ω in the system

S is at most $d(S)\,|S|$. The length of binary representation of such a number is at most $\log_2(d(S)\,|S|) + 1$. The length of each rule from S is at most $d(S)$. One can show that the length of word representing each rule (including the sign ";" after it) is at most $10d(S)(\log_2(d(S)\,|S|) + 1)$ and

$$size(S) \le 10\,|S|\,d(S)(\log_2(d(S)\,|S|) + 1). \tag{13.1}$$

Lemma 13.9 *Let $n, k, d \in \omega\backslash\{0\}$, $d \ge 2$, and $k \ge 2$. Then there exists a reduced decision rule system S such that $n(S) = 2n + d$, $k(S) = k$, $d(S) = d$, $|S| = 2n + k - 1$, and $L_{SR}(S) \ge 2^n$.*

Proof Let us consider a system of decision rules $S = S_1 \cup S_2 \cup S_3$, where $S_1 = \{(a_{2i-1} = 0) \wedge (a_{2i} = 0) \to 0, (a_{2i-1} = 1) \wedge (a_{2i} = 1) \to 0 : i = 1, \dots, n\}$, $S_2 = \{(a_1 = 2) \wedge (a_2 = 2) \to 0, \dots, (a_1 = k - 1) \wedge (a_2 = k - 1) \to 0\}$, and $S_3 = \{(a_{2n+1} = 0) \wedge \cdots \wedge (a_{2n+d} = 0) \to 0\}$. If $k = 2$, then $S_2 = \emptyset$. It is clear that S is reduced, $n(S) = 2n + d$, $k(S) = k$, $d(S) = d$, and $|S| = 2n + k - 1$. Denote by Δ the set of tuples $(\delta_1, \dots, \delta_{2n+d}) \in \{0, 1\}^{2n+d}$ such that $\delta_{2i-1} + \delta_{2i} = 1$ for $i = 1, \dots, n$ and $\delta_{2n+1} = \cdots = \delta_{2n+d} = 1$. It is clear that $|\Delta| = 2^n$ and, for any $\bar{\delta} \in \Delta$, there is no a rule from S that is realizable for the tuple $\bar{\delta}$.

Let Γ be a decision tree over S, which solves the problem $SR(S)$ and for which $L(\Gamma) = L_{SR}(S)$. Let $\bar{\delta} = (\delta_1, \dots, \delta_{2n+d}) \in \Delta$. It is clear that there exists a complete path ξ in Γ such that $K(\xi) \subseteq K(S, \bar{\delta})$. Evidently, the terminal node of this path is labeled with the empty set. Let us show that $\{a_1 = \delta_1, \dots, a_{2n} = \delta_{2n}\} \subseteq K(\xi)$. Let us assume the contrary. Then there exists $j \in \{1, \dots, 2n\}$ such that the equation $a_j = \delta_j$ does not belong to $K(\xi)$. Let $j \in \{2i - 1, 2i\}$, where $i \in \{1, \dots, n\}$. Denote by r_0 the rule $(a_{2i-1} = 0) \wedge (a_{2i} = 0) \to 0$ and by r_1 – the rule $(a_{2i-1} = 1) \wedge (a_{2i} = 1) \to 0$. It is clear that $r_0, r_1 \in S$ and at least one of the equation systems $K(r_0) \cup K(\xi)$ and $K(r_1) \cup K(\xi)$ is consistent but this is impossible. Therefore $\{a_1 = \delta_1, \dots, a_{2n} = \delta_{2n}\} \subseteq K(\xi)$. From this relation and the inclusion $K(\xi) \subseteq K(S, \bar{\delta})$ it follows that in Γ there are at least 2^n pairwise different complete paths. Therefore $L(\Gamma) \ge 2^n$ and $L_{SR}(S) \ge 2^n$. □

Lemma 13.10 *Let $n, d \in \omega\backslash\{0\}$ and $d \ge 2$. Then there exists a reduced decision rule system S such that $n(S) = 2n + d$, $k(S) = 1$, $d(S) = d$, $|S| = n + 1$, and $L_{ESR}(S) \ge 2^n$.*

Proof Let us consider a system of decision rules $S = S_1 \cup S_2$, where $S_1 = \{(a_{2i-1} = 0) \wedge (a_{2i} = 0) \to 0 : i = 1, \dots, n\}$ and $S_2 = \{(a_{2n+1} = 0) \wedge \cdots \wedge (a_{2n+d} = 0) \to 0\}$. It is clear that S is reduced, $n(S) = 2n + d, d(S) = d, k(S) = 1$, and $|S| = n + 1$. Denote by Δ the set of tuples $(\delta_1, \dots, \delta_{2n+d}) \in \{0, *\}^{2n+d}$ such that $\{\delta_{2i-1}, \delta_{2i}\} = \{0, *\}$ for $i = 1, \dots, n$ and $\delta_{2n+1} = \cdots = \delta_{2n+d} = *$. It is clear that $|\Delta| = 2^n$ and, for any $\bar{\delta} \in \Delta$, there is no a rule from S that is realizable for the tuple $\bar{\delta}$. Let Γ be a decision tree over S, which solves the problem $ESR(S)$ and for which $L(\Gamma) = L_{ESR}(S)$. Let $\bar{\delta}, \bar{\sigma} \in \Delta$ and $\bar{\delta} \ne \bar{\sigma}$. It is clear that there exist complete paths ξ, τ in Γ such that $K(\xi) \subseteq K(S, \bar{\delta})$ and $K(\tau) \subseteq K(S, \bar{\sigma})$. Let us show that $\xi \ne \tau$. Let us assume the contrary: $\xi = \tau$. It is clear that the terminal node of the path ξ is labeled with the empty set. Since $\bar{\delta} \neq \bar{\sigma}$, there exists $i \in \{1, \ldots, n\}$ such that $(2i - 1)$th and $(2i)$th digits of the tuples $\bar{\delta}$ and $\bar{\sigma}$ are different. Therefore the attributes a_{2i-1} and a_{2i} are not attached to any node of the path ξ. Hence the system of equations $K(r) \cup K(\xi)$ is consistent, where r is the rule $(a_{2i-1} = 0) \wedge (a_{2i} = 0) \rightarrow 0$ from S, but this is impossible. Therefore $\xi \neq \tau$. Thus, in the tree Γ, there are at least 2^n pairwise different complete paths. As a result, we have $L(\Gamma) \geq 2^n$ and $L_{ESR}(S) \geq 2^n$. □

Lemma 13.11 (a) *Let $n, d \in \omega \setminus \{0\}$. Then there exists a reduced decision rule system S such that $n(S) = n + d$, $k(S) = 1$, $d(S) = d$, $|S| = n + 1$, and $T_{EAD}(S) \geq 2^{n+1}$.*

(b) *Let $n, d, k \in \omega \setminus \{0\}$ and $k \geq 2$. Then there exists a reduced decision rule system S such that $n(S) = n + d$, $k(S) = k$, $d(S) = d$, $|S| = 2n + k - 1$, and $T_{AD}(S) \geq 2^n$.*

Proof (a) Let us consider a system of decision rules $S = \{(a_1 = 0) \rightarrow 1, (a_2 = 0) \rightarrow 2, \ldots, (a_n = 0) \rightarrow n, (a_{n+1} = 0) \wedge \cdots \wedge (a_{n+d} = 0) \rightarrow n + 1\}$. It is clear that S is reduced, $n(S) = n + d$, $k(S) = 1$, $d(S) = d$, $|S| = n + 1$, and the problem $EAD(S)$ has 2^{n+1} pairwise different solutions. Therefore any decision tree solving the problem $EAD(S)$ has at least 2^{n+1} terminal nodes that are labeled with pairwise different sets of decision rules. Hence $T_{EAD}(S) \geq 2^{n+1}$.

(b) Let us consider a decision rule system $S = S_1 \cup S_2 \cup S_3$, where $S_1 = \{(a_i = 0) \rightarrow 2i - 1, (a_i = 1) \rightarrow 2i : i = 1, \ldots, n\}$, $S_2 = \{(a_{n+1} = 0) \wedge \cdots \wedge (a_{n+d} = 0) \rightarrow 0\}$, and $S_3 = \{(a_1 = 2) \rightarrow 0, \ldots, (a_1 = k - 1) \rightarrow 0\}$. If $k = 2$, then $S_3 = 0$. It is clear that S is reduced, $n(S) = n + d$, $k(S) = k$, $d(S) = d$, and $|S| = 2n + k - 1$. It is also clear that the problem $AD(S)$ has at least 2^n pairwise different solutions. Therefore any decision tree solving the problem $AD(S)$ has at least 2^n terminal nodes that are labeled with pairwise different sets of decision rules. Hence $T_{AD}(S) \geq 2^n$. □

Let $k, d \in \omega \setminus \{0\}$. Denote $\Sigma(k, d) = \{S \in \Sigma : k(S) = k, d(S) = d\}$. We will consider only sequential algorithms for the construction of decision trees and will evaluate their time complexity depending on the size of decision rule systems on the algorithm input.

Lemma 13.12 *Let $d \in \omega \setminus \{0\}$ and $C \in \{SR, AD, AR\}$. Then there exists a polynomial time algorithm that, for a given decision rule system $S \in \Sigma(1, d)$, constructs a decision tree solving the problem $C(S)$.*

Proof Let $S \in \Sigma(1, d)$, $A(S) = \{a_{i_1}, \ldots, a_{i_m}\}$, $i_1 < \cdots < i_m$, and

$$V(S) = \{(\delta_1, \ldots, \delta_m)\}.$$

We construct a decision tree Γ that consists of one complete path $\xi = v_1, d_1, \ldots, v_m, d_m, v_{m+1}$ such that, for $j = 1, \ldots, m$, the node v_j is labeled with the attribute a_{i_j}, the edge d_j is labeled with the number δ_j, and the node v_{m+1} is labeled with the set of decision rules S. It is clear that Γ solves the problem $C(S)$ and the considered algorithm has polynomial time complexity. □

Lemma 13.13 *Let $k \in \omega\setminus\{0\}$ and $C \in \{SR, ESR\}$. Then there exists a polynomial time algorithm that, for a given decision rule system $S \in \Sigma(k, 1)$, constructs a decision tree solving the problem $C(S)$.*

Proof Let $S \in \Sigma(k, 1)$ and S contain a rule r of the length 0. Construct a decision tree Γ_0 that contains only one node labeled with the set $\{r\}$. It is clear that Γ_0 solves the problem $C(S)$ and the considered algorithm has polynomial time complexity.

Let us assume now that S does not contain rules of the length 0 and $A(S) = \{a_{i_1}, \ldots, a_{i_n}\}$. Let $j \in \{1, \ldots, n\}$ and $\delta \in V_S(a_{i_j})$. It is clear that in the system S there exists a decision rule for which the left-hand side is equal to $a_{i_j} = \delta$. Denote this rule by $r(j, \delta)$.

Let us construct a decision tree Γ_1. Let $V_S(a_{i_1}) = \{\delta_1, \ldots, \delta_m\}$. The decision tree Γ_1 consists of the nodes $v_0, v_1, \ldots, v_m$ and the edges $d_1, \ldots, d_m$. The node v_0 is labeled with the attribute a_{i_1}, and the node v_j is labeled with the set $\{r(1, \delta_j)\}$, $j = 1, \ldots, m$. For $j = 1, \ldots, m$, the edge d_j is labeled with the number δ_j, the edge d_j leaves the node v_0 and enters the node v_j. It is clear that Γ_1 solves the problem $SR(S)$ and the considered algorithm has polynomial time complexity.

Let us construct a decision tree Γ_2. The tree Γ_2 contains a complete path $\xi = v_1, d_1, \ldots, v_n, d_n, v_{n+1}$. For $j = 1, \ldots, n$, the node v_j is labeled with the attribute a_{i_j} and the edge d_j is labeled with the symbol $*$. The node v_{n+1} is labeled with the empty set. Let $j \in \{1, \ldots, n\}$ and $V_S(a_{i_j}) = \{\delta_1, \ldots, \delta_m\}$. Besides the edge d_j, also the edges $d_{j1}, \ldots, d_{jm}$ leave the node v_j. These edges are labeled with the numbers $\delta_1, \ldots, \delta_m$, respectively. The edges $d_{j1}, \ldots, d_{jm}$ enter the nodes $v_{j1}, \ldots, v_{jm}$ that are labeled with the sets $\{r(j, \delta_1)\}, \ldots, \{r(j, \delta_m)\}$, respectively. The tree Γ_2 does not contain any other nodes and edges. It is clear that Γ_2 solves the problem $ESR(S)$ and the considered algorithm has polynomial time complexity. □

13.4 On Construction of Decision Trees

As it was mentioned above, we consider only sequential algorithms for the construction of decision trees and evaluate their time complexity depending on the size of decision rule systems on the algorithm input. We assume that, during each unit of time, the algorithm can add to the constructing decision tree at most one node.

Let $C \in \{SR, ESR, AD, EAD, AR, EAR\}$. Denote by $P_{DT}(C)$ the set of pairs $(k, d) \in (\omega\setminus\{0\})^2$ satisfying the following condition: there exists a polynomial time algorithm that, for an arbitrary decision rule system $S \in \Sigma(k, d)$, constructs a decision tree solving the problem $C(S)$.

Theorem 13.1 (a) $P_{DT}(ESR) = \{(k, 1) : k \in \omega\setminus\{0\}\}$, (b) $P_{DT}(SR) = \{(k, 1), (1, d) : k, d \in \omega\setminus\{0\}\}$, (c) $P_{DT}(EAD) = P_{DT}(EAR) = \emptyset$, *and* (d) $P_{DT}(AD) = P_{DT}(AR) = \{(1, d) : d \in \omega\setminus\{0\}\}$.

Proof Let $C \in \{SR, ESR, AD, EAD, AR, EAR\}$, $(k, d) \in (\omega\setminus\{0\})^2$ and let us assume that there exists a polynomial time algorithm, which, for a given decision rule system $S \in \Sigma(k, d)$, constructs a decision tree solving the problem $C(S)$. Then there exists a polynomial π such that, for any $S \in \Sigma(k, d)$, $L_C(S) \leq \pi(size(S))$. If S is reduced, then, by (13.1), $size(S) \leq 10\,|S|\,d(S)(\log_2(d(S)\,|S|) + 1)$. We know that, for systems $S \in \Sigma(k, d)$, $d(S) \leq d$. Therefore there exists a polynomial ϱ such that $L_C(S) \leq \varrho(|S|)$ for any reduced system $S \in \Sigma(k, d)$.

(a) Using Lemma 13.13, we obtain that $\{(k, 1) : k \in \omega\backslash\{0\}\} \subseteq P_{DT}(ESR)$. Let $k, d \in \omega\backslash\{0\}$ and $d > 1$. We now show that $(k, d) \notin P_{DT}(ESR)$. Let us assume the contrary: $(k, d) \in P_{DT}(ESR)$. Then there exists a polynomial ϱ such that $L_{ESR}(S) \leq \varrho(|S|)$ for any reduced system $S \in \Sigma(k, d)$. Let $k = 1$. Using Lemma 13.10, we obtain that, for any $n \in \omega\backslash\{0\}$, there exists a reduced decision rule system $S \in \Sigma(k, d)$ such that $|S| = n + 1$ and $L_{ESR}(S) \geq 2^n$. Therefore $2^n \leq \varrho(n + 1)$ for any $n \in \omega\backslash\{0\}$ but this is impossible. Let $k > 1$. Using Lemma 13.9, we obtain that, for any $n \in \omega\backslash\{0\}$, there exists a reduced decision rule system $S \in \Sigma(k, d)$ such that $|S| = 2n + k - 1$ and $L_{SR}(S) \geq 2^n$. By Lemma 13.8, $L_{ESR}(S) \geq 2^n$. Therefore, $2^n \leq \varrho(2n + k - 1)$ for any $n \in \omega\backslash\{0\}$ but this is impossible since k is a constant for the considered systems of decision rules. Hence $(k, d) \notin P_{DT}(ESR)$ if $d > 1$. Thus, $P_{DT}(ESR) = \{(k, 1) : k \in \omega\backslash\{0\}\}$.

(b) Using Lemmas 13.12 and 13.13, we obtain that $\{(k, 1), (1, d) : k, d \in \omega\backslash\{0\}\} \subseteq P_{DT}(SR)$. Let $k, d \in \omega\backslash\{0, 1\}$. We now show that $(k, d) \notin P_{DT}(SR)$. Let us assume the contrary: $(k, d) \in P_{DT}(SR)$. Then there exists a polynomial ϱ such that $L_{SR}(S) \leq \varrho(|S|)$ for any reduced system $S \in \Sigma(k, d)$. Using Lemma 13.9, we obtain that, for any $n \in \omega\backslash\{0\}$, there exists a reduced decision rule system $S \in \Sigma(k, d)$ such that $|S| = 2n + k - 1$ and $L_{SR}(S) \geq 2^n$. Therefore, $2^n \leq \varrho(2n + k - 1)$ for any $n \in \omega\backslash\{0\}$ but this is impossible since k is a constant for the considered systems of decision rules. Hence $(k, d) \notin P_{DT}(SR)$ if $k > 1$ and $d > 1$. Thus, $P_{DT}(SR) = \{(k, 1), (1, d) : k, d \in \omega\backslash\{0\}\}$.

(c) Let $k, d \in \omega\backslash\{0\}$. We now show that $(k, d) \notin P_{DT}(EAD)$. Let us assume the contrary: $(k, d) \in P_{DT}(EAD)$. Then there exists a polynomial ϱ such that $L_{EAD}(S) \leq \varrho(|S|)$ for any reduced system $S \in \Sigma(k, d)$. Let $k = 1$. Using Lemma 13.11, we obtain that, for any $n \in \omega\backslash\{0\}$, there exists a reduced decision rule system $S \in \Sigma(k, d)$ such that $|S| = n + 1$ and $T_{EAD}(S) \geq 2^{n+1}$. It is clear that $L_{EAD}(S) \geq T_{EAD}(S)$ and $L_{EAD}(S) \geq 2^{n+1}$. Therefore, $2^{n+1} \leq \varrho(n + 1)$ for any $n \in \omega\backslash\{0\}$ but this is impossible. Let $k > 1$. Using Lemma 13.11, we obtain that, for any $n \in \omega\backslash\{0\}$, there exists a reduced decision rule system $S \in \Sigma(k, d)$ such that $|S| = 2n + k - 1$ and $T_{AD}(S) \geq 2^n$. By Lemma 13.8, $T_{EAD}(S) \geq T_{AD}(S)$. It is clear that $L_{EAD}(S) \geq T_{EAD}(S)$ and $L_{EAD}(S) \geq 2^n$. Therefore, $2^n \leq \varrho(2n + k - 1)$ for any $n \in \omega\backslash\{0\}$ but this is impossible since k is a constant for the considered systems of decision rules. Hence $(k, d) \notin P_{DT}(EAD)$. Thus, $P_{DT}(EAD) = \emptyset$.

Let $k, d \in \omega\backslash\{0\}$. We now show that $(k, d) \notin P_{DT}(EAR)$. Let us assume the contrary: $(k, d) \in P_{DT}(EAR)$. Then there exists a polynomial ϱ such that $L_{EAR}(S) \leq \varrho(|S|)$ for any reduced system $S \in \Sigma(k, d)$. Let $k = 1$. Using Lemma 13.11, we obtain that, for any $n \in \omega\backslash\{0\}$, there exists a reduced decision rule system $S \in \Sigma(k, d)$ such that $|S| = n + 1$ and $T_{EAD}(S) \geq 2^{n+1}$. It is clear that $L_{EAD}(S) \geq T_{EAD}(S)$. By Lemma 13.8, $L_{EAR}(S) \geq L_{EAD}(S)$ and $L_{EAR}(S) \geq 2^{n+1}$. Therefore, $2^{n+1} \leq \varrho(n + 1)$ for any $n \in \omega\backslash\{0\}$ but this is impossible. Let $k > 1$. Using

Lemma 13.11, we obtain that, for any $n \in \omega\backslash\{0\}$, there exists a reduced decision rule system $S \in \Sigma(k, d)$ such that $|S| = 2n + k - 1$ and $T_{AD}(S) \geq 2^n$. By Lemma 13.8, $T_{EAR}(S) \geq T_{AD}(S)$. It is clear that $L_{EAR}(S) \geq T_{EAR}(S)$ and $L_{EAR}(S) \geq 2^n$. Therefore, $2^n \leq \varrho(2n + k - 1)$ for any $n \in \omega\backslash\{0\}$ but this is impossible since k is a constant for the considered systems of decision rules. Hence $(k, d) \notin P_{DT}(EAR)$. Thus, $P_{DT}(EAR) = \emptyset$.

(d) Using Lemma 13.12, we obtain that $\{(1, d) : d \in \omega\backslash\{0\}\} \subseteq P_{DT}(AD)$ and $\{(1, d) : d \in \omega\backslash\{0\}\} \subseteq P_{DT}(AR)$.

Let $k, d \in \omega\backslash\{0\}$ and $k \geq 2$. We now show that $(k, d) \notin P_{DT}(AD)$. Let us assume the contrary: $(k, d) \in P_{DT}(AD)$. Then there exists a polynomial ϱ such that $L_{AD}(S) \leq \varrho(|S|)$ for any reduced system $S \in \Sigma(k, d)$. Using Lemma 13.11, we obtain that, for any $n \in \omega\backslash\{0\}$, there exists a reduced decision rule system $S \in \Sigma(k, d)$ such that $|S| = 2n + k - 1$ and $T_{AD}(S) \geq 2^n$. It is clear that $L_{AD}(S) \geq T_{AD}(S)$ and $L_{AD}(S) \geq 2^n$. Therefore, $2^n \leq \varrho(2n + k - 1)$ for any $n \in \omega\backslash\{0\}$ but this is impossible since k is a constant for the considered systems of decision rules. Hence $(k, d) \notin P_{DT}(AD)$. Thus, $P_{DT}(AD) = \{(1, d) : d \in \omega\backslash\{0\}\}$.

Let $k, d \in \omega\backslash\{0\}$ and $k \geq 2$. We now show that $(k, d) \notin P_{DT}(AR)$. Let us assume the contrary: $(k, d) \in P_{DT}(AR)$. Then there exists a polynomial ϱ such that $L_{AR}(S) \leq \varrho(|S|)$ for any reduced system $S \in \Sigma(k, d)$. Using Lemma 13.11, we obtain that, for any $n \in \omega\backslash\{0\}$, there exists a reduced decision rule system $S \in \Sigma(k, d)$ such that $|S| = 2n + k - 1$ and $T_{AD}(S) \geq 2^n$. It is clear that $L_{AD}(S) \geq T_{AD}(S)$. By Lemma 13.8, $L_{AR}(S) \geq L_{AD}(S)$ and $L_{AR}(S) \geq 2^n$. Therefore, $2^n \leq \varrho(2n + k - 1)$ for any $n \in \omega\backslash\{0\}$ but this is impossible since k is a constant for the considered systems of decision rules. Hence $(k, d) \notin P_{DT}(AR)$. Thus, $P_{DT}(AR) = \{(1, d) : d \in \omega\backslash\{0\}\}$. □

13.5 On Construction of Acyclic Decision Graphs

Besides decision trees, we will also consider *acyclic decision graphs* that are defined similarly to the decision trees, but instead of finite directed trees with roots, finite directed graphs with roots that do not have directed cycles are considered. Note that any decision tree is an acyclic decision graph. Let $C \in \{SR, ESR, AD, EAD, AR, EAR\}$. For a decision rule system S, we denote by $L_C^{DG}(S)$ the minimum number of nodes in an acyclic decision graph, which solves the problem $C(S)$. One can show that $L_C^{DG}(S) \geq T_C(S)$.

First, we describe a construction that will be used in this and in the next section. Let $C \in \{SR, ESR, AD, EAD, AR, EAR\}$, $S \in \Sigma$, and $r \in S$ be a decision rule of the form $(a_{i_1} = \delta_1) \wedge \cdots \wedge (a_{i_m} = \delta_m) \rightarrow \sigma$, where $m > 0$. For $j = 1, \ldots, m$, let $V_S^C(a_{i_j}) = V_S(a_{i_j})$ if $C \in \{SR, AD, AR\}$ and $V_S^C(a_{i_j}) = EV_S(a_{i_j})$ if $C \in \{ESR, EAD, EAR\}$. We now describe an acyclic decision graph $G_S^C(r)$. The graph $G_S^C(r)$ contains $m + 2$ nodes $v_1, \ldots, v_m, v_{m+1}, v_{m+2}$. The node v_1 is the root of $G_S^C(r)$. For $j = 1, \ldots, m$, the node v_j is labeled with the attribute a_{i_j}, the node

v_{m+1} is labeled with the set $\{r\}$, and the node v_{m+2} is labeled with the empty set. For $j = 1, \ldots, m$, exactly $|V_S^C(a_{i_j})|$ edges leave the node v_j. These edges are labeled with pairwise different elements from the set $V_S^C(a_{i_j})$. The edge labeled with the number δ_j enters the node v_{j+1}. All other edges enter the node v_{m+2}. The graph $G_S^C(r)$ does not contain other nodes and edges.

We consider only sequential algorithms for the construction of acyclic decision graphs and evaluate their time complexity depending on the size of decision rule systems on the algorithm input. We assume that, during each unit of time, the algorithm can add to the constructing acyclic decision graph at most one node.

Let $C \in \{SR, ESR, AD, EAD, AR, EAR\}$. By $P_{DG}(C)$ we denote the set of pairs $(k, d) \in (\omega \backslash \{0\})^2$ satisfying the following condition: there exists a polynomial time algorithm that, for an arbitrary decision rule system $S \in \Sigma(k, d)$, constructs an acyclic decision graph solving the problem $C(S)$.

Theorem 13.2 (a) $P_{DG}(SR) = P_{DG}(ESR) = (\omega \backslash \{0\})^2$, (b) $P_{DG}(AD) = P_{DG}(AR) = \{(1, d) : d \in \omega \backslash \{0\}\}$, *and* (c) $P_{DG}(EAD) = P_{DG}(EAR) = \emptyset$.

Proof Let $C \in \{SR, ESR, AD, EAD, AR, EAR\}$ and $(k, d) \in (\omega \backslash \{0\})^2$. Let us assume that there exists a polynomial time algorithm, which, for a given decision rule system $S \in \Sigma(k, d)$, constructs an acyclic decision graph solving the problem $C(S)$. Then there exists a polynomial π such that, for any $S \in \Sigma(k, d)$, $L_C^{DG}(S) \leq \pi(size(S))$. If S is reduced, then, by (13.1), $size(S) \leq 10\,|S|\,d(S)(\log_2(d(S)\,|S|) + 1)$. We know that, for systems $S \in \Sigma(k, d)$, $d(S) \leq d$. Therefore there exists a polynomial ϱ such that $T_C(S) \leq L_C^{DG}(S) \leq \varrho(|S|)$ for any reduced system $S \in \Sigma(k, d)$.

(a) Let $C \in \{SR, ESR\}$. We now show that $P_{DG}(C) = (\omega \backslash \{0\})^2$. Moreover, we show that there exists a polynomial time algorithm, which, for a given decision rule system $S \in \Sigma$, constructs an acyclic decision graph solving the problem $C(S)$.

Let $S \in \Sigma$. If the system S contains a decision rule r of the length 0, then the acyclic decision graph consisting of one node that is labeled with the set $\{r\}$ solves the problem $C(S)$. Let the system S do not contain rules of the length 0 and $S = \{r_1, \ldots, r_t\}$. We connect the graphs $G_S^C(r_1), \ldots, G_S^C(r_t)$ (the description of the graph $G_S^C(r)$ for a decision rule $r \in S$ can be found at the beginning of Sect. 13.5). To this end, for $j = 1, \ldots, t-1$, replace the node of the graph $G_S^C(r_j)$ labeled with the empty set with the root of the graph $G_S^C(r_{j+1})$. Denote by G_S^C the obtained graph. One can show that the graph G_S^C solves the problem $C(S)$. It is clear that the considered algorithm has polynomial time complexity. Thus, $P_{DG}(C) = (\omega \backslash \{0\})^2$.

(b) Using Lemma 13.12, we obtain that $\{(1, d) : d \in \omega \backslash \{0\}\} \subseteq P_{DG}(AD)$ and $\{(1, d) : d \in \omega \backslash \{0\}\} \subseteq P_{DG}(AR)$.

Let $k, d \in \omega \backslash \{0\}$ and $k \geq 2$. We now show that $(k, d) \notin P_{DG}(AD)$. Let us assume the contrary: $(k, d) \in P_{DG}(AD)$. Then there exists a polynomial ϱ such that $T_{AD}(S) \leq \varrho(|S|)$ for any reduced system $S \in \Sigma(k, d)$. Using Lemma 13.11, we obtain that, for any $n \in \omega \backslash \{0\}$, there exists a reduced decision rule system $S \in \Sigma(k, d)$ such that $|S| = 2n + k - 1$ and $T_{AD}(S) \geq 2^n$. Therefore, $2^n \leq \varrho(2n +$

$k-1)$ for any $n \in \omega\setminus\{0\}$ but this is impossible since k is a constant for the considered systems of decision rules. Hence $(k,d) \notin P_{DG}(AD)$. Thus, $P_{DG}(AD) = \{(1,d) : d \in \omega\setminus\{0\}\}$.

Let $k,d \in \omega\setminus\{0\}$ and $k \geq 2$. We now show that $(k,d) \notin P_{DG}(AR)$. Let us assume the contrary: $(k,d) \in P_{DG}(AR)$. Then there exists a polynomial ϱ such that $T_{AR}(S) \leq \varrho(|S|)$ for any reduced system $S \in \Sigma(k,d)$. Using Lemma 13.11, we obtain that, for any $n \in \omega\setminus\{0\}$, there exists a reduced decision rule system $S \in \Sigma(k,d)$ such that $|S| = 2n+k-1$ and $T_{AD}(S) \geq 2^n$. By Lemma 13.8, $T_{AR}(S) \geq T_{AD}(S)$ and $T_{AR}(S) \geq 2^n$. Therefore, $2^n \leq \varrho(2n+k-1)$ for any $n \in \omega\setminus\{0\}$ but this is impossible since k is a constant for the considered systems of decision rules. Hence $(k,d) \notin P_{DT}(AR)$. Thus, $P_{DG}(AR) = \{(1,d) : d \in \omega\setminus\{0\}\}$.

(c) Let $k,d \in \omega\setminus\{0\}$. We now show that $(k,d) \notin P_{DG}(EAD)$. Let us assume the contrary: $(k,d) \in P_{DG}(EAD)$. Then there exists a polynomial ϱ such that $T_{EAD}(S) \leq \varrho(|S|)$ for any reduced system $S \in \Sigma(k,d)$. Let $k=1$. Using Lemma 13.11, we obtain that, for any $n \in \omega\setminus\{0\}$, there exists a reduced decision rule system $S \in \Sigma(k,d)$ such that $|S| = n+1$ and $T_{EAD}(S) \geq 2^{n+1}$. Therefore, $2^{n+1} \leq \varrho(n+1)$ for any $n \in \omega\setminus\{0\}$ but this is impossible. Let $k > 1$. Using Lemma 13.11, we obtain that, for any $n \in \omega\setminus\{0\}$, there exists a reduced decision rule system $S \in \Sigma(k,d)$ such that $|S| = 2n+k-1$ and $T_{AD}(S) \geq 2^n$. By Lemma 13.8, $T_{EAD}(S) \geq T_{AD}(S)$ and $T_{EAD}(S) \geq 2^n$. Therefore, $2^n \leq \varrho(2n+k-1)$ for any $n \in \omega\setminus\{0\}$ but this is impossible since k is a constant for the considered systems of decision rules. Hence $(k,d) \notin P_{DG}(EAD)$. Thus, $P_{DG}(EAD) = \emptyset$.

Let $k,d \in \omega\setminus\{0\}$. We now show that $(k,d) \notin P_{DG}(EAR)$. Let us assume the contrary: $(k,d) \in P_{DG}(EAR)$. Then there exists a polynomial ϱ such that $T_{EAR}(S) \leq \varrho(|S|)$ for any reduced system $S \in \Sigma(k,d)$. Let $k=1$. Using Lemma 13.11, we obtain that, for any $n \in \omega\setminus\{0\}$, there exists a reduced decision rule system $S \in \Sigma(k,d)$ such that $|S| = n+1$ and $T_{EAD}(S) \geq 2^{n+1}$. By Lemma 13.8, $T_{EAR}(S) \geq T_{EAD}(S)$ and $T_{EAR}(S) \geq 2^{n+1}$. Therefore, $2^{n+1} \leq \varrho(n+1)$ for any $n \in \omega\setminus\{0\}$ but this is impossible. Let $k > 1$. Using Lemma 13.11, we obtain that, for any $n \in \omega\setminus\{0\}$, there exists a reduced decision rule system $S \in \Sigma(k,d)$ such that $|S| = 2n+k-1$ and $T_{AD}(S) \geq 2^n$. By Lemma 13.8, $T_{EAR}(S) \geq T_{AD}(S)$ and $T_{EAR}(S) \geq 2^n$. Therefore, $2^n \leq \varrho(2n+k-1)$ for any $n \in \omega\setminus\{0\}$ but this is impossible since k is a constant for the considered systems of decision rules. Hence $(k,d) \notin P_{DG}(EAR)$. Thus, $P_{DG}(EAR) = \emptyset$. □

13.6 On Construction of Acyclic Decision Graphs with Writing

Difficulties associated with a large number of pairwise different solutions to a problem can be circumvented by considering *acyclic decision graphs with writing*, which, in addition to working nodes labeled with attributes, have *writing* nodes and one terminal node. One of the nodes of the graph is distinguished as the root. Each writing node is labeled with a decision rule and has only one leaving edge. This edge is not labeled. The only terminal node is labeled with the letter W, denoting a set of decision rules W. This set can be changed during the work of the acyclic decision graph with writing. At the start of the work (when we are at the root of the graph), $W = \emptyset$. If during the work we come to a writing node that is labeled with a decision rule, we add this rule to the set W. When we reach the terminal node, the set W formed by this moment is the result of the work of the considered acyclic decision graph with writing.

We consider only sequential algorithms for the construction of acyclic decision graphs with writing and evaluate their time complexity depending on the size of decision rule systems on the algorithm input. We assume that, during each unit of time, the algorithm can add to the constructing acyclic decision graph with writing at most one node.

Let $C \in \{SR, ESR, AD, EAD, AR, EAR\}$. Denote by $P_{DGW}(C)$ the set of pairs $(k, d) \in (\omega \backslash \{0\})^2$ satisfying the following condition: there exists a polynomial time algorithm that, for an arbitrary decision rule system $S \in \Sigma(k, d)$, constructs an acyclic decision graph with writing solving the problem $C(S)$.

Theorem 13.3 $P_{DGW}(SR) = P_{DGW}(AD) = P_{DGW}(AR) = P_{DGW}(ESR) = P_{DGW}(EAD) = P_{DGW}(EAR) = (\omega \backslash \{0\})^2$.

Proof Let $C \in \{SR, ESR, AD, EAD, AR, EAR\}$, $S \in \Sigma$, and $r \in S$. We now describe an acyclic decision graph with writing $D_S^C(r)$. Let the length of the decision rule r be equal to 0. Then $D_S^C(r)$ contains two nodes v_1 and v_2 and an edge that leaves the node v_1 and enters the node v_2. The node v_1 is the root of $D_S^C(r)$. The node v_1 is labeled with the decision rule r and the node v_2 is labeled with the letter W. Let r be a decision rule of the form $(a_{i_1} = \delta_1) \wedge \cdots \wedge (a_{i_m} = \delta_m) \rightarrow \sigma$, where $m > 0$. Then the graph $D_S^C(r)$ is obtained from the graph $G_S^C(r)$ (this graph is defined at the beginning of Sect. 13.5) in the following way. Instead of the set $\{r\}$, we label the node v_{m+1} with the rule r. Instead of the empty set, we label the node v_{m+2} with the letter W. We add an edge leaving the node v_{m+1} and entering the node v_{m+2}.

We now show that $P_{DGW}(C) = (\omega \backslash \{0\})^2$. In fact, we show that there exists a polynomial time algorithm, which, for a given decision rule system $S \in \Sigma$, constructs an acyclic decision graph with writing G that solves the problem $C(S)$. Let $S = \{r_1, \ldots, r_t\}$. First, we construct graphs $D_S^C(r_1), \ldots, D_S^C(r_t)$. Then we connect these graphs. To this end, for $j = 1, \ldots, t-1$, replace the node of the graph $D_S^C(r_j)$ labeled with the letter W with the root of the graph $D_S^C(r_{j+1})$. Denote by G the obtained acyclic decision graph with writing. One can show that the graph G solves the problem $C(S)$. It is clear that the considered algorithm has polynomial time complexity. □

In connection with the difficulties arising in the construction of decision trees, it would be possible to move on to the study of acyclic decision graphs and acyclic decision graphs with writing. However, when constructing them, one should simultaneously try to make the depth as small as possible without allowing an excessive increase in the number of nodes. The possibilities of such bi-criteria optimization are the subject of a special study in the future. In several subsequent sections, we will consider another approach based on the simulation of the work of a decision tree on a given tuple of attribute values.

13.7 Bounds Based on Node Covers

In this section, we will study bounds on the minimum depth of decision trees solving problems. These bounds depend on the maximum length of the rule and parameters based on node covers for hypergraphs corresponding to the rule systems. In the next section, we will consider polynomial time algorithms for modeling the operation of a decision tree on a given tuple of attribute values that are based on the ideas proposed in this section.

Let S be a decision rule system with $n(S) > 0$. We denote by S^+ the subsystem of S containing only rules of the length $d(S)$. Denote $\beta^+(S) = \beta(S^+)$, where $\beta(S^+)$ is the minimum cardinality of a node cover of the hypergraph $G(S^+)$. It is clear that $\beta(S^+) \leq \beta(S)$. Let $C \in \{EAR, EAD, ESR\}$. We denote by $E_C(S)$ the set of consistent equation systems $\{a_{i_1} = \delta_1, \ldots, a_{i_m} = \delta_m\}$ such that $a_{i_j} \in A(S)$ and $\delta_j \in EV_S(a_{i_j})$ for $j = 1, \ldots, m$. Let $C \in \{AR, AD, SR\}$. We denote by $E_C(S)$ the set of consistent equation systems $\{a_{i_1} = \delta_1, \ldots, a_{i_m} = \delta_m\}$ such that $a_{i_j} \in A(S)$ and $\delta_j \in V_S(a_{i_j})$ for $j = 1, \ldots, m$. Denote $I_{EAD}(S) = I_{AD}(S)$ and $I_{ESR}(S) = I_{SR}(S)$.

Let $C \in \{AR, AD, SR, EAR\}$. Denote

$$\beta_C(S) = \max\{\beta(S_\alpha) : \alpha \in E_C(S)\},$$
$$\beta_C^+(S) = \max\{\beta^+(S_\alpha) : \alpha \in E_C(S)\}.$$

It is clear that $\beta_C^+(S) \leq \beta_C(S)$.

Let $C \in \{EAD, ESR\}$. Denote

$$\beta_C(S) = \max\{\beta(I_C(S_\alpha)) : \alpha \in E_C(S)\},$$
$$\beta_C^+(S) = \max\{\beta^+(I_C(S_\alpha)) : \alpha \in E_C(S)\}.$$

It is clear that $\beta_C^+(S) \leq \beta_C(S)$.

Lemma 13.14 *Let S be a decision rule system with $n(S) > 0$ and $C \in \{AR, EAR, EAD, ESR\}$. Then $h_C(S) \geq \beta_C(S) \geq \beta_C^+(S)$.*

Proof Let $C \in \{AR, EAR\}$, $\alpha \in E_C(S)$, and $\beta_C(S) = \beta(S_\alpha)$. Using Lemma 13.4, we obtain $h_C(S) \geq h_C(S_\alpha)$. By Lemma 13.5, $h_C(S_\alpha) \geq \beta(S_\alpha) = \beta_C(S)$. As we already mentioned, $\beta_C(S) \geq \beta_C^+(S)$.

Let $C \in \{EAD, ESR\}$, $\alpha \in E_C(S)$, and $\beta_C(S) = \beta(I_C(S_\alpha))$. Using Lemma 13.4, we obtain $h_C(S) \geq h_C(S_\alpha)$. By Lemma 13.5, $h_C(S_\alpha) \geq \beta(I_C(S_\alpha)) = \beta_C(S)$. As we already mentioned, $\beta_C(S) \geq \beta_C^+(S)$. □

Lemma 13.15 *Let S be a decision rule system with $n(S) > 0$ and $C \in \{AR, EAR\}$. Then*

$$h_C(S) \leq d(S)\beta_C^+(S).$$

Proof Let $V_C = V(S)$ if $C = AR$ and $V_C = EV(S)$ if $C = EAR$. We now describe the work of a decision tree Γ on a tuple from V_C. This work consists of rounds.

First round. We construct a node cover B_1 of the hypergraph $G(S^+)$ that has the minimum cardinality, i.e., $|B_1| = \beta(S^+)$. The decision tree Γ sequentially computes values of the attributes from B_1. As a result, we obtain a system α_1 consisting of $|B_1|$ equations of the form $a_{i_j} = \delta_j$, where $a_{i_j} \in B_1$ and δ_j is the computed value of the attribute a_{i_j}. If $S_{\alpha_1} = \emptyset$ or all rules from S_{α_1} have the empty left-hand side, then the tree Γ finishes its work. The result of this work is the set of decision rules r from S for which the system of equations $K(r) \cup \alpha_1$ is consistent. Otherwise, we move on to the second round of the decision tree Γ work.

Second round. We construct a node cover B_2 of the hypergraph $G((S_{\alpha_1})^+)$ that has the minimum cardinality, i.e., $|B_2| = \beta((S_{\alpha_1})^+)$. The decision tree Γ sequentially computes values of the attributes from B_2. As a result, we obtain a system α_2 consisting of $|B_2|$ equations. If $S_{\alpha_1 \cup \alpha_2} = \emptyset$ or all rules from $S_{\alpha_1 \cup \alpha_2}$ have the empty left-hand side, then the tree Γ finishes its work. The result of this work is the set of decision rules r from S for which the system of equations $K(r) \cup \alpha_1 \cup \alpha_2$ is consistent. Otherwise, we move on to the third round of the decision tree Γ work, etc., until we obtain empty system of rules or system in which all rules have empty left-hand side.

It is clear that $d(S) > d(S_{\alpha_1}) > d(S_{\alpha_1 \cup \alpha_2}) > \cdots$. Therefore, the number of rounds is at most $d(S)$. The number of attributes values of which are computed by Γ during each round is at most $\beta_C^+(S)$. Therefore $h(\Gamma) \leq d(S)\beta_C^+(S)$. It is easy to check that Γ solves the problem $C(S)$. Thus, $h_C(S) \leq d(S)\beta_C^+(S)$. □

Lemma 13.16 *Let S be a decision rule system with $n(S) > 0$ and $C \in \{EAD, ESR\}$. Then*

$$h_C(S) \leq d(I_C(S))\beta_C^+(S).$$

Proof Denote $V_C = EV(S)$. We now describe the work of a decision tree Γ on a tuple from V_C. If all rules from $I_C(S)$ have the empty left-hand side, then the tree Γ finishes its work. The result of this work is the set of decision rules $I_C(S)$. Otherwise, we move on to the first round of the decision tree Γ work.

We construct a node cover B_1 of the hypergraph $G(I_C(S)^+)$ that has the minimum cardinality, i.e., $|B_1| = \beta(I_C(S)^+)$. The decision tree Γ sequentially computes values of the attributes from B_1. As a result, we obtain a system α_1 consisting of $|B_1|$ equations of the form $a_{i_j} = \delta_j$, where $a_{i_j} \in B_1$ and δ_j is the computed value of the attribute a_{i_j}. If $I_C(S_{\alpha_1}) = \emptyset$ or all rules from $I_C(S_{\alpha_1})$ have the empty left-hand side, then the tree Γ finishes its work. The result of this work is the set of decision rules r from S for which the system of equations $K(r) \cup \alpha_1$ is consistent and $A(r) \subseteq B_1$. Otherwise, we move on to the second round of the decision tree Γ work.

We construct a node cover B_2 of the hypergraph $G(I_C(S_{\alpha_1})^+)$ that has the minimum cardinality, i.e., $|B_2| = \beta(I_C(S_{\alpha_1})^+)$. The decision tree Γ sequentially computes values of the attributes from B_2. As a result, we obtain a system α_2 consisting of $|B_2|$ equations. If $I_C(S_{\alpha_1 \cup \alpha_2}) = \emptyset$ or all rules from $I_C(S_{\alpha_1 \cup \alpha_2})$ have the empty left-hand side, then the tree Γ finishes its work. The result of this work is the set of decision rules r from S for which the system of equations $K(r) \cup \alpha_1 \cup \alpha_2$ is consistent and $A(r) \subseteq B_1 \cup B_2$. Otherwise, we move on to the third round of the decision tree Γ work, etc., until we obtain empty system of rules or system in which all rules have empty left-hand side.

One can show that $d(I_C(S)) > d(I_C(S_{\alpha_1})) > d(I_C(S_{\alpha_1 \cup \alpha_2})) > \cdots$. Therefore the number of rounds is at most $d(I_C(S))$. The number of attributes values of which are computed by Γ during each round is at most $\beta_C^+(S)$. Therefore $h(\Gamma) \le d(I_C(S))\beta_C^+(S)$. One can show that Γ solves the problem $C(S)$. Thus, $h_C(S) \le d(I_C(S))\beta_C^+(S)$. □

Theorem 13.4 *Let S be a decision rule system with $n(S) > 0$. Then*
(a) $\max\{d(S), \beta_{AR}(S)\} \le h_{AR}(S) \le d(S)\beta_{AR}^+(S)$.
(b) $\max\{d(S), \beta_{EAR}(S)\} \le h_{EAR}(S) \le d(S)\beta_{EAR}^+(S)$.
(c) $h_{AD}(S) \le d(S)\beta_{AD}^+(S)$.
(d) $\max\{d(S'), \beta_{EAD}(S')\} \le h_{EAD}(S) \le d(S')\beta_{EAD}^+(S')$, *where* $S' = R_{AD}(S)$.
(e) $h_{SR}(S) \le d(S)\beta_{SR}^+(S)$.
(f) $\max\{d(S'), \beta_{ESR}(S')\} \le h_{ESR}(S) \le d(S')\beta_{ESR}^+(S')$, *where* $S' = R_{SR}(S)$.

Proof (a) The lower bounds on $h_{AR}(S)$ follow from Lemmas 13.6 and 13.14. The upper bound on $h_{AR}(S)$ follows from Lemma 13.15.

(b) The lower bounds on $h_{EAR}(S)$ follow from Lemmas 13.6 and 13.14. The upper bound on $h_{EAR}(S)$ follows from Lemma 13.15.

(c) From Lemma 13.15 it follows that $h_{AR}(S) \le d(S)\beta_{AR}^+(S)$. The inequality $h_{AD}(S) \le h_{AR}(S)$ follows from Lemma 13.3. It is clear that $\beta_{AD}^+(S) = \beta_{AR}^+(S)$. Therefore $h_{AD}(S) \le d(S)\beta_{AD}^+(S)$.

(d) From Lemma 13.7 it follows that $h_{EAD}(S) = h_{EAD}(S')$. The bounds

$$\max\{d(S'), \beta_{EAD}(S')\} \le h_{EAD}(S')$$

follow from Lemmas 13.6 and 13.14. The bound $h_{EAD}(S') \le d(I_{EAD}(S'))\beta_{EAD}^+(S')$ follows from Lemma 13.16. It is clear that $I_{EAD}(S') = S'$.

(e) From Lemma 13.15 it follows that $h_{AR}(S) \le d(S)\beta_{AR}^+(S)$. The inequality $h_{SR}(S) \le h_{AR}(S)$ follows from Lemma 13.3. It is clear that $\beta_{SR}^+(S) = \beta_{AR}^+(S)$. Therefore $h_{SR}(S) \le d(S)\beta_{SR}^+(S)$.

(f) From Lemma 13.7 it follows that $h_{ESR}(S) = h_{ESR}(S')$. The bounds

$$\max\{d(S'), \beta_{ESR}(S')\} \le h_{ESR}(S')$$

follow from Lemmas 13.6 and 13.14. The bound $h_{ESR}(S') \le d(I_{ESR}(S'))\beta_{ESR}^+(S')$ follows from Lemma 13.16. It is clear that $I_{ESR}(S') = S'$. □

13.8 Algorithms Based on Node Covers

In this section, for each of the considered six problems, we design polynomial time algorithms that, for a given tuple of attribute values, describe the work on this tuple of a decision tree, which solves the problem. These algorithms are based on node covers for hypergraphs corresponding to decision rule systems.

13.8.1 Auxiliary Statements

In this section, we prove three lemmas.

Let S be a decision rule system and S^+ be the set of rules of the length $d(S)$ from S. Two decision rules r_1 and r_2 from S^+ are called *equivalent* if $K(r_1) = K(r_2)$. This equivalence relation provides a partition of the set S^+ into equivalence classes. We denote by $S^{\max}$ the set of rules that contains exactly one representative from each equivalence class and does not contain any other rules. It is clear that a set of attributes is a node cover for the hypergraph $G(S^+)$ if and only if this set is a node cover for the hypergraph $G(S^{\max})$.

Lemma 13.17 *Let S be a decision rule system with $n(S) > 0$. Then*

$$h_{EAR}(S) \geq \ln |S^{\max}| / \ln(k(S) + 1).$$

Proof Let $r \in S^{\max}$ and the rule r be equal to $(a_{i_1} = \delta_1) \wedge \cdots \wedge (a_{i_m} = \delta_m) \to \sigma$. We now define a tuple $\bar{\delta}(r) \in EV(S)$. For $j = 1, \ldots, m$, the tuple $\bar{\delta}(r)$ in the position corresponding to the attribute a_{i_j} contains the number δ_j. All other positions of the tuple $\bar{\delta}(r)$ are filled with the symbol $*$. We denote by $S^{\max}(\bar{\delta}(r))$ the set of rules from $S^{\max}$ that are realizable for the tuple $\bar{\delta}(r)$. One can show that $S^{\max}(\bar{\delta}(r)) = \{r\}$. From here it follows that any decision tree solving the problem $EAR(S)$ has at least $|S^{\max}|$ terminal nodes.

Let Γ be a decision tree, which solves the problem $EAR(S)$ and for which $h(\Gamma) = h_{EAR}(S)$. It is easy to show that the number of terminal nodes in Γ is at most $(k(S) + 1)^{h(\Gamma)}$. Therefore $(k(S) + 1)^{h(\Gamma)} \geq |S^{\max}|$ and $h(\Gamma) \geq \ln |S^{\max}| / \ln(k(S) + 1)$. Thus, $h_{EAR}(S) \geq \ln |S^{\max}| / \ln(k(S) + 1)$. □

Lemma 13.18 *Let S be an SR-reduced decision rule system with $n(S) > 0$. Then*

$$h_{ESR}(S) \geq \ln |S^{\max}| / \ln(k(S) + 1).$$

Proof Let $r \in S^{\max}$. We define the tuple $\bar{\delta}(r) \in EV(S)$ just as in the proof of Lemma 13.17. We denote by $S(\bar{\delta}(r))$ the set of rules from S that are realizable for the tuple $\bar{\delta}(r)$. Using the fact that the system S is SR-reduced, one can show that the set $S(\bar{\delta}(r))$ consists of the rule r and all rules from S^+, which are equivalent to r.

From here it follows that any decision tree solving the problem $ESR(S)$ has at least $|S^{\max}|$ terminal nodes. Further, just as in the proof of Lemma 13.17, we show that $h_{ESR}(S) \geq \ln |S| / \ln(k(S) + 1)$. □

Lemma 13.19 *Let S be an AD-reduced decision rule system with $n(S) > 0$. Then*

$$h_{EAD}(S) \geq \ln |S^{\max}| / \ln(k(S) + 1).$$

Proof Let $r \in S^{\max}$ and the right-hand side of r be equal to σ. We define the tuple $\bar{\delta}(r) \in EV(S)$ just as in the proof of Lemma 13.17. We denote by $S(\bar{\delta}(r))$ the set of rules from S that are realizable for the tuple $\bar{\delta}(r)$. The set $S^{\max}(\bar{\delta}(r))$ is defined in a similar way. Using the fact that the system S is AD-reduced, one can show that in $S(\bar{\delta}(r))$ only rule r and rules equal to r have right-hand sides equal to σ and r is the only rule in $S^{\max}(\bar{\delta}(r))$. From here it follows that any decision tree solving the problem $EAD(S)$ has at least $|S^{\max}|$ terminal nodes. Further, just as in the proof of Lemma 13.17, we show that $h_{EAD}(S) \geq \ln |S| / \ln(k(S) + 1)$. □

13.8.2 Algorithms for Construction of Node Covers

In this section, we consider two algorithms for construction of node covers.

13.8.2.1 Algorithm $\mathcal{N}_{cover}$

Let S be a decision rule system with $n(S) > 0$. We now describe a polynomial time algorithm $\mathcal{N}_{cover}$ for the construction of a node cover B for the hypergraph $G(S^+)$ such that $|B| \leq \beta(S^+)d(S)$.

Algorithm $\mathcal{N}_{cover}$

Set $B = \emptyset$. We find in S^+ the rule r_1 with the minimum identifier and add all attributes from $A(r_1)$ to B. We remove from S^+ all rules r such that $A(r_1) \cap A(r) \neq \emptyset$. Denote the obtained system by S_1^+. If $S_1^+ = \emptyset$, then B is a node cover of $G(S^+)$. If $S_1^+ \neq \emptyset$, then we find in S_1^+ the rule r_2 with the minimum identifier and add all attributes from $A(r_2)$ to B. We remove from S_1^+ all rules r such that $A(r_2) \cap A(r) \neq \emptyset$. Denote the obtained system by S_2^+. If $S_2^+ = \emptyset$, then B is a node cover of $G(S^+)$. If $S_2^+ \neq \emptyset$, then we find in S_2^+ the rule r_3 with the minimum identifier, and so on.

Lemma 13.20 *Let S be a decision rule system with $n(S) > 0$. The algorithm $\mathcal{N}_{cover}$ constructs a node cover B for the hypergraph $G(S^+)$ such that $|B| \leq \beta(S^+)d(S)$.*

Proof Let $B = \bigcup_{i=1}^{t} A(r_i)$. Since the sets $A(r_1), \ldots, A(r_t)$ are pairwise disjoint and the length of each of the rules $r_1, \ldots, r_t$ is equal to $d(S)$, we obtain $|B| = t \cdot d(S)$ and $\beta(S^+) \geq t$. Therefore $|B| \leq \beta(S^+)d(S)$. □

13.8.2.2 Algorithm $\mathcal{N}_{greedy}$

Let S be a decision rule system with $n(S) > 0$. We now describe a polynomial time algorithm $\mathcal{N}_{greedy}$ for the construction of a node cover B for the hypergraph $G(S^+)$ such that $|B| \le \beta(S^+) \ln |S^{\max}| + 1$. We will say that an attribute a_i covers a rule $r \in S^{\max}$ if $a_i \in A(r)$.

Algorithm $\mathcal{N}_{greedy}$

Set $B = \emptyset$. During each step, the algorithm chooses an attribute $a_i \in A(S^{\max})$ with the minimum index i, which covers the maximum number of rules from $S^{\max}$ uncovered during previous steps, and add it to the set B. The algorithm will finish the work when all rules from $S^{\max}$ are covered.

The considered algorithm is essentially a well-known greedy algorithm for the set cover problem – see Sect. 4.1.1 of the book [8].

Lemma 13.21 (follows from Theorem 4.1 [8]) *Let S be a decision rule system with $n(S) > 0$. The algorithm $\mathcal{N}_{greedy}$ constructs a node cover B for the hypergraph $G(S^+)$ such that* $|B| \le \beta(S^{\max}) \ln |S^{\max}| + 1 = \beta(S^+) \ln |S^{\max}| + 1$.

13.8.3 Algorithm $\mathcal{A}^C_{\mathcal{N}}$, $C \in \{AR, EAR\}$, $\mathcal{N} \in \{\mathcal{N}_{cover}, \mathcal{N}_{greedy}\}$

Let S be a decision rule system with $n(S) > 0$, $C \in \{AR, EAR\}$, $\mathcal{N} \in \{\mathcal{N}_{cover}, \mathcal{N}_{greedy}\}$, $V_C = V(S)$ if $C = AR$, and $V_C = EV(S)$ if $C = EAR$. We now describe a polynomial time algorithm $\mathcal{A}^C_{\mathcal{N}}$ that, for a given tuple of attribute values $\bar{\delta}$ from the set V_C, describes the work on this tuple of a decision tree Γ, which solves the problem $C(S)$.

Algorithm $\mathcal{A}^C_{\mathcal{N}}$

Set $Q := S$.

Step 1. If $Q = \emptyset$ or all rules from Q have an empty left-hand side, then the tree Γ finishes its work. The result of this work is the set of decision rules from S that correspond to the rules with an empty left-hand side from the set Q. Otherwise, we move on to Step 2.

Step 2. Using the algorithm $\mathcal{N}$, we construct a node cover B of the hypergraph $G(Q^+)$. The decision tree Γ sequentially computes values of the attributes from the set B ordered by ascending attribute indices. As a result, we obtain a system α consisting of $|B|$ equations of the form $a_{i_j} = \delta_j$, where $a_{i_j} \in B$ and δ_j is the computed value of the attribute a_{i_j} (value of this attribute from the tuple $\bar{\delta}$). Set $Q := Q_\alpha$ and move on to Step 1.

We now consider bounds on accuracy for the algorithm $\mathcal{A}^C_{\mathcal{N}}$, $\mathcal{N} \in \{\mathcal{N}_{cover}, \mathcal{N}_{greedy}\}$, $C \in \{AR, EAR\}$.

Theorem 13.5 *Let $C \in \{AR, EAR\}$ and S be a decision rule system with $n(S) > 0$. Then the algorithm $\mathscr{A}^{C}_{\mathscr{N}_{cover}}$ describes the work of a decision tree Γ, which solves the problem $C(S)$ and for which $h(\Gamma) \leq h_C(S)^3$.*

Proof Let us consider Step 2 of the algorithm. It is clear that $d(Q_\alpha) \leq d(Q) - 1$. Therefore the number of repetitions of Step 2 is at most $d(S)$. By Lemma 13.6, $d(S) \leq h_C(S)$. It is clear that $Q = S_\gamma$ for some subsystem γ of the equation system $K(S, \bar{\delta})$. From Lemma 13.20 it follows that the number of attributes values of which are computed by Γ during Step 2 is at most $\beta((S_\gamma)^+)d(S_\gamma)$. Evidently, $d(S_\gamma) \leq d(S) \leq h_C(S)$ and $\beta((S_\gamma)^+) \leq \beta(S_\gamma)$. By Lemmas 13.4 and 13.5, $\beta(S_\gamma) \leq h_C(S_\gamma) \leq h_C(S)$. Therefore the number of attributes values of which are computed by Γ during Step 2 is at most $h_C(S)^2$. The number of repetitions of Step 2 is at most $h_C(S)$. Thus, the depth of the decision tree Γ is at most $h_C(S)^3$. □

Theorem 13.6 *Let S be a decision rule system with $n(S) > 0$.*

(a) *The algorithm $\mathscr{A}^{EAR}_{\mathscr{N}_{greedy}}$ describes the work of a decision tree Γ, which solves the problem $EAR(S)$ and for which $h(\Gamma) \leq h_{EAR}(S)^3 \ln(k(S) + 1) + h_{EAR}(S)$.*

(b) *The algorithm $\mathscr{A}^{AR}_{\mathscr{N}_{greedy}}$ describes the work of a decision tree Γ, which solves the problem $AR(S)$ and for which $h(\Gamma) \leq h_{AR}(S)^2 \ln(|S|) + h_{AR}(S)$.*

Proof (a) Let us consider Step 2 of the algorithm. It is clear that $d(Q_\alpha) \leq d(Q) - 1$. Therefore the number of repetitions of Step 2 is at most $d(S)$. By Lemma 13.6, $d(S) \leq h_{EAR}(S)$. It is clear that $Q = S_\gamma$ for some subsystem γ of the equation system $K(S, \bar{\delta})$. From Lemma 13.21 it follows that the number of attributes values of which are computed by Γ during Step 2 is at most $\beta((S_\gamma)^{\max}) \ln |(S_\gamma)^{\max}| + 1$. Evidently, $\beta((S_\gamma)^{\max}) \leq \beta(S_\gamma)$. By Lemmas 13.4 and 13.5, $\beta(S_\gamma) \leq h_{EAR}(S_\gamma) \leq h_{EAR}(S)$. Therefore $\beta((S_\gamma)^{\max}) \leq h_{EAR}(S)$. Evidently, $k(S_\gamma) \leq k(S)$. By Lemmas 13.4 and 13.17, $\ln |(S_\gamma)^{\max}| \leq h_{EAR}(S_\gamma) \ln(k(S_\gamma) + 1) \leq h_{EAR}(S) \ln(k(S) + 1)$. Therefore the number of attributes values of which are computed by Γ during Step 2 is at most $h_{EAR}(S)^2 \ln(k(S) + 1) + 1$. The number of repetitions of Step 2 is at most $h_{EAR}(S)$. Thus, the depth of the decision tree Γ is at most $h_{EAR}(S)^3 \ln(k(S) + 1) + h_{EAR}(S)$.

(b) Let us consider Step 2 of the algorithm. It is clear that $d(Q_\alpha) \leq d(Q) - 1$. Therefore the number of repetitions of Step 2 is at most $d(S)$. By Lemma 13.6, $d(S) \leq h_{AR}(S)$. It is clear that $Q = S_\gamma$ for some subsystem γ of the equation system $K(S, \bar{\delta})$. From Lemma 13.21 it follows that the number of attributes values of which are computed by Γ during Step 2 is at most $\beta((S_\gamma)^{\max}) \ln |(S_\gamma)^{\max}| + 1$. Evidently, $\beta((S_\gamma)^{\max}) \leq \beta(S_\gamma)$. By Lemmas 13.4 and 13.5, $\beta(S_\gamma) \leq h_{AR}(S_\gamma) \leq h_{AR}(S)$. Therefore $\beta((S_\gamma)^{\max}) \leq h_{AR}(S)$. Evidently, $|(S_\gamma)^{\max}| \leq |S|$. Therefore the number of attributes values of which are computed by Γ during Step 2 is at most $h_{AR}(S) \ln |S| + 1$. The number of repetitions of Step 2 is at most $h_{AR}(S)$. Thus, the depth of the decision tree Γ is at most $h_{AR}(S)^2 \ln(|S|) + h_{AR}(S)$. □

13.8.4 Algorithm $\mathscr{A}_{\mathscr{N}}^{C}$, $C \in \{SR, ESR, AD, EAD\}$, $\mathscr{N} \in \{\mathscr{N}_{cover}, \mathscr{N}_{greedy}\}$

Let S be a decision rule system with $n(S) > 0$, $C \in \{SR, ESR, AD, EAD\}$, and $\mathscr{N} \in \{\mathscr{N}_{cover}, \mathscr{N}_{greedy}\}$. Let $V_C = V(S)$ if $C \in \{SR, AD\}$, and $V_C = EV(S)$ if $C \in \{ESR, EAD\}$. Let $R_C = R_{SR}$ if $C \in \{SR, ESR\}$ and $R_C = R_{AD}$ if $C \in \{AD, EAD\}$. We now describe a polynomial time algorithm $\mathscr{A}_{\mathscr{N}}^{C}$ that, for a given tuple of attribute values $\bar{\delta}$ from the set V_C, describes the work on this tuple of a decision tree Γ, which solves the problem $C(S)$.

Algorithm $\mathscr{A}_{\mathscr{N}}^{C}$

Set $Q := S$.

Step 1. Set $P := R_C(Q)$. If $P = \emptyset$ or all rules from P have an empty left-hand side, then the tree Γ finishes its work. The result of this work is the set of decision rules from S that correspond to the rules with an empty left-hand side from the set P. Otherwise, we move on to Step 2.

Step 2. Using the algorithm $\mathscr{N}$, we construct a node cover B of the hypergraph $G(P^+)$. The decision tree Γ sequentially computes values of the attributes from the set B ordered by ascending attribute indices. As a result, we obtain a system α consisting of $|B|$ equations of the form $a_{i_j} = \delta_j$, where $a_{i_j} \in B$ and δ_j is the computed value of the attribute a_{i_j} (value of this attribute from the tuple $\bar{\delta}$). Set $Q := P_\alpha$ and move on to Step 1.

We now consider bounds on accuracy for the algorithm $\mathscr{A}_{\mathscr{N}}^{C}$, $C \in \{SR, ESR, AD, EAD\}$, $\mathscr{N} \in \{\mathscr{N}_{cover}, \mathscr{N}_{greedy}\}$.

Theorem 13.7 *Let S be a decision rule system with $n(S) > 0$.*

(a) *If $C \in \{ESR, EAD\}$, then the algorithm $\mathscr{A}_{\mathscr{N}_{cover}}^{C}$ describes the work of a decision tree Γ, which solves the problem $C(S)$ and for which $h(\Gamma) \le h_C(S)^3$.*

(b) *If $C \in \{SR, AD\}$, then the algorithm $\mathscr{A}_{\mathscr{N}_{cover}}^{C}$ describes the work of a decision tree Γ, which solves the problem $C(S)$ and for which $h(\Gamma) \le d(S)^2 \beta_C(S)$.*

Proof (a) Let us consider Step 2 of the algorithm. Note that $P := R_C(Q)$. Using Lemmas 13.4 and 13.7, we obtain $h_C(Q) = h_C(P) \ge h_C(P_\alpha)$. If we apply the same relations to the previous repetitions of Step 2, we obtain $h_C(S) \ge h_C(Q)$.

One can show that $d(R_C(S)) \ge d(P) > d(P_\alpha)$. From these inequalities it follows that the number of repetitions of Step 2 is at most $d(R_C(S))$. Using Lemmas 13.6 and 13.7, we obtain $h_C(S) = h_C(R_C(S)) \ge d(R_C(S))$. Thus, the number of repetitions of Step 2 is at most $h_C(S)$ and $d(P) \le h_C(S)$.

From Lemma 13.20 it follows that the number of attributes values of which are computed by Γ during Step 2 is at most $\beta(P^+)d(P)$. We know that $d(P) \le h_C(S)$. Evidently, $P^+ \subseteq P = R_C(Q) \subseteq I_C(Q)$. Therefore $\beta(P^+) \le \beta(I_C(Q))$. By Lemma 13.5, $\beta(I_C(Q)) \le h_C(Q)$. We know that $h_C(Q) \le h_C(S)$. Thus, $\beta(P^+) \le$

$h_C(S)$. As a result, we obtain that the number of attributes values of which are computed by Γ during Step 2 is at most $h_C(S)^2$. As we know, the number of repetitions of Step 2 is at most $h_C(S)$. Thus, the depth of the decision tree Γ is at most $h_C(S)^3$.

(b) Let us consider Step 2 of the algorithm. Note that $P := R_C(Q)$. It is clear that $d(S) \geq d(P) > d(P_\alpha)$. From these inequalities it follows that the number of repetitions of Step 2 is at most $d(S)$. From Lemma 13.20 it follows that the number of attributes values of which are computed by Γ during Step 2 is at most $\beta(P^+)d(P)$. One can show that $P = S'_\gamma$ for some subsystem S' of the decision rule system S and some subsystem γ of the equation system $K(S, \bar{\delta})$. It is clear that $P^+ \subseteq P = S'_\gamma \subseteq S_\gamma$. Therefore $\beta(P^+) \leq \beta(S_\gamma) \leq \beta_C(S)$. As a result, the depth of the decision tree Γ is at most $d(S)^2\beta_C(S)$. □

Theorem 13.8 *Let S be a decision rule system with $n(S) > 0$.*

(a) *If $C \in \{ESR, EAD\}$, then the algorithm $\mathcal{A}^C_{\mathcal{N}_{greedy}}$ describes the work of a decision tree Γ, which solves the problem $C(S)$ and for which $h(\Gamma) \leq h_C(S)^3 \ln(k(S) + 1) + h_C(S)$.*

(b) *If $C \in \{SR, AD\}$, then the algorithm $\mathcal{A}^C_{\mathcal{N}_{greedy}}$ describes the work of a decision tree Γ, which solves the problem $C(S)$ and for which $h(\Gamma) \leq d(S)\beta_C(S) \ln |S| + d(S)$.*

Proof (a) Let us consider Step 2 of the algorithm. Note that $P := R_C(Q)$. Using Lemmas 13.4 and 13.7, we obtain $h_C(Q) = h_C(P) \geq h_C(P_\alpha)$. If we apply the same relations to the previous repetitions of Step 2, we obtain $h_C(S) \geq h_C(Q) = h_C(P)$.

It is clear that $d(R_C(S)) \geq d(P) > d(P_\alpha)$. From these inequalities it follows that the number of repetitions of Step 2 is at most $d(R_C(S))$. Using Lemmas 13.6 and 13.7, we obtain $h_C(S) = h_C(R_C(S)) \geq d(R_C(S))$. Thus, the number of repetitions of Step 2 is at most $h_C(S)$.

From Lemma 13.21 it follows that the number of attributes values of which are computed by Γ during Step 2 is at most $\beta(P^{\max}) \ln |P^{\max}| + 1$. Evidently, $P^{\max} \subseteq P = R_C(Q) \subseteq I_C(Q)$. Therefore $\beta(P^{\max}) \leq \beta(I_C(Q))$. By Lemma 13.5, $\beta(I_C(Q)) \leq h_C(Q)$. We know that $h_C(Q) \leq h_C(S)$. Thus, $\beta(P^{\max}) \leq h_C(S)$. Evidently, $k(P) \leq k(S)$. By Lemmas 13.18 and 13.19, $\ln |P^{\max}| \leq h_C(P) \ln(k(P) + 1) \leq h_C(S) \ln(k(S) + 1)$. Therefore the number of attributes values of which are computed by Γ during Step 2 is at most $h_C(S)^2 \ln(k(S) + 1) + 1$. The number of repetitions of Step 2 is at most $h_C(S)$. Thus, the depth of the decision tree Γ is at most $h_C(S)^3 \ln(k(S) + 1) + h_C(S)$.

(b) Let us consider Step 2 of the algorithm. Note that $P := R_C(Q)$. It is clear that $d(S) \geq d(P) > d(P_\alpha)$. From these inequalities it follows that the number of repetitions of Step 2 is at most $d(S)$. From Lemma 13.21 it follows that the number of attributes values of which are computed by Γ during Step 2 is at most $\beta(P^+) \ln |P^{\max}| + 1$. One can show that $P = S'_\gamma$ for some subsystem S' of the decision rule system S and some subsystem γ of the equation system $K(S, \bar{\delta})$. It is clear that $P^+ \subseteq P = S'_\gamma \subseteq S_\gamma$. Therefore $\beta(P^+) \leq \beta(S_\gamma) \leq \beta_C(S)$. It is clear that $|P^{\max}| \leq |S|$. As a result, the depth of the decision tree Γ is at most $d(S)\beta_C(S) \ln |S| + d(S)$. □

13.9 Completely Greedy Algorithms

In this section, for each of the considered six problems, we design a polynomial time algorithm that, for a given tuple of attribute values, describes the work on this tuple of a decision tree, which solves the problem. This algorithm is a completely greedy algorithm by nature.

It will be convenient for us to use the following notation: for any decision rule system S, let $R_{AR}(S) = S$.

Let S be a decision rule system with $n(S) > 0$ and $C \in \{AR, EAR, SR, ESR, AD, EAD\}$. Let $V_C = V(S)$ if $C \in \{AR, SR, AD\}$, and $V_C = EV(S)$ if $C \in \{EAR, ESR, EAD\}$. Let $R_C = R_{AR}$ if $C \in \{AR, EAR\}$, $R_C = R_{SR}$ if $C \in \{SR, ESR\}$ and $R_C = R_{AD}$ if $C \in \{AD, EAD\}$. We now describe a polynomial time algorithm $\mathscr{A}_G^C$ that, for a given tuple of attribute values $\bar{\delta}$ from the set V_C, describes the work on this tuple of a decision tree Γ, which solves the problem $C(S)$.

Algorithm $\mathscr{A}_G^C$

Set $Q := S$.

Step 1. Set $P := R_C(Q)$. If $P = \emptyset$ or all rules from P have an empty left-hand side, then the tree Γ finishes its work. The result of this work is the set of decision rules from S that correspond to the rules with an empty left-hand side from the set P. Otherwise, we move on to Step 2.

Step 2. We choose an attribute $a_i \in A(P)$ with the minimum index i, which covers the maximum number of rules from P. The decision tree Γ computes the value of this attribute. As a result, we obtain a system of equations $\{a_i = \delta\}$, where δ is the computed value of the attribute a_i (value of this attribute from the tuple $\bar{\delta}$). Set $Q := P_{\{a_i=\delta\}}$ and move on to Step 1.

One can show that the decision tree Γ described by the algorithm $\mathscr{A}_G^C$ solves the problem $C(S)$. We have no bounds on the accuracy of the algorithm $\mathscr{A}_G^C$.

13.10 Dynamic Programming Algorithms

In this section, for each of the considered six problems, we describe a dynamic programming algorithm that computes the minimum depth of a decision tree solving the problem. This algorithm uses a directed acyclic graph (DAG), which nodes are systems of decision rules.

13.10.1 Construction of DAG $\Delta_C(S)$

Let S be a decision rule system with $n(S) > 0$ and $C \in \{AR, EAR, AD, EAD, SR, ESR\}$. We now describe an algorithm $\mathscr{A}^C_{DAG}$ that constructs a DAG $\Delta_C(S)$. For an attribute $a_i \in A(S)$, let $V^C_S(a_i) = V_S(a_i)$ if $C \in \{AR, AD, SR\}$ and $V^C_S(a_i) = EV_S(a_i)$ if $C \in \{EAR, EAD, ESR\}$.

Algorithm $\mathscr{A}^C_{DAG}$

Step 1. Construct a DAG Δ containing only one node S that is not marked as processed and move on to Step 2.

Step 2. If all nodes in Δ are marked as processed, then return this DAG as $\Delta_C(S)$. The work of the algorithm is finished.

Otherwise, choose a node Q that is not marked as processed. If the set of rules $R_C(Q)$ is empty or contains only rules with an empty left-hand side, then mark the node Q as processed and move on to Step 2.

Otherwise, for each attribute $a_i \in A(Q)$ and each number $\delta \in V^C_S(a_i)$, add to Δ an edge that leaves the node Q, enters the node $Q_{\{a_i=\delta\}}$, and is labeled with the equation $a_i = \delta$. If the node $Q_{\{a_i=\delta\}}$ does not belong to Δ, then add it to Δ and do not mark it as processed. Mark the node Q as processed and move on to Step 2.

A node Q of the DAG $\Delta_C(S)$ will be called *terminal* if this node has no leaving edges.

13.10.2 Dynamic Programming Algorithm $\mathscr{A}^C_{DP}$

Let S be a decision rule system with $n(S) > 0$ and $C \in \{AR, EAR, AD, EAD, SR, ESR\}$. We now describe a dynamic programming algorithm A^C_{DP}, which returns the minimum depth $h_C(S)$ of a decision tree that solves the problem $C(S)$. This algorithm is used the DAG $\Delta_C(S)$. During each step, the algorithm labels a node Q of the DAG $\Delta_C(S)$ with a number $H_C(Q)$.

Algorithm $\mathscr{A}^C_{DP}$

Step 1. If the node S of the DAG $\Delta_C(S)$ is labeled with a number $H_C(S)$, then return this number as $h_C(S)$. The algorithm finishes its work. Otherwise, move on to Step 2.

Step 2. Choose a node Q in $\Delta_C(S)$ that is not labeled with a number and such that either Q is a terminal node or all children of Q are labeled with numbers.

Let node Q be a terminal node. Then label this node with the number $H_C(Q) = 0$ and move on to Step 1.

Let all children of the node Q be labeled with numbers. For each $a_i \in A(Q)$, compute the number $H_C(Q, a_i) = 1 + \max\{H_C(Q_{\{a_i=\delta\}}) : \delta \in V^C_S(a_i)\}$, label the node Q with the number $H_C(Q) = \min\{H_C(Q, a_i) : a_i \in A(Q)\}$ and move on to Step 1.

Let $C \in \{AR, EAR, AD, EAD, SR, ESR\}$, S be a decision rule system with $n(S) > 0$, and Q be a node of the DAG $\Delta_C(S)$. Let Γ be a decision tree over S (see Definition 13.10) in which each terminal node is labeled with a subset of the set Q. We will say that this tree *solves* the problem $C(Q)$ if it satisfies conditions from Definition 13.12 for $C \in \{AR, EAR\}$, Definition 13.13 for $C \in \{AD, EAD\}$, and Definition 13.14 for $C \in \{SR, ESR\}$ in which the rule system S is replaced with the rule system Q. We denote by $h_C^S(Q)$ the minimum depth of a decision tree over S in which each terminal node is labeled with a subset of the set Q and which solves the problem $C(Q)$. It is clear that $h_C^S(S) = h_C(S)$.

Lemma 13.22 *Let $C \in \{AR, EAR, AD, EAD, SR, ESR\}$, S be a decision rule system with $n(S) > 0$, and Q be a node of the DAG $\Delta_C(S)$, which is not terminal. Then $h_C^S(Q) = \min\{h_C^S(Q, a_i) : a_i \in A(Q)\}$, where $h_C^S(Q, a_i) = 1 + \max\{h_C^S(Q_{\{a_i=\delta\}}) : \delta \in V_S^C(a_i)\}$ for any $a_i \in A(Q)$.*

Proof It is easy to see that, for any $a_i \in A(Q)$, $h_C^S(Q, a_i)$ is the minimum depth of a decision tree over S in which each terminal node is labeled with a subset of the set Q, which solves the problem $C(Q)$ and in which the root is labeled with the attribute a_i. One can show that in a decision tree over S in which each terminal node is labeled with a subset of the set Q, which solves the problem $C(Q)$ and which depth is equal to $h_C^S(Q)$, the root cannot be labeled with an attribute from the set $A(S) \setminus A(Q)$. Therefore $h_C^S(Q) = \min\{h_C^S(Q, a_i) : a_i \in A(Q)\}$. □

Theorem 13.9 *Let S be a decision rule system with $n(S) > 0$ and $C \in \{AR, EAR, AD, EAD, SR, ESR\}$. Then the number $H_C(S)$ returned by the algorithm $\mathscr{A}_{DP}^C$ is equal to the minimum depth $h_C(S)$ of a decision tree solving the problem $C(S)$.*

Proof We prove by induction on nodes of the DAG $\Delta_C(S)$ that, for any node Q of $\Delta_C(S)$, $h_C^S(Q) = H_C(Q)$. If Q is a terminal node, then the rule system $R_C(Q)$ is empty or contains only rules with an empty left-hands side. In this case, $h_C^S(Q) = 0$ and $h_C^S(Q) = H_C(Q)$.

Let Q be a nonterminal node such that, for each $a_i \in A(Q)$ and each $\delta \in V_S^C(a_i)$, $h_C^S(Q_{\{a_i=\delta\}}) = H_C(Q_{\{a_i=\delta\}})$. Then, for each $a_i \in A(Q)$, $h_C^S(Q, a_i) = H_C(Q, a_i)$. By Lemma 13.22, $h_C^S(Q) = \min\{h_C^S(Q, a_i) : a_i \in A(Q)\}$. Therefore $h_C^S(Q) = H_C(Q)$.

As a result, we obtain $h_C^S(S) = H_C(S)$. Since $h_C^S(S) = h_C(S)$, the algorithm $\mathscr{A}_{DP}^C$ returns the number $h_C(S)$. □

References

1. Durdymyradov, K., Moshkov, M.: Complexity of transforming decision rule systems into decision trees and acyclic decision graphs. In: Nguyen, N.T., Chbeir, R., Fujita, H., Hong, T.-P., Le Nguyen, M., Wojtkiewicz, K. (eds.). Recent Challenges in Intelligent Information and Database Systems – 16th Asian Conference on Intelligent Information and Database Systems, ACIIDS

2024, Ras Al Khaimah, UAE, April 15–18, 2024, Proceedings, Part I. Communications in Computer and Information Science. vol. 2144, pp. 153–163. Springer (2024)
2. Durdymyradov, K., Moshkov, M.: Modeling the functioning of decision trees based on decision rule systems by greedy algorithm. In: Nguyen, N.T., Franczyk, B., Ludwig, A., Núñez, M., Treur, J., Vossen, G., Kozierkiewicz, A. (eds.). Computational Collective Intelligence – 16th International Conference, ICCCI 2024, Leipzig, Germany, September 9–11, 2024, Proceedings, Part II. Lecture Notes in Artificial Intelligence. vol. 14811, pp. 153–162. Springer (2024)
3. Durdymyradov, K., Moshkov, M.: Simulating functioning of decision trees for tasks on decision rule systems. In: Hu, M., Cornelis, C., Zhang, Y., Lingras, P., Slezak, D., Yao, J. (eds.). Rough Sets – International Joint Conference, IJCRS 2024, Halifax, NS, Canada, May 17–20, 2024, Proceedings, Part I. Lecture Notes in Computer Science. vol. 14839, pp. 188–200. Springer (2024)
4. Durdymyradov, K., Moshkov, M.: Construction of decision trees and acyclic decision graphs from decision rule systems (2023). arXiv:2305.01721
5. Durdymyradov, K., Moshkov, M.: Greedy algorithm for inference of decision trees from decision rule systems (2024). arXiv:2401.06793
6. Moshkov, M.: Some relationships between decision trees and decision rule systems. In: Polkowski, L., Skowron, A. (eds.) Rough Sets and Current Trends in Computing, First International Conference, RSCTC'98, Warsaw, Poland, June 22–26, 1998, Proceedings, Lecture Notes in Computer Science, vol. 1424, pp. 499–505. Springer (1998)
7. Moshkov, M.: On transformation of decision rule systems into decision trees. In: Proceedings of the Seventh International Workshop Discrete Mathematics and its Applications, Moscow, Russia, January 29 – February 2, 2001, Part 1, pp. 21–26. Center for Applied Investigations of Faculty of Mathematics and Mechanics, Moscow State University (2001). (in Russian)
8. Moshkov, M., Zielosko, B.: Combinatorial Machine Learning—A Rough Set Approach. Studies in Computational Intelligence, vol. 360. Springer (2011)

Final Remarks

The main goal of this book was to understand the relationships between deterministic decision trees and systems of decision rules, often represented by nondeterministic decision trees.

To this end, we studied the complexity of deterministic and nondeterministic decision trees for problems over information systems, for decision tables from closed classes, and for problems related to formal languages. We also studied the complexity of deterministic decision trees derived from decision rule systems, and the complexity of algorithms for constructing these trees or their representations in the form of acyclic decision graphs.

Future research will be devoted to further generalization of the results obtained for problems over information systems to the case of decision tables from closed classes, as well as a more in-depth theoretical and experimental study of algorithms for transforming systems of decision rules into deterministic decision trees.

K. Durdymyradov et al., *Decision Trees Versus Systems of Decision Rules*, Studies in Big Data 160, https://doi.org/10.1007/978-3-031-71586-0

Index

K. Durdymyradov et al., *Decision Trees Versus Systems of Decision Rules*, Studies in Big Data 160, https://doi.org/10.1007/978-3-031-71586-0

The manufacturer's authorised representative in the EU is Springer Nature Customer Service Centre GmbH, Europaplatz 3, 69115 Heidelberg, Germany. If you have any concerns regarding our products, please contact ProductSafety@springernature.com

Printed and bound by CPI Group (UK) Ltd, Croydon, CR0 4YY

30/09/2026

02225570-0009